BREAKTHROUGH
science

BRENDAN CASSERLY
AND BERNARD HORGAN

Gill & Macmillan

Published in Ireland by
Gill & Macmillan Ltd
Goldenbridge
Dublin 8
with associated companies throughout the world
© Brendan Casserly and Bernard Horgan 1994
© Artwork Gill & Macmillan

Design: Identikit Design Consultants, Dublin
Artwork: Maureen Kelly, Shayne Fox, Alanna Corballis
Editing: Brendan O'Brien
Photo Research: Anne-Marie Ehrlich

0 7171 2165 8
Print origination by Identikit Design Consultants, Dublin
and Typeform Ltd, Dublin
Printed in Ireland by ColourBooks Ltd, Dublin

All rights reserved.
No part of this publication may be reproduced, copied or transmitted in any form or by any means without written permission of the publishers or else under the terms of any licence permitting limited copying issued by the Irish Copyright Licensing Agency, The Writers' Centre, Parnell Square, Dublin 1.

ACKNOWLEDGMENTS

For permission to reproduce transparencies and photographs grateful acknowledgment is made to the following:
Science Photo Library; Stephen O'Reilly; A.C. Crookes; Slide File; E.T. Archive; Mansell Collection; Harry Smith Collection; Archiv für Kunst und Geschichte; Redferns; Hulton Deutsch; Zefa; National Portrait Gallery, London; Trinity College Dublin; Giraudon; Novosti; National Museum of Ireland; Preston-Mafham/Premaphoto Wildlife; Holt Studios International; Royal College of Physicians; Image Selection; Royal College of Surgeons of England; National Dairy Council; A.C. Cooper; Meteorological Service; J.F.P. Galvin; Butter Council.

CONTENTS

Introduction	iv
Safety	v

PHYSICS
1. Measurement — 1
2. Matter — 5
3. Energy — 7
4. Energy Sources — 13
5. Mass and Density — 19
6. Force, Weight & Mass — 24
7. Motion — 30
8. Levers — 34
9. Pressure — 40
10. Heat — 46
11. Movement of Heat — 53
12. Changes of State — 63
13. Electricity — 69
14. Ohm's Law — 76
15. Uses of Electricity — 82
16. Magnetism — 91
17. Static Electricity — 98
18. Waves and Sound — 102
19. Light — 110

CHEMISTRY
20. Starting Chemistry — 121
21. How to Examine a Substance — 129
22. Different Sorts of Changes — 133
23. Elements and Compounds—Atoms and Molecules — 139
24. Compounds and Mixtures — 144
25. Separating Mixtures — 154
26. The Air We Breathe — 164
27. Carbon Dioxide — 174
28. Water — 182
29. What are Atoms Made of? — 191
30. Elements Can Be Grouped Together — 197
31. Organising the Elements — 205
32. How Do Elements Form Compounds? — 208
33. Valency, Equations and Energy — 219
34. Acids and Bases — 226
35. Hard and Soft Water — 238
36. Metals — 245
37. How Reactive Are Metals? — 251
38. Electricity from Chemicals — 258
39. Oxidation and Reduction — 265

BIOLOGY
40. Starting Biology — 270
41. Cells — 275
42. Food — 280
43. The Digestive System — 283
44. Respiration and Breathing — 289
45. The Circulatory System — 295
46. Excretion — 303
47. Support, Protection and Movement — 306
48. Sensitivity and Co-ordination — 310
49. Reproduction — 314
50. Genetics — 319
51. Plants — 321
52. Photosynthesis — 324
53. Transport in Plants — 331
54. Responsiveness in Plants — 335
55. Plant Reproduction — 337
56. Ecology — 344
57. Conservation and Pollution — 350
58. Habitat—Field Study — 354
59. Soil — 362
60. Micro-organisms — 368

APPLIED SCIENCE

EARTH SCIENCE
61. The Universe — 374
62. Weather — 382

FOOD
63. Food — 394
64. Food Preservation — 402
65. Processing our Food — 409

ELECTRONICS
66. Electronics — 416
67. Diodes — 423
68. Transistors — 428

ENERGY CONVERSION
69. Energy Conversion — 432
70. Electromagnets — 436

HORTICULTURE
71. Horticulture — 441
72. Growing Plants — 459

MATERIALS
73. Materials Science — 477
74. Types of Materials — 489

Index — 517

INTRODUCTION

So you are going to study science over the next few years. Good for you!

Is the teacher going to show you how to build a space shuttle to go into space? Well, no. But remember that the scientists and engineers who built the Apollo space ship that took the first men to the moon began by studying the same things that you are going to study in your science course.

So, watch what happens in your experiments, listen carefully to your teacher, ask questions about what is happening and enjoy learning science.

WRITING REPORTS ABOUT YOUR EXPERIMENTS

Scientists have always written reports of the experiments they did. There are two reasons for this. A scientist who does not write a report may not, some time later, remember exactly what happened. Scientists also write reports so that they can share their discoveries with others. When you write about an experiment you should use the following headings when possible to help you write a clear report of your work.

Date: The date you did the experiment.
Aim: This is the purpose of the experiment—what you set out to measure or find out.
Method: A step-by-step explanation of how you did the experiment, with a clear and simple labelled diagram of the equipment you used.
Result: A clear description of what happened or the measurements you took when you did the experiment.
Conclusion: Your explanation of what happened when you did the experiment.

THE COURSE

Your Junior Certificate science course consists of five sections:
- physics
- chemistry
- biology
- applied science (choose two units out of six)
- local studies (available to only a few schools).

The material in each section is divided into
(a) core
(b) ordinary level
(c) higher level.

All students must study core material from entire course. Ordinary level students must study **ALL** the core material plus the ordinary level material only from any three sections. Higher level students must study **ALL** the core material plus **ALL** the other material from physics, chemistry and biology and two of the applied science units.

In this book the core material is marked with a blue border and the higher material with a red border. The ordinary level material is unmarked:
i.e. ■ = core material
     ■ = higher level material.

Definitions are highlighted with a black border: e.g.

The volume of a body is the amount of space it takes up.

INTRODUCTION

SAFETY

A science laboratory can be a very dangerous place for you and your friends if you are careless. Chemicals, electricity, hot liquids and broken glassware can cause serious injuries.

Your teacher will have safety rules for the school laboratories to keep you and your classmates safe. You must understand and follow these rules when doing experiments.

SAFETY RULES

- Do not enter the laboratory without permission.
- Put your bag and any other personal things safely out of the way.
- Do not eat or drink in the laboratory.
- Make sure you know the position of all safety equipment—fire extinguishers, fire blankets, eye-wash bottles, first-aid kit.

- Do not leave the place assigned to you without permission.
- Do not interfere with laboratory equipment.
- Keep your work area clean and tidy.
- Keep all equipment away from the edge of the bench.
- Make sure the equipment you use is clean and in good condition.

- Listen to your teacher's instructions carefully.
- Do not start an experiment until you are sure you understand these instructions.
- Wear safety glasses during all experiments with chemicals or hot liquids (even hot water).
- Report any cracked or damaged glassware to your teacher.
- Report all accidents, however slight, to your teacher.

- Do not touch, taste or smell any chemical.
- Never handle chemicals with your bare hands.
- Do not return chemicals to jars once you have removed them.
- Use pipette fillers to fill pipettes.
- Dry up any spillage immediately.

- Be careful with flammable vapours of substances like alcohol.
- Do not overfill test-tubes.
- Point test-tubes away from yourself and everybody else when heating substances.
- Clean and put away all equipment.
- Wash your hands at the end of each laboratory session.

SECTION I

PHYSICS

CHAPTER 1 MEASUREMENT

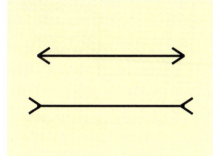

One of the reasons we study science is to make ourselves more aware of the world around us. To help us to do this we have five senses, **sight**, **hearing**, **smell**, **taste** and **touch**. But how reliable are these senses? Can we rely on them alone or do they need some help? Let's find out.

Which of the two lines in the diagram is the longer? Measure them and see.

Is the hat taller than it is wide? Measure it and find out.

Don't depend on your senses alone; use measuring instruments to help them.

LENGTH

Length is measured in **metres** (m) or **centimetres** (cm). 1 m = 100 cm. Long distances are measured in **kilometres** (km). 1 km = 1000 m.

To measure a straight length we use a ruler or metrestick.

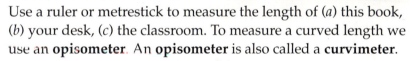

Use a ruler or metrestick to measure the length of (*a*) this book, (*b*) your desk, (*c*) the classroom. To measure a curved length we use an **opisometer**. An **opisometer** is also called a **curvimeter**.

Large opisometers, called **trundle wheels**, are used to measure distances along roads, athletic tracks, etc. Each time the trundle wheel does one turn it clicks. Each click tells you that the wheel has travelled one metre.

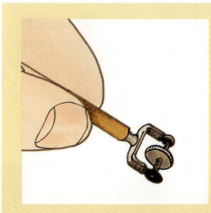

EXPERIMENT 1.1
AIM: TO MEASURE THE LENGTH OF A CURVE
Method
1. Turn the wheel until it sticks at the pointer.
2. Roll the wheel along the curve from one end to the other.
3. Now place the wheel at zero on a metrestick and roll it in the opposite direction until it sticks at the pointer again.

Result: The reading on the metrestick is the length of the curved line. To get an accurate result, repeat the experiment and take an average of the readings.

BREAKTHROUGH SCIENCE

Your metrestick or ruler can measure only to the nearest millimetre. To get a more accurate measurement we use a **vernier callipers**. Your teacher will show you how to use one.

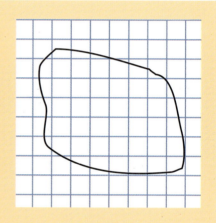
vernier callipers

AREA = 14 cm × 9 cm = 126 cm²

AREA

The unit of area is the **square metre** (m²) or **square centimetre** (cm²).

To calculate a regular rectangular area simply multiply length by breadth.

EXPERIMENT 1.2
AIM: TO MEASURE AN IRREGULAR AREA

Method

1. Draw the outline of an irregular area, for example your hand, on a sheet of graph paper.
2. Count the number of squares that are completely inside the outline.
3. Go round the edge and count the number of squares that are halfway, or more, inside the outline.
4. Add the two counts and you have the approximate area.

VOLUME

The volume of a body is the amount of space it takes up.

The unit of volume is the **cubic metre** (m³), but a more convenient unit is the **cubic centimetre** (cm³).

The volume of a liquid is measured in **litres** (1 litre = 1000 cm³)

In the case of a regular rectangular solid, volume = length × breadth × height. The volume of the block in the diagram is 24 cm³.

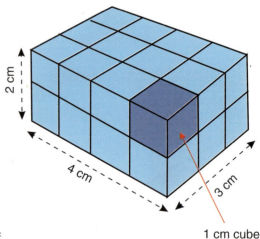
1 cm cube

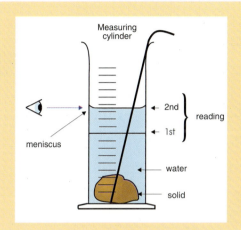

EXPERIMENT 1.3
AIM: TO FIND THE VOLUME OF AN IRREGULAR SOLID

Method
1. Pour some water into the graduated cylinder and note the volume.
2. Slide the solid gently into the water and note the new volume.

Result: The difference between the two volumes is the volume of the solid.

Note: the water surface is not flat but is curved downwards. This is called a **meniscus**. The readings should be taken from the lowest point (the bottom of the meniscus). In the case of a body that is too big to fit into a graduated cylinder, we can do the following experiment.

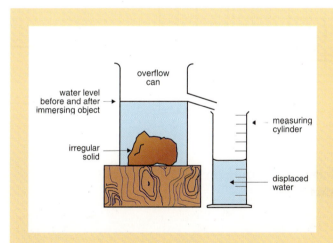

EXPERIMENT 1.4
AIM: TO FIND THE VOLUME OF A LARGE SOLID

Method
1. Fill the overflow can with water and wait until it stops dripping from the spout.
2. Place a cylinder under the spout and lower the solid gently into the water.
3. Wait until the water has stopped overflowing into the cylinder.

Result: The volume of water collected in the cylinder gives the volume of the body.

SUMMARY

Quantity	Standard unit	Symbol	Other units
Length	Metre	m	Kilometre, centimetre, millimetre
Area	Square metre	m^2	Square centimetre
Volume	Cubic metre	m^3	Cubic centimetre, litre

- The volume of a body is the amount of space it takes up.

QUESTIONS

Section A
1. The standard unit of length is the _____ .
2. The length of a page of this book would be measured in _____ , but the distance from Cork to Dublin would be measured in _____ .
3. To measure the length of a road on a map you would use an _____ . To measure the length of a football pitch you would use a _____ _____ .
4. One metre equals _____ centimetres.
5. Name an instrument that may be used to measure the diameter of a metal cylinder. _____ _____
6. One kilometre equals _____ metres.
7. Underline the area in cm^3 shown in the diagram: 4·4, 5·8, 6·5, 7·0.
8. 1 litre equals _____ cm^3.
9. 1 millilitre equals _____ cm^3.
10. A box of length 30 cm, width 20 cm and height 15 cm has a volume of _____ .
11. In the diagram below, X is an _____ _____ ; Y is a _____ _____ . The volume of the stone is _____ .

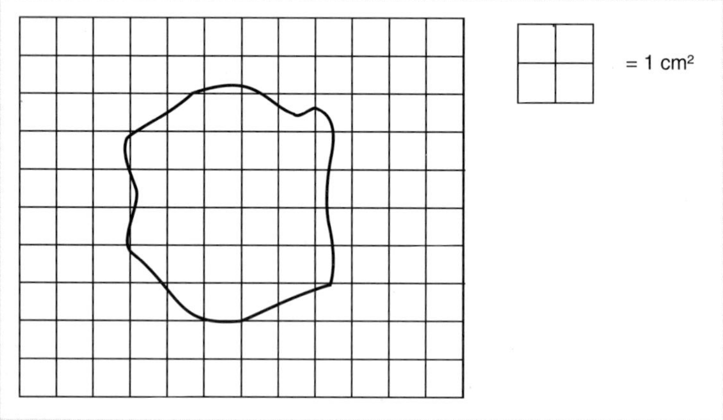

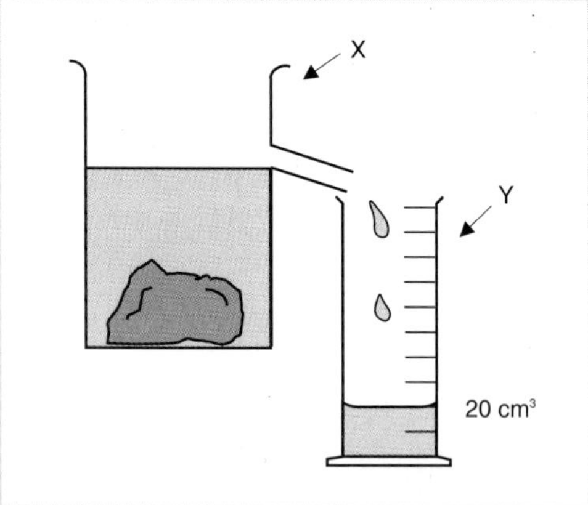

12. What is the reading on the vernier scale in the diagram below? _____

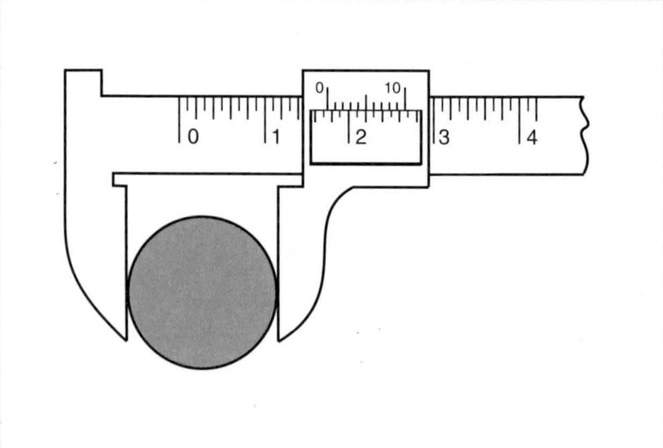

Section B

1. Which is the greatest: 50 cm, 3 m or 0·01 km?
2. Name the instrument shown in the diagram. Give one practical use of the instrument.
3. Using a map of Ireland, find the distance (a) by air, (b) by road from Cork to Galway.
4. Work out the area of one page of this book. What area of paper is in the whole book?
5. Describe an experiment to find the area of your hand.
6. What is the curved surface on the top of the water in the diagram called? What is the volume of the stone? If the stone were too big to fit in the cylinder, how would you find its volume?
7. Define the word "volume". Why might it be difficult to find the volume of one paper clip? How would you solve this problem?
8. Describe an experiment to find the volume of a cork. Why might your answer be slightly inaccurate?

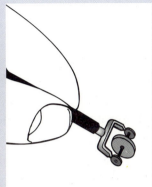

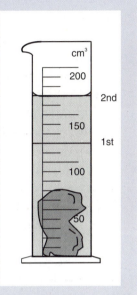

CHAPTER 2 MATTER

You probably know the names of hundreds of substances, such as concrete, glass, timber, plastic, milk, water and air. In science all of these substances are known as matter.

Matter is anything that occupies space.

Even though they are all called matter, they are not all the same, because matter can exist in three different states: solid, liquid and gas.

Solid	**Liquid**	**Gas**
Definite volume	Definite volume	No definite volume
Definite shape	No definite shape	No definite shape
Cannot flow	Can flow	Can flow
Cannot be compressed	Cannot be compressed	Can be compressed

A substance may change from one form to another, as you can see if you do the following experiment.

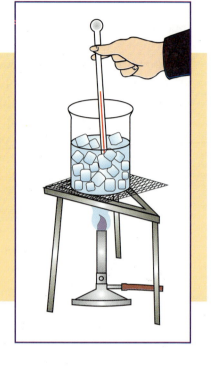

EXPERIMENT 2.1
AIM: TO SHOW THAT A SUBSTANCE CAN CHANGE STATE
Method
1. Cover the bulb of a thermometer with some ice in a beaker.
2. Note the temperature of the ice.
3. Heat the ice gently and note the temperature when it begins to melt.
4. When all the ice has turned to water, continue heating and note the temperature at which the water begins to boil.

Conclusion : A substance can change from solid (ice) to liquid (water) to gas (steam) if it is heated.

If a number of students do this experiment you will see that the windows of the science room become covered with condensation, because when the steam loses heat it turns back to water. If you take the water remaining in the beaker, pour it into an ice tray and put it in the freezer compartment of a fridge it will turn back to ice.

Conclusion: A substance can change from gas to liquid to solid if it loses heat.

On a rainy day the road becomes wet. If it is not raining the following day the road will soon dry up. So where did the water that was on the road go? It changed to gas and rose into the air. It did not boil because it was not hot enough. It evaporated.
A liquid can change to gas by evaporation even if it is not hot enough to boil.

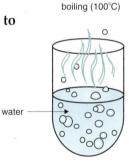
boiling (100°C)
boiling takes place throughout the liquid

evaporation (20°C)
evaporation takes place at the surface only

SUMMARY
- Matter is anything that takes up space.
- There are three states of matter: solid, liquid and gas.
- A substance can change from solid, to liquid, to gas if it gains heat.
- It can change back if it loses heat.

QUESTIONS

1. Matter is anything that _____ _____ .
2. Matter exists in three states: _____ _____ and _____ .
3. A _____ has a definite volume but no definite shape.
4. A _____ has no definite volume and no definite shape.
5. A _____ has a definite volume and a definite shape.
6. Underline the liquids in this list: bread, butter, milk, water, chalk, lemonade.
7. At what temperature does ice melt? _____
8. At what temperature does water boil? _____
9. Fill in the missing words in the diagram.
10. A change in state involves the loss or gain of _____
11. Choose words from the following—melting, freezing, evaporation, condensation—to describe these changes: (a) liquid to solid, (b) gas to liquid.

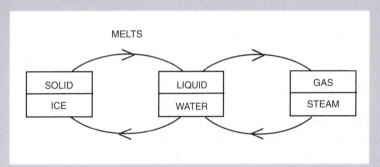

CHAPTER 3 — ENERGY

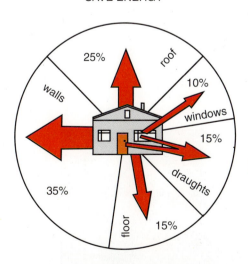

SAVE ENERGY — WITH INSULATION!

Some years ago people throughout the world became worried about energy. The cost of oil, coal and gas was going up. Supplies of oil, coal and gas were going down. We were all urged to play our part in energy conservation.

Posters began to appear with slogans like "SAVE ENERGY, SWITCH IT OFF". Other posters told us how much energy we were wasting by not having our homes insulated.

Energy was needed to heat homes and offices and to keep buses, trucks and factory machinery going. It was even needed to keep our bodies going, because *energy is the ability to do work*.

Energy is measured in **joules** (J). The joule is a very small unit, so we use the kilojoule (kJ). 1 kilojoule = 1000 joules.

In science, doing work means making something move. There are many different forms (kinds) of energy.

BREAKTHROUGH SCIENCE

STORED ENERGY
Potential energy
The water behind the dam has energy **stored** in it because of its **position**. It can run down through the channel and cause the turbine to spin. This can be used to make electricity.

The elastic band has energy stored in it because of its **condition** (it is stretched). It can make the paper move.

The kind of energy that the water and the elastic band have stored in them is called **potential energy**.

> *Potential energy is energy something has because of its position or condition.*

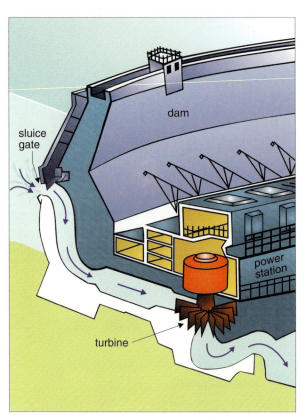

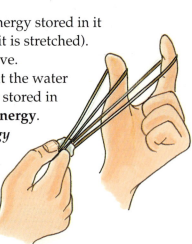

Chemical energy
All the things in the diagram have chemical energy stored in them. How much chemical energy is stored in a Mars bar? Examine the wrapper and find out.

petrol

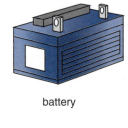
battery

food

Nuclear energy
The substances in the diagram are **radioactive**. They have energy called nuclear energy stored in them. This energy can be released in a nuclear power station and used to make electricity.

PHYSICS · ENERGY

OTHER FORMS OF ENERGY (NOT STORED)

Kinetic energy (movement energy)

The car and the hammer have energy because they are moving. This kind of energy is called kinetic energy or movement energy.

| *Kinetic energy is energy something has because it is moving.*

Heat energy

The water in the kettle has heat energy. The hotter it gets, the more heat energy it will have. If it gets hot enough it will turn to steam. Steam can make things move.

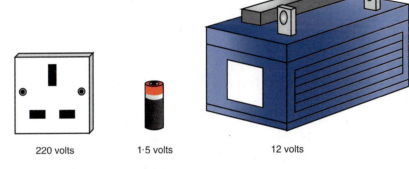

220 volts   1·5 volts   12 volts

Electrical energy

The ESB supplies us with electrical energy. We can also get electrical energy from a cell or from a battery.

Have you ever used any of these in your personal stereo?

Light energy

The bulb is giving out light energy.

Sound energy

The radio is giving out sound energy.

Magnetic energy

The magnet has magnetic energy. It makes the pins move.

9

WHERE ENERGY COMES FROM

All our energy came in the first place from the sun. Millions of years ago, dense tropical forests grew in many parts of the world. The trees took in energy from the sun and changed it into chemical energy. Other smaller plants also took in energy from the sun. Animals got their energy by eating the plants. As time went by, many changes occurred on the surface of the earth and the remains of plants and animals became buried under the ground and the seas. The pressure changed them into coal, oil and natural gas. The energy stored in coal, oil and natural gas is being used today to keep cars, trucks, trains and factories going.

It is important to realise that when we eat our food we are gaining energy but **we are not creating energy**. Neither are we destroying the energy stored in the food. We are simply changing it from one form to another.

Energy cannot be created or destroyed, it can only change from one form to another.

This is known as the **law of conservation of energy**. Many of the things that are most useful to us in our daily lives change energy from one form to another. Electrical cookers change electrical energy to heat energy. Light bulbs change electrical energy to heat and light energy.

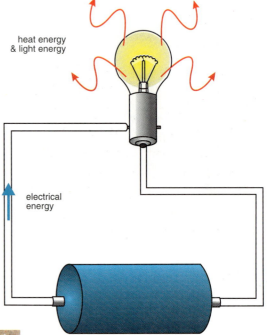

ENERGY CONVERSION

Rub your hands together. You are changing **kinetic energy** into **heat energy**.

Rattle a few coins in your pocket. You are changing **kinetic energy** into **sound energy**.

The English experimentalist William Crookes (1832–1919) invented the Crookes radiometer. He also investigated electrical discharges through vacuum tubes. These led to Thomson's discovery of the electron. Crookes investigated radioactivity and invented an alpha particle detector.

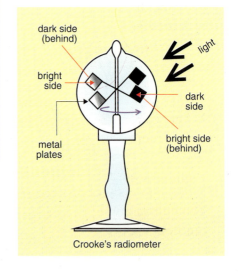

Crooke's radiometer

Place the Crookes radiometer in a sunny position. The plates will begin to spin round. Crookes' radiometer converts light energy to kinetic energy.

The bulb converts electrical energy to heat and light energy.

PHYSICS · ENERGY

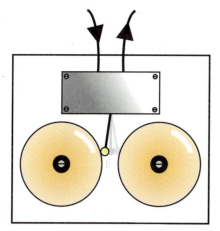

electric bell

The bell converts electrical energy to sound energy.

The electric motor converts electrical energy to kinetic energy.

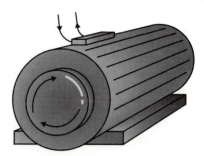

Electric Motor

SUMMARY

- Energy is the ability to do work.
- Energy is measured in joules.
- All our energy comes in the first place from the sun.
- **The law of conservation of energy**: energy cannot be created or destroyed, it can only change from one form to another.
- There are many different types of energy, e.g. kinetic energy, potential energy, heat energy, light energy, sound energy, chemical energy, electrical energy and nuclear energy.
- Chemical energy, nuclear energy and potential energy are examples of stored energy.
- Many things have been invented to help us to change energy from one form to another.

QUESTIONS

Section A
1. Define energy. _____
2. Energy is measured in units called _____.
3. All our energy comes, or has come from _____.
4. Name three different forms of energy. _____
5. A moving car has _____ energy.
6. A wound-up alarm clock has _____ energy.
7. A solar calculator can be powered by light because _____.
8. Chemical energy, nuclear energy and potential energy are all examples of _____ energy.
9. The energy a body has because of its position or condition is called _____ _____.
10. The energy a body has because it is moving is called _____ _____.
11. Energy cannot be _____ or _____, it can only be _____ from one _____ to another.
12. Why does a nail become hot when struck a number of times by a hammer? _____.

Section B
1. Give three ways in which the energy bill for your home could be reduced.
2. Name some appliance in your home that changes
   (a) electrical energy to light
   (b) electrical energy to heat
   (c) electrical energy to sound.
3. The sun's energy is responsible for world stocks of coal, oil and natural gas. Explain this.
4. Fill in the missing words in the diagram on the right.

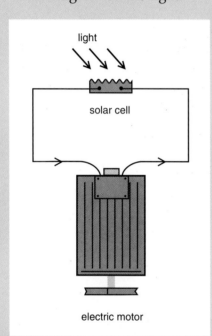
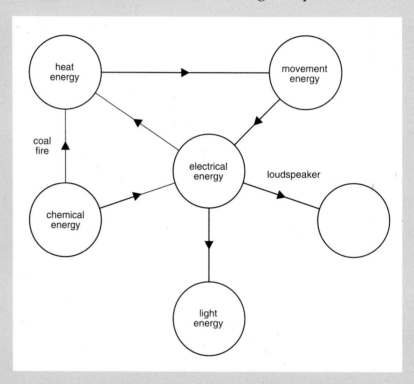

5. In the diagram (above left), what does the solar cell do to the light energy? What energy change takes place in the electric motor?
6. What energy changes take place when water falls from the top of a waterfall to the bottom? Will the water at the bottom be different from the water at the top in any way?
7. What energy changes take place when a girl throws a basketball into the air and it drops into the basket?
8. State what energy conversions take place in (i) a television, (ii) a Bunsen burner, (iii) a washing machine, (iv) a car, (v) an electric hairdrier.

CHAPTER 4 ENERGY SOURCES

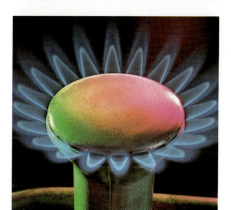

A fuel is any substance that can be used as a source of energy. Fossil fuels are the remains of plants and animals that lived millions of years ago. Coal, peat, oil and gas are all fossil fuels. Coal and peat were formed from plants; oil and gas were formed from tiny plants and animals that lived in the sea. Coal, oil and gas are **hydrocarbons** because they contain hydrogen and carbon.

There are two problems with fossil fuels.

1. Fossil fuels cause air pollution.

To study this, try the following experiment. Note: carbon dioxide gas turns limewater milky. Water changes cobalt chloride paper from blue to pink. You will probably need to dry the cobalt chloride paper in an oven before using it.

EXPERIMENT 4.1
AIM: TO EXAMINE THE PRODUCTS OF COMBUSTION (BURNING)

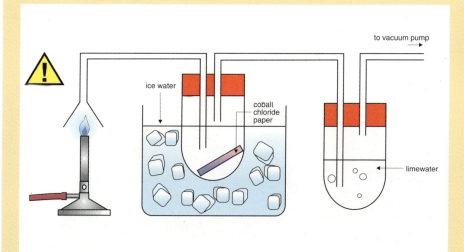

Method
1. Set up the experiment as shown and allow it to run until you see a change in the limewater and the cobalt chloride paper.
2. **Result:** the cobalt chloride paper turns pink and the limewater turns milky.
3. **Conclusion:** water and carbon dioxide have been produced.
4. Repeat the experiment with a candle or a burning peanut.

We learn two things from this experiment. First, natural gas is a clean fuel which produces carbon dioxide gas and water. (We breathe out the same things ourselves when we burn food in our bodies.) Second, other fuels are not so clean! Oil can produce a gas called sulphur dioxide. Household coal produces smoke and tarry fumes as well as sulphur dioxide. Even smokeless fuel, such as coalite, produces sulphur dioxide.

BREAKTHROUGH SCIENCE

2. The supply of fossil fuels is limited.
We have been burning fossil fuels for a long time now, and scientists are worried because the supply will eventually run out. They say that fossil fuels are **non-renewable**.
"Non-renewable" means that the supply will eventually run out.

ALTERNATIVE SOURCES OF ENERGY
Since the supply of non-renewable energy will run out, we need to look for alternatives. The alternatives in the table below have the advantage of being renewable, but they also have disadvantages.
"Renewable" means that the supply will not run out.

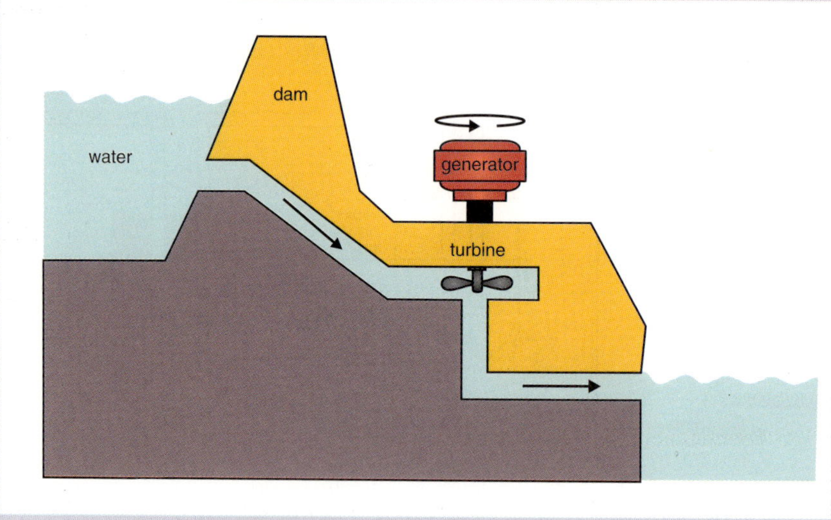

TYPE OF ENERGY	HOW IT IS PRODUCED	DISADVANTAGES
Hydroelectricity	A dam built across a river forms a lake. Water flowing through the dam turns a generator to produce electricity.	Not all rivers and lakes are suitable.
Wind energy	Large windmills are used to turn generators.	Expensive; low output; only a few areas are suitable. Can be unsightly if windmills are located in scenic areas.

14

TYPE OF ENERGY	HOW IT IS PRODUCED	DISADVANTAGES
Tidal energy	A dam built across an estuary forms a lake behind it, which fills and empties with the tides. The fast-flowing water turns generators.	Expensive to build. Few areas are suitable.

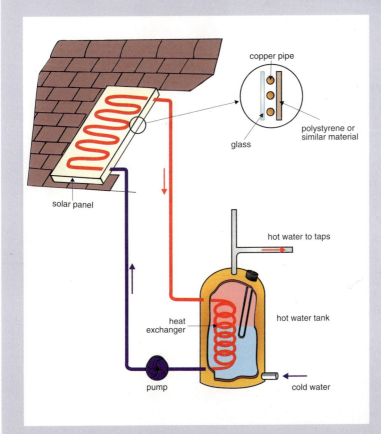

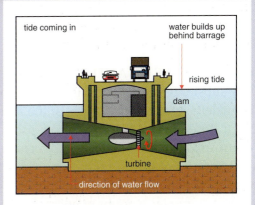

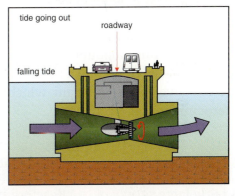

Solar energy	The sun's energy is trapped by solar panels and used to heat water. Another kind of solar panel contains cells that convert solar energy into electricity.	The energy is most needed in winter, when the sunshine is weakest.
Biomass	Fast-growing plants are used to make alcohol, which can be used instead of petrol.	Very large areas of land are needed.

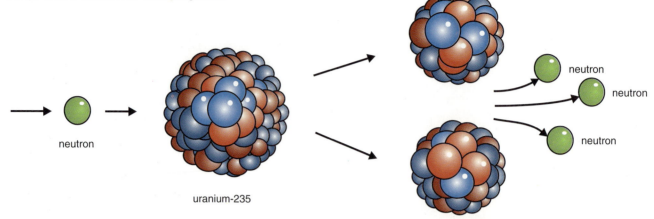

a single atom of uranium-235 undergoing fission

NUCLEAR ENERGY AND RADIOACTIVITY

Every substance in the world is made up of atoms. The nucleus, which is the centre of the atom, has a lot of energy stored in it. Some substances such as **uranium**, **plutonium** and **radon** have so much energy that they have to get rid of some of it by sending out radiation. These are called **radioactive substances**.

A much greater amount of energy can be released by a process called **nuclear fission**. This means that tiny particles called neutrons strike the nucleus and split it in two, releasing a huge amount of energy in the form of heat. This is what happens in a nuclear reactor. The heat is used to convert water to steam, which turns a generator and makes electricity.

Advantages	Disadvantages
A fairly large supply of fuel is available. Huge amounts of energy can be released from a tiny amount of fuel.	Reactors are very expensive to build and maintain. Dangerous nuclear waste is produced.

Nuclear energy is the energy stored in the nucleus of an atom.

The question of whether nuclear energy should be regarded as renewable or non-renewable is far too complicated to be considered in this book.

Marie Curie (1867–1934), a Polish physicist, made a lot of discoveries on radioactivity. She discovered the elements radium and polonium in 1898, and shared the 1903 Nobel prize for physics with her French husband, Pierre. She received the prize for chemistry in 1911, and is the only person to have won two Nobel science prizes. Her daughter Irene Joliot-Curie (jointly with her husband Frederic) was awarded the 1935 Nobel prize for chemistry. The unit of radioactivity, the curie, is named in honour of Marie Curie.

PHYSICS · ENERGY SOURCES

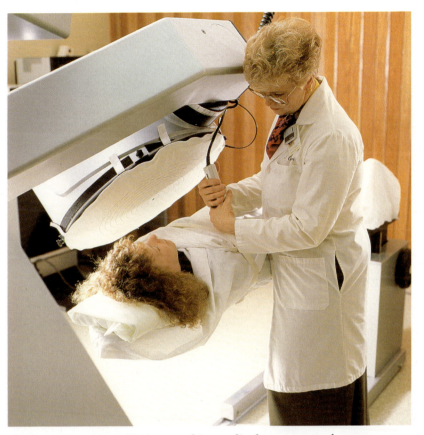
Radioactive material being used in medical treatment of cancer

USES OF RADIOACTIVE SUBSTANCES
Radioactive substances are used:
- in nuclear power stations, to generate electricity
- to preserve food by irradiating it
- like X-rays to detect faults in machinery
- to make nuclear weapons
- in medicine, to treat cancer
- in medicine, to trace the flow of blood in the body
- to sterilise medical equipment.

SUMMARY
- A fuel is any substance that can be used as a source of energy.
- Fossil fuels are the remains of dead plants and animals.
- Fossil fuels such as coal, peat, oil and gas are non-renewable, which means that the supply will eventually run out.
- Coal, oil and gas are hydrocarbons, because they contain hydrogen and carbon.
- Uranium, plutonium and radon are radioactive substances.
- Nuclear energy is energy stored in the nucleus of an atom.
- Nuclear fission means that neutrons split the nucleus of an atom in two, releasing a large amount of energy.

Renewable	**Non-renewable**
Hydro	Coal
Wind	Peat
Tidal	Oil
Solar	Gas
Biomass	

QUESTIONS

Section A
1. What is a fuel? _____
2. What are fossil fuels? _____ Name three. _____
3. _____ _____ and _____ are called hydrocarbons because they contain _____.
4. When natural gas burns, it produces the gas _____ _____ and the liquid _____.
5. There are two problems with fossil fuels. What are they? _____
6. Non-renewable means _____.
7. Renewable means _____.
8. Name two renewable sources of energy. _____.
9. Name two non-renewable sources. _____.
10. What is meant by biomass? _____
11. Name three radioactive substances. _____, _____ and _____
12. State two uses of radioactive substances.
13. Nuclear energy is the energy _____.
14. Nuclear fission means that _____.

Section B
1. Try to find out the following.
   (a) What is the chemical name for natural gas?
   (b) What sort of gas was used in the home before natural gas?
   (c) How is natural gas better than what went before it?
   (d) Why did so many of the pipes have to be replaced when natural gas was introduced?
2. Why is electricity not regarded as a source of energy?
   What energy changes take place in a hydroelectric station?
   Why is hydroelectricity used more in Norway than in Ireland?
3. Wind energy has one great advantage—it is renewable. What does this mean?
   Why might people object to the erection of wind generators in their area?
4. What is a nucleus?
   What is nuclear fission?
   How is the energy released in a nuclear reactor used to make electricity?
   Give as many uses as you can for radioactive substances.
5. Give as many benefits of nuclear energy as you can. A government plan to build a nuclear power station at Carnsore Point, County Wexford was scrapped because of opposition from certain sections of the public. Why do you think they opposed it?
6. Visit your school library or local public library and do a project on solar energy.

CHAPTER 5 MASS AND DENSITY

A standard bag of sugar has a mass of one kilogram. If you add some sugar to the bag, will this change the mass? If you take some sugar out, will the mass change?

The mass of an object is the amount of matter in it.
The standard unit of mass is the kilogram (kg). When we are dealing with smaller masses we use the gram (g). A kilogram is 1000 grams.

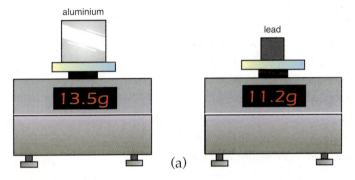
(a)

In diagram (a) we are comparing the mass of aluminium and lead. Is this a fair comparison? We should compare the same volume of each substance, as in diagram (b).

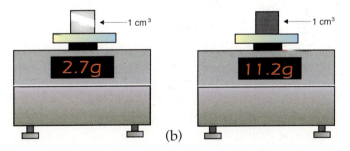
(b)

Scientists say that the density of lead is 11·2 g/cm^3 and the density of aluminium is 2·7 g/cm^3. This can also be written as 2·7 g cm^{-3}.

The density of a substance is the mass of 1 cm^3 of that substance.

PROBLEM
A piece of metal has a mass of 72 g and a volume of 18 cm^3. What is the density of this metal?

18 cm^3 has a mass of 72 g => 1 cm^3 has a mass of 4 g. The density of the metal is 4 g/cm^3. From this we learn that

Density = mass/volume

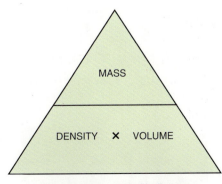

Cover what you want in this diagram and you will see how to find it.

BREAKTHROUGH SCIENCE

PROBLEM

A piece of metal has a mass of 37 g. If the density of the metal is 7·2 g/cm³, what is the volume of the piece of metal?

From the triangle

$$\text{Volume} = \frac{\text{mass}}{\text{density}} = \frac{37 \text{ g}}{7\cdot2 \text{ g/cm}^3} = 5 \text{ cm}^3$$

EXPERIMENT 5.1
AIM: TO FIND THE DENSITY OF AN IRREGULAR SOLID

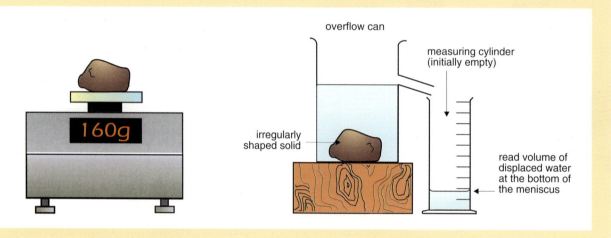

Method

1. Use a balance to find the mass of the solid.
2. Fill the overflow can with water and wait until it stops dripping.
3. Place a graduated cylinder under the spout.
4. Lower the solid gently into the water.

Result: The volume of the water that overflows equals the volume of the solid.

$$\text{Density} = \frac{\text{mass}}{\text{volume}}$$ can be used to calculate your answer.

Archimedes (298–212 BC) was the greatest mathematician of ancient times. He lived at Syracuse in Sicily and was killed during its capture by the Romans in the Second Punic War. He designed an enormous catapult, and a concave mirror to focus the heat of the sun on the sails of ships attacking Syracuse. He invented the Archimedes screw, which is still used to transfer grain, and made many original contributions to geometry. Archimedes is best known for his discovery on the weight of a body immersed in a liquid: Archimedes' Principle.

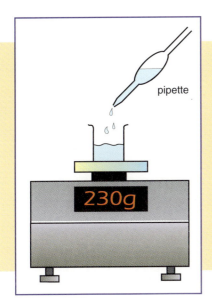

EXPERIMENT 5.2
AIM: TO FIND THE DENSITY OF A LIQUID

Method
1. Find the mass of an empty beaker.
2. Using a pipette, place 100 cm^3 of the liquid in the beaker.
3. Find the mass of the beaker plus the liquid.
4. Subtract the mass of the beaker from this to get the mass of the liquid.

Result: Density = $\dfrac{\text{mass}}{100 \text{ cm}^3}$

This table gives the densities of some common substances.

SUBSTANCE	DENSITY (g/cm^3)	SUBSTANCE	DENSITY (g/cm^3)
Aeroboard	0·02	Aluminium	2·7
Oak	0·65	Steel	7·9
Methylated spirits	0·8	Copper	8·9
Ice	0·92	Lead	11·2
Water	1·0	Mercury	13·6
Ebony	1·1	Gold	19

FLOATING AND DENSITY

When the air in a balloon is heated it spreads out (expands) and takes up a lot more space. As a result it becomes less dense. Since the hot air inside the balloon is less dense than the cold air outside, the balloon floats. Cork is less dense than water, so it floats in water. A stone is more dense than water, so it sinks.

A body will float in a fluid if it is less dense than the fluid.

SUMMARY

- The mass of a body is the amount of matter in it.
- The density of a substance is the mass of 1 cm³ of it.
- Density = $\dfrac{\text{mass}}{\text{volume}}$
- A body will float in a fluid if it is less dense than the fluid.

QUESTIONS

Section A

1. What is this piece of apparatus called? _____ What is it used for? _____
2. Define mass. _____
   The standard unit of mass is the _____
3. Define density. _____
4. Water has a density of 1 g cm⁻³. What does this mean?
   _____
5. A coin has a mass of 12 g and a volume of 1·5 cm³. The density of the coin is _____.
6. A stone has a density of 2·5 g/cm³ and a volume of 30 cm³. The mass of the stone is _____.
7. The mass of A shown in the diagram is 50 g. What is its density? _____ 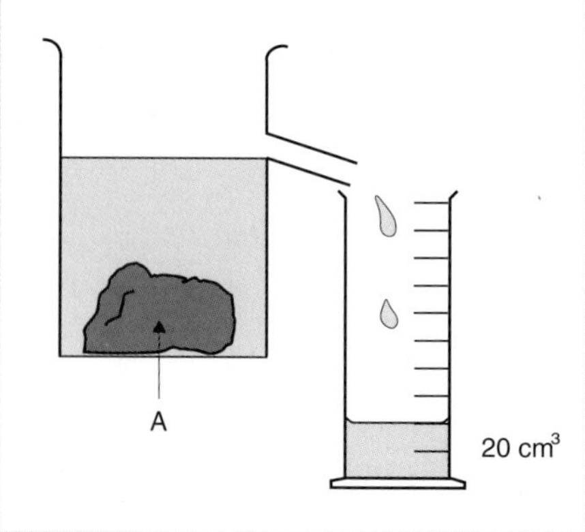
8. Why does a copper coin sink in water?
   _____
9. Why does a match float in water?
   _____
10. Three liquids that do not mix are shaken together and allowed to settle as in the diagram. Which liquid has the lowest density? ___ If liquid Y is water, suggest a possible density for liquid Z. _____

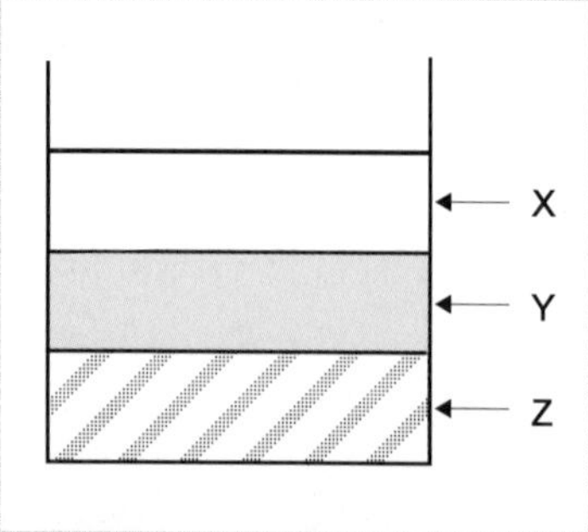

11. A body cannot float in a fluid whose _____ is less than that of the body.
12. In the diagram, the loaded test tube floats vertically in water with 18 cm³ of the tube below the surface. How much further will it sink if placed in a liquid of density 0·9 g/cm³? ___

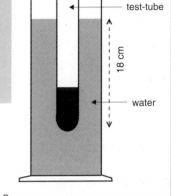

Section B
1. In the diagram, what is A called? What is B called? What happens when the stone is lowered gently into the water? What is the density of the stone if it has a mass of 25 g and a volume of 10 cm³?

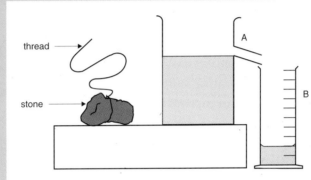

2. How would you find the mass of water needed to fill a beaker to the brim?
3. You have a box of matches. When you place a match on a balance, its mass is so small that it does not register. How would you find the mass of a single match?
4. Convert 8 kg to g and 350 g to kg.
5. In the diagram, what is the mass of the stone? What is its volume? What is its density?
6. A stone of mass 120 g displaces 40 cm³ of water when it is dropped into an overflow can. What is the density of the stone?
7. 30 ballbearings displace 45 cm³ of water. What is the volume of one ballbearing?
8. Complete the table.

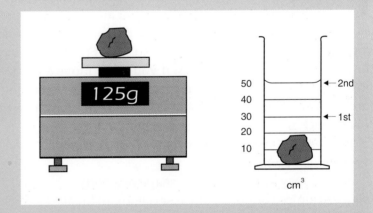

Object	Volume	Mass	Density
A	5 cm³	40 g	—
B	50 cm³	—	2 g cm⁻³
C	—	18 g	3 g cm⁻³

9. Explain why cream floats on top of milk. Name a substance that would not float in mercury.
10. Why does a hot air balloon float up in the air? Explain how the submarine in the diagram can be made to sink. How can it be made to rise again?
11. Describe an experiment to find the density of a cork.
12. Describe an experiment to find the density of sugar. (Note: sugar dissolves in water.)
13. Describe a simple experiment to show that cooking oil is less dense than water.
14. An object has a volume of 60 cm^{-3} and a mass of 54 g. What is its density? Will it float in water? Will it float in methylated spirits? (The density of methylated spirits is 0·8 g cm^{-3}.)

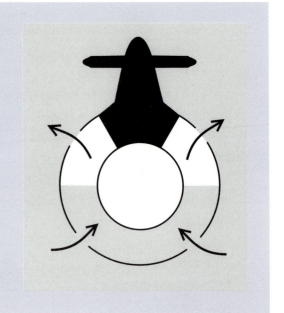

FORCE, WEIGHT & MASS — CHAPTER 6

If you cycle your bicycle you are exerting a force. This causes the bicycle to move. If you pedal harder the bicycle moves faster. If you turn the handlebars you are exerting a force that causes the bicycle to change direction. By applying the brakes you apply a force that causes the bicycle to stop. So a force can change the speed or direction of a body. If you try to move a bus by pushing it, you are hardly likely to succeed. But even though the force you are exerting is not big enough, **you are still exerting a force** which is trying to change the motion of the bus.

A force is anything that changes, or tries to change, the motion of a body.

A force may also change the shape of a body (squeeze a piece of putty) or cause a body to break (snap a pencil).

Force is measured in units called **newtons (N)**. For example, when you lift a mass of 100 g you are applying a force of 1 newton.

PHYSICS · FORCE, WEIGHT & MASS

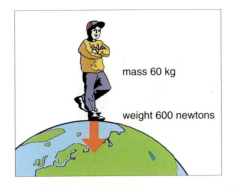

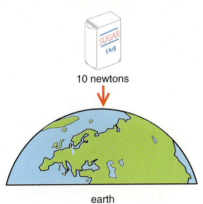

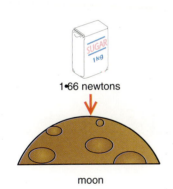

GRAVITY AND WEIGHT

When a hammer thrower starts to swing a hammer, why does it keep going round in a circle? Why doesn't it fly off? Because the hammer thrower is exerting a force which keeps it in place. As soon as he releases, it it will fly off.

Why does the moon keep spinning round the earth? Same reason. There is a force keeping it in place. The earth exerts a force on everything within a certain distance of it. This force is called gravity. The force it exerts on me is called my weight.

The weight of a body is the force pulling it towards the centre of the earth.

Like all other forces, weight is measured in newtons. In the diagram, the amount of sugar in the bag is the same in both cases, but the pull of gravity is less on the moon. Remember that the weight of a body will change if the pull of gravity on it changes, but the mass of a body will not change unless we take something out of it or put something into it.

Weight (in newtons) = mass (in kilograms) × 10

This means that a mass of 1 kilogram will have a weight of approximately 10 newtons on earth.

WORK

Work is done when a force moves an object. In which of the diagrams is more work being done?

BREAKTHROUGH SCIENCE

Work = force × distance
The unit of work is the **joule (J)**, which is also the unit of energy.

Problem
If you used a force of 500 N to push a trolley a distance of 10 m, how much work would you do?
Work = force × distance = 500 N × 10 m = 5000 J = 5 kJ

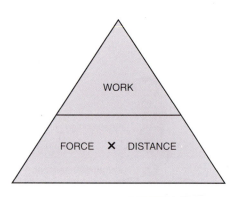

POWER

Mowing a lawn is hard work. A person who has a large lawn might use a motor mower (petrol or electric). An electric mower changes electrical energy into movement energy. This makes the blade spin round and cut the grass. The mower is very powerful, which means it can do a great deal of work in a short time.

Power is the amount of work done per second.

Power is measured in units called **watts (W)**.
1 watt means 1 joule per second.
1000 watts = 1 kilowatt

The mower is marked 3·5 kW (3·5 thousand watts), so it can do 3500 joules of work per second! When we do work we are changing energy from one form to another. In fact, we could define work as the amount of energy changed from one form to another in one second.

PAIRS OF FORCES

If you pushed the wall of your school, would it push you back? Yes it would. Try it while wearing roller skates.

Blow up a balloon, hold it by the neck and then let it go.

Hook two spring balances together and pull from each end. What do you notice about the two readings?

Simple experiments like this make it clear what is meant by Newton's third law.

Newton's third law: To every action there is an equal and opposite reaction.

PHYSICS · FORCE, WEIGHT & MASS

FRICTION
Rub your hands together very quickly. Can you feel them heating up? This heat is caused by friction.

Friction is a force that tries to stop things from sliding over each other.

Put some soap solution on your hands. Now rub them together very quickly. Do they heat up? Not very much? That's because there is very little friction. You used soap as a lubricant to reduce the friction between your hands.

Lubricants are substances that reduce friction.

EXPERIMENT 6.1
AIM: TO INVESTIGATE FRICTION

Method
1. Attach a spring balance to a block of wood and pull until the block just begins to move. Note the reading on the balance.

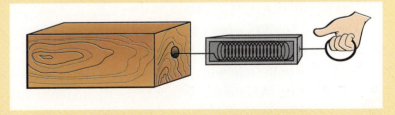

2. Attach some sandpaper to the block and repeat step 1. A greater force is needed.
3. Remove the sandpaper and put some grease on the bottom of the block. A smaller force is needed.

Conclusion: Lubricants such as grease reduce friction.

EFFECTS OF FRICTION

If there were no friction between your bicycle tyre and the ground, what would happen? Could you light a match if the edge of the matchbox was perfectly smooth? What would happen to a car engine if it ran out of oil?

Now make a list of ways in which friction can be (i) helpful, (ii) troublesome.

SUMMARY
....
- A force is anything that changes, or tries to change, the motion of a body.
- The weight of a body is the force pulling it towards the centre of the earth.
- Weight (newtons) = Mass (kilograms) × 10
- Work = Force × Distance
- Power is the amount of work done per second.
- To every action there is a equal and opposite reaction.
- Friction tries to prevent surfaces from sliding over each other..

QUESTIONS
....

Section A
1. Define force. _____ The unit of force is the _____.
2. Give two effects that force can have on a body. _____ and _____
3. The weight of a body is the force _____.
4. Weight is measured in _____.
5. The pull of gravity is _____ on the moon than on earth.
6. The weight of a body depends on its _____ and its _____.
7. The _____ of a body is the same on the moon as on earth, but its _____ is less.
8. A car of mass 1000 kg has a weight of _____.
9. Underline in the following list the mass of the hydrogen atom: 3.4×10^2 kg, 2.4×10^{24} kg, 1.7×10^{-27} kg. What is the weight of the hydrogen atom? _____
10. A force of 5 N moves an object a distance of 4 m. The work done is _____.
11. Power is defined as _____. The unit of power is the _____.
12. Give two examples of friction being useful. _____ and _____
13. State one method by which friction can be reduced. _____
14. _____ are substances that reduce friction.

Section B

1. If you are cycling to school, what forces do you apply and what effects do they have? Are you being affected by friction? Explain. How do you stop? Is friction involved here?
2. What is the weight of an object that has a mass of (i) 2 kg, (ii) 5 kg, (iii) 500 g?
3. What is the mass of an object that has a weight of (i) 30 N, (ii) 4·5 N?
4. A man pushes a pram weighing 250 N a distance of 200 m. How much work has he done?
5. Complete the table.

FORCE	DISTANCE	WORK
20 N	30 m	—
50 N	—	750 N
—	15 m	600 N

6. A man raising money for charity pulls a trailer weighing 1000 N a distance of 5 kilometres. How much work has he done?
7. Find out your own mass. Can you now calculate your weight?
   If you walk up a stairs, what mass are you carrying?
   What force are you exerting?
   If the upstairs is 3 m higher than the downstairs, how much work have you done?
8. A girl has a mass of 40 kg. What weight is she?
   If she climbs up onto a platform that is 10 m high, how much work has she done? How much energy has she used?
   If she took 30 seconds to climb up onto the platform, what is her average power?

MOTION CHAPTER 7

Work is done when a force is used to move something. The rate at which something moves is called its **speed**.

$$Speed = \frac{distance}{time}$$

Speed is measured in metres per second (written as m/s or m s^{-1})

Speed is not the same thing as velocity. To know the velocity of a body we must know its speed and the direction in which it is moving. For example, a fisherman listening to the weather forecast is interested in both the speed and the direction of the wind. Indeed, the direction could be more important than the speed.

Velocity is speed in a given direction.

For example, a runner could be said to have a velocity of 5 metres per second, due east.

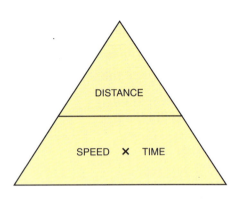

ACCELERATION

A woman is sitting in a car. The car is not moving, so its velocity is zero. She starts the car and presses the accelerator. One second later the car is moving at 5 metres per second. After another second it is moving at 10 metres per second. Its velocity is changing by 5 metres per second each second. So we can say that it has an acceleration of 5 metres per second each second, or 5 metres per second squared.

Acceleration is the change of velocity per second, or

$$Acceleration = \frac{change\ in\ velocity}{time\ taken}$$

Problem

A car takes 5 seconds to increase its velocity from 10 metres per second to 25 metres per second. What is the acceleration of the car?

$$Acceleration = \frac{change\ in\ velocity}{time\ taken} = \frac{(25-10)\ m/s}{5\ s} = \frac{15\ m/s}{5\ s} = 3\ m/s/s.$$

PHYSICS · MOTION

Note the units. This means that the car increases its velocity by **3 metres per second each second**. The acceleration can be written as 3 m/s/s or 3 m/s² or 3 m s⁻².

DISTANCE–TIME GRAPH

This table gives the figures for a girl running 50 metres.

Time (s)	0	1	2	3	4	5	6	7	8	9	10
Distance (m)	0	5	10	15	20	25	30	35	40	45	50

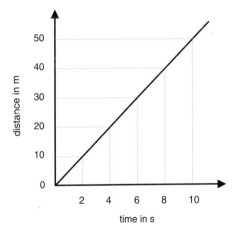

We can now plot a graph of distance against time, putting time on the horizontal axis and distance on the vertical axis.

Since velocity = $\frac{\text{distance}}{\text{time}}$

we can find the velocity from the graph. Can you see how to do this?

The velocity is $\frac{50 \text{ m}}{10 \text{ s}}$ = 5 m/s.

VELOCITY–TIME GRAPH

The table gives the velocity of a coin dropping through the air.

Velocity (m/s)	0	10	20	30	40	50	60	70	80	90	100
Time (s)	0	1	2	3	4	5	6	7	8	9	10

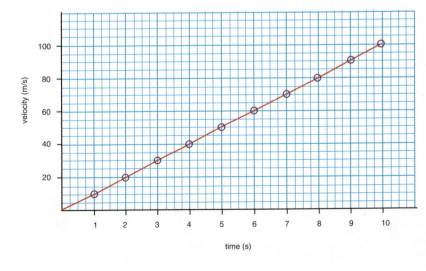

We can now plot a graph of the velocity against the time.

Since acceleration is $\frac{\text{change in velocity}}{\text{time taken}}$

we can find the acceleration from the graph.

The acceleration is $\frac{100 \text{ m s}^{-1}}{10 \text{ s}}$ = 10 m s⁻².

MOMENTUM

Would it be easy to stop a golf ball travelling at 120 kilometres per hour? No! Why? Because its velocity is so big.

Would it be easy to stop a bus travelling at 60 kilometres per hour? No! Why? Because its mass is so big. To stop something that is moving you must destroy its momentum.

Momentum = mass × velocity

Problem
A body of mass 60 kg is moving at 3 m/s. What is its momentum?
Momentum = mass × velocity = 60 kg × 3 m s^{-1} = 180 kilograms.metres per second.

SUMMARY

- Speed = $\dfrac{\text{distance}}{\text{time}}$ (units m s^{-1})

- Velocity is speed in a given direction.

- Acceleration = $\dfrac{\text{change in velocity}}{\text{time taken}}$ (units m s^{-2})

- Momentum = mass × velocity (units kilogram metres per second)

QUESTIONS

Section A
1. Define velocity. _____
2. A car travels 100 m in 4 s. What is its average velocity? _____
3. Define acceleration. _____ What are the units of acceleration? _____
4. A car increases its velocity from 10 m/s to 25 m/s in 5 s. What is its acceleration? _____
5. A car starting from rest accelerates at 3 m/s for 4 s. Its velocity at the end of this time is _____.
6. The velocity of a car changes from 25 m s^{-1} to 40 m s^{-1} in 5 seconds. What is its acceleration?
7. The velocity of a car changes from 24 m s^{-1} to zero in 3 seconds. What is its acceleration?
8. The diagram shows a graph of distance against time. What is the distance travelled after 3 s? _____ What is the velocity? _____

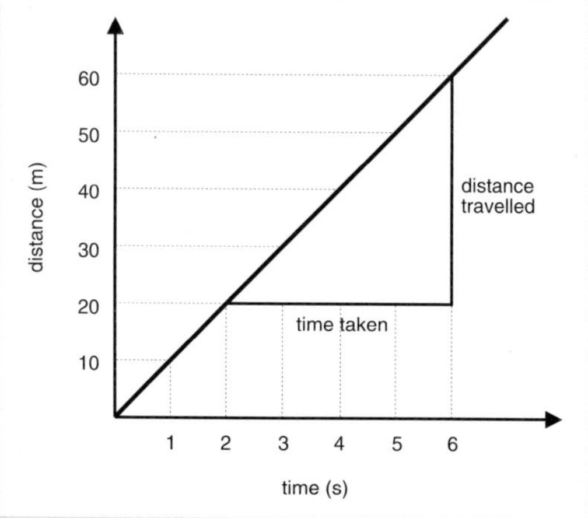

9. From the graph of velocity against time, find (i) the velocity after 3·5 s; _____, (ii) the acceleration. _____
10. Define momentum. _____
11. A body has a mass of 5 kg and a velocity of 3 m/s. Its momentum is _____.
12. If a body of mass 4 kg has a momentum of 16 kilograms.metres per second, its velocity is _____.

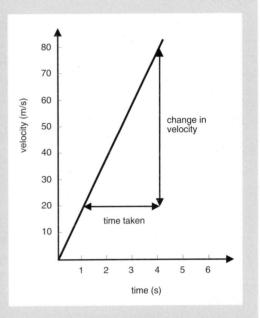

Section B
1. What is speed? What are the units of speed? A cyclist travels 100 m in 20 seconds. What is the average speed of the cyclist?
2. A car starts from rest with a constant acceleration of 5 m/s^2. What is its speed after 4 s? How long would it take to travel 2 km at this speed?
3. Complete the table.

After	1 s	2 s	3 s	4 s	5 s	6 s
Speed	5 m s^{-1}	10 m s^{-1}				30 m s^{-1}

What is the acceleration?

4. A motorist travelling at 20 m s^{-1} has 4 seconds in which to stop her car to avoid crashing. If she just succeeds, what is her acceleration? Scientists use two words to describe minus acceleration. Try to find out what they are.
5. A trolley moved along a track in a straight line. The distance travelled in metres by the trolley from the start was measured every second. The results are shown in the table.

Time (s)	1	2	3	4	5	6
Distance (m)	2	4	6	8	10	12

Draw a graph of the results, putting time on the horizontal axis (x-axis) and distance on the vertical axis (y-axis).
(i) On the graph, mark the distance travelled in 2·5 s.
(ii) Mark the time taken to travel 9 m.
(iii) What is the velocity of the trolley when it has travelled 9 m?
(iv) Is the velocity different when it has travelled 12 m?
(v) State whether or not the trolley is accelerating. Explain your answer.

6. A parachutist freefalls from an aircraft. The table shows her speed during the first 10 seconds.

Time (s)	0	1	2	3	4	5	6	7	8	9	10
Speed (m/s)	0	10	20	30	40	25	18	14	10	10	10

Draw a graph of speed against time. What is the acceleration at the beginning? What is causing this acceleration? What is this acceleration called? When did she open her parachute?

7. The graph represents a man walking along a straight road.

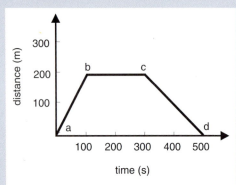

(i) Between what two points on the graph is he travelling at a steady speed away from his starting point? What is his speed at this stage?
(ii) What is happening between points (b) and (c)?
(iii) What is happening between points (c) and (d)? What is his speed at this stage?

LEVERS CHAPTER 8

CENTRE OF GRAVITY

A metrestick can be balanced on one finger if the finger is placed at the correct point. In a box of chocolates, each chocolate has its own weight and the chocolates are all in different positions, but the box of chocolates can be balanced on one finger.

The point at which all the weight of a body appears to act is called its centre of gravity.

EXPERIMENT 8.1
AIM: TO FIND THE CENTRE OF GRAVITY OF A SHEET OF CARDBOARD

Method
1. Set up the apparatus as shown in the diagram. Make sure the cardboard can swing freely on the pin.
2. When the cardboard comes to rest, draw a line on the cardboard along the plumb line.

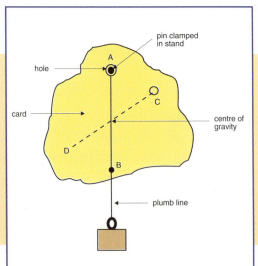

PHYSICS · LEVERS

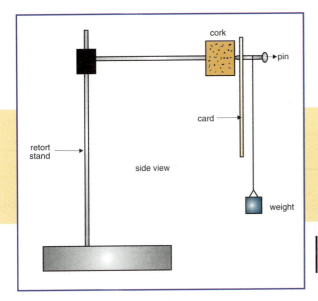

3. Repeat step 2 with the pin at a different corner of the cardboard.
Result: The point where the lines meet is the centre of gravity of the cardboard.

Equilibrium: a body is in equilibrium when all the forces acting on it cancel each other.

STATES OF EQUILIBRIUM

Stable
Tilt the flask slightly. This **raises its centre of gravity.** Now let it go: it will not fall over.

Unstable
Tilt the flask slightly. This **lowers its centre of gravity.** Now let it go: it will fall over.

Neutral
Move the flask slightly. There is **no change in the height of its centre of gravity.** Now let it go: it will stay in its new position.

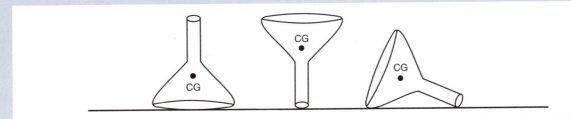

STABLE DESIGN

There is a difference between stable equilibrium and stable design. Did you notice that when you tilted the flask in diagram (b) the vertical line through its centre of gravity moved outside its base? This means that it is unstable. It is unstable because its base is too narrow and its centre of gravity is too high. It is clear from the diagrams that the principle of stable design is to have a **low centre of gravity** and a **wide base**.

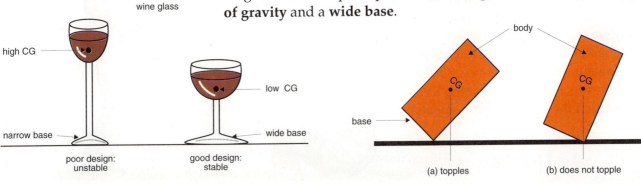

35

BREAKTHROUGH SCIENCE

LEVERS

If the centre of gravity of the leaning tower of Pisa were not directly above the base, the tower's weight would create a turning effect which would cause the tower to topple over. However, not all turning effects are bad. Without turning effects we could not use the many levers that make our lives so much easier.

For example, the door in the diagram is a lever. When it is pushed it turns on its hinges and opens. The force we use to push the door open is called the **effort**. The hinge is called the **fulcrum**.

A lever is a rigid body which is free to turn about a fixed point called the fulcrum.

Here are some more common levers.

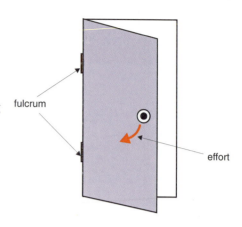

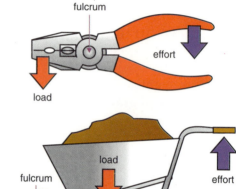

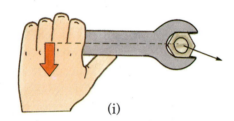

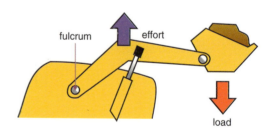

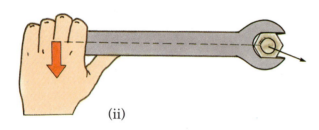

In diagram (i) the nut cannot be loosened.

In diagram (ii) the nut is loosened by applying the same force at a greater distance from the fulcrum.

The turning effect of a force is called the moment of the force.

The moment of a force = force × perpendicular distance from the fulcrum to the line of action of the force.

Problem

What is the moment of the force being applied in the diagram?

  Moment = force × distance = 500 N × 2 m = 1000 N m.

  All levers obey the law of the lever (also called the principle of moments).

PHYSICS · LEVERS

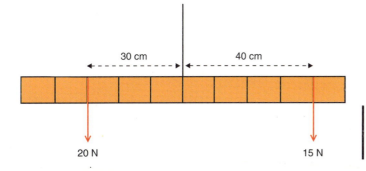

Law of the lever: when a lever is balanced, the sum of the clockwise moments is equal to the sum of the anticlockwise moments.

EXPERIMENT 8.2
AIM: TO VERIFY THE LAW OF THE LEVER

Method
1. Hang a metrestick from a stand.
2. Adjust the thread until the metrestick is balanced. The thread is now at the centre of gravity.
3. Hang three masses on the metrestick and adjust their positions until the metrestick balances.

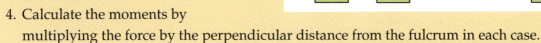

4. Calculate the moments by multiplying the force by the perpendicular distance from the fulcrum in each case.
5. The sum of the clockwise moments should equal the sum of the anticlockwise moments.
6. If you have time, repeat steps 1–5 with three forces on one side and two forces on the other.

We can use the law of the lever to find the weight of an object.

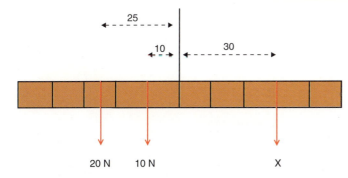

Problem
The lever in the diagram is balanced under the action of the forces shown. Find the value of X.
$(20 \times 25) + (10 \times 10) = X \times 30$
$500 + 100 = 30X$
$600 = 30X$
$20 = X$.

SUMMARY
- The point at which all the weight of a body appears to act is called its centre of gravity.
- A body is in stable equilibrium if tilting it slightly raises its centre of gravity.
- A body is in unstable equilibrium if tilting it slightly lowers its centre of gravity.
- A body is in neutral equilibrium if moving it slightly does not change the height of its centre of gravity.
- The principle of stable design is to have a low centre of gravity and a wide base.
- A lever is a rigid body that is free to turn about a fixed point called a fulcrum.
- The moment of a force means the turning effect of the force.

- The moment of a force = the force × the perpendicular distance from the fulcrum to the line of action of the force.
- The law of the lever: when a lever is balanced, the sum of the clockwise moments equals the sum of the anticlockwise moments.

QUESTIONS

Section A

1. The point at which the weight of a body appears to act is called the _____.
2. The diagram shows a piece of cardboard that has been hung freely from points A, B and C. What name is given to the point X? _____
   What is the purpose of the plumb line AW? _____ 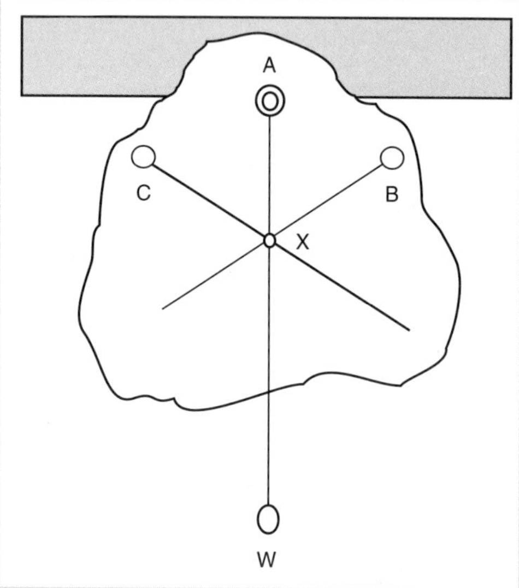
3. Name the three states of equilibrium. _____
4. What is the state of equilibrium of a rectangular piece of wood hanging freely from a nail as shown in the diagram? _____ 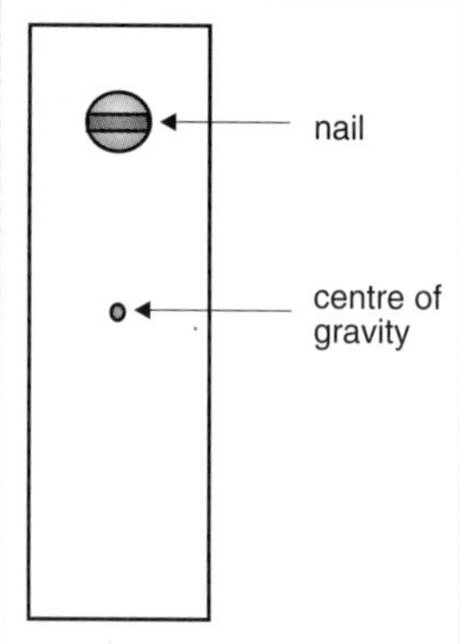
5. The stability of a body can be increased by _____ its base or _____ its centre of gravity.
6. Name the state of equilibrium of a spherical marble at rest on a clockglass as shown in the diagram. _____ _____ 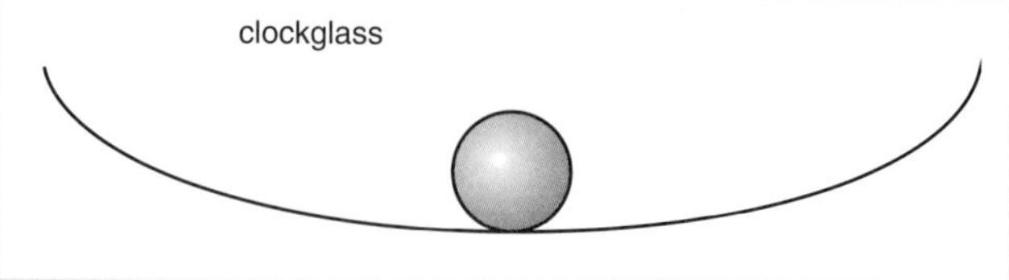
7. Why are passengers allowed to stand downstairs but not upstairs in a bus? _____
8. A lever is a _____ body that is free to _____ about a fixed point called the _____.
9. Name three common levers. _____
10. Give two common examples of a lever. _____ and _____
11. The metrestick in the diagram is balanced. At what mark does the weight of the metrestick appear to act? _____

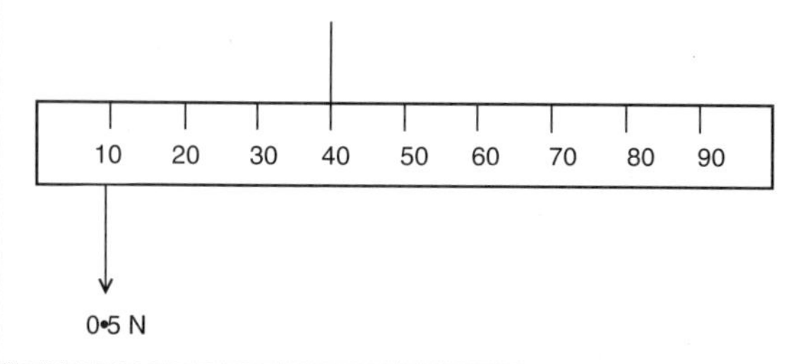

12. To find the weight of a bowl of berries, a girl set up the apparatus in the diagram. What result did she get? _____

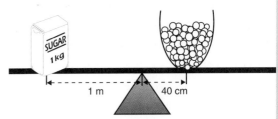

13. When a force of 12 N is applied at right angles to a door at a distance of 0·5 m from the hinge, the moment of the force about the hinge is _____.

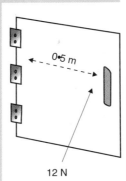

14. Draw a diagram of a wheelbarrow in your copy and mark the fulcrum.

Section B
1. Describe, with the aid of a diagram, how to find the centre of gravity of an irregularly shaped piece of cardboard.
   How would you check that the point you get is correct?
   What has the centre of gravity got to do with stability?
2. Vehicles must have a stable design; otherwise they would turn over easily. Explain, with the aid of simple diagrams, how the design of (a) a racing car, (b) a bus is made stable.
3. The diagram shows an example of a lever. What is a lever? Give two more everyday examples of a lever.
4. What is meant by the moment of a force?
   State the law of the lever.
   A uniform metrestick is balanced as shown in the diagram. At what point of the metrestick is the 25 N weight?

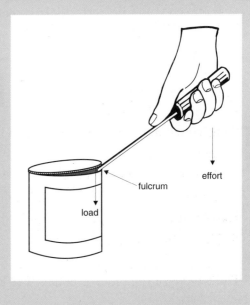

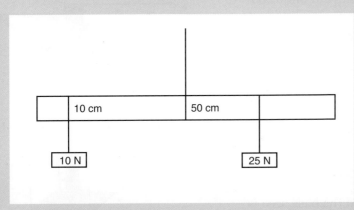

5. Study these diagrams carefully. The arrows indicate the position of the load, effort or fulcrum. Label each one correctly.
6. A uniform metrestick is suspended at its midpoint. A mass of 30 g is suspended at the 15 cm mark. Where must a 25 g mass be suspended to balance the metrestick?
7. A uniform metrestick is suspended at its midpoint. A mass of 20 g suspended at the 5 cm mark is balanced by a mass of 25 g at the X cm mark. Find the value of X.
8. A uniform metrestick, supported at its centre of gravity, has a 50 g mass suspended at the 20 cm mark, a 20 g mass at the 40 cm mark and a 10 g mass at the 70 cm mark. Where must a 100 g mass be placed so that the metrestick will be perfectly balanced?

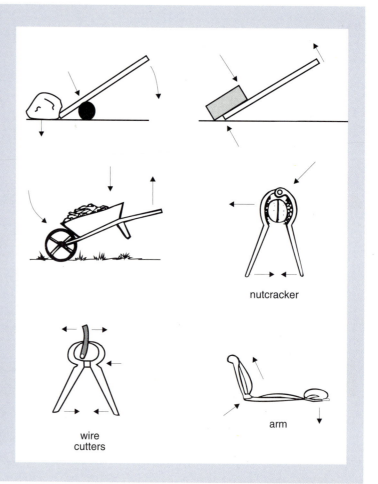

nutcracker

wire cutters

arm

PRESSURE CHAPTER 9

Fill your schoolbag with books and hang it from your shoulder by a strap. Now hang it from your shoulder by a length of string. Feel the difference? The weight is the same but the force is now pressing on a smaller area, so the pressure is greater because:

$$pressure = \frac{force}{area}$$

Pressure is force per unit area.

The weight of the block is 600 N. The base area is 3 cm × 4 cm = 12 cm².

The pressure = $\frac{force}{area}$ = $\frac{600 \text{ N}}{12 \text{ cm}^2}$ = 50 N/cm²

But when the block is standing on its end the base area is 3 cm × 2 cm = 6 cm².

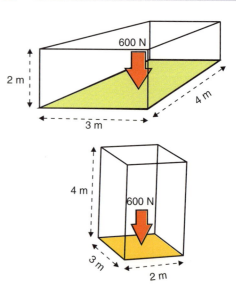

The pressure = $\dfrac{\text{force}}{\text{area}}$ = $\dfrac{600 \text{ N}}{6 \text{ cm}^2}$ = 100 N/cm^2 or 100 Ncm^{-2}.

It is clear from this example that the pressure depends not just on the force (the weight of the block), but also on the area over which the force is spread. The smaller the area, the greater is the pressure.

Small area => large pressure

Large area => small pressure

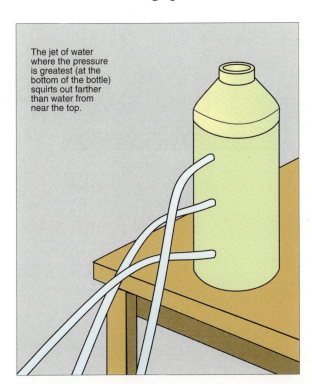

The jet of water where the pressure is greatest (at the bottom of the bottle) squirts out farther than water from near the top.

For mathematical problems on pressure we can use the triangle in the diagram.

A block that has a weight of 40 N exerts a pressure of 8 N/m^2 = on the ground. What is the base area of the block?

The area = $\dfrac{\text{force}}{\text{pressure}}$ = $\dfrac{40 \text{ N}}{8 \text{ N/m}^2}$ = 5 m^2

PRESSURE IN FLUIDS (LIQUIDS AND GASES)

Try the simple experiment shown in the diagram. Note that the nearer the hole is to the bottom, the further out the water shoots. In other words, the greater the depth, the greater is the pressure.

This experiment can be done very simply by filling a tall plastic bottle with water, screwing the top on, making three holes in the side of the bottle and removing the top.

BREAKTHROUGH SCIENCE

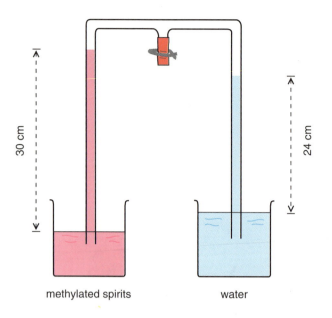

methylated spirits   water

In the diagram, the atmospheric pressure supports 30 cm of methylated spirits but only 24 cm of water. This is because water is more dense than methylated spirits.

AIR

Air has weight. To prove this, weigh a soft basketball or football. Pump air into the ball until it is very hard. Weigh the ball again. The increase in weight is caused by the air that was pumped into the ball.

ATMOSPHERIC PRESSURE

We live at the bottom of a sea, but it is a sea of air, not a sea of water. Nevertheless, the air has weight and since it has weight it exerts pressure. Usually we are not aware of atmospheric pressure because, strange as it may seem, the atmosphere does not just press down, **it presses equally in all directions**. To illustrate this, try the following experiment.

EXPERIMENT 9.1
AIM: TO DEMONSTRATE ATMOSPHERIC PRESSURE I

Method
1. Fill a glass jar to the brim with water.
2. Cover it with a sheet of cardboard.
3. Holding the cardboard in place, turn the jar upside down.
4. Remove your hand. The cardboard stays in place. It is supported by the atmospheric pressure **acting upwards**.

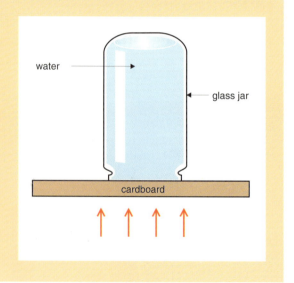

The following experiment shows that atmospheric pressure also works sideways.

PHYSICS · PRESSURE

EXPERIMENT 9.2
AIM: TO DEMONSTRATE ATMOSPHERIC PRESSURE II

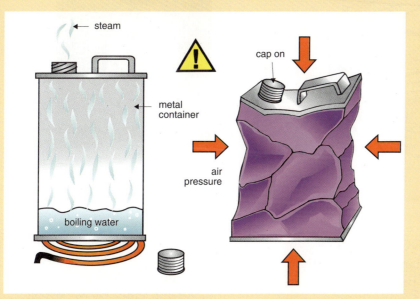

Method
1. Place a little water in the bottom of a tin can.
2. Boil the water until the steam has driven all the air out of the can.
3. Carefully seal the can tightly and cool it until the steam condenses, leaving a partial vacuum inside the can.

Result: Watch how the atmospheric pressure crushes the can.

Note: This experiment can be done very effectively using an empty mineral can and some plasticine, but wear a padded glove to protect your hand from being burned.

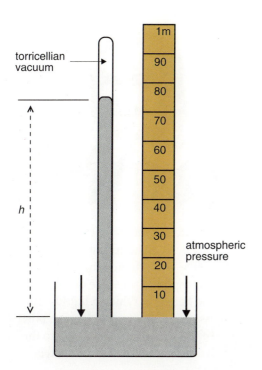

THE BAROMETER

One way of measuring atmospheric pressure is to see how much mercury it will support in a closed tube. Pressure is measured in newtons per square metre (N/m^2). The newton per square metre is also called the **pascal (Pa)**, so $1\,N/m^2 = 1$ Pa.

Normal atmospheric pressure is sufficient to support a column of mercury 76 cm high. Another way of saying this is to say that normal atmospheric pressure is 1000 hectopascals. High atmospheric pressure generally indicates dry weather; low pressure indicates wet weather.

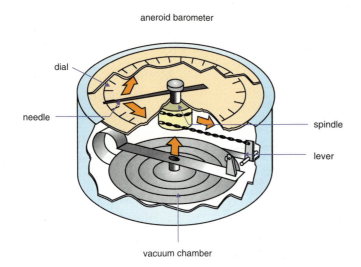

THE ANEROID BAROMETER (NO LIQUID)

This consists of a corrugated box that has been partly evacuated. When the atmospheric pressure increases it presses down the top of the box. This, through a system of levers, causes a needle to move over a scale.

Pressure in a fluid depends on depth and density.
Since pressure in fluids depends on depth, the pressure changes as we go higher into the atmosphere. An instrument called an **altimeter** is used by pilots to measure altitude (height). It does this by measuring pressure, so it is based on the barometer.

SUMMARY

- Pressure = $\dfrac{\text{force}}{\text{area}}$
- The unit of pressure is 1 N/m² or 1 pascal.
- Pressure in a liquid depends on depth and density.
- Standard atmospheric pressure is 76 cm of mercury or 1000 hectopascals.
- An aneroid barometer contains no liquid.

The French philosopher and scientist Blaise Pascal (1623–1662) made many contributions to mathematics and physics. Pascal invented the first calculating machine. He studied hydrostatics and invented the syringe and the hydraulic press. He worked on probability theory by calculating the odds involved in various gambling games. The unit of pressure, the pascal, is named in his honour.

The Italian mathematician and physicist, Evangelista Torricelli (1608–1647) demonstrated that atmospheric pressure determines the height to which a liquid will rise in a tube inverted over a bowl of the same liquid. This led to the development of the barometer.

QUESTIONS

Section A
1. Pressure is defined as _____.
2. Pressure is measured in _____ or _____.
3. A cube of side 2 metres rests on a table and exerts a force of 20 newtons. What is the pressure? _____
4. A rectangular block is 5 m long, 3 m wide and 3 m high. The weight of the block is 180 newtons. Calculate the pressure exerted by the block
   (a) when it is resting on one of its square faces _____
   (b) when it is resting on one of its rectangular faces. _____
5. Pressure in a fluid depends on _____ and _____.
6. A _____ is used to measure atmospheric pressure.
7. What property of mercury makes it suitable as a liquid in barometers? _____
8. The atmosphere can normally support a column of mercury _____ centimetres high or a column of water _____ metres high.
9. X is a liquid often found in a thermometer but never found in a barometer. X is _____.
10. A type of barometer that does not contain liquid is _____.
11. An instrument used to measure altitude is called _____ and is based on _____.
12. On a wet day the atmospheric pressure would tend to be _____ than normal.
13. The base of a dam is thicker than the top because _____.
14. Why is it necessary to punch two small holes or one larger hole in the top of a can of liquid in order to get the liquid to flow freely? _____
15. The diagram shows the weather forecast chart for a certain day. What type of weather would you have expected in Ireland on that day?

Section B
1. Define pressure. A box weighing 48 newtons has a length of 2 m, a breadth of 2 m and a height of 3 m. It is standing on a table as shown in the diagram. Calculate the pressure exerted by the box on the table
   (i) when one of its square faces is resting on the table as shown
   (ii) when it is turned so that one of its rectangular faces is resting on the table.

2. Describe a simple experiment to show that the atmosphere exerts pressure. Comment on the atmospheric pressure on top of a mountain and at the bottom of a mine.
3. What change, if any, would you expect in the pressure of the atmosphere if the amount of water vapour in the air increased? Explain briefly why it is possible to use a barometer to measure altitude.
4. On a day when the height of the mercury barometer was 76 mm, what would be the atmospheric pressure in hectopascals?
5. What is the aim of the experiment in the diagram? What happens to the air in the can when the water boils? The cap is screwed back on the can and the burner is turned off. What happens to the steam in the can when it cools? What happens to the can then?
6. A block of metal weighing 200 N exerts a pressure of 2 N m^{-2} on a floor. What is the area of the base of the block?

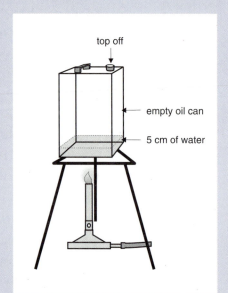

HEAT CHAPTER 10

In an earlier chapter we learned that **heat is a form of energy.** We shall now study the effects of heat on solids, liquids and gases.

EXPERIMENT 10.1
AIM: TO SHOW THAT SOLIDS EXPAND WHEN HEATED

Method
1. Check that the cold ball will just pass through the ring.
2. Heat the ball strongly. The ball expands. (This means that its volume increases.) It will not now pass through the ring.
3. Allow the ball to cool. It contracts. (This means that its volume decreases.) It can pass through the ring once again.

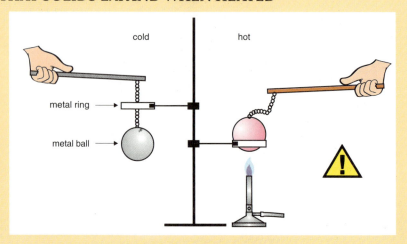

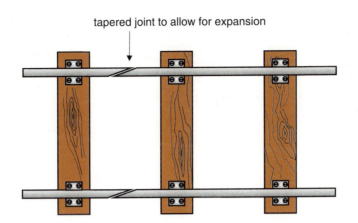

The ends of each length of rail are tapered to overlap with the next length of rail. This allows the rails to expand while at the same time allowing the train to move smoothly along the track.

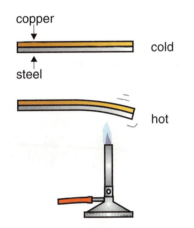

THE BIMETALLIC STRIP

Different substances expand by different amounts when they are heated. One application of this is the bimetallic strip. When the strip is heated the copper expands more than the iron. This causes the strip to bend. The strip straightens out again when it cools. We use this to make a **thermostat** (a device for keeping the temperature fairly steady). When the temperature reaches a preset value the strip bends and cuts off the electricity. The temperature at which this happens can be set by turning the screw. When the strip cools it straightens out and switches the heater back on. A bimetallic strip can also be used to set off a fire alarm. Can you see from the diagram how it works?

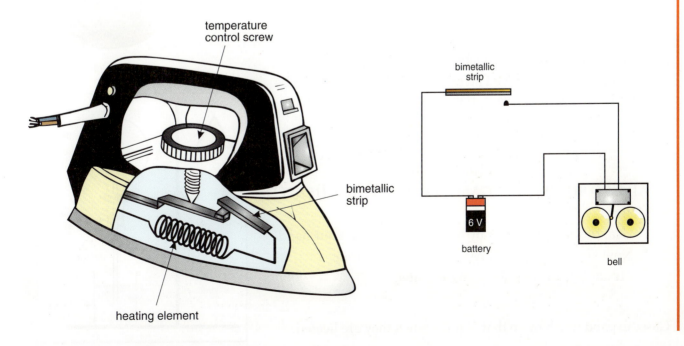

BREAKTHROUGH SCIENCE

EXPERIMENT 10.2
AIM: TO SHOW THAT LIQUIDS EXPAND WHEN HEATED
Method
1. Mark the level of the liquid with an elastic band.
2. Heat the liquid. It expands up the tube.
3. Cool the liquid. It contracts.

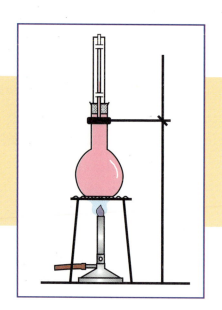

Liquids expand more than solids when they are heated.

WATER
Most substances expand (take up more space) when they are heated, and contract (take up less space) when they are cooled. Water is somewhat different.

When water is cooled it contracts, but when the temperature reaches 4°C it begins to expand again. So water has its greatest density at 4°C. From 4°C to 0°C the water expands and becomes less dense. As a result, ice floats in water.

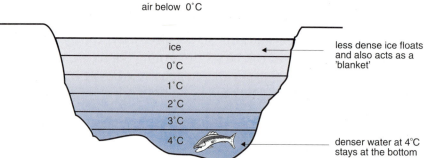

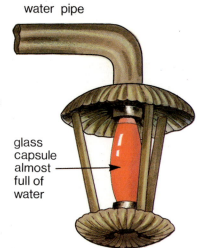

a sprinkler

SPRINKLERS
In the event of a fire the water will expand. This causes the glass to burst, and water is sprayed all over the fire. Sprinklers can be seen on the ceilings of supermarkets, carpet showrooms, etc.

EXPERIMENT 10.3
AIM: TO SHOW THAT GASES EXPAND WHEN HEATED
Method
1. Heat the air in the flask by holding it between your hands.
2. Now heat the flask gently with a Bunsen burner.
   The air bubbles out through the water.

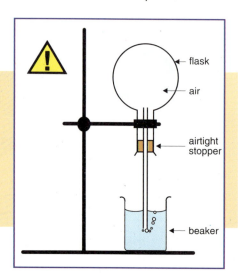

Gases expand much more than liquids when they are heated.

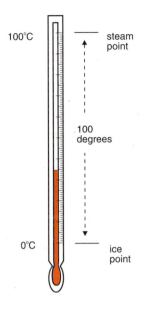

EFFECTS OF HEAT
When we heat a substance we give its molecules more energy, so they move further apart and the substance expands. As we give the molecules more energy the temperature rises. An instrument for measuring temperature is called a thermometer. The temperature of a substance is a measure of how hot the substance is.

THERMOMETERS
The most simple type of thermometer contains mercury in a glass tube. This thermometer works on the principle that liquids expand when they are heated. The scale used on this thermometer is called the Celsius scale.

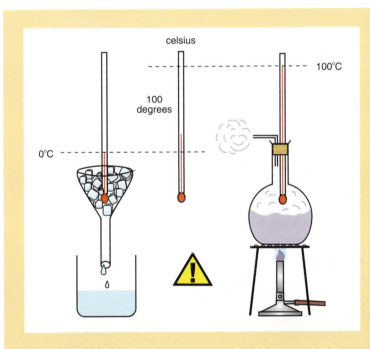

EXPERIMENT 10.4
AIM: TO MARK THE TWO FIXED POINTS ON AN UNMARKED THERMOMETER

Method
1. Place the thermometer in a funnel of melting ice. Mark the mercury level as 0 °C.
2. Place the thermometer above some water in a flask and heat the water until it boils. Mark the mercury level as 100 °C.
3. The length between the two marks can now be divided into 100 equal parts, each representing 1 °C.

ALCOHOL THERMOMETERS
Alcohol is also used in thermometers because it has a lower freezing point than mercury (112 °C), it expands more than mercury and it is cheaper.

Mercury	Alcohol
1. Easily seen	1. Must be coloured
2. Measures from −39 to 357 °C	2. Measures from −112 to 78 °C
3. Does not wet (stick to the glass)	3. Wets the glass
4. Expensive	4. Cheap

Other liquids were tried in thermometers. Water was not suitable because it froze at 0 °C, boiled at 100 °C, did not expand evenly and stuck to the glass when it expanded up the tube.

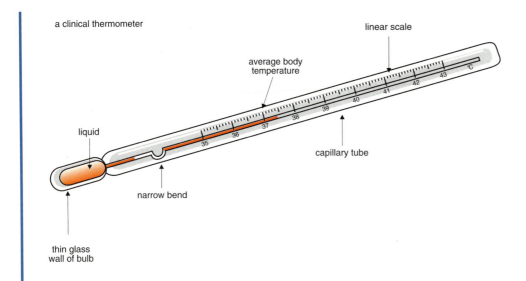

a clinical thermometer

THE CLINICAL THERMOMETER

If you have ever had your temperature taken, you have seen a clinical thermometer. The scale reads from 34 °C to 43 °C. The normal temperature of the human body is about 37 °C, but if you are ill your temperature may be higher. When the thermometer is put into your mouth the mercury expands up the tube. When it is taken out of your mouth the mercury would be expected to contract fairly quickly: however, the constriction in the tube causes the mercury thread to break. This gives plenty of time to read the temperature. The thermometer must then be shaken to return the mercury to the bulb before the thermometer is used again. Ordinary thermometers should not be shaken.

TEMPERATURE AND HEAT

It is important to realise that, although heat affects the temperature of a body, heat and temperature are not the same thing.

Temperature indicates the level of heat in a body, not the amount of heat.

The amount of water in these two vessels is the same, but the level of water is not. If you open the tap, water flows from the higher level to the lower level until the two levels are the same.

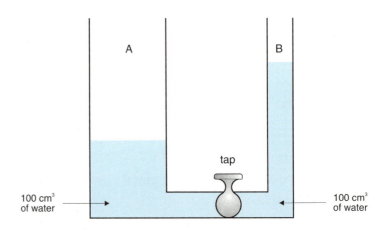

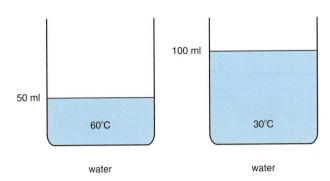

The beakers have been placed on two identical heaters for the same amount of time.

In the same way, heat will flow from the higher temperature (level) to the lower temperature if it can until the two temperatures are the same. The difference between temperature and heat can be illustrated as shown in the diagram. The amount of heat that has been put into these two beakers of water is the same, but their temperature is not the same.

Temperature is a measure of how hot a body is.

SUMMARY

- Heat is a form of energy. Heat is measured in joules.
- Temperature indicates the level of heat in a body.
- Temperature is measured in degrees Celsius.
- The fixed points on the Celsius scale are 0 °C and 100 °C.
- Solids, liquids and gases all expand when heated.
- The advantages of mercury as a liquid in thermometers are that
  (1) it is easy to see
  (2) it can measure up to 357 °C
  (3) it does not wet (stick to) the glass.
- The advantages of alcohol as a liquid in thermometers are that
  (1) it expands more than mercury
  (2) it can measure down to −112 °C
  (3) it is cheap.
- Water is an unsuitable liquid for use in thermometers because
  (1) it freezes at 0 °C
  (2) it boils at 100 °C
  (3) it does not expand evenly
  (4) it wets the glass.

QUESTIONS

Section A

1. Heat is a form of _____.
2. Solids, liquids and gases all _____ when they are heated.
3. A _____ is used to measure temperature.
4. The freezing point of water is _____ degrees Celsius.
5. The boiling point of water is _____ degrees Celsius.
6. The two liquids commonly used in thermometers are _____ and _____.
7. Give two reasons why water is an unsuitable liquid for use in thermometers.
   _____
8. Give two advantages of alcohol over mercury in thermometers. _____
9. Give two advantages of mercury over alcohol in thermometers. _____
10. A _____ thermometer is used to take your temperature.
11. What is the principle on which the mercury thermometer is based? _____
12. Water has its greatest density at a temperature of _____.
13. A bimetallic strip _____ when it is heated.
14. A thermostat switches a device _____ or _____ when a certain _____ is reached.
15. How does this flashing bulb work? _____
16. Why are gaps left in railway lines? _____
17. Usually when a body is heated its volume increases and its _____ decreases.
18. Most substances _____ when heated and _____ when cooled. Water is different. When water is cooled it _____ until a temperature of _____ is reached. Below that temperature it begins to expand again. As it expands it becomes less _____. This is why _____ floats in water.

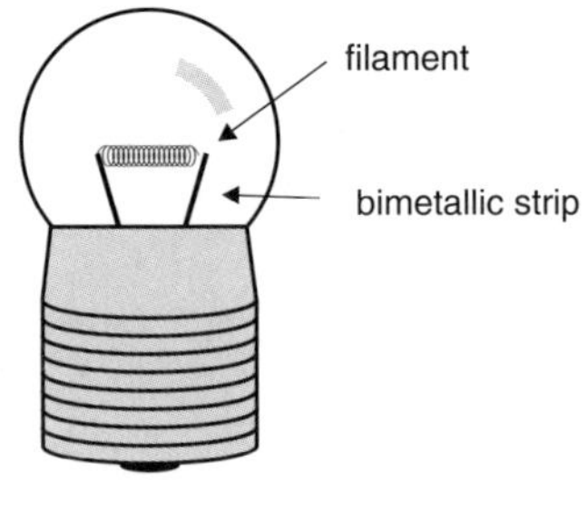

Section B

1. Describe experiments to show that (i) solids, (ii) liquids, (iii) gases expand when they are heated.
2. If your warm hands are placed around the flask in the diagram, what will be observed? What will be observed if you then remove your hands? What do these results suggest about the effect of temperature on the volume of a gas?

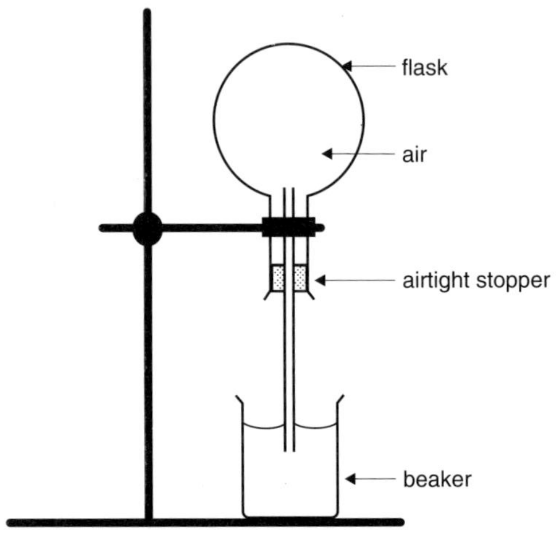

a 'flashing bulb'
filament
bimetallic strip

flask
air
airtight stopper
beaker

3. Explain the following. (*a*) When telephone wires are being strung between poles in hot weather, they are left hanging slack. (*b*) Gaps are left between concrete slabs in a footpath.
4. Name and draw a diagram of the type of thermometer used to take the temperature of the human body.
   What is the normal temperature of the human body?
   Why is this thermometer shaken after being used?
5. Give as many practical applications as you can of the fact that solids and liquids expand when they are heated.
   Are there any practical applications of the fact that gases expand when they are heated?
6. When the flask in experiment 10.4 is heated, the level of the liquid falls **very slightly** at first before it begins to rise. Why?
7. What is temperature?
   How would you show that heat and temperature are not the same?
   Which of the following contains the greater amount of heat: 200 cm^3 of water at 50 °C or 300 cm^3 of water at 50 °C? Explain your answer.

CHAPTER 11 — MOVEMENT OF HEAT

Heat travels in rays from the sun to the earth

Many forms of energy would not be of much use to us if they could not travel from one place to another. Imagine what would happen if light could not travel from the sun to the earth, or electricity could not travel from the generating station to your home, or sound could not travel from a radio to your ear. Light, electricity and sound can travel. Heat always travels from a higher temperature to a lower temperature. Heat travels in three ways.

HEAT CONDUCTION
Conduction is the movement of heat through a substance without any movement of the substance.

What happens is that heat travels from one particle to the next, but **the particles themselves do not move**.

EXPERIMENT 11.1
AIM: TO COMPARE THE HEAT CONDUCTION OF COPPER, IRON, ALUMINIUM AND GLASS

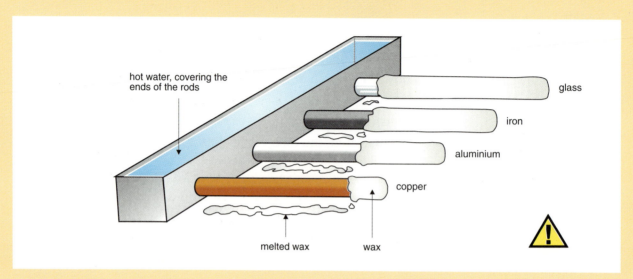

Method
1. Pour the boiling water into the container.
2. The ballbearing under the wax will fall from the copper first, then from the aluminium.
3. The ballbearing will not fall from the glass.
   Glass is a poor conductor.

EXPERIMENT 11.2
AIM: TO SHOW THAT COPPER IS A BETTER HEAT CONDUCTOR THAN WOOD

Method
1. Shove a piece of wood into one end of a copper pipe.
2. Wrap paper around the joint and heat the joint **gently**.
3. The paper covering the wood becomes charred but the paper covering the copper does not. This is because the copper pipe conducts (carries) the heat away from the paper. The wood does not.

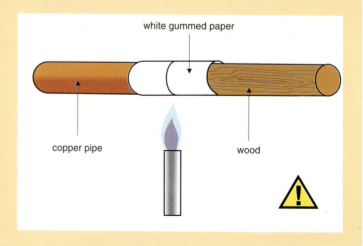

CONDUCTORS AND INSULATORS OF HEAT

Generally speaking, metals are good conductors of heat and non-metals are bad conductors. A poor conductor of heat is called an **insulator**.

Most liquids are poor conductors of heat.

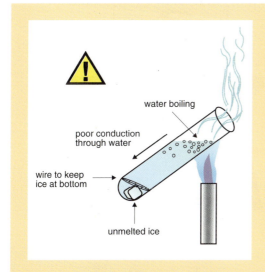

EXPERIMENT 11.3
AIM: TO SHOW THAT WATER IS A POOR CONDUCTOR OF HEAT

Method

1. Heat the water at the top of the test tube.
2. Even when the water at the top is boiling, the ice at the bottom has not melted.
3. This shows that water is a poor conductor of heat.

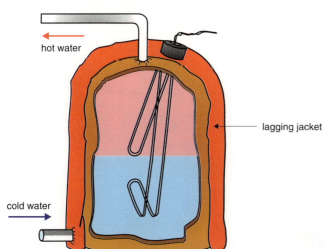

If you heat the water in the top part of the hot-water cylinder in your hot press, the heat is not conducted down into the water below. So the water on top remains hot.

Gases are poor conductors of heat. This is because the particles in a gas are too far apart for the heat to travel from one particle to the next.

CONVECTION

It is very important for us to be able to heat water. We need hot water for cooking, washing, etc. But water is a very poor conductor of heat, so how do we heat it? The answer is that it heats by **convection**. The particles nearest the heater heat up and move away, carrying the heat with them.

Convection means that heat is carried through liquids and gases by the movement of particles.

EXPERIMENT 11.4
AIM: TO SHOW CONVECTION CURRENTS IN WATER

Method
1. Drop a crystal of potassium permanganate down a funnel to the bottom of a beaker of water.
2. Heat the water **gently** just below the crystal.
3. Watch the warm water rising and the cold water coming down to take its place. You are watching **convection currents**. Without the coloured crystal the convection currents would still be there, but you would not be able to see them. Convection can occur only in fluids (liquids or gases).

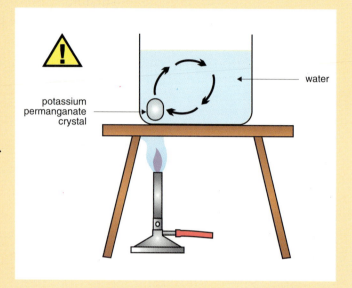

EXPERIMENT 11.5
AIM: TO SHOW CONVECTION CURRENTS IN AIR

Method
1. Light the candle and place it under the chimney.
2. Hold a smoking string over the other chimney.
3. Normally we would expect the smoke from the string to float upwards, but in this case it goes down the chimney.
   The reason for this is that hot air from the candle rises up the first chimney and cold air comes down the other chimney to take its place, pulling the smoke down with it.

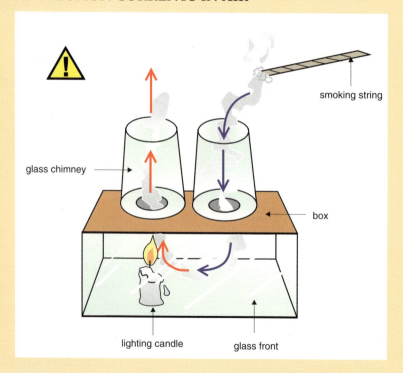

There are many examples of convection in everyday life.

PHYSICS · MOVEMENT OF HEAT

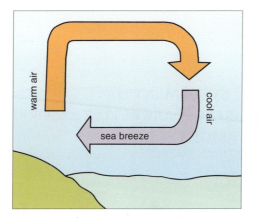
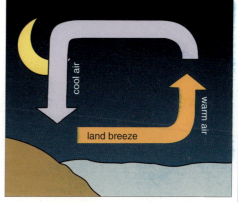

Convection currents cause land and sea breezes.

RADIATION

Have you ever cooked meat under a grill? The heat travels downwards from the grill to the meat. This heat cannot travel by conduction, since air is a bad conductor. Neither can it travel by convection, since hot air rises. In fact the heat travels by radiation.

> *Heat radiation means that heat travels in invisible rays without needing a substance to travel through.*

The heat that reaches us from the sun travels by radiation. Unlike conduction and convection, radiated heat can travel through a vacuum. Have you ever burned a piece of paper using a magnifying glass, as shown in the diagram? It is not the visible light that burns the paper. It is the invisible heat rays which, like the light rays, are focused on one spot by the magnifying glass.

EXPERIMENT 11.6
AIM: TO SHOW THAT DARK SURFACES ARE BETTER RADIATORS THAN BRIGHT SURFACES

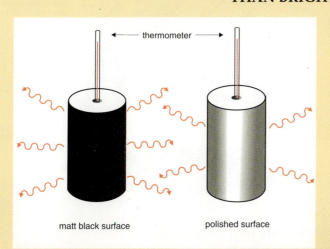

Method

1. Get two similar metal cans. Rub the surface of one of them with fine sandpaper until it is clean and shiny.
2. Blacken the other can with matt black paint.
3. Fill each of the cans with boiling water. Cover them as shown in the diagram.
4. Record the temperature of each can every 2 minutes.

Result: The temperature falls faster in the blackened can.

Conclusion: Dark surfaces are better radiators of heat than bright surfaces.

BREAKTHROUGH SCIENCE

EXPERIMENT 11.7
AIM: TO SHOW THAT DARK SURFACES ARE BETTER ABSORBERS OF HEAT THAN BRIGHT SURFACES

Method
1. Set up the apparatus as shown in the diagram.
2. The coin on the black plate falls off first. This shows that dark surfaces are better at absorbing heat.

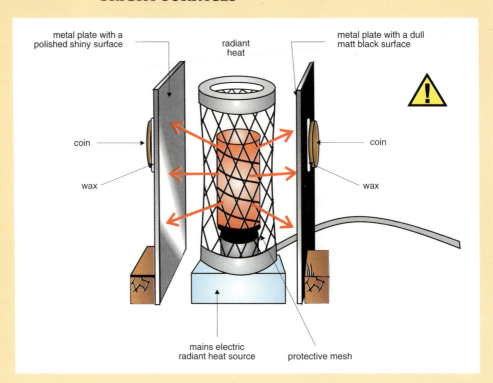

Dark surfaces are better at radiating and absorbing heat than bright surfaces.

To put this another way, good absorbers are good radiators and bad absorbers are bad radiators.

SHORT WAVES AND LONG WAVES

Bodies at very high temperatures give out short-wave heat radiation and bodies at lower temperatures give out long-wave heat radiation. Short-wave radiation is more penetrating than long-wave radiation. This is how a greenhouse heats up. The short-wave radiation comes in through the glass and causes the plants and soil to heat up. These now give out long-wave radiation because they are not nearly as hot as the sun. The long-wave radiation cannot penetrate the glass, and so the heat is trapped in the greenhouse.

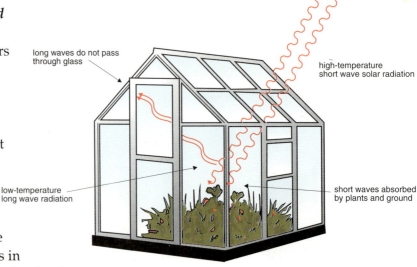

Cavity wall insulation

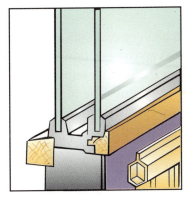
Double glazing

Attic insulation

INSULATION

You can keep yourself warm by wearing a string vest beneath your clothes. The pockets of air trapped in the string vest are a very good insulator, and prevent your body from losing heat. Loose woollen clothes are good insulators because of the air trapped in them. Remember the posters telling us how much energy we are wasting by not insulating our homes? One answer to this is to insulate the attic with fibreglass, which has a lot of air trapped in it, and to insulate the walls with aeroboard, which also has air trapped in it.

EXPERIMENT 11.8
AIM: TO STUDY THE EFFECTS OF INSULATION

Method
1. Set up two identical metal cans as shown in the diagram.
2. Pour an equal amount of boiling water into each and cover.
3. Take the temperature in each can every three minutes for half an hour.
4. On the same axis and using the same scales, plot graphs showing the rate of fall of temperature in each can.

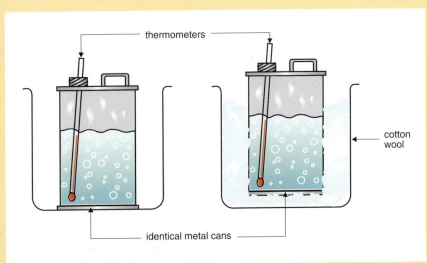

Result: The temperature falls faster in the can with the insulation, especially at the beginning.
Conclusion: Cotton wool is a good insulator.

BREAKTHROUGH SCIENCE

TOG VALUES

Shopping for continental quilts (duvets) can be very confusing. One quilt can cost twice as much as another even though they look exactly the same on the outside. If you look at the labels you will see why. The dearer quilt will have a higher tog value. The higher the tog value, the better the insulation. Tog values run from about 1 for light summer clothes to 13·5 for quilts filled with feathers and down.

The tog value tells us how good an insulator the material is.

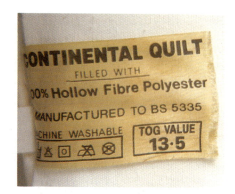

SUMMARY

- Heat can be transferred in three ways: conduction, convection and radiation.
- Conduction is the movement of heat through a substance without any movement of the particles of the substance.
- Metals are good conductors; non-metals are bad conductors.
- As a general rule liquids and gases are poor conductors.
- Convection means that heat is carried through liquids and gases by the movement of particles.
- Heat radiation means that heat travels in invisible rays without needing a substance to travel through.
- Dark surfaces are better at radiating and absorbing heat than bright surfaces.
- Generally speaking, good absorbers are good radiators and bad absorbers are bad radiators.
- A tog value is a measure of insulation.

QUESTIONS

Section A

1. Heat can be transferred in three ways:
   (a) _____; (b) _____; (c) _____.
2. _____ are good conductors of heat.
3. The diagram shows three rods of different material being heated by hot water.
   What is the purpose of this experiment?
   _____

   Is this a fair test? _____ Why?
   _____
4. Substances that do not conduct heat well are called _____.
5. Fluids are not good conductors because
   _____

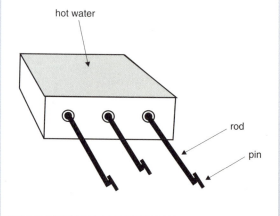

6. In the diagram, which piece of ice, A or B, will melt first? _____ Give a reason for your answer. _____
7. On a cold day the bare metal handlebars of a bicycle feel colder than the plastic handgrips, because _____
8. Heat travels through fluids by _____.
9. Solids are not heated by convection because _____.
10. Heat travels from the sun to the earth by _____.
11. Dark surfaces are better at _____ heat than bright surfaces.
12. Dark surfaces are also better at _____ heat than bright surfaces.
13. Bright clothing is worn in hot countries because _____.
14. Bright clothing is also worn by Eskimos because _____.
15. What is meant by the tog value of a material? _____
16. Underline the insulators in the following list: aluminium, wool, glass, tin, plastic, air.

Section B

1. In the diagram, there is wax on the top of each of the rods. How does the heat travel along the rods? On which rod will the wax melt first?
2. In which way is heat transferred in a liquid? In which way is heat carried from the sun to the earth? Name one method used to prevent heat loss in our homes.

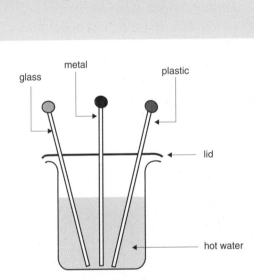

3. In the diagram, by what method is the heat from the candle flame transferred to outside the box? Give one everyday example of the transfer of heat by this method.

4. In which container in the diagram does the temperature drop faster? Explain your answer. Give one everyday application of the results of this experiment.
5. Three metal rods of equal diameter are shown in the lower diagram. How would you carry out a fair test to see which metal is the best conductor?
6. Name three ways in which a house could be insulated. Two houses, A and B, are equally well heated, but snow melts much more quickly from the roof of house B than from the roof of house A. Which house has its attic insulated? Explain your answer.
7. What is meant by convection? Describe an experiment to show convection currents in water or air. Why is the heating element in an electric kettle placed so near the bottom of the kettle? Explain how a dual immersion heater works in a domestic hot water tank.
8. What is meant by "radiation"? Describe experiments to show that dull surfaces are better radiators and absorbers of heat than bright surfaces. Domestic heaters are often called radiators. Why is this name not quite correct?
9. On a day when it is calm inland you may feel a breeze at the seaside. Why is this? Would this breeze blow in the same direction at night as during the day? Why?
10. A beaker of boiling water is placed on a bench in a science lab.
    Name three ways in which it will lose heat. The temperature of the water is taken every two minutes, and recorded as follows.

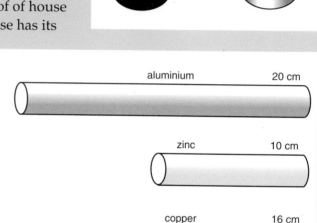

Time (minutes)	0	2	4	6	8	10	12
Temperature (°C)	100	70	54	42	32	25	20

(i) Draw a graph of temperature against time.
(ii) From your graph, what was the temperature after 5 minutes?
(iii) If the temperature in the lab is 17°C, what will the final temperature of the water be?

CHAPTER 12 CHANGES OF STATE

When a piece of ice is melting, it is changing from the solid state (ice) to the liquid state (water). When water is boiling it is changing from the liquid state (water) to the gas state (water vapour).

We can learn a lot about changes of state from this simple experiment.

EXPERIMENT 12.1
AIM: TO DEMONSTRATE LATENT (HIDDEN) HEAT

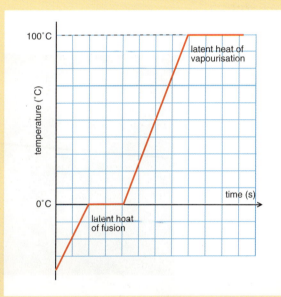

Method
1. Place about 250 g of crushed ice in a beaker and take the temperature.
2. Heat the ice gently and record the temperature every minute until all the ice has melted.
3. Continue to heat and to take the temperature until the water begins to boil.
4. Take readings for 5 more minutes.
Draw a graph of temperature against time. Your graph should be like the one in the diagram.
Result: The temperature does not change while the ice is melting; neither does it change while the water is boiling.

There is no change of temperature during a change of state.
Even though there is no change of temperature during a change of state, heat is still being taken in: it simply does not show up on the thermometer. This heat is known as **latent** (hidden) heat.
Heat taken in during melting or boiling is called latent heat.

THE KINETIC THEORY
Scientists have a theory, called the kinetic theory, to explain what is happening in Experiment 12.1. All matter is made up of atoms, or groups of atoms called molecules. In a solid such as ice the molecules are tightly packed, and can only vibrate. As the ice is heated the temperature rises to 0 °C. Then the ice begins to melt,

Breakthrough Science

but the temperature does not change: instead, the heat energy is used to break some of the bonds that hold the molecules together. Layers of molecules can now slide over each other, and the solid has become a liquid. As the water is heated, the temperature again rises until it reaches 100 °C and the water begins to boil. Once again the temperature does not change as the heat energy is used to break the remaining bonds and turn the liquid into a gas. When the temperature does not change the heat involved is called hidden heat or latent heat.

Latent heat is the heat involved when a substance changes state without changing temperature.

It takes 3.3×10^5 (330,000) joules of heat to change 1 kg of ice to water without a change in temperature. This is called the latent heat of fusion of ice.

It takes 2.3×10^6 joules to change 1 kg of water to steam without a change in temperature. This is called the latent heat of vaporisation of water.

Example
How much heat is required just to change 5 kg of ice at 0 °C to 5 kg of water at 0 °C?

$5 \times (3.3 \times 10^5) = 16.5 \times 10^5 = 1.65 \times 10^6$ joules

SUBLIMATION
If we heat some ammonium chloride as shown in the diagram, it will change directly from solid to gas. The gas will change back to solid when it hits the cold flask.

Sublimation means that a substance changes directly from solid to gas when heated.

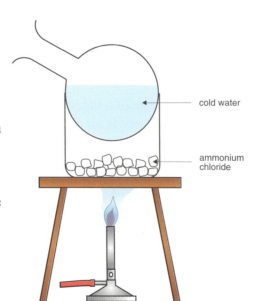

The British physicist James Prescott Joule (1818–1889) is known for his discoveries about heat. In studying electrical, chemical and mechanical energy, Joule found that the amount of heat produced by each form of energy is proportional to the energy used— conservation of energy. The unit of energy (joule) is named in his honour.

PHYSICS · CHANGES OF STATE

At very low temperatures carbon dioxide is solid, and is called "dry ice". As its temperature rises towards room temperature it changes directly to gas. This is used on stage, particularly in rock concerts and videos, to create a "mist" effect. Iodine also sublimes.

BOILING AND PRESSURE

When you boiled water in Experiment 6.1, did it boil at exactly 100 °C? Maybe it boiled at 98 °C or 102 °C. It depends on the atmospheric pressure at the time. When water boils, molecules escape from the surface. The atmospheric pressure opposes this. If the pressure is high, the water may have to reach a temperature of 102 °C before it can boil. At low pressure it may boil at 98 °C.

In the pressure cooker, the water vapour cannot escape so the pressure above the water becomes so great that it will not allow the water to boil until a temperature of 120 °C is reached. At this temperature the food cooks much faster.

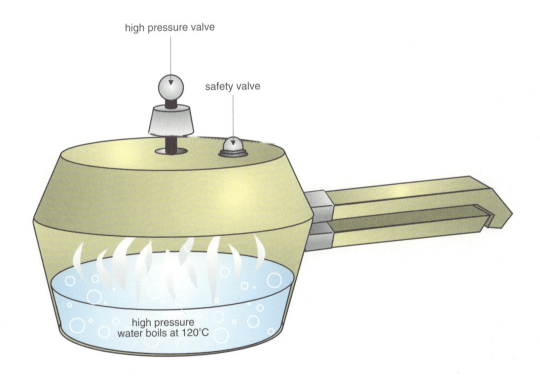

BREAKTHROUGH SCIENCE

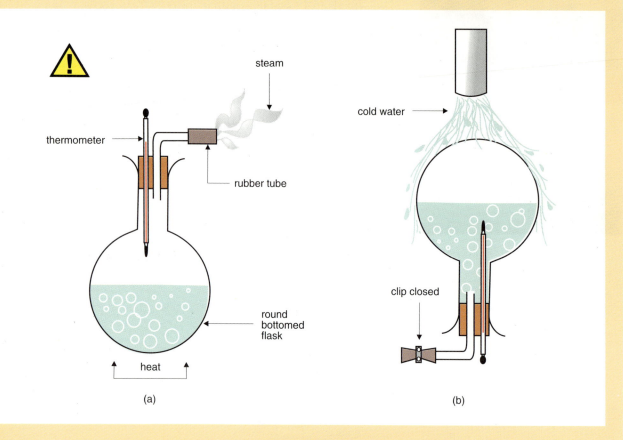

EXPERIMENT 12.2
AIM: TO SHOW THE EFFECT OF PRESSURE ON THE BOILING POINT

Method
1. With the clip open, heat the water until it boils. Note the temperature.
2. Close the clip for a few seconds and note the temperature.
3. Turn the flask upside down and pour cold water over it. The water begins to boil again. Can you explain what is happening in this experiment?

PRESSURE AND MELTING

When you take some snow and press it between your two hands, the increased pressure raises the melting point of the snow and causes some of it to melt. When the pressure is removed, this snow freezes again and holds the rest of the snow together in a ball.

PHYSICS · CHANGES OF STATE

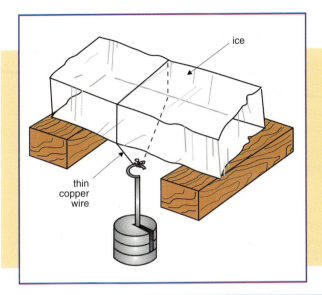

EXPERIMENT 12.3
AIM: TO SHOW THE EFFECT OF PRESSURE ON THE MELTING POINT

Method
1. Place thin wire across the block of ice and hang a large mass from it.
2. The ice under the wire melts, allowing the wire to sink down through it.
3. As the wire moves down, the water above the wire freezes again.

SUMMARY

- The main difference between the three states of matter is the freedom of movement of the molecules.
- There is no change of temperature during a change of state.
- Latent heat is the heat involved when a substance changes state without changing temperature.
- Sublimation means that a substance changes directly from liquid to gas when heated.
- Increased pressure raises the boiling point of a liquid.
- Increased pressure raises the melting point of a solid.

QUESTIONS

Section A
1. What changes would you notice on a thermometer if you placed it in a beaker of ice and began to heat it gently? _____
2. There is no change of _____ during a change of _____.
3. Latent heat means _____.
4. Why does your hand feel cold if you spill methylated spirits or perfume on it?
   _____
5. Which would be more effective in cooling a drink, 20 g of water at 0 °C or 20 g of ice at 0 °C? _____ Give a reason for your answer. _____
6. Why are you likely to catch a cold if you wear wet clothes? _____
8. How much heat does it take to melt 2 kg of ice at 0 °C if the specific latent heat of fusion of ice is $3\cdot 3 \times 10^5$ joules? _____
9. What is meant by sublimation? _____
   Name two substances that sublime. _____ and _____
10. Increased pressure _____ the melting point of a solid and _____ the boiling point of a liquid.

Section B
1. When heat is supplied to a substance its temperature usually rises, but it is possible to supply heat to a substance without raising its temperature. Give two examples of this.
2. If a small drop of perfume or methylated spirits is applied to the back of your hand, it dries very quickly but your hand feels cold. What exactly has happened?
3. When you add a few lumps of ice to a drink, the drink gets colder and the ice begins to melt. Why exactly is this?
4. Which would give you a worse burn, boiling water or steam? Why?
5. The temperatures at which water boils at sea level, down a mine and on top of a mountain are 105 °C, 100 °C, 96 °C, but not in that order. Say which is which, and explain why.
6. If you look at the back of a fridge you will see a series of very narrow pipes. What is inside in these pipes? Why are they painted black?
   What is the compressor for?
   How does the fridge work?
   Why should old fridges be disposed of very carefully?

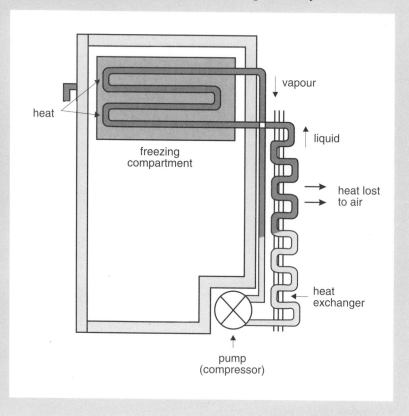

PHYSICS · ELECTRICITY

CHAPTER 13 ELECTRICITY

We use electricity in the home because it has the ability to do work for us.
Electricity is a form of energy.

All the electrical devices we use in the home change energy from one form to another. But what do we do if we are not in the home? If we want to go outside and use a personal stereo or a flashlamp, we use a cell or a battery (a number of cells joined together).

POTENTIAL DIFFERENCE

If your tyre is flat, it is because there is no air inside the tube. There is plenty of air outside, but it will not go in unless the pressure outside is greater than the pressure inside. Your pump provides the difference in pressure necessary to drive air into the tube.

A battery is another type of pump. It provides the difference in electric pressure necessary to drive electricity round a circuit. This difference in electric pressure is called **potential difference**, and is measured in **volts**.

The Italian physicist Alessandro Volta (1745–1827) was the inventor of the first electric battery. Volta began his experiments after Galvani's experiments with "animal electricity" suggested wrongly that electricity was a living substance in animals. Volta also discovered methane gas. The volt is named in his honour.

BREAKTHROUGH SCIENCE

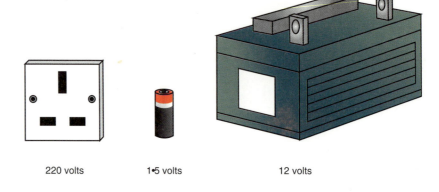

220 volts   1·5 volts   12 volts

A flashlamp cell provides a potential difference of 1·5 volts. A car battery provides a potential difference of 12 volts. The ESB provides a potential difference of 220 volts!

In the diagram, will the bulb light if the switch is open? Will the bulb light if the battery is removed?

Electricity will flow if there is a potential difference and a complete circuit.

CONDUCTORS AND INSULATORS

Anything that allows electricity to flow through it is an electrical conductor. Anything that does not allow electricity to flow through it is an electrical insulator.

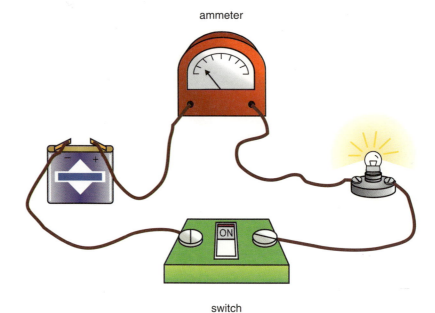

EXPERIMENT 13.1
AIM: TO SEE WHETHER A SUBSTANCE IS A CONDUCTOR OR AN INSULATOR

Method
1. Set up the apparatus as shown.
2. Place the substance you want to test between the clips, and see if the bulb lights.
3. Repeat with different substances, and draw up a table of conductors and insulators.

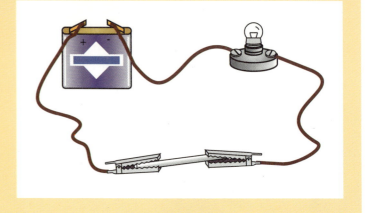

So now you know! Some things are conductors, others are insulators. But are all conductors the same? Let's find out.

PHYSICS · ELECTRICITY

EXPERIMENT 13.2
AIM: TO COMPARE SOME CONDUCTORS

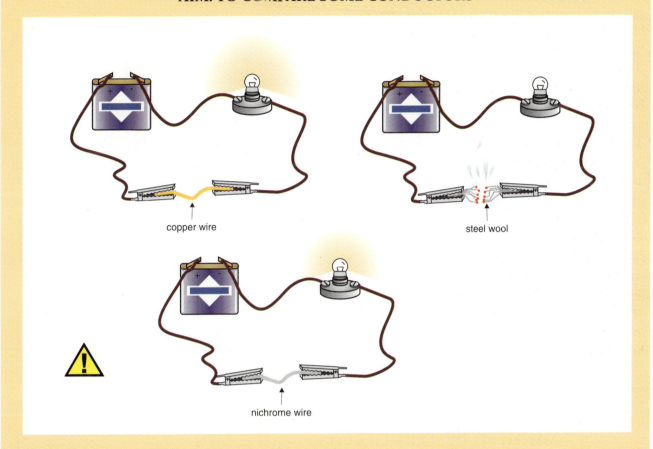

Method
1. Set up the circuit shown in the diagram.
2. Place the following between the clips and see how brightly the bulb lights: copper wire, thin nichrome wire, a few strands of steel wool.

Electricity passes through the copper wire easily. It does not pass through the nichrome wire so easily. In other words, the nichrome wire has greater **resistance** to electricity than copper wire has. A lot of energy has to be used to push the electricity through the nichrome wire. This energy turns to heat, and the wire gets hot. The steel wool gets so hot that it melts.

If energy is used in passing electricity through an object, that object is a resistor.

The unit of resistance is the **ohm**.

Electrical energy is changed to heat energy in a resistor.

CURRENT

We use the word "current" to describe the flow of a river. We use the same word in electricity.

An electric current is a flow of electric charge.

Electric current is measured in units called **amperes**. To measure electric current we use an **ammeter**.

ammeter

symbol

Andre Ampere (1775–1836) was a French physicist who demonstrated that electric currents produce magnetic fields. He also showed that the direction of the magnetic field is determined by the direction of the current. He defined the unit of current later named in his honour, the ampere.

SERIES AND PARALLEL CIRCUITS

An electrical circuit can be connected in either of two ways, depending on what results you want.

Series circuits

Bulbs connected in series are connected one after another.

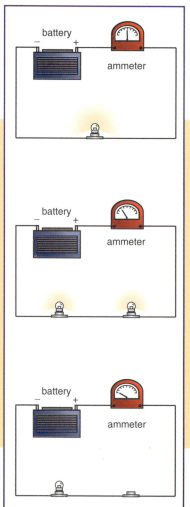

EXPERIMENT 13.3
AIM: TO STUDY SERIES CIRCUITS

Method
1. Set up a circuit as shown in the top diagram. Note the current.
2. Connect a second bulb in series with the first. What happens to the current?
3. Connect a third bulb in series. Note the current.
4. Remove one of the bulbs from its holder. What happens?

Result: For a fixed voltage, two bulbs in series are dimmer than one bulb on its own. If one bulb blows, the circuit is broken.

Conclusion: The only advantage of connecting bulbs in series is that the current is smaller, and therefore the battery lasts longer.

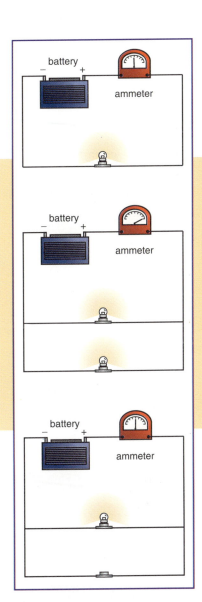

Parallel circuits
Bulbs connected in parallel are side by side in a circuit.

EXPERIMENT 13.4
AIM: TO STUDY PARALLEL CIRCUITS

Method
1. Set up the circuit shown in the top diagram in the previous experiment. Note the current.
2. Connect a second bulb in parallel with the first, as shown in the second diagram on the left. What happens to the current?
3. Remove one of the bulbs from its holder. What happens?
   Result: For a fixed voltage, two bulbs in parallel are brighter than one bulb on its own. Also, if one bulb blows the circuit is **not** broken.
   Conclusion: The only disadvantage is that the battery runs down faster.

In future when drawing circuit diagrams we shall use the following symbols.

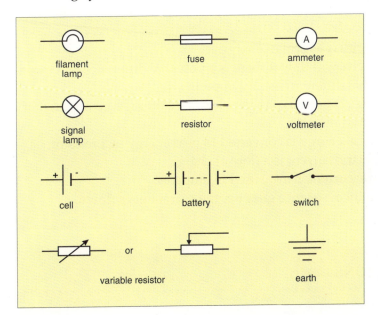

Note that
(i) conventional current flows from the positive of a battery or cell to the negative
(ii) in the symbol for a cell or battery (group of cells), the long line is positive (+) and the short line is negative (−)
(iii) an **ammeter** is always connected into a circuit in **series**; a **voltmeter** is always connected into a circuit in **parallel**.

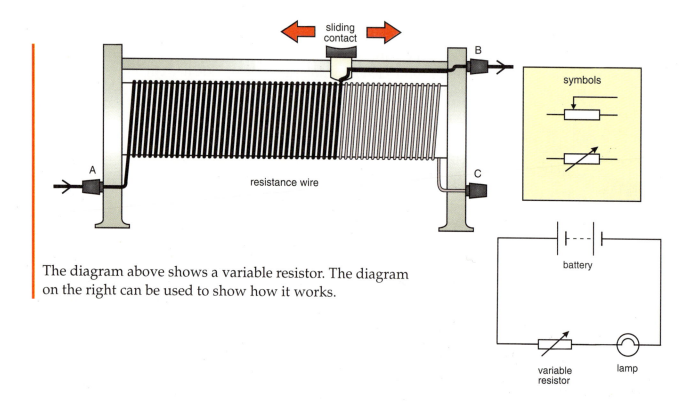

The diagram above shows a variable resistor. The diagram on the right can be used to show how it works.

SUMMARY

- Electricity is a form of energy.
- Electricity will flow if there is a potential difference and a complete circuit.
- Potential difference is measured in volts.
- If energy is used in passing electricity through an object, that object is a resistor. Resistance is measured in ohms.
- Electrical energy is converted to heat energy in a resistor.
- An electric current is a flow of charge.
- Electric current is measured in amperes.
- When the resistance goes up, the current goes down.

QUESTIONS

Section A
1. Electricity is a form of _____.
2. Electricity will flow if there is a _____ _____ and a _____ _____.
3. A light bulb converts _____ energy to _____ _____.
4. Potential difference is measured in _____.
5. An electric current is a flow of _____.
6. The unit of current is the _____.

7. The potential difference of
   (a) a flashlamp cell is _____
   (b) a car battery is _____
   (c) the ESB domestic supply is _____

8. What happens in the diagram when A is joined to B by
   (a) wood? _____
   (b) metal? _____
   (c) rubber? _____

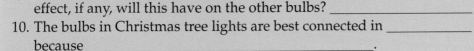
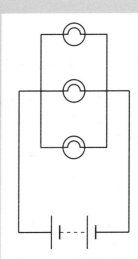

9. If one of the bulbs in the diagram is removed from its socket, what effect, if any, will this have on the other bulbs? _____

10. The bulbs in Christmas tree lights are best connected in _____ because _____.

Section B

1. Set up the apparatus shown in the diagram. Touch the two clips together. What happens? What does this tell you? Place pieces of the following materials between the clips and fill in the table: copper, wool, plastic, aluminium, iron, wood, paper.
   Conductor
   Insulator
   Of the materials you have tested, which would be suitable for use in electric wires? Why are wires that carry electric current coated? Which of the materials you have tested would you choose to coat electric wires? Why?

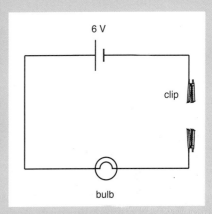

2. Name the type of circuit in diagram A. Name the type of circuit in diagram B. What happens in each circuit if the switch is closed? What happens in each circuit if one of the bulbs blows? In which circuit is the greater current being drawn from the battery? In which circuit will the battery last longer?

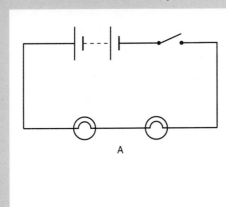
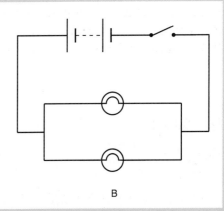

3. The diagram shows a battery, a meter and two lamps in a circuit. What type of meter is this?
   Draw a circuit diagram for this circuit using the usual symbols.
   In what way are the lamps connected? Would you connect lamps in this way in your house? Why?
4. Give two advantages of connecting bulbs in parallel.
5. Set up the circuit shown in the diagram. Slide the contact on the variable resistor to the left. What happens to the bulb? Slide it to the right. What happens to the bulb? Now explain how a variable resistor works.

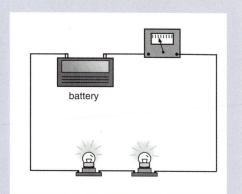

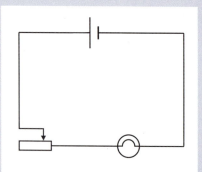

OHM'S LAW — CHAPTER 14

A 1·5 V cell will cause electricity to flow through a conductor at a certain rate. Two such cells connected in series will cause the electricity to flow faster (greater current). If we use a conductor with greater resistance, the current will be reduced. Clearly, voltage, current and resistance are all related to each other. This relationship was studied by a German schoolteacher called George Ohm, who arrived at Ohm's Law.

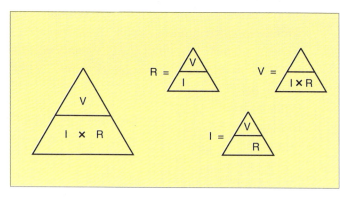

Ohm's law: For a metal resistor at constant temperature,

$$resistance = \frac{voltage}{current}$$

In symbols, $R = \frac{V}{I}$ *or* $V = I.R$

Resistance is measured in ohms.
When using Ohm's law to solve problems, we can use the triangles in the diagram.

The German physicist Georg Simon Ohm (1787–1854) discovered the relationship between the current, voltage and resistance in a circuit: Ohm's law. The unit of electrical resistance, the ohm, is named in his honour.

Problem

A potential difference of 12 volts is applied to a 3 ohm resistor. What current flows?

$$\text{Current} = \frac{\text{voltage}}{\text{resistance}} = \frac{12 \text{ volts}}{3 \text{ ohms}} = 4 \text{ amps}$$

EXPERIMENT 14.1
AIM: TO VERIFY OHM'S LAW

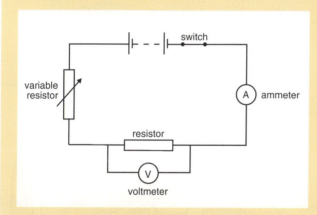

Method

1. Set up the apparatus as shown.
2. Draw up a table for voltage and current.
3. Set the variable resistor to give a small current. Note the voltage.
4. Adjust the variable resistor to give a slightly larger current. Note the voltage again.
5. Repeat step 4 four or five times.
6. Draw a graph of voltage (y-axis) against current (x-axis).

Result: The current is directly proportional to the voltage.

What kind of graph did you get? How could you use this graph to find the resistance of the resistor?

Problem

In a laboratory experiment, a pupil made the following measurements of voltage and current using a metallic conductor.

Voltage (volts)	0	3	6	9	12
Current (amps)	0	0·5	1	1·5	2

BREAKTHROUGH SCIENCE

Plot a graph of voltage against current and use it to find the resistance of the conductor.

Resistance = $\dfrac{12 \text{ volts}}{2 \text{ amps}}$ = 6 ohms

Study the following circuits and see if you can understand what is happening.

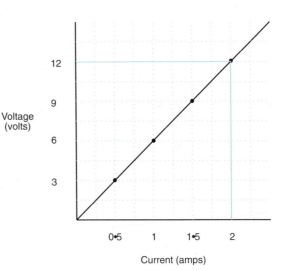

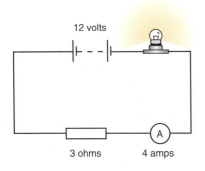

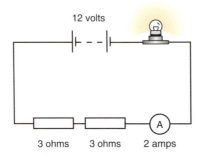

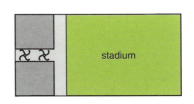

If the crowd has to pass through two turnstiles, one after the other, it makes it twice as difficult to get in.

When resistors are connected in series, the total resistance is obtained by adding the individual resistances ($R = R_1 + R_2$)

POWER

ESB power stations supply us with electrical energy. We use this energy for heat and for light. The rate at which the heater uses electrical energy is known as its power, and is measured in watts (W).

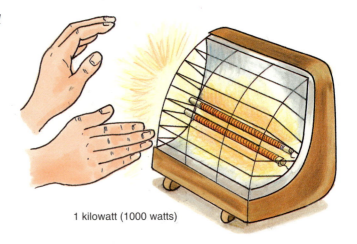

1 kilowatt (1000 watts)

James Watt (1736–1819), a Scottish engineer and inventor, contributed to the development of the steam engine as a practical power source. Although Watt did not invent the steam engine, his improved engine was the first practical device to convert heat efficiently into useful work. The unit of power (watt) is named in his honour.

HOW TO FIND THE COST OF RUNNING AN ELECTRICAL APPLIANCE

A device that uses 1 joule of energy per second has a power of 1 watt.

The watt is a very small unit, so we use the kilowatt (1000 watts). The ESB charges you for the amount of its electrical energy you use. It uses a unit of energy called the kilowatt hour. One kilowatt for one hour is one kilowatt hour.

Kilowatt hours = kilowatts × hours

Problem
A 2 kilowatt heater is left on for 4 hours. How many units of electricity does it use?

2 kilowatts × 4 hours = 8 kilowatt hours. Answer: 8 units.

Problem
Seven 100 watt lamps are left on for 12 hours. How much does this cost at 8p per unit?

700 watts = 0·7 kilowatts.   0·7 × 12 = 8·4 units
8·4 units @ 8p each = 67·2p

A useful equation for dealing with electrical power is

watts = amperes × volts

Most electrical calculations can be dealt with by using one or both of these triangles.

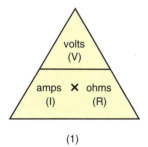

(1)

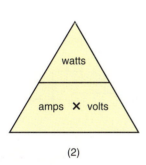

(2)

Problem
An electrical appliance is marked 12 volts, 60 watts. What current does it take, and what is its resistance?

Using triangle (2) Amps = $\dfrac{\text{watts}}{\text{volts}} = \dfrac{60 \text{ W}}{12 \text{ V}} = 5$ amps

Using triangle (1) R = $\dfrac{V}{I} = \dfrac{12 \text{ V}}{5 \text{ A}} = 2\cdot 4$ ohms

SUMMARY

- Ohm's law: $\dfrac{\text{voltage}}{\text{current}}$ = resistance
- Resistance is measured in ohms.
- A power of 1 joule per second = 1 watt.
- For resistors in series, $R = R_1 + R_2$.
- The unit used by the ESB is the kilowatt hour.
- kilowatt hours = kilowatts × hours
- watts = amperes × volts

	Units	**Symbol**
Voltage	Volts	V
Current	Amps	I
Resistance	Ohms	R
Power	Watts	W
ESB	Kilowatt hours	kW h

QUESTIONS

Section A

1. According to Ohm's law, _____ divided by _____ equals _____.
2. The unit of resistance is the _____.
3. A current of 0·5 amperes flows through a wire when the potential difference is 12 volts. What is the resistance of the wire? _____ 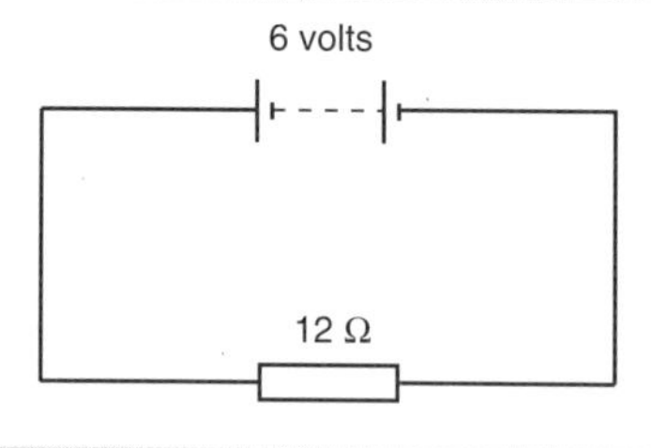
4. The potential difference across the resistor in the circuit on the right is 6 volts. What is the current in the resistor? _____
5. What mistake has been made in this circuit (below right)? _____
   What is the purpose of this circuit? _____
6. What is the total resistance in the diagram below? _____ 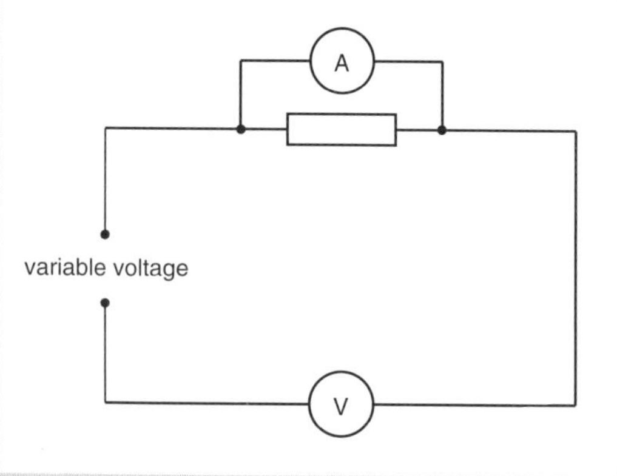 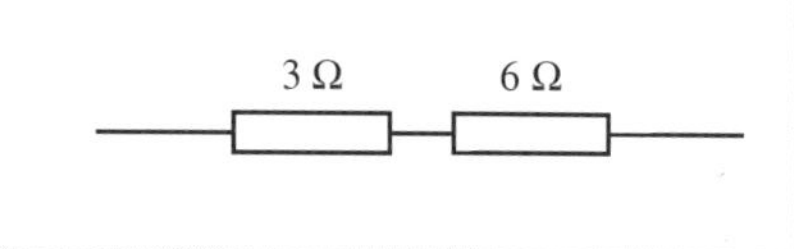 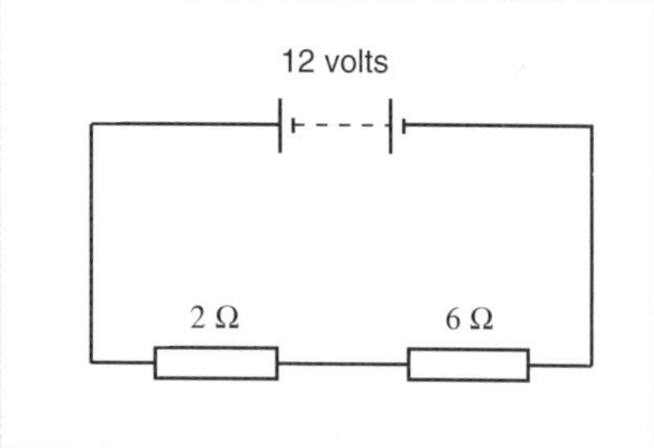
7. What current flows in this circuit (above)? _____
8. A device that uses 1 joule of energy per second has a power of _____.
9. The unit of electrical energy used by the ESB is the _____ _____.

10. Give the symbol for each of the following. Ampere: _____. Volt: _____. Kilowatt hour: _____.
11. A bulb has 60 W stamped on it. What does this tell us? _____
12. Calculate the cost of using a 100 watt bulb for 8 hours per day, 5 days per week at 7p per unit. _____

Section B

1. A consumer received the following information on her ESB bill.
   Present reading 94612. Previous reading 93908. How many units had she used? How much did this cost at 8·2 pence per unit?
2. State Ohm's law. In a laboratory experiment, a pupil made the following measurements of voltage and current using a metallic conductor.

Voltage (volts)	0	2	4	6	8
Current (amps)	0	0·6	1·2	1·8	2·4

   Plot a graph of voltage against current, and use the graph to find the resistance of the conductor.
3. In the circuit in the diagram, what is the total resistance?
   What is the current?
   What is the potential across the 3 ohm resistor? 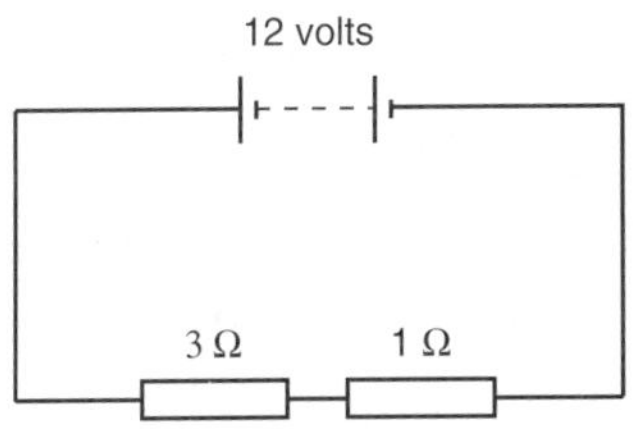
4. What is the relationship between potential difference, current and resistance? Describe, with the aid of a labelled circuit diagram, an experiment to verify this relationship.
5. A piece of nichrome wire is kept at a steady temperature. Different voltages are applied across it and the current is measured each time. Copy and fill in the table.

Voltage	Current	Resistance
12 V	2 A	
6 V		
	1·5 A	

6. An electrical device is marked 500 W, 250 V. What does this mean?
   If this device is connected to a 250 volt supply, what current does it take?
   How much would it cost to run this device for 8 hours at 2 pence per unit?
7. Name the units of electric current, potential difference and resistance, and state the relationship between them.
   Calculate the current flowing through a 3 kilowatt electric heater when it is connected to a 250 volt supply. How much would it cost to run this electric heater for 10 hours a day, 5 days a week at 7 pence per kilowatt hour?

8. Give one example to show that electricity is a form of energy.
   Household electricity bills are based on the number of units of electrical energy used. What is the correct name for these units?
   If the cost of one of these units is 8 pence, and the total cost of using a 2·5 kW heater for X hours is 60 pence, calculate the value of X.
9. An electric soldering iron uses a current of 3 amps from a 240 volt supply. What is the resistance of the iron? What is the power of the iron in watts? How much would it cost to run this iron for a total of 2 hours at 8p per unit?
10. An electric fire has a resistance of 80 ohms and is plugged into a 240 V supply. If the fire is left on for 10 hours at 8p per kilowatt hour, how much will it cost?

USES OF ELECTRICITY — CHAPTER 15

Electricity is useful to us because it can be converted to other forms of energy.

EXPERIMENT 15.1

AIM: TO DEMONSTRATE THE EFFECTS OF ELECTRICITY

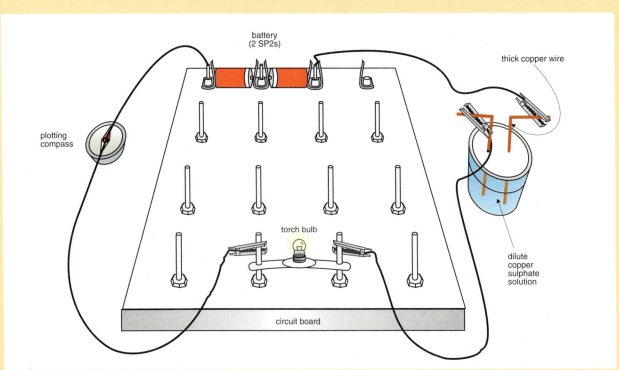

Method
1. Set up the apparatus as shown in the diagram, and close the switch.
2. The filament in the bulb heats up and the bulb lights (**heating effect**).
3. The compass needle moves as if there were a magnet near it (**magnetic effect**).
4. The cathode becomes coated (**chemical effect**). The cathode is the thick copper wire connected to the minus of the battery.

Alternatively, experiment 15.1 can be done as three separate experiments.

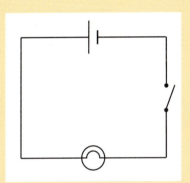

EXPERIMENT 15.2
AIM: TO DEMONSTRATE THE HEATING EFFECT OF ELECTRICITY

Method
1. Set up the apparatus as shown in the diagram.
2. Close the switch.
   Result: The filament heats up and the bulb lights.
   Conclusion: Electricity has a heating effect.

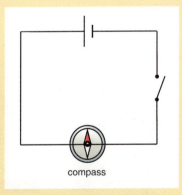

compass

EXPERIMENT 15.3
AIM: TO DEMONSTRATE THE MAGNETIC EFFECT OF ELECTRICITY

Method
1. Set up the apparatus as shown in the diagram.
2. Close the switch.
   Result: The compass needle moves as if there were a magnet near it.
   Conclusion: Electricity has a magnetic effect.

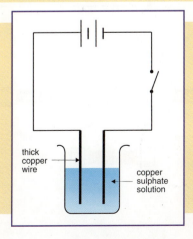

thick copper wire
copper sulphate solution

EXPERIMENT 15.4
AIM: TO DEMONSTRATE THE CHEMICAL EFFECT OF ELECTRICITY

Method
1. Set up the apparatus as shown in the diagram.
2. Close the switch.
   Result: The cathode becomes coated with copper.
   Conclusion: Electricity has a chemical effect.

BREAKTHROUGH SCIENCE

FUSES

When a current flows through a resistor the resistor gets hot. If the current were too big it could cause a fire. A fuse prevents this. A fuse is a short piece of wire with a low melting point. *If the current exceeds a certain value, the fuse will melt and break the circuit.*

Close the switch in this simple circuit and you will see how the fuse works. Look at the plug in the diagram. If you buy a plug it usually comes fitted with a 13 amp fuse. Most people take the plug home and fit it to an electrical device such as a kettle or a table lamp. It is clear from the diagram that a 13 amp fuse is the most suitable fuse available for the kettle, but not for the lamp. Of course the lamp will work with the 13 amp fuse, but if something goes wrong the wires may overheat and cause a fire without blowing the fuse.

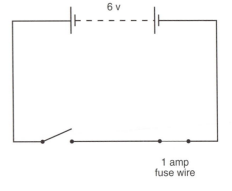
1 amp fuse wire

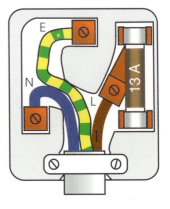

Which of the fuses in the diagram should be used with (i) the TV, (ii) the lamp?
Always use a fuse with a value greater than the correct current but as close to it as possible.

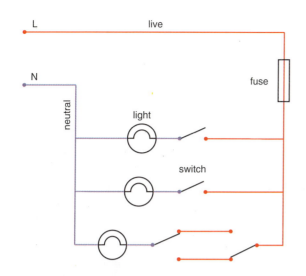

DOMESTIC CIRCUITS

The ESB cable coming into your house contains two wires, a live and a neutral. There are two types of circuit in a house, a lighting circuit and a **ring main** circuit.

Lighting circuit

This is a simple circuit in which lights are connected in parallel, so that if one light is switched off or a bulb blows, the others will continue to light.

Ring main circuit

In this case the live and neutral wires each form a ring or loop. A third loop is formed by the earth wire. When you plug a kettle into a socket, the live and neutral pins connect with the live and neutral wires so that current can flow from one wire to the other through the heating element of the kettle. No current should flow in the earth wire, unless of course a fault develops.

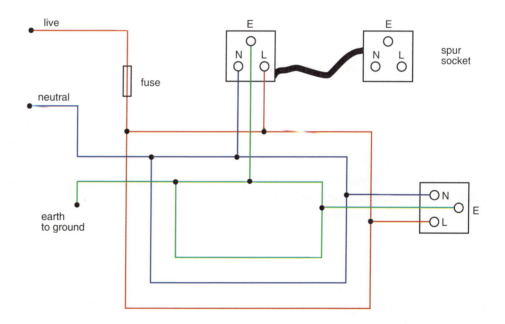

Spur sockets

A spur socket is an extra socket attached to one of the existing sockets. Spur sockets should be avoided if possible, although they can sometimes be convenient. If you use them, remember the rule: **one spur per socket**, no more.

BREAKTHROUGH SCIENCE

Switch and fuse
Did you notice that the switch and fuse are always on the live wire? Can you see why from the diagram?

EARTHING
Electricity will flow to earth if there is something it can flow through. This can be very dangerous.

In diagram (A), current flows through the heating element of the kettle. The resistance of the heating element keeps the current down to 10 amps. In diagram (B) the live wire is badly insulated at the point where it enters the kettle. The electricity has an easier path than it should have (short circuit). How will this affect the current?

What will happen to the fuse?

If there were no earth wire and you touched the case of the kettle, what would happen?

Now can you understand what the earth wire in a plug is for?

In some cases, circuit breakers are used instead of fuses. They switch off the current if it goes above a certain level. It can be switched on again when the fault has been dealt with.

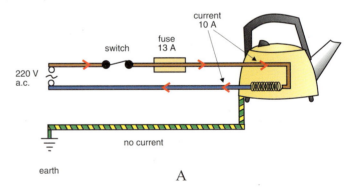

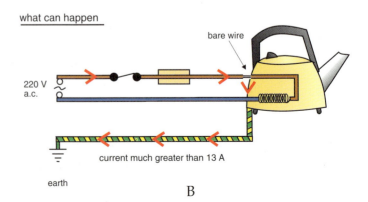

a.c./d.c.

When a battery is used in a circuit, the current flows in the same direction all the time. This is direct current, or d.c. for short. When you plug in a kettle to the ESB mains you are using 220 volt alternating current, or a.c. This current changes direction 100 times per second. Lots of things, such as cassette players, can be run off batteries or off the mains, but to do this a.c. must be converted to d.c.: in other words, it must be rectified. One way to do this is to use a diode, which allows current to flow in one direction only.

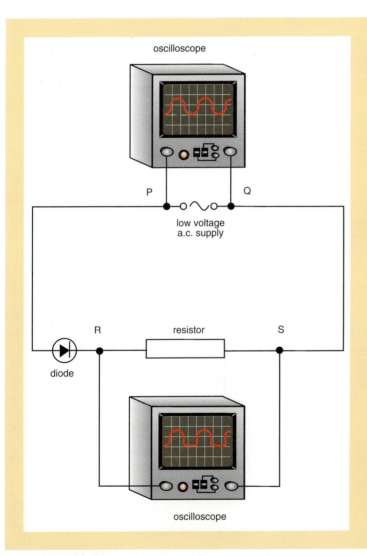

EXPERIMENT 15.5
AIM: TO CONVERT a.c. TO d.c.
Method
1. Set up the apparatus as shown.
2. Connect an oscilloscope between P and Q.
   Result (i) should appear on the screen.
3. Connect the oscilloscope between R and S.
   Result (ii) should appear on the screen.

Conclusion: A diode rectifies alternating current.

Televisions, radios, computers and battery chargers need d.c., so each must have a built-in rectifier.

| *A rectifier changes a.c. to d.c.*

BREAKTHROUGH SCIENCE

SUMMARY

- Electricity has a heating, chemical and magnetic effect.
- A fuse is a device that melts and breaks the circuit if the current exceeds a certain value.
- The purpose of earthing is to make sure that even if the metal body of an electrical device becomes live the person using it will not get a shock.
- If the metal casing of a device becomes live, a large current will flow to earth through the copper earth wire and the fuse will blow.
- Direct current (d.c.) flows in the same direction all the time.
- Alternating current (a.c.) changes direction 100 times per second.
- a.c. can be converted to d.c. using a rectifier.

QUESTIONS

Section A

1. Give three effects of an electric current. _____
2. A _____ contains a short piece of wire of low melting point.
3. In a three-pin plug, the brown wire is connected to _____, the blue wire is connected to _____ and the green/yellow wire is connected to _____.
4. The fuse and the switch should always be on the _____ wire.
5. What effect of an electric current is being shown in the diagram? _____ Give two applications of this effect in the home. _____ 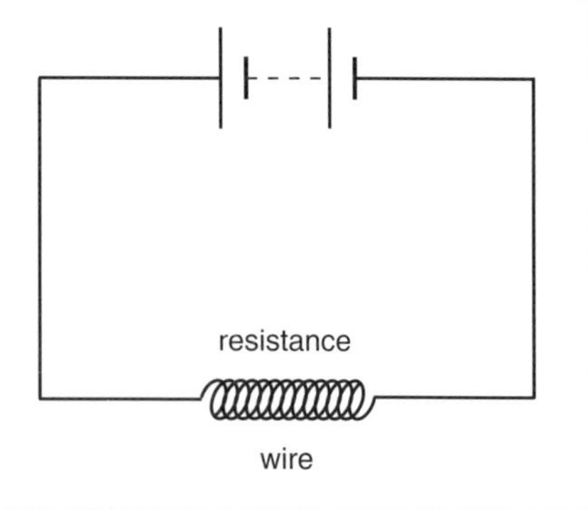
6. In the diagram, name A, B and C.

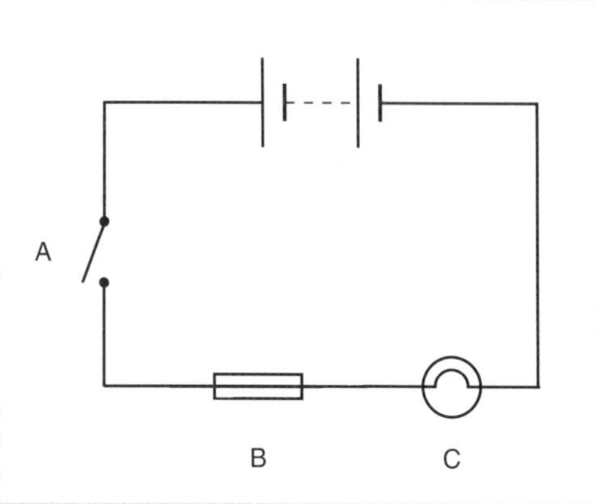

7. In the diagram of the kettle, which is
   (a) the live wire?
   (b) the earth wire?
   (c) the neutral wire?
8. What is meant by
   (i) a.c.? _____ _____
   (ii) d.c? _____ _____
   Give one difference between them.
   _____
9. A _____ can be used to change _____ to d.c.
10. In some cases ___ _____ _____ are used instead of fuses.

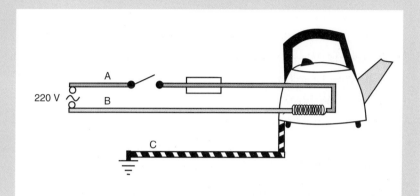

Section B
1. What are the two faults in this circuit?
2. Name the parts of the electric circuit labelled P, Q, R. What will happen if the current exceeds 3 amps? What will then happen to the bulb? Why is a fuse important in an electric circuit?

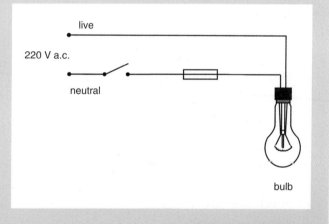

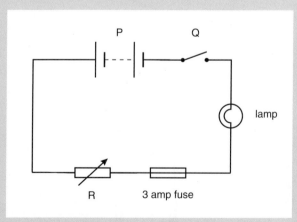

3.

	Conventional oven	Microwave oven
Voltage (V)	250	250
Current (A)		4
Power (kW)	2·5	
Cooking time (hours)	2	0·5
Units (kW hours)		
Cost @ 8p per unit		

The table refers to the cooking of a 2 kg chicken. Fill in the missing details.
Some people are concerned about the use of microwave ovens.
Do some research to find out why.

4. The diagram shows a domestic lighting circuit.
   What is A?
   Name the wires B and C.
   In what way are the lights connected?
   Give two advantages of connecting them this way.
   Where in your home would you find the arrangement at D?

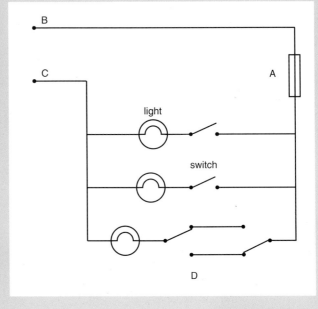

5. In the diagram, what happens when the hairdryer is plugged in?
   What colour are the wires L, N and E?
   What part of the dryer should E be connected to?
   Many hairdryers have no earth wire. Why?

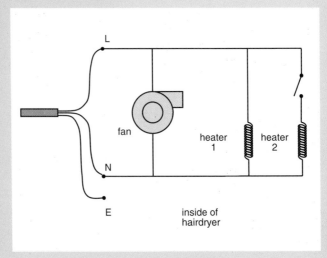

PHYSICS · MAGNETISM

CHAPTER 16 # MAGNETISM

Most people know that a magnet can attract certain metals. If you put a bar magnet into a box of small nails, you will find that the force of attraction is strongest at the two ends.

Hang a magnet from a non-metal stand and wait for it to stop swinging. It will stop with one end pointing north. Mark this end with an N. Swing the magnet slightly. When it stops the same end will be pointing north.

This end is called the north pole of the magnet. The other end is called the south pole.

| *When a magnet is free to swing it settles in a north-south direction.*

A COMPASS

Almost 2000 years ago, the Chinese dug an iron ore called lodestone out of the ground and found that if a piece of it was suspended from a wooden pole it would come to rest with one end pointing north. They used this as a primitive compass. A simple modern compass contains a magnetic needle which does the same thing.

compass

OTHER PROPERTIES

Take two bar magnets and suspend one of them from a wooden stand.

Bring the two north poles close together. Bring the two south poles close together. Bring a north pole close to a south pole.

| *Like magnetic poles repel. Unlike magnetic poles attract.*

This type of compass is used in navigation

91

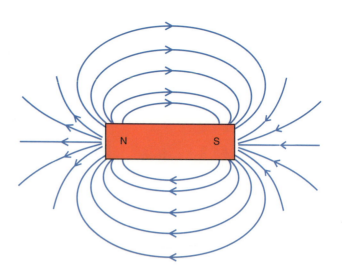

MAGNETIC FIELDS

If a small nail is placed too far away from a magnet, the magnet cannot move it. The magnet has an effect only within a certain space.

The space within which a magnet has an effect is called a magnetic field.

EXPERIMENT 16.1
AIM: TO PLOT THE MAGNETIC FIELD ROUND A BAR MAGNET

Method

1. Place a bar magnet on a sheet of paper and draw its outline.
2. Place a small plotting compass close to the north pole of the magnet.
3. Draw a dot in front of the needle.
4. Move the compass until the dot is behind the needle.
5. Mark another dot in front of the needle.
6. Continue in this way to produce a magnetic field map.

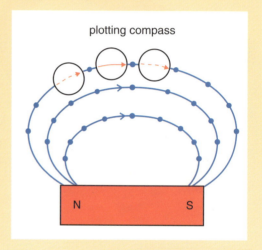

A magnetic field line is the path along which a north pole would move if it were free to do so.

MAKING MAGNETS

The most efficient way to make a magnet is to place a metal bar inside a solenoid (long coil of wire) and pass a large direct current through the wire. Only a few substances can be magnetised.

Iron, nickel, cobalt and some of their alloys can be magnetised.

Steel, which is an alloy of iron and carbon, can be magnetised.

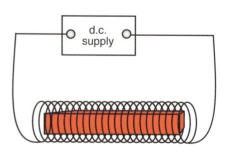

metal bar being magnetised

STORING MAGNETS

Magnets can lose their magnetism if they are not stored properly as shown in the diagram. The two keepers along with the magnets form a closed system, and the magnetism is preserved.

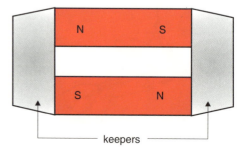

PHYSICS · MAGNETISM

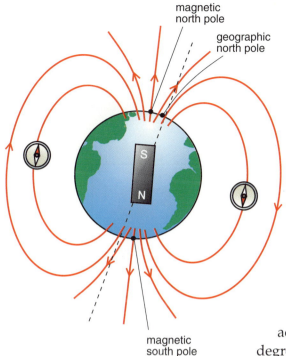

If you attach some steel nails to a bar magnet, the nails become magnetised temporarily. After you do this a number of times, steel nails become permanently magnetised. Iron nails do not.

THE EARTH'S MAGNETISM

Because there are magnetic materials in the earth, the earth has a magnetic field. The earth's magnetic field is the same as the field that would be caused by a bar magnet in the centre of the earth. There is, of course, no bar magnet in there.

The earth's magnetic poles are not the same as its geographic poles. Anyone using a compass for navigational purposes must take the difference into account. In Ireland the difference is an angle of about 10·5 degrees. The magnetic poles are drifting slowly all the time. Look this up in your geography book.

ELECTROMAGNETISM

Switch on the current. The nail becomes a magnet and attracts the pins. Switch off the current: the nail is no longer a magnet. The nail is a magnet only when the current is flowing. This type of magnet is called an electromagnet.

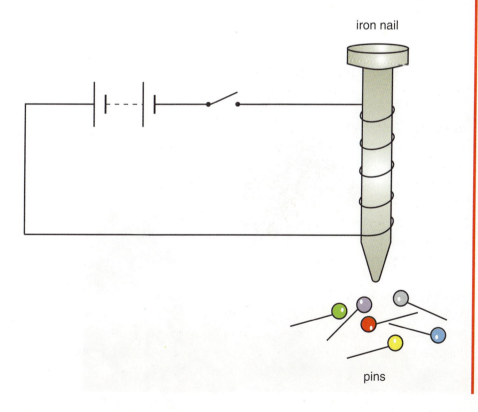

93

BREAKTHROUGH SCIENCE

MAGNETIC FIELD DUE TO A STRAIGHT WIRE

Sprinkle some iron filings on the cardboard and close the switch. Tap the cardboard gently. The lines of force should be circles. You can also plot the magnetic field, using a plotting compass. (You need a low voltage supply, as it is difficult to get this to work without a fairly large current.)

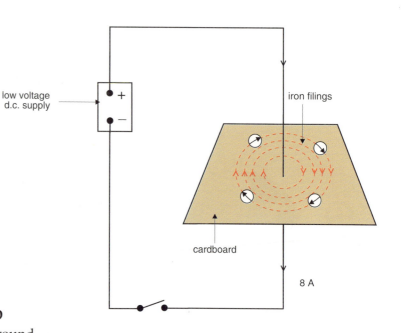

MAGNETIC FIELD DUE TO A SOLENOID

Close the switch and plot the magnetic field round the solenoid using a plotting compass. The field is similar to the field of a bar magnet.

USES

Magnets are used as catches on cupboard doors. Magnetic seals are used on fridge doors. Magnets are also used in electric motors, loudspeakers and dynamos, as we shall see later.

Electromagnets are used on cranes in scrapyards and in electric doorbells.

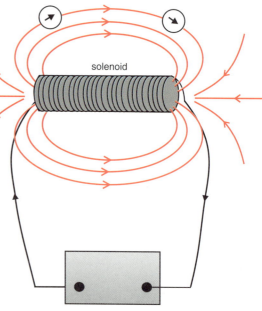

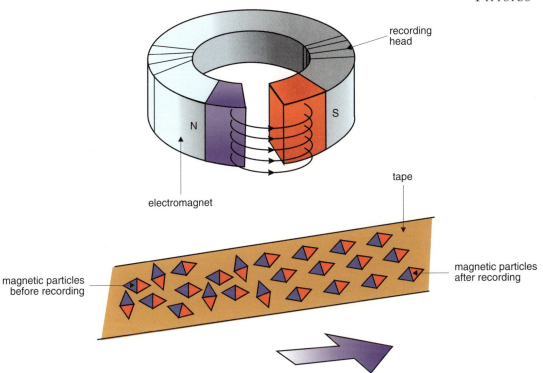

MAGNETIC TAPES

Audio and video tapes are widely used for storing information (music, films, etc.). The plastic tape is coated with iron oxide particles. The recorder converts the information into electric pulses. The pulses magnetise the iron oxide particles into a pattern which represents the information. When the tape is played the pattern is converted back to electric pulses, which are then converted to sound (and pictures).

SUMMARY

- A freely suspended magnet will settle in a north–south direction.
- Like magnetic poles repel; unlike magnetic poles attract.
- Iron, nickel, cobalt and steel can be magnetised.
- The space within which a magnet has an effect is called its magnetic field.
- A magnetic line of force is the path along which a north pole would move if it were free to do so.

QUESTIONS

Section A

1. A freely suspended magnet comes to rest pointing _____.
2. Like magnetic poles _____; unlike magnetic poles _____.
3. The elements _____, _____, _____ and some of their alloys can be magnetised.

4. In the diagram, place the letter N in the box at the north pole of the magnet.
5. What is a magnetic field line? _____

6. The diagram shows two bar magnets with keepers attached. Draw one magnetic field line passing through each keeper.

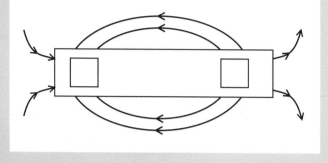

7. The space within which a magnet has an effect is called _____.
8. The core of an electromagnet is usually made of _____. Steel would not be suitable for use in the core of an electromagnet because _____.
9. Give two uses of magnets. _____
10. If you have a music system, why should you not store cassette tapes near the speakers? _____ _____

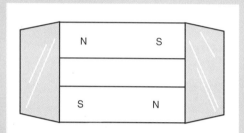

Section B

1. The magnet shown in the drawing is swinging freely. In what direction will the north pole point when it comes to rest?
   In which of the pairs A, B, C in the diagram will the magnets be attracted to each other? Describe a simple experiment to show the field of a magnet.

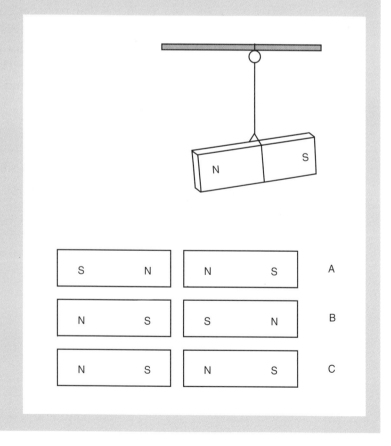

2. In the diagram, in which direction will the magnet move in X? In which direction will the magnet move in Y? In which direction will the magnet move in Z? Based on these results, state the law of attraction and repulsion for magnets.

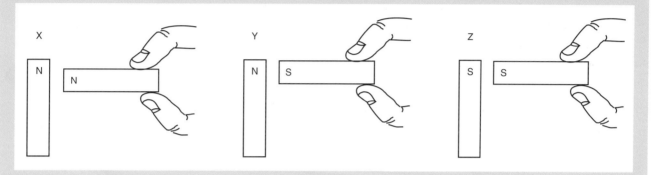

3. You are given three metal bars, A, B and C, which have been painted black and look exactly alike. One of them is a magnet, another is made of steel but is not magnetised, and the third one is made of copper.
   Write an account of an experiment, using another bar magnet, to find out which is which.
   Use the headings Aim, Method, Result and Conclusions.

4. The diagram shows the apparatus required to make an electromagnet.
   Identify the items labelled A and B.
   Explain how the items could be used to make an electromagnet.

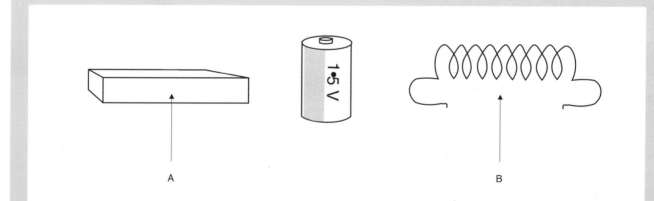

5. Have you ever seen anyone running a magnet over the body of a second-hand car before deciding whether or not to buy it? Why do they do this?

STATIC ELECTRICITY CHAPTER 17

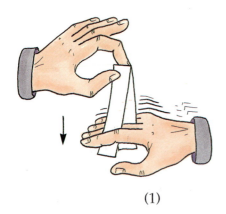

(1)   (2)

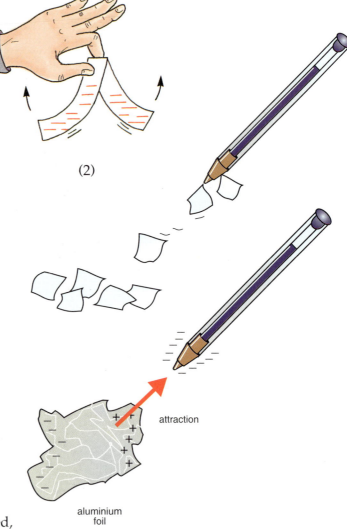

Cut two strips of polythene from a plastic bag and hold them together at one end. Pull them between your fingers. The strips now repel each other.

Put some small pieces of paper or aluminium foil on your desk. Rub a plastic biro against your jumper. The biro now attracts the pieces of paper.

EXPLANATION

Everything is made of atoms. Normally an atom has the same number of protons (positively charged) as of electrons (negatively charged). Electrons can be removed by friction (rubbing). Electrons moved from your fingers to the polythene strips. Both strips were now negatively charged, and like charges repel each other.

| *A body becomes negatively charged when it gains electrons.*
| *A body becomes positively charged when it loses electrons.*

When you rub the biro against your jumper, the biro gains electrons from your jumper and becomes negatively charged. It repels electrons in the aluminium foil (like charges repel). This leaves a positive charge on the top of the foil which is then attracted by the biro (unlike charges attract).

| *Like charges repel; unlike charges attract*.

It is important to remember that **electric charges are always caused by the movement of electrons**, never by the movement of protons.

POLYTHENE AND PERSPEX

When you rubbed a plastic biro against your jumper, electrons moved from your jumper to the biro. Does this always happen? Does the cloth always lose electrons, no matter what type of plastic you use? Let us see.

EXPERIMENT 17.1
AIM: TO SHOW THAT LIKE CHARGES REPEL AND UNLIKE CHARGES ATTRACT

Method
1. Charge a polythene (white plastic) rod by rubbing it against your jumper, and hang it from a non-metal stand.
2. Charge another polythene rod the same way, and hold it close to the first rod. Note how they **repel** each other.
3. Repeat steps 1 and 2 with two perspex (clear plastic) rods, and note how they also **repel** each other.
4. Now hang a polythene rod from the stand and bring a perspex rod close to it. Note how they **attract** each other.

   Result: The polythene and the perspex have unlike charges.
   Conclusion: The polythene gained electrons; the perspex lost electrons.

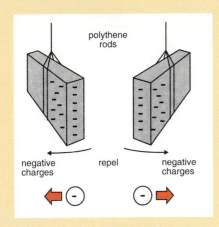
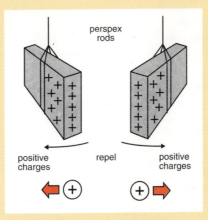
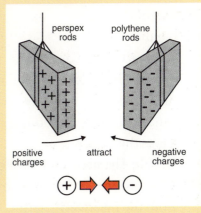

Breakthrough Science

This diagram shows what happens to the polythene. Can you draw a similar diagram to show what happens to the perspex?

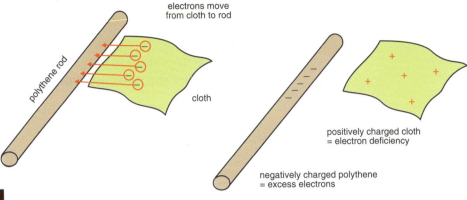

Benjamin Franklin (1706–1790), American scientist, inventor and writer, was prominent in the foundation of the USA. Franklin showed with his famous kite experiment that lightning was in fact a form of electricity. He invented lightning rods to protect buildings. His later work included a theory of heat absorption, designing ships, tracking storm paths and inventing bifocal lenses.

EXPERIMENT 17.2
AIM: TO SHOW THAT CURRENT WILL FLOW FROM A POSITIVELY CHARGED BODY TO A NEGATIVELY CHARGED BODY

Method
1. Set up the apparatus as shown in the diagram.
2. Turn on the Van de Graaff generator and let it run for a few minutes so that a positive charge can build up on the dome.
3. Close the switch.

Result: The neon bulb flashes.

Conclusion: Current has flowed from the positively charged dome to the earth, which is negatively charged relative to the dome.

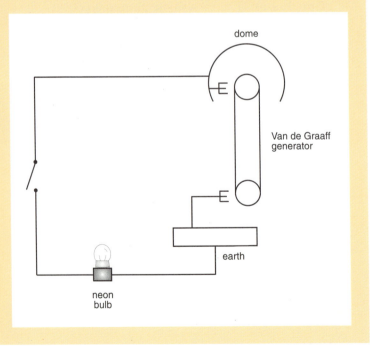

PHYSICS · STATIC ELECTRICITY

EVERYDAY EXAMPLES

There are many everyday examples of the effects of static electricity. One of these is lightning. A cloud becomes charged by friction with the wind. The negative charge on the bottom of the cloud causes a positive charge on the ground directly underneath the cloud in much the same way as the negative charge on the biro caused a positive charge on the top of the aluminium foil. If the force of attraction between the two charges is big enough, charge will flow through the air. This is what we call lightning. Lightning heats the air, causing it to expand very quickly and produce a sound called thunder.

N.B. An electric charge will always run to earth if it can.

SUMMARY

- A positive charge is caused by losing electrons.
- A negative charge is caused by gaining electrons.
- Like charges repel, unlike charges attract.
- A perspex rod becomes positively charged when rubbed with a dry cloth.
- A polythene rod becomes negatively charged.

QUESTIONS

1. Static electricity is always caused by the movement of _____.
2. A neutral body has the same number of _____ as of _____.
3. A positively charged body has lost _____.
4. A negatively charged body has _____ _____.
5. When a polythene rod is rubbed with a duster the rod becomes _____ charged and the duster becomes _____ charged.
6. Explain what happens in question 5 in terms of electrons. _____

7. It is not possible to hold a _____ in your bare hand and charge it by friction. Why is this? _____
8. Like charges _____; unlike charges _____.
9. Give an everyday example of static electricity. _____
10. Have you ever seen a strap like this hanging from the back of a car?
    What should it be made of? _____
    What is it for? _____

WAVES AND SOUND — CHAPTER 18

As we saw in chapter 3, sound is a form of energy. In order to produce sound, something must vibrate.

The strings of the guitar vibrate.
The cone of the loudspeaker vibrates.
The prongs of the tuning fork vibrate.

When the prongs of the tuning fork vibrate, they cause the layer of air molecules beside them to vibrate. These cause the next layer of molecules to vibrate, and so on until the vibrations reach your eardrum. This is how you hear the sound of the tuning fork. Another way of saying this is to say that sound waves travel from the tuning fork to your ear.

A wave is a means of transferring energy from one place to another.

PHYSICS · WAVES AND SOUND

EXPERIMENT 18.1
AIM: TO ILLUSTRATE THE WAVE NATURE OF SOUND

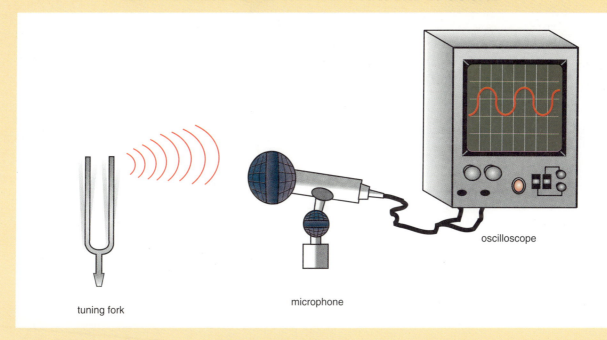

Method
1. Set up the apparatus as shown in the diagram.
2. Hold a vibrating tuning fork in front of the microphone.
3. A wave pattern will appear on the screen of the oscilloscope. This shows that sound has a wave nature.

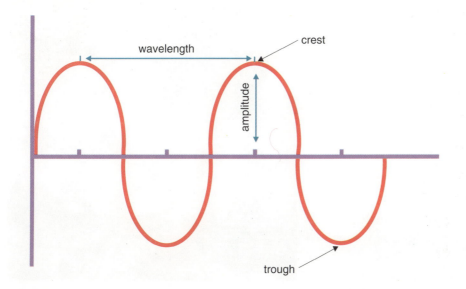

PROPERTIES OF WAVES
The amplitude is the greatest displacement from rest.

The wavelength is the distance from one crest to the next (in metres).

The frequency is the number of vibrations per second (hertz).

103

Breakthrough Science

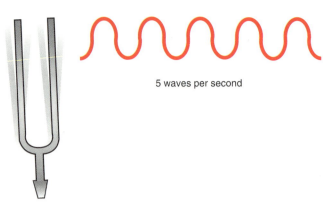

5 waves per second

One hertz means one vibration per second. If an object has a frequency of 5 hertz (Hz), it will send out 5 waves per second.

5 vibrations per second

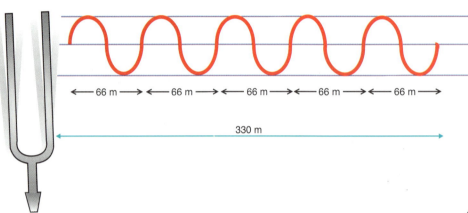

five waves in one second

← 66 m → ← 66 m → ← 66 m → ← 66 m → ← 66 m →

330 m

If each wave is 66 m long, the front of the first wave will have travelled 330 m in one second. $330 = 5 \times 66$

| *Velocity = frequency × wavelength*

This equation applies to all waves, including light waves, radio waves, etc.

Problem
If a sound wave has a velocity of 330 m s^{-1} and a frequency of 220 hertz, what is its wavelength?

$$\text{Wavelength} = \frac{\text{velocity, 330 m s}^{-1}}{\text{frequency, 220 hertz}} = 1\cdot 5 \text{ m}$$

Alexander Graham Bell (1847–1922), a Scottish-American scientist, invented the telephone. He was trained in public speaking and in teaching the deaf to speak, and studied anatomy and physiology, which gave him some ideas for the telephone. He also invented a number of other devices, including one that transmitted sound using light waves.

PHYSICS · WAVES AND SOUND

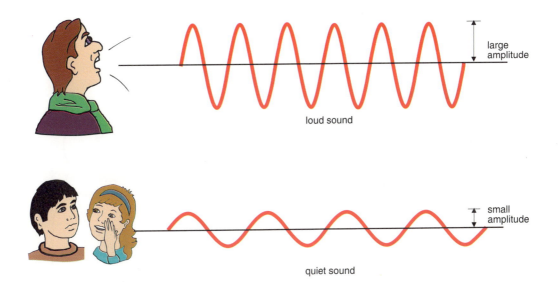

loud sound

quiet sound

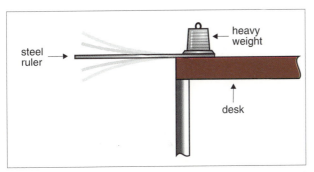

AMPLITUDE AND FREQUENCY

With waves, the amplitude determines the amount of energy being transmitted. The brightness of light depends on the amplitude of the light wave.

The **loudness** of a sound depends on the **amplitude** of the sound wave.

The frequency determines the pitch. The **pitch** of a sound depends on the **frequency** of the sound wave.

This can be illustrated by causing a steel ruler to vibrate as shown in the diagram. When a short piece of the ruler is protruding over the edge of the desk, we get a high frequency of vibration and a high-pitched sound. When a larger piece is protruding we get a low frequency and a low-pitched sound.

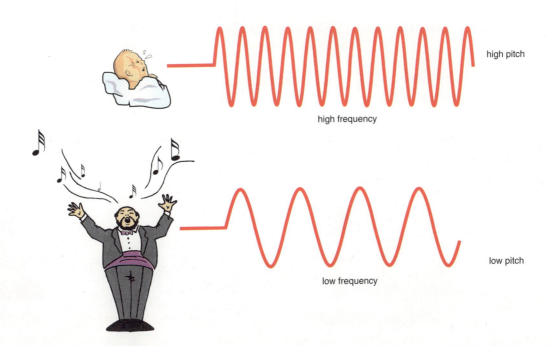

EXPERIMENT 18.2
AIM: TO SHOW THAT SOUND CANNOT TRAVEL THROUGH A VACUUM

Method
1. Set up the apparatus as shown in the diagram.
2. Start the bell ringing.
3. Slowly evacuate the air from the jar.
   Result: The sound will fade until you can hear no sound at all, even though you can still see the hammer striking the gong.
   Conclusion: This shows that sound needs a medium in which it can travel.

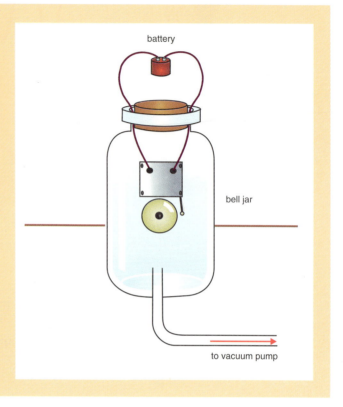

SPEED OF SOUND AND ECHOES
An echo means that sound is reflected (bounces back) off a surface.

EXPERIMENT 18.3
AIM: TO DEMONSTRATE REFLECTION OF SOUND

Method
1. Set up the apparatus as shown in the diagram.
2. Adjust the position of tube B until the sound is at its loudest.
   Result: Angle X is equal to angle Y.

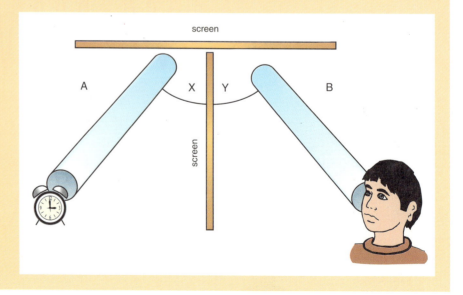

Physics · Waves and Sound

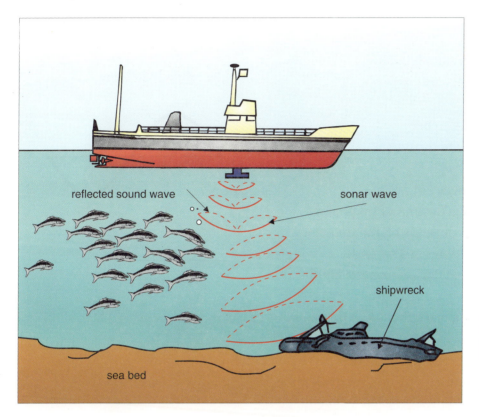

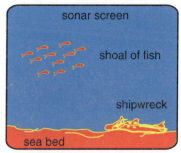

The speed of sound in air is about 340 m s^{-1}. The speed of sound in water is over four times as great. (Can you say why?)

The process being used in the diagrams is called echo-sounding. Can you see how it works? Ask your geography teacher about the use of echo-sounding in geological surveys.

Problem

A girl standing 100 m from a wall gives a shout. She hears the echo 0·6 of a second later. What is the speed of sound in air?

$$\text{Velocity} = \frac{\text{distance to wall and back}}{\text{time}} = \frac{200 \text{ m}}{0.6 \text{ s}} = 333 \text{ m s}^{-1}$$

SUMMARY

- Sound is a form of energy.
- Sound travels in waves.
- The number of vibrations per second is called the frequency.
- The amplitude is the greatest displacement from rest.
- The wavelength is the distance from one crest to the next.
- Velocity = frequency × wavelength
- The loudness depends on the amplitude; the pitch depends on the frequency.
- The velocity of sound in air is about 340 metres per second.

QUESTIONS

Section A
1. Sound is a form of _____.
2. In the diagram of a wave, A is the _____.
3. The distance from one crest to the next is called the _____. 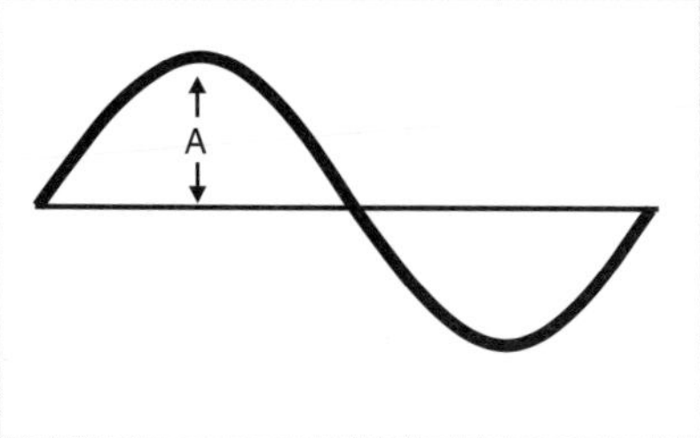
4. What is meant by the frequency of a wave? _____ What is the unit of frequency? _____
5. The _____ of a sound depends on the frequency.
6. A tuning fork vibrates 800 times in 4 seconds. Its frequency is _____.
7. The _____ of a sound depends on the amplitude.
8. _____ = _____ x wavelength.
9. What is the wavelength of a wave that has a frequency of 15 Hz and a velocity of 330 m s^{-1}? _____
10. A wave of wavelength 20 m travels at 340 m s^{-1}. What is its frequency? _____
11. Sound cannot travel through a _____; it needs a _____.
12. What is an echo? _____
13. The speed of sound in water is _____ than the speed of sound in air.
14. Dogs can hear a dog whistle but humans cannot, because _____.

Section B
1. The girl in the drawing shouts loudly and hears an echo. What causes the echo?

If air is removed from the belljar, why can the alarm not be heard when it rings?

2. A girl standing in front of a cliff gives a shout and hears the echo 4 seconds later. If the velocity of sound in air is 340 m s^{-1}, how far is she from the cliff?

3. Describe a simple experiment to show that sound cannot travel through a vacuum. Can light travel through a vacuum? Give an example.

4. What is the velocity of sound in air? Which would sound travel through fastest: a solid, a liquid or a gas? Explain.

5. Fill in the gaps in this table.

Velocity	Frequency	Wavelength
330 m s^{-1}	55 Hz	
	20 Hz	17 m
340 m s^{-1}		15 m
335 m s^{-1}	5 Hz	

6. In the diagram, what is the boat doing? What type of sound waves are usually used for this job?
If the echo is received after 0·1 seconds, how deep are the fish?
(Velocity of sound in water = 1500 m s^{-1})

7. How would you demonstrate in a laboratory: (i) reflection of sound, (ii) that sound is a wave motion? A tuning fork emits sound waves of frequency 256 Hz. Calculate the wavelength of the sound waves if the velocity of sound in air is 340 m s^{-1}.

BREAKTHROUGH SCIENCE

LIGHT CHAPTER 19

In chapter 3 we saw that the Crookes radiometer converts light energy to movement energy. The solar calculator and the solar cell both convert light energy to electrical energy.
Light is a form of energy.

One common application of the solar cell is in the solar calculator

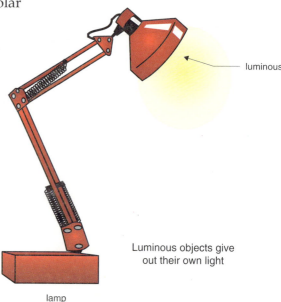

Luminous objects give out their own light

lamp

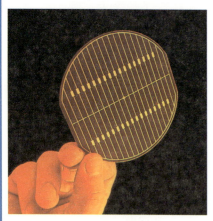

Cell converts light energy to electrical energy

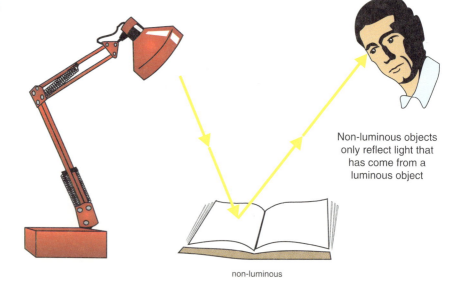

Non-luminous objects only reflect light that has come from a luminous object

non-luminous

EXPERIMENT 19.1
AIM: TO SHOW THAT LIGHT TRAVELS IN STRAIGHT LINES

Method
1. Look at a light (a small bulb or a candle) through a drinking straw.
2. Bend the straw. (Can you see the light now?) This shows that light travels in straight lines.

110

PHYSICS · LIGHT

THE SUN

If you look up at the night sky, you see stars. These stars are held in their positions by gravitational forces.

A group of stars held together by gravitational forces is called a galaxy.

Our galaxy is called **the milky way**. The nearest star to our earth is the sun. It takes the earth 365·25 days to orbit the sun.

The sun and the nine planets that orbit around it make up the **solar system**. The earth is the only planet that can support life as we know it, because it has enough water, it has oxygen in its atmosphere and it is neither too hot nor too cold.

ECLIPSES

A solar eclipse (eclipse of the sun) occurs when the moon comes between the sun and the earth.

total eclipse of the sun

A lunar eclipse occurs when the moon is in the earth's shadow. It takes the moon 28 days to orbit the earth.

An eclipse provides further evidence that light travels in straight lines.

total eclipse of the moon

Breakthrough Science

REFLECTION

If there is a very small child who is at the crawling stage in your family, why not have a little fun? Hold a large mirror in front of the child and watch while it stops, looks in the mirror, puts its hand on the front of the mirror to try and touch the "other child", crawls round to the back of the mirror to see where the "other child" is, and stops puzzled when it finds no one there. What the child does not realise is that the image it sees in the mirror is caused by reflection (light bouncing off the mirror).

In the diagram you can see how this works. The rays of light entering the eye seem to be coming from behind the mirror, so that is where the image is seen.

Another application of reflection is the simple periscope. You can make a simple periscope as shown in the diagram.

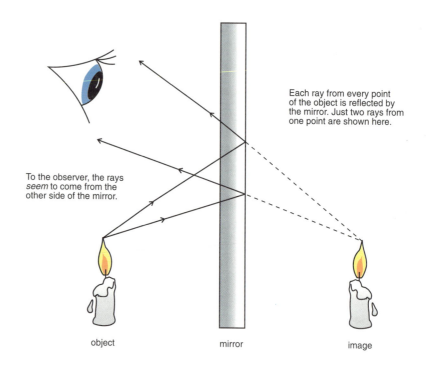

Each ray from every point of the object is reflected by the mirror. Just two rays from one point are shown here.

To the observer, the rays *seem* to come from the other side of the mirror.

object   mirror   image

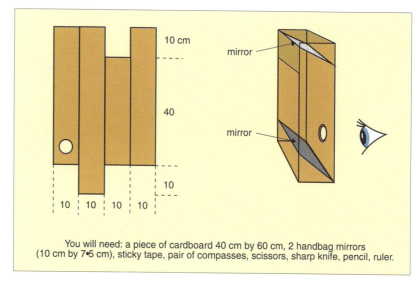

You will need: a piece of cardboard 40 cm by 60 cm, 2 handbag mirrors (10 cm by 7•5 cm), sticky tape, pair of compasses, scissors, sharp knife, pencil, ruler.

Sir Isaac Newton (1643–1727), one of the most important persons in the history of science, made discoveries in physics, astronomy and mathematics. He developed calculus and a reflecting telescope, and did work on optics. He discovered the laws of motion and universal gravitation, and used these to explain the movements of the planets. The unit of force (newton) is named in his honour.

PHYSICS · LIGHT

A convex mirror can be used to "see around corners"

Curved mirrors

If you look into the front of a highly polished spoon, you will see an image of yourself upside down. You are now looking into a **concave** mirror. Curved mirrors like this are used as reflectors behind the bulb of a flashlamp or a car headlamp.

If you look into the back of the spoon you will see an image of yourself the right way up. You are now looking into a **convex** mirror. These are used at dangerous exits or road junctions so that drivers can see if anything is coming.

REFRACTION

Light also bends when it travels from one transparent substance into another. This can have some peculiar effects.

EXPERIMENT 19.2
AIM: TO DEMONSTRATE REFRACTION

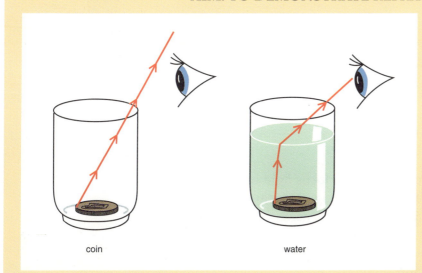

Method
1. Place a coin in a mug.
2. Move back until you cannot see the coin over the rim of the mug.
3. Now get someone to pour water into the mug until you can see the coin again.

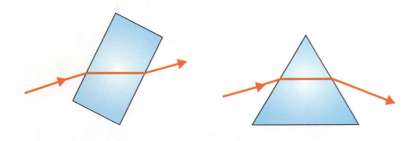

The diagram shows the path followed by a ray of light as it travels through a glass block. Try this in the science lab. Mark the outline of the block and the path of the ray of light with a pencil, so that you have a drawing of the path when the block and the light are removed.

113

BREAKTHROUGH SCIENCE

Some other effects of refraction are shown in the diagrams. In the diagram on the right, the light bends and makes the water appear less deep than it is. Could this effect be dangerous for non-swimmers?
Refraction is the bending of light when it travels from one transparent substance into another.

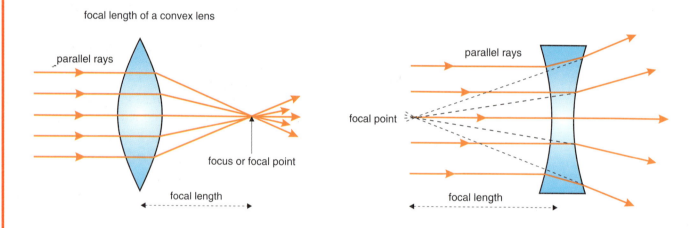

LENSES
Refraction can also be useful. Convex or converging lenses are used in microscopes, telescopes, etc. A convex lens brings light rays together at a point (focus). A concave lens causes light rays to spread out.

Stand near the window and use a convex lens to focus light on a sheet of white paper. You should get a very clear image. Concave and convex lenses are used in spectacles and contact lenses.

DISPERSION OF LIGHT
When the English scientist Isaac Newton passed light through a glass prism, the white light split up into seven bands of colour: red, orange, yellow, green, blue, indigo and violet.
The breaking up of white light into different colours is called dispersion.

A rainbow is caused by the dispersion of white light passing through raindrops. White light can also be dispersed by reflection from the surface of a compact disc.

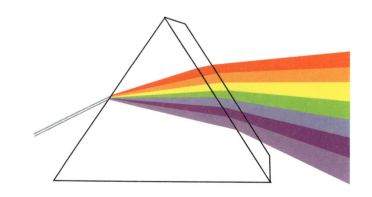

PHYSICS · LIGHT

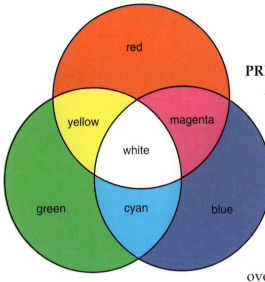

PRIMARY AND SECONDARY COLOURS

Although the spectrum contains seven colours, there are in fact only three basic or primary colours.

| *The primary colours of light are red, green and blue.*

When we mix two primary colours we get a secondary colour.

| *The secondary colours of light are cyan (turquoise), yellow, and magenta.*

The mixing of primary and secondary colours can be demonstrated by using three coloured spotlights, as shown in the diagram or by using three flashlamps with coloured filters over them.

Blue is a primary colour. Yellow is a secondary colour. If we mix blue and yellow we get white. Blue and yellow are complementary colours.

| *Complementary colours are a primary colour and a secondary colour that together give white.*

Mixing colours of paint is not the same as mixing colours of light.

ELECTROMAGNETIC RADIATION

Light is a member of a group of waves called the electromagnetic spectrum. They are all fundamentally the same. They all travel at the same speed (3×10^8 metres per second), but their frequencies and wavelengths are different. Some, such as X-rays, have high frequencies and short wavelengths. Others, such as radio waves, have low frequencies and long wavelengths. Light is the only part of the electromagnetic spectrum that is visible.

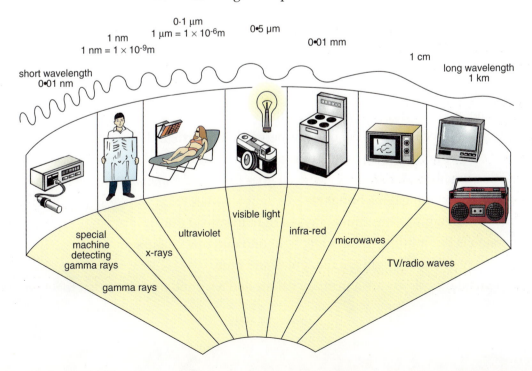

Radiation	Uses and effects
Radio waves	Radio and TV transmission
Microwaves	Cooking, communications, radar
Infra-red	Given out by hot bodies. Heating, treatment of rheumatism and injured muscles, photography in foggy and dark conditions
Ultra-violet	Causes suntan, causes some washing powders to glow, is absorbed by glass
X-rays	Used in medicine for X-ray photography and cancer treatment
Gamma rays	Come from the nucleus of an atom. Cancer treatment

UHF and VHF

You probably listen to FM radio. VHF (very high frequency) waves are used to give high-quality stereo reception. UHF (ultra-high frequency) waves are used for television. One problem with high frequency waves is that they have short wavelengths and so do not bend round obstacles such as hills. This can cause poor reception in valleys etc.

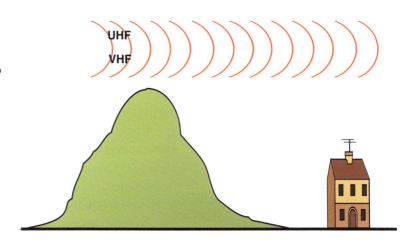

SUMMARY

- It takes the earth 365·25 days to orbit the sun.
- It takes the moon 28 days to orbit the earth.
- A solar eclipse occurs when the moon is between the earth and the sun.
- A lunar eclipse occurs when the earth is between the moon and the sun.
- Refraction means that light bends when it goes from one transparent substance to another.
- A convex (converging) lens brings rays of light together.
- A concave (diverging) lens causes rays of light to spread out.
- Dispersion is the splitting up of white light into different colours
- The primary colours of light are red, green and blue.
- Complementary colours are a primary colour and a secondary colour that together give white.

PHYSICS · LIGHT

QUESTIONS

Section A

1. Light is a form of _____.
2. A solar cell converts _____ energy to _____ energy.

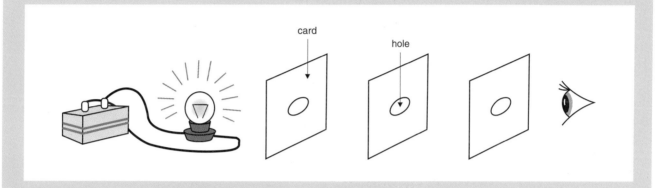

3. What property of light is being shown in the diagram? _____
4. What is a galaxy? _____ Name our galaxy. _____
5. Name the planet nearest the sun. _____
6. Refraction means that light _____.
7. The diagram shows a ray of light XY reaching the surface of water from underneath. The direction of the ray of light in air is given by _____.

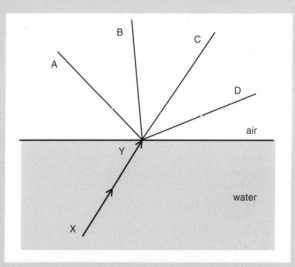

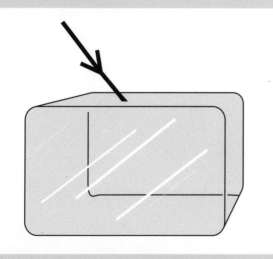

8. Trace the path of the ray of light through the glass block.
9. What is meant by dispersion of light? _____
10. What are complementary colours? _____

11. In the diagram, what colours are X _____ and Y? _____
12. What happens when a thermometer with a blackened bulb is held just beyond the red end of the spectrum? _____ Why? _____
13. Name the radiation that lies just beyond the violet end of the visible spectrum. _____
14. Can you get a suntan in a glasshouse? _____ Why? _____

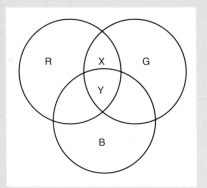

15. Apart from infra-red and ultra-violet, name two invisible members of the electromagnetic spectrum. _____

Section B

1. The diagram shows the separation of white light into seven colours. List these colours in the correct order beginning with red. Name the three primary colours. What happens to light so that you can see this page?

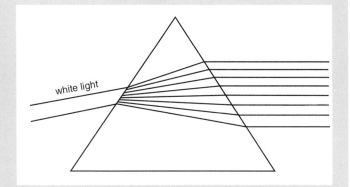

2. Show, by means of a diagram, the dispersion of light by a glass prism. How is a rainbow formed?
3. In an experiment to detect infra-red radiation, why is the bulb of the thermometer blackened? List the properties of infra-red radiation.
   Give two uses and name one source of this radiation.
4. Give an everyday example to show that light travels faster than sound.
   Give another difference between light and sound.
   Which travel faster, sound waves or radio waves? Explain.
5. Match each item in column X with one in column Y.

X	Y
blue + green	white
blue + yellow	cyan
red + green	yellow
blue + red	white
red + cyan	magenta
green + magenta	white

6. What are complementary colours? In question 5, which pairs of colours in column X are complementary colours?

SECTION 2

CHEMISTRY

COMMON CHEMICAL APARATUS

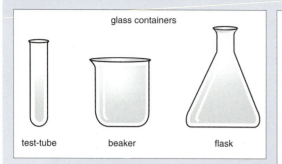

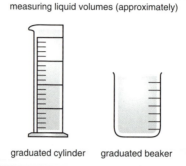

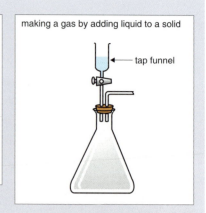

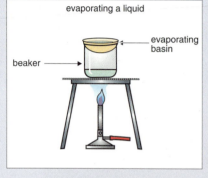

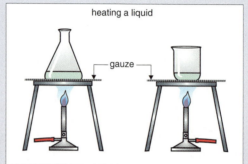

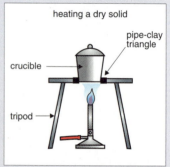

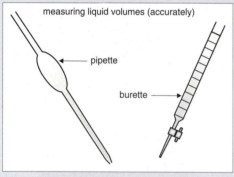

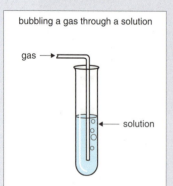

HOW THE BUNSEN BURNER WORKS

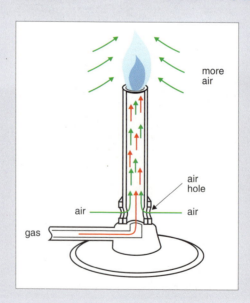

The Bunsen burner burns a mixture of gas and air. The gas is controlled by the gas tap. The air is controlled by the air hole. The flame is yellow and visible when the air hole is closed. The flame is blue and almost invisible when the air hole is fully open. The blue flame burns a gas/air mixture and is hot. The yellow flame is cooler.

CHAPTER 20 STARTING CHEMISTRY

How many different chemicals do you use every day? They could include plastics, detergents, paints, oils, medicines, metals, fuels and fertilisers.

You already know the names of many hundreds of these substances: concrete, glass, different plastics, metals, timber, milk, water, air, etc. All these substances are known as **matter**.

Matter is anything that takes up space and has mass.

Every substance is a solid, a liquid or a gas. These are the **three states of matter**. Iron is a solid, water is a liquid and oxygen is a gas.

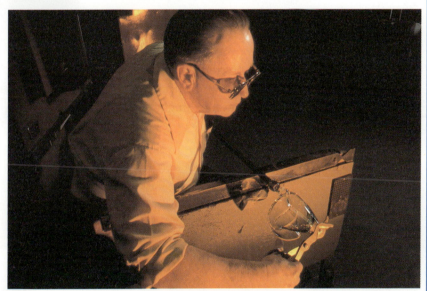

SOLIDS

A solid has a definite shape. It also takes up a fixed amount of space: a definite volume. A piece of iron cannot be squeezed into a smaller space: it cannot be **compressed**. A solid cannot flow. If you leave a piece of iron on the bench, it will not flow over the edge of the bench and onto the floor. We call these the **properties of solids**.

Breakthrough Science

LIQUIDS

A liquid has no definite shape. If you pour water into a bottle it will take the shape of the bottle. If you pour it into a beaker it will take the shape of the beaker.

A liquid does have a definite volume. If you pour a litre of water from the beaker back into the bottle, it will still be a litre of water and will take up the same amount of space as before. A liquid cannot be compressed. A liquid can flow. If you pour some water onto the bench it will flow over the edge and onto the floor.

GASES

If you turn on the gas tap at one end of your school laboratory **for a few seconds (and no longer!)** you will soon smell the gas at the other end of the room. The gas has spread all over the room, therefore it can flow. The gas is now in every part of the room: it occupies the volume of the room. It has no definite volume and no definite shape. A gas can be compressed: when you pump up your bicycle tyre you are compressing air (a gas) into the bicycle tube.

Joseph Priestley (1733–1804) was one of the founders of chemistry. Priestley discovered several gases, including oxygen (which was named by Lavoisier). He also studied plant respiration.

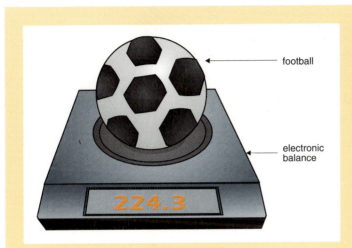

EXPERIMENT 20.1

AIM: TO SHOW THAT A GAS HAS MASS

Method

1. Find the mass of a deflated football.
2. Pump air (a gas) into the football until it is hard.
3. Find the mass of the pumped-up football.

Result: The football is heavier.

Conclusion: A gas has mass.

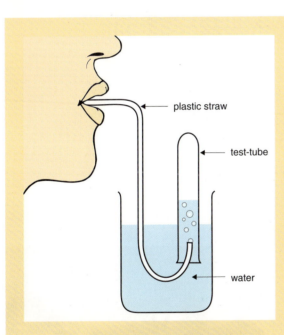

EXPERIMENT 20.2

AIM: TO SHOW THAT A GAS HAS VOLUME

Method

1. Fill a test-tube with water. Put your finger over the open end of the tube.
2. Invert the test-tube (turn it upside down) into a bowl of water.
3. Blow into the tube with a plastic straw.

Result: The water is pushed out by the air you blow in: the space occupied by the water is taken by the air.

Conclusion: A gas has volume.

Solid	Liquid	Gas
Definite volume	Definite volume	No definite volume
Definite shape	No definite shape	No definite shape
Cannot flow	Can flow	Can flow
Cannot be compressed	Cannot be compressed	Can be compressed
Support from below only	Support from below and at sides	Support from all directions

CHANGE OF STATE

A substance can change from solid to liquid to gas, as you will see if you do the following experiment.

EXPERIMENT 20.3
AIM: TO SHOW THAT A SUBSTANCE CAN CHANGE STATE

Method
1. Cover the bulb of a thermometer with some ice in a beaker.
2. Note the temperature of the ice.
3. Heat the ice gently and note the temperature when it begins to melt. This is the **melting point of water**: 0° Celsius.
4. When all the ice has turned to water, continue the heating and note the temperature at which the water boils. This is the **boiling point of water**: 100°C.

Conclusion: A substance can change from solid (ice) to liquid (water) to gas (steam) when heated.

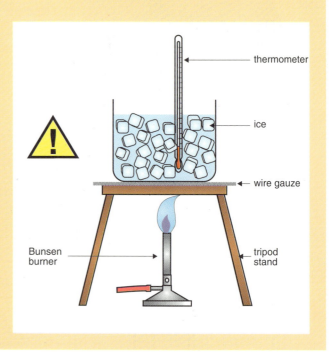

If a number of students do this experiment at the same time (on a cold day), you will see that the windows of the laboratory become covered with condensation. This happens when the steam loses heat and turns back to water. If you take the water remaining in the beaker, pour it into an ice tray and put it into the freezer compartment of a fridge it will turn back into ice.

Conclusion: A substance can change from a gas (steam) to a liquid (water) to a solid (ice) when it loses heat.

Usually when we think of a substance we think of it in one state only. We think of iron as a solid, but when iron is heated sufficiently it turns into a liquid and can be poured into moulds to make tools, machine parts and wheels. Plumbers and electricians use a metal alloy (called solder) that melts easily (at a low temperature) to connect pipes and wires together. They first melt the solder and then allow it to cool and turn solid.

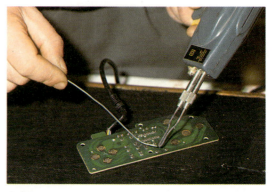

Soldering

We think of water as a liquid, but we can turn it into a solid (ice) by cooling and into a gas (steam) by heating. There are even some substances that can change directly from solid to gas when heated: this is **sublimation**. At some rock concerts, solid carbon dioxide (called "dry ice", and very cold) is allowed to warm up and change directly into a gas. This causes water vapour to form thick white clouds on stage.

Sublimation means that a substance changes directly from a solid to a gas when heated.

THINGS DON'T HAVE TO BOIL TO CHANGE STATE

On a rainy day the road is wet. If the following day is dry, the road will soon dry up. Where did the water go?

The water changed to a gas and rose into the air. It did not boil because it was not hot enough: it **evaporated**. A liquid can change to a gas by evaporation at temperatures much lower than its boiling temperature.

EVERYTHING IS MADE UP OF PARTICLES

All matter is made up of particles. We cannot see these individual particles, as they are very small. The tiniest speck of dust or smoke you see contains about a million million million of the particles from which the substance is made.

In chemistry we want to know what happens inside the substance when changes take place. Why does a substance change from solid to liquid to gas when heated?

The particles that make up a solid are tightly packed. They cannot move about but they can vibrate. They need more energy if they are to move about. If we give them heat energy they will eventually be able to slide over each other and move about. When this happens the substance can flow, and has changed from a solid to a liquid. If we continue to heat the liquid, the particles will eventually have enough energy to move about freely and escape from the liquid. When this happens, the liquid has changed to a gas.

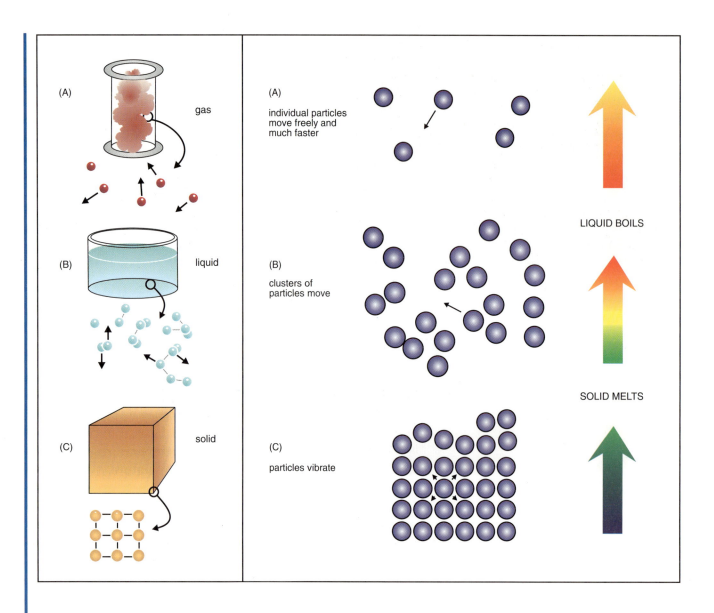

EXPERIMENT 20.4
AIM: TO SHOW THAT SUBSTANCES CONTAIN MANY PARTICLES

Method
1. Half-fill a beaker with water.
2. Drop a small crystal of potassium permanganate (potassium manganate VII) into the water through a plastic straw.

Result: The purple colour spreads slowly throughout the water.

Conclusion: Potassium permanganate contains many particles.

CHEMISTRY · STARTING CHEMISTRY

> *Whether a substance is a solid, a liquid or a gas depends on the freedom of movement of its particles.*
> *The temperature at which a solid changes to a liquid is its melting point.*
> *The temperature at which a liquid changes to a gas is its boiling point.*

Robert Brown (1773–1858)

BROWNIAN MOVEMENT

Movement of particles was first noticed by the Scottish botanist Robert Brown (1773–1858). He saw pollen grains darting about in water. He tested other substances, including coal dust, and found the same movement. Brown thought this was caused by the particles themselves. Scientists have since discovered that this movement—called **Brownian movement**—is due to these small particles being struck by moving water particles. The theory that explains how the differences between solid, liquid and gas are due to the differences in the movement of their particles is called the kinetic theory.

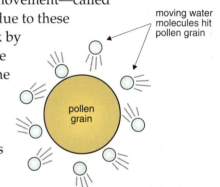

Solid	Liquid	Gas
Particles tightly packed	Particles move	Particles move fast
Little movement	about slowly	and freely all the time

SUMMARY

- Matter is anything that takes up space and has mass.
- The three states of matter are solid, liquid and gas.
- A substance can change from solid (ice) to liquid (water) to gas (steam) when heated.
- All matter is made up of particles.
- A substance is a solid, a liquid or a gas depending on the freedom of movement of its particles.
- Sublimation means that a substance changes directly from a solid to a gas when heated.
- The temperature at which a solid changes to a liquid is its melting point.
- The temperature at which a liquid changes to a gas is its boiling point.

QUESTIONS

Section A

1. Complete the following table.

Solid	Liquid	Gas
Definite volume		No definite volume
	No definite shape	
Cannot be compressed		Can flow

2. Name the change of state taking place in each of the following:
   (a) SOLID → GAS
   (b) GAS → LIQUID
   (c) LIQUID → SOLID
   (d) LIQUID → GAS
3. The temperature at which a solid turns to a liquid is its _____ point.
4. The temperature at which a liquid turns to a gas is its _____ point.
5. Brownian movement is due to _____.
6. Name a substance that sublimes on heating.
7. The spreading out of a gas due to the movement of its particles is called _____.
8. Why can a gas be compressed into a container? _____
9. Matter is anything that _____.
10. What is a change of state? _____
11. Name the three states of matter. _____
12. Condensation is _____.
13. What do you mean when you say that solid carbon dioxide sublimes?

14. Why are liquids and gases particularly suitable for use as fuels in cars?

15. When a beam of sunlight passes through a room you sometimes see small specks of dust moving about in many different directions. Why do you think this happens?

Section B

1. Describe a simple experiment to show that particles move about in a liquid. Use the headings Aim, Method, Result and Conclusion. Draw a simple diagram of the apparatus you used.
2. Solid carbon dioxide (dry ice) is used at rock concerts to produce vapour clouds on the stage. Explain the changes of state that take place to produce this effect.

3. Condensation of water vapour is a problem in some homes in winter. Explain why condensation happens mainly in winter. Suggest a number of things you could do to reduce the problem.
4. Liquids cannot be compressed. This fact is essential for car brakes to work. Why is this so? What do you think might happen when you pressed on the brakes if air had got into the brake fluid pipes?
5. Describe two simple experiments to show that a gas has mass and occupies space. Draw a simple diagram and use the headings Aim, Method, Result and Conclusion.

CHAPTER 21 HOW TO EXAMINE A SUBSTANCE

As you study chemistry you will examine many new substances. It is important to learn to examine each substance in a **systematic** way.

To examine the **physical properties** of any substance, you must answer the following questions:

State: Is it a solid, liquid or gas?
Colour: What colour is it, if any?
Smell: Wave the smell towards your nose by moving your hand over the substance. What kind of smell does it have?
Solubility in water: Does it dissolve in water?

SAFETY WARNING: DO NOT SMELL OR SNIFF A SUBSTANCE DIRECTLY. DO NOT UNDER ANY CIRCUMSTANCES TASTE ANYTHING IN A CHEMISTRY LABORATORY.

Salt seen under a microscope

SOLID
Examine the solid with a magnifying glass. Is it made of small shiny blocks, like grains of salt or sugar (i.e. **crystalline**)?
Is it non-crystalline, like baking flour (i.e. **amorphous**)?

BREAKTHROUGH SCIENCE

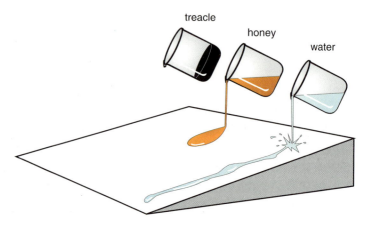
test for viscosity

LIQUID

Pour the liquid.

If it flows quickly we say that it has a **low viscosity**.

If it flows slowly (like honey or treacle) we say that it has a **high viscosity** or that it is a **viscous** liquid.

GAS

Put a gas jar of air on top of the jar of gas you are examining.

Remove the glass cover plates for a few seconds. Test each jar for the gas you are examining.

If the gas is **less dense** ("lighter") **than air** it will be mainly in the upper jar.

If the gas is **more dense** ("heavier") **than air** it will be mainly in the lower jar.

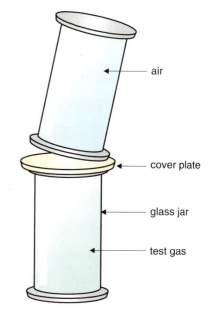
test for density

A substance also has **chemical properties**. Chemical properties tell us how a substance behaves when we add it to other substances. An example of a chemical property of iron is that it rusts: it forms a brown flaky substance when left in damp air.

We shall come across many other examples of chemical properties in our study of chemistry.

Does it burn? Test a **small** quantity carefully by putting a lighted taper (a taper is like a very thin wax candle) to it. Does it "catch fire"? A substance that "catches fire" is **flammable**.

Does it help other substances to burn: does the taper burn more brightly when put in the substance? A substance that helps other substances to burn **supports combustion**.

Does it extinguish ("put out") a flame?

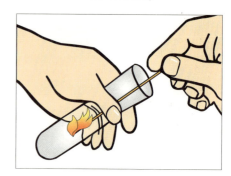
test for combustion

Is the substance an **acid?** Sulphuric acid is a strong acid that can burn your skin. Orange juice (like some other juices with a sharp taste) is a weak acid. We test for acids by using **blue litmus** paper: a blue substance that changes colour to **red** when put in an acid solution. Litmus is an **indicator**.

Is the substance a **base**? A base is a substance that has a soapy feel when rubbed with your fingers: soap is a base. Sodium hydroxide (caustic soda or oven cleaner) is a strong base that can burn your skin. We use **red litmus** paper to tell us whether or not a substance is a base. Red litmus turns **blue** when put in a basic solution. If the base does not dissolve easily in water, wet the litmus paper and touch it to the base.

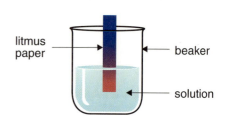
test for acid or base

CHEMISTRY · HOW TO EXAMINE A SUBSTANCE

A bee sting is acid, and so you should treat it with a mild base such as milk of magnesia. This changes the acid into a harmless substance and helps stop the irritation caused by the sting. A wasp sting is a base, so you should treat it with a mild acid such as vinegar.

EXPERIMENT 21.1
AIM: TO EXAMINE SOME COPPER SULPHATE

1. State: Solid
2. Colour: Bright blue—Copper sulphate is also called **bluestone**.
3. Smell: No noticeable smell.
4. Solubility: It dissolves in water to give a blue solution.
5. Is it crystalline? Examine some solid copper sulphate with a magnifying glass. Copper sulphate is made of shiny crystals—copper sulphate is crystalline.
6. Is it an acid? Put some **blue litmus** paper into copper sulphate solution. The colour of the paper does not change—copper sulphate is not an acid.

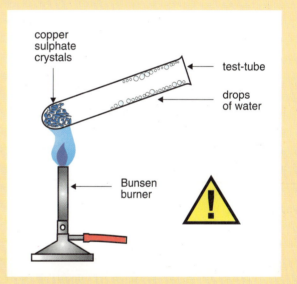

7. Is it a base? Put some **red litmus** paper into copper sulphate solution. The colour of the paper does not change—copper sulphate is not a base.
8. Put a small amount of the copper sulphate into a test-tube. Point the tube at a window or a wall (**never** towards yourself or another person) and heat gently with a **Bunsen** burner. The copper sulphate loses its blue colour and turns into a pale whitish-blue powder. Allow the test-tube to cool. Add a few drops of water to the powder: it turns back to the bright blue colour of the copper sulphate.

Conclusion: Crystals of copper sulphate contain water.

You should now examine a number of other common chemical substances in the same way. Can you describe the properties of copper, iron oxide (rust), sulphur, dilute ammonia liquid, sodium chloride (common table salt)?

131

Breakthrough Science

The German chemist Robert Wilhelm Bunsen (1811–1899) invented and improved many pieces of laboratory equipment, including a battery, the grease spot photometer, the spectroscope and the Bunsen burner. He also invented new methods of identifying, separating and measuring quantities of inorganic substances. Using the spectroscope, Bunsen (with Kirchhoff) discovered the elements caesium and rubidium.

SUMMARY

- Physical properties: solid, liquid or gas; coloured or colourless; a distinctive smell or not; dissolves in water or not; crystalline or amorphous; high or low viscosity; denser or less dense than air (if a gas)?
- Chemical properties: does it burn; does it help other things to burn faster; is it acid, base or neutral (test with litmus paper)?

QUESTIONS

Section A

1. What type of solid is salt? _____
2. Flour is an _____ solid.
3. A substance that catches fire easily is said to be _____.
4. A substance that turns red litmus blue is a _____.
5. A bee sting is _____. To treat it you should _____.
6. A wasp sting should be treated with vinegar because it is _____.
7. When copper sulphate crystals are heated, what change occurs? _____
8. A liquid that pours very slowly is a _____ liquid.
9. Salt is soluble in water. This means that salt _____ in water.
10. A gas that sinks down in air is _____ than air.
11. A match burns very brightly in oxygen. This tells us that oxygen _____.
12. Blue litmus turns red when put into vinegar. You can conclude from this that vinegar is _____.
13. When carbon dioxide gas is poured over a lighted match it quenches the match. This tells us that carbon dioxide _____.
14. Wax does not dissolve in water: wax is _____ in water.

Section B

1. List the properties of copper sulphate under the following headings.
   (Hint: see experiment 21.1)

   Physical properties   Chemical properties
   ....

2. Describe a simple experiment you could do to show that carbon dioxide gas is denser than air. Draw a diagram and use the headings Aim, Method, Result and Conclusion.
3. Sugar is a chemical that is used in food—in tea, coffee, sweets and cakes. List as many properties as you can of sugar, using the headings "Physical properties" and "Chemical properties".

CHAPTER 22 — DIFFERENT SORTS OF CHANGES

We see many different changes in substances: ice melting to give water, iron turning into rust and coal burning to give heat and leave cinders and ash. Is there any difference between these changes? You will carry out some experiments to discover that there are different sorts of changes.

EXPERIMENT 22.1

AIM: TO EXAMINE THE CHANGES THAT TAKE PLACE WHEN ICE IS HEATED

Method

1. Find the mass of an empty clean dry plastic bag.
2. Find the mass of the plastic bag with some ice in it. Calculate the mass of the ice.
3. Seal the plastic bag and allow the ice to melt.
4. Find the mass of the plastic bag with the water. Calculate the mass of water.
5. Put the plastic bag into the freezer compartment of a fridge. After a few hours you will find that you have ice once again.

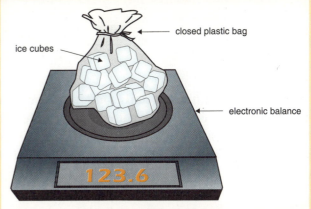

Result: There is no change in mass when ice melts to form water.
Conclusion: Water and ice are different states of the same substance: ice is the solid state and water is the liquid state.

Water and ice are chemically the same substance.

EXPERIMENT 22.2
AIM: TO EXAMINE THE CHANGES THAT TAKE PLACE WHEN AN IRON NAIL IS MAGNETISED

Method
1. Find the mass of the nail.
2. Dip the nail into some iron filings. The iron filings do not cling to it.
3. Rub the nail a few times in one direction with a strong bar magnet.
4. Dip the nail into the iron filings. The iron filings now cling to the ends of the nail.
5. Find the mass of the nail again after removing the iron filings.
   Result: The mass did not change when the nail was magnetised.
   Do you notice any other changes in the properties of the iron nail?
6. Hold the nail with a long tongs over a Bunsen flame until it is red hot. Dip it into cold water to cool it. Dry the nail.
7. Dip the nail into iron filings. The filings do not stick to it now.
8. Find the mass of the nail.

Demagnetising the nail did not change its mass.

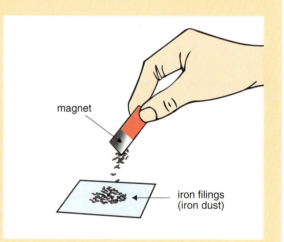

Result: The magnetised nail is *not* a new substance. It has all the properties of the original nail with the single added property of magnetism. This property of magnetism is easily removed.

Physical change is a change in which no new substance is formed, but the original substance gains new properties.

There is no change in mass in a physical change; it is easy to reverse a physical change.

Now try this experiment.

EXPERIMENT 22.3
AIM: TO INVESTIGATE THE CHANGES THAT HAPPEN WHEN PAPER IS BURNED

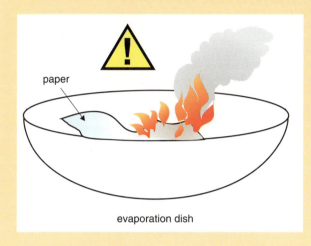

Method
1. Place a piece of paper in an evaporating dish and light it.
2. When it has stopped burning, examine what it left in the dish.

Result: Is this the same substance that you started with? Of course not! Can you get the piece of paper back? Not a chance.

Conclusion: This kind of change is a **chemical change**.

We shall find out more about chemical changes in the following experiments.

EXPERIMENT 22.4
AIM: TO EXAMINE THE CHANGES THAT TAKE PLACE WHEN MAGNESIUM IS BURNED IN AIR

Method
1. Examine a piece of magnesium and list its properties.
2. Find the mass of a small amount of magnesium (about 0·1 g) accurately.
3. Place the magnesium in a crucible and weigh the lot. Place the crucible in a fireclay triangle on a tripod stand.
4. Heat the crucible carefully with the lid slightly raised.
5. When the magnesium stops burning, allow the crucible to cool.
6. Find the mass of the crucible again.

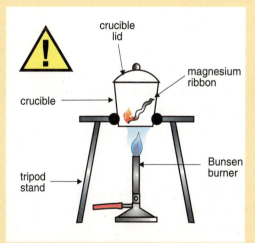

Result: There is an increase in the mass.

7. Examine the contents of the crucible. The substance in the crucible is a white powdery substance. When you test it with moist red litmus, it turns the litmus blue. This shows it is a base.

Conclusion: This white powder is a *new* substance called magnesium oxide.

When we burn **magnesium** it joins with a gas in the air called **oxygen** to form a white powdery substance called **magnesium oxide**. Magnesium oxide has a greater mass than magnesium because it contains oxygen particles as well as magnesium particles.

EXPERIMENT 22.5
AIM: TO EXAMINE THE CHANGES THAT TAKE PLACE WHEN IRON FILINGS RUST

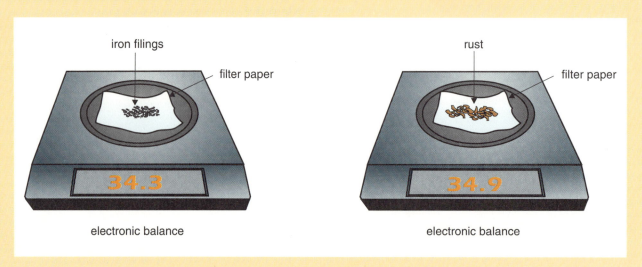

Method
1. Examine some iron filings and list the properties of iron.
2. Find the mass of about 1 g of iron filings.
3. Spread the iron filings thinly on a piece of filter paper.
4. Leave the iron filings in a damp airy place. After some days the iron filings will have changed to a reddish-brown substance. Place the filter paper and contents in a cool oven to dry.
5. Find the mass of this substance.
   Result: The mass is greater than that of the iron filings.
6. Examine the substance and list its properties.
   Conclusion: This substance is clearly not iron. It is what we call rust: its correct chemical name is **iron oxide**.

In this experiment **iron** joined with **oxygen** in the air to form the reddish-brown powder **iron oxide**. This reaction happens only in moist or damp air. The iron oxide formed has a greater mass than the iron originally present.

CHEMISTRY · DIFFERENT SORTS OF CHANGES

In both these experiments a new substance was formed, and the substance we started with gained mass in the chemical reaction. The chemical change in each case is extremely difficult to reverse.

A chemical change is a change in which at least one new substance is formed.

When a chemical change takes place we call what happens a chemical reaction.

In these experiments we ended up with a heavier mass of substance than we started with. Did we create matter? The following experiment will help to answer this question.

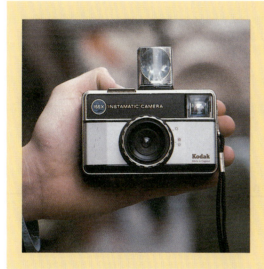

EXPERIMENT 22.6
AIM: TO SHOW THAT THERE IS NO OVERALL CHANGE IN MASS IN A CHEMICAL REACTION

Method
1. Find the mass of a flash bulb carefully.
2. Set off the bulb (use a camera or connect to a battery).
3. Find the mass of the bulb again after the reaction.

Result: There is no change in mass.

Conclusion: When a chemical reaction takes place **in a sealed container** there is no change of mass.

The total mass of the substances at the start of the reaction (the **reactants**) equals the total mass of the substances produced (the **products**). This shows that in a chemical reaction matter is neither created nor destroyed.

Law of conservation of matter: matter is neither created nor destroyed in a chemical reaction.

SUMMARY

- **Physical change**: A change in which no new substance is formed (but the original substance gains new properties).
- **Chemical change**: a change in which at least one new substance is formed.

Chemical change	Physical change
• One or more new substances formed	• No new substances formed
• A change in mass	• No change in mass
• Very difficult to reverse reaction	• Can be reversed easily
• Heat usually involved in reaction	• Often no heat involved

- **Law of conservation of matter**: matter is neither created nor destroyed in a chemical reaction.

QUESTIONS

Section A

1. Underline the physical changes in the following list: expansion, bleaching, melting, burning.
2. A change in which a new substance is formed is a _____ change.
3. Give two differences between a physical change and a chemical change.
   _____
4. Iron joins with _____ in damp air to form rust (iron oxide).
5. When magnesium burns in air, it joins with _____ to form a white powdery substance called _____.
6. A physical change is one in which _____.
7. Underline the chemical changes in the following list: melting ice, burning paper, magnetising iron, burning phosphorus.
8. Magnetising a bar of iron is a _____ change.
9. When a change in a substance results in a change in mass, the change must be _____.
10. When iron forms rust, the mass _____. This tells us that rusting is a _____ change.

Section B
1. The mass of a flash bulb does not change when the bulb is used. Does this mean that this is a physical change? Explain your answer.
2. Describe a simple experiment you could do to show that iron rusting is a chemical change. Draw a diagram and use the headings Aim, Method, Result and Conclusion.
3. Magnesium burns in air to form a white powder which has a greater mass than the magnesium. Compare the properties of magnesium and magnesium oxide using the following table.

	Magnesium	Magnesium oxide
Physical		
Chemical		

CHAPTER 23 ELEMENTS AND COMPOUNDS—ATOMS AND MOLECULES

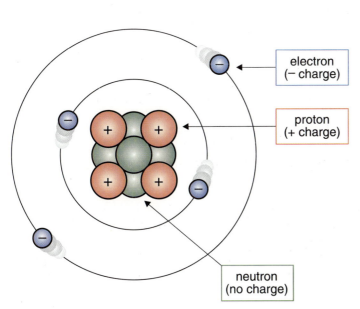

Everything you see around you is made up of tiny particles called **atoms**. However, you cannot see individual atoms. They are so small that a million million million atoms would fit on the full stop at the end of this sentence. Even the smallest speck of dust you see contains billions of atoms. Even though we cannot see individual atoms, scientists have discovered that all atoms are made of three different smaller particles: **protons**, **neutrons** and **electrons**.

The protons and neutrons are in a cluster at the centre of the atom: the **nucleus**. The electrons spin around the nucleus in much the same way as the planets spin around the sun. If we picture the nucleus as a golfball, an atom on this scale is the size of a playing field.

There are over a hundred different kinds of atoms. Ninety-two of these are found naturally on the earth: **the natural atoms**.

> The Irish physicist E.T.S. Walton (1903–) shared the 1951 Nobel Prize for physics with J.D. Cockcroft for the first artificial changing of one element into another (in 1932).

The rest are made artificially using nuclear reactors.

Substances usually contain groups made up of two or more atoms joined together. These groups of atoms joined together chemically are **molecules**.

A molecule is two or more atoms combined chemically.

ELEMENTS

When all the atoms in a substance are alike, the substance is an element. There are over 100 different kinds of atom, so there are over 100 different elements.

COMMON ELEMENTS

Hydrogen, carbon, nitrogen, oxygen, silicon, aluminium, sulphur, chlorine, iron, copper and silver.

An element is a substance that cannot be broken down chemically into simpler substances.

All the atoms of an element are alike and have the same chemical properties.

Even though there are just over 100 elements in the world, there are more than a million different substances. What are all these substances made of? The answer is that they are all made of elements. Elements are the building blocks of nature. Nature uses elements in the same way that a builder uses different arrangements of simple blocks to build many different buildings.

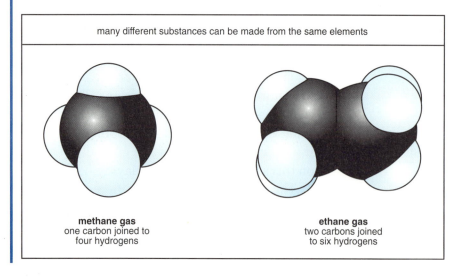

many different substances can be made from the same elements

methane gas
one carbon joined to four hydrogens

ethane gas
two carbons joined to six hydrogens

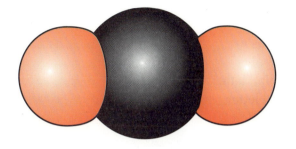

carbon dioxide molecule, CO_2

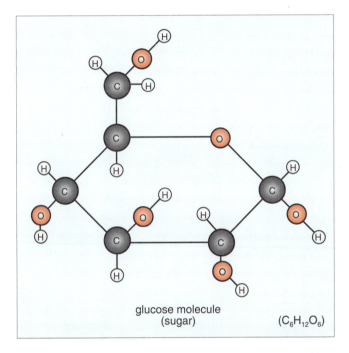

glucose molecule (sugar)   ($C_6H_{12}O_6$)

COMPOUNDS

Hydrogen gas is an element. Oxygen gas is another common element. When these two elements combine chemically, they form the liquid we call "water". Each molecule of water contains two atoms of hydrogen combined with one atom of oxygen. This is why scientists call water H_2O. Water is a compound.

> *A compound is formed when two or more elements combine chemically.*

In other words, if the molecules of a substance contains different kinds of atom, that substance is a compound.

Carbon dioxide gas is a common compound, formed when the element carbon combines with the element oxygen when fuels are burned. Each molecule of carbon dioxide contains one atom of carbon combined with two atoms of oxygen. A scientist would say that the chemical formula for carbon dioxide is CO_2.

Sugar is another compound you know, but it is a bit more complicated than water or carbon dioxide. A sugar molecule contains 6 atoms of carbon, 12 atoms of hydrogen and 6 atoms of oxygen. Sugar is used as a sweetener and as a source of energy.

EXPERIMENT 23.1

AIM: TO SHOW SOME COMMON LABORATORY ELEMENTS AND COMPOUNDS

1. Examine the following substances. Elements: carbon, magnesium, calcium, copper and iron. Compounds: water, calcium carbonate, magnesium sulphate and copper sulphate.
2. List their physical and chemical properties.

You will deal with many other elements and compounds later in your work.

SYMBOLS

When scientists write about a substance, they use symbols as a short way of telling us what the substance is. A short symbol is used for each element. The symbol consists of a single capital letter such as C, for carbon, or a capital letter followed by a small letter, such as Ca, for calcium. This table gives the symbols for the first 20 elements.

Number	Element	Symbol	Number	Element	Symbol
1	Hydrogen	H	11	Sodium	Na
2	Helium	He	12	Magnesium	Mg
3	Lithium	Li	13	Aluminium	Al
4	Beryllium	Be	14	Silicon	Si
5	Boron	B	15	Phosphorus	P
6	Carbon	C	16	Sulphur	S
7	Nitrogen	N	17	Chlorine	Cl
8	Oxygen	O	18	Argon	Ar
9	Fluorine	F	19	Potassium	K
10	Neon	Ne	20	Calcium	Ca

Some other elements with their symbols are:

Number	Element	Symbol	Number	Element	Symbol
25	Manganese	Mn	47	Silver	Ag
26	Iron	Fe	50	Tin	Sn
27	Cobalt	Co	78	Platinum	Pt
29	Copper	Cu	79	Gold	Au
30	Zinc	Zn	80	Mercury	Hg

As you can see, not all these symbols are taken from the English name for the element: some, like gold (Latin—**Au**rum) are taken from the Latin name. The symbol for sodium is taken from its Latin name **Na**trium, and the symbol for potassium is from its Latin name **K**alium. The symbol for each element, with its atomic number, is given in the periodic table on page 206 of this book (and on page 44 of the Mathematical Tables).

CHEMISTRY · ELEMENTS AND COMPOUNDS—ATOMS AND MOLECULES

SUMMARY

- Everything is made up of atoms.
- All atoms are made up of three different kinds of particles: protons, neutrons and electrons.
- A molecule is two or more atoms combined chemically.
- An element is a substance that cannot be broken down chemically into simpler substances.
- A compound is formed when two or more elements combine chemically.
- All atoms of an element are alike and have the same chemical properties.

QUESTIONS

Section A

1. All matter is made of particles called _____.
2. What is a compound? _____
3. Carbon is an _____. Carbon dioxide is a _____.
4. Underline the compounds in the following list: air, oxygen, water, iron, magnesium, rust.
5. Which of the following are elements? Water, hydrogen, oxygen, magnesium, carbon dioxide, iron.
6. Why is the smallest unit of water called a molecule? _____
7. Atoms are made up of small particles. Name the three types of particles.
   _____
8. What is an element? _____
9. Name the elements that have the symbols Na _____, K _____, C _____, Fe _____.
10. The core of an atom is called the _____.

Section B

1. Most fuels are hydrocarbons—compounds of just two elements, hydrogen and carbon. Explain how so many different substances can be made from just two elements. Draw simple diagrams of two hydrocarbons.
2. Fill in the following table.

	Location	Charge
Proton		
Neutron		
Electron		

143

3. Hydrogen, helium, nitrogen, oxygen, neon and chlorine are all gases used in different ways. Find out about these elements (look them up in an encyclopedia or other science book in the library—**hint**, use the index of a book to find the page the information is on). Fill in the following table.

	Uses
Hydrogen	
Helium	
Nitrogen	
Oxygen	
Neon	
Chlorine	

COMPOUNDS AND MIXTURES CHAPTER 24

Do you like sweets? You probably know a shop where you can buy mixtures of sweets. You can pick up whatever mixture you like, and there are many other mixtures of sweets you could buy.

You have probably seen fertiliser bags with numbers written on them, e.g. 10, 10, 20. This means that the bag contains 10 parts of nitrates, 10 parts of phosphates and 20 parts of potassium. What do you think 0, 7, 30 means? This is a different mixture of fertiliser. Farmers use different mixtures of fertilisers for different crops.

What is the difference between a mixture and a compound? One difference is that the composition of a mixture is variable. This means that you can put as much as you wish of each substance into the mixture.

There are other differences between a compound and a mixture, as the following experiment shows.

CHEMISTRY · COMPOUNDS AND MIXTURES

EXPERIMENT 24.1
AIM: TO PREPARE A COMPOUND OF IRON AND SULPHUR

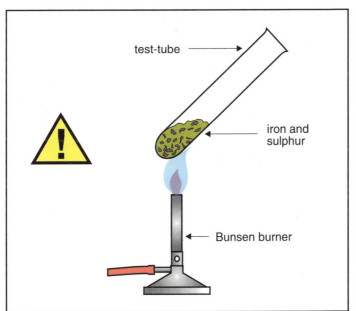
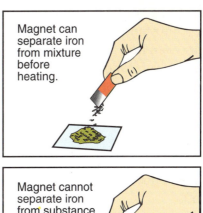
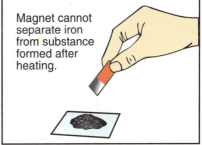

Method
1. Examine some iron (**Fe**) filings. List the properties: iron is a grey metallic solid; does not dissolve in water; is attracted to a magnet; does not react vigorously with dilute acids.
2. Examine some sulphur (**S**). List the properties. Sulphur is a yellow powdery substance; does not dissolve in water; does not react with dilute acids.
3. Weight out 5 g each of iron filings and of sulphur. Mix together in a test-tube. Divide the mixture into two parts.
4. Put one part into a test-tube. Heat in a fume cupboard until the test-tube glows red. Continue to heat in order to burn off any excess sulphur. Allow to cool.
5. Compare the resulting substance with the mixture you did not heat.
   Result: The yellow colour of sulphur and grey specks of iron can be seen in the mixture, but not in the part you heated. A magnet can be used to separate the iron filings from the mixture, but not from the part you heated.
   Additional test: Teacher demonstration experiment only. **Safety warning! Do this is a fume cupboard.**
6. Add one drop of dilute hydrochloric acid (**HCl**) to the black solid. The foul-smelling gas given off is hydrogen sulphide (**H_2S**) (the smell of rotten eggs). Hydrogen sulphide is very toxic, and in large amounts can kill.

The substance in the test-tube is clearly a **new substance**—iron (II) sulphide (**FeS**)—a **compound** of iron and sulphur. It does not have the properties of iron or the properties of sulphur. It has properties of its own, different from those of either sulphur or iron. It would also be very difficult to separate the iron from the sulphur in this new substance. We cannot easily reverse the process that happened when iron and sulphur joined together chemically to form this new substance. The molecules that make up iron sulphide each consist of one atom of iron chemically joined to one atom of sulphur.

A compound is formed when two or more elements are chemically combined.

You cannot put as much as you like of each substance into a compound.

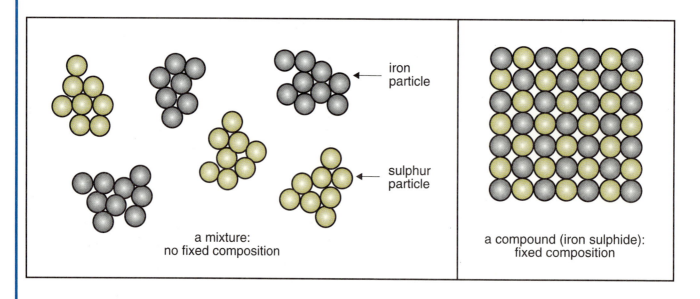

a mixture: no fixed composition

a compound (iron sulphide): fixed composition

Compounds have a fixed composition: the proportions of the elements from which they are made are fixed.

The iron and sulphur that you did not heat are obviously **NOT** a compound. The sulphur and iron retain all their own separate properties: they are not chemically combined, but simply mixed together. You could separate the iron by using a strong magnet. This would leave the sulphur.

CHEMISTRY · COMPOUNDS AND MIXTURES

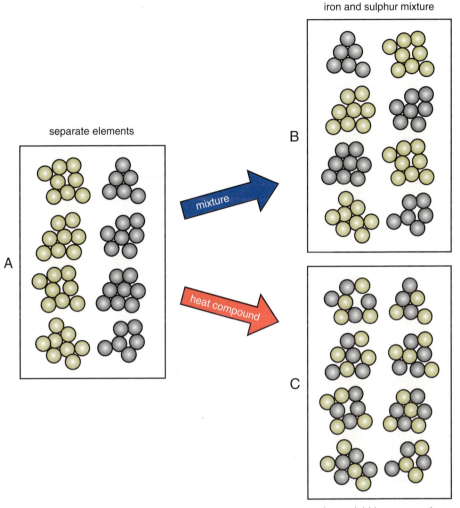

A mixture is formed when two or more substances are put together but are not combined chemically.

SOLUTION

A very common type of mixture is a **solution**. When you put some sugar into your tea, the sugar seems to disappear. But you know from the taste that the sugar is still there. When you taste the tea it is sweet: the sugar is now mixed throughout the tea. We say the sugar has **dissolved** in the tea, and has formed a **solution** of sugar in tea.

tea (solvent)   sugar (soluble substance)   cup of tea (solution)

BREAKTHROUGH SCIENCE

Not all substances dissolve in water. If you get tar on your feet at a beach, the tar will not dissolve in water. Another solvent such as butter or white spirits is needed to dissolve tar. You must choose a suitable solvent for each different substance. Shops have many different solvents for removing different stains.

A solution is made up of two parts: the substance you are dissolving and the **solvent** you dissolve it in (usually a liquid).

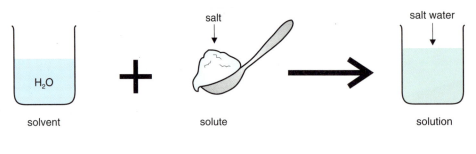

Solution	=	Soluble substance (solute)	+	Solvent
Salt water		Salt		Water

A solution is a mixture of a solute and a solvent.

DILUTE OR CONCENTRATED

If you look at the label on a bottle of orange squash, it says "dilute to taste". That means you should add water to the squash before drinking it. This is because the solution in the bottle is a **concentrated** orange squash solution: it is a strong solution. The solution in your glass is a **dilute** solution: it is a weak solution.

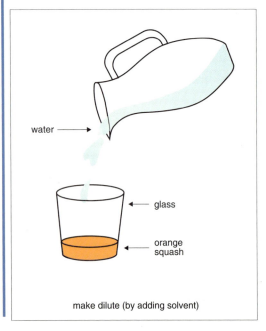

make dilute (by adding solvent)

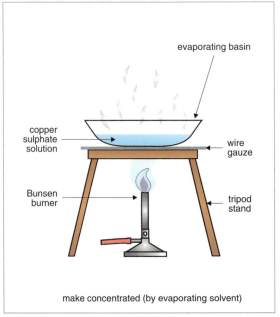

make concentrated (by evaporating solvent)

SOLUBILITY

If someone puts three or four teaspoons of sugar into a cup of tea, you probably find a lot of sugar left in the bottom of the cup when you go wash it. Why is this? Is there a limit to the amount that can be dissolved?

The following experiment will help to answer this question.

EXPERIMENT 24.2
AIM: TO INVESTIGATE THE SOLUBILITY OF COPPER SULPHATE

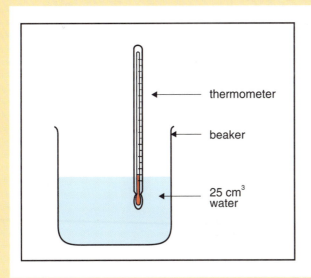
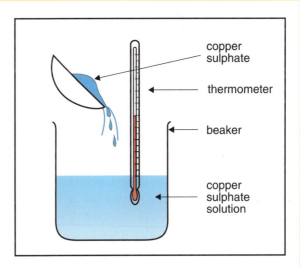

Method
1. With a **graduated cylinder**, measure out 25 cm³ of water into a beaker. Find the mass of the beaker.
2. Take the temperature of the water with a thermometer.
3. Add copper sulphate (**CuSO₄**) to the water, a small amount at a time. Stir the mixture to ensure that the copper sulphate dissolves.
4. Stop adding copper sulphate when it will no longer dissolve but stays undissolved at the bottom of the beaker. Find the mass of the beaker again. Calculate the mass of copper sulphate that dissolved in 25 cm³ of water.
   As the water has dissolved all it can at this temperature, we say that the solution is **saturated**.
5. Heat the solution to 60 °C. Add more copper sulphate. The hotter the water is, the more copper sulphate it can dissolve. Keep your copper sulphate solution for experiment 24.3.

> *A saturated solution is a solution that has dissolved as much solute as it can hold (at that temperature).*

20 g of copper sulphate will dissolve in 100 g of water at 20 °C.
40 g of copper sulphate will dissolve in 100 g of water at 60 °C.

EXPERIMENT 24.3
AIM: TO PRODUCE A CRYSTAL OF COPPER SULPHATE FROM A COPPER SULPHATE SOLUTION

Method
1. Allow your copper sulphate solution from experiment 24.2 to cool.
2. To speed it up, put the beaker into some cold water.
   Result: You will soon notice crystals of copper sulphate falling to the bottom of the beaker.

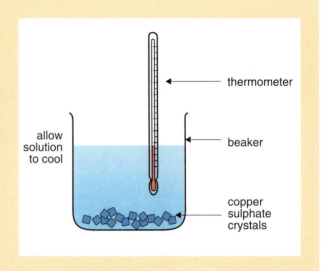

This is because more copper sulphate dissolves in water at 60 °C than at 20 °C. As the temperature drops, not all the copper sulphate can remain dissolved, and some comes out of solution and falls to the bottom of the beaker. If this is allowed to happen slowly, large crystals of copper sulphate are formed. This process of forming crystals of a compound by cooling a saturated solution is called **crystallisation**.

In general, the hotter the solution the more substance it can dissolve, and the lower the temperature the less substance it can dissolve.

EXPERIMENT 24.4
AIM: TO INVESTIGATE SUPERSATURATION WITH SODIUM THIOSULPHATE

Method
1. Half-fill a test-tube with crystals of **sodium thiosulphate**.
2. Heat the test-tube gently. This drives out the water locked up in the crystals. This water dissolves the sodium thiosulphate.
3. Allow the solution to cool slowly.
4. Add a small crystal of sodium thiosulphate. All the sodium thiosulphate comes out of solution, and the crystals again fill the test-tube.

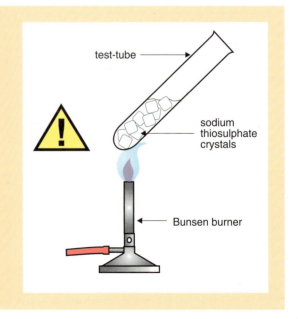

When you cool a hot solution slowly without shaking, the substance does not crystallise out of solution. The solution now contains more substance than a saturated solution normally would at this temperature. This solution is known as a **supersaturated solution**. This solution is **unstable**, and if you suspend a small crystal in it the excess substance will crystallise out onto this crystal, forming a larger crystal.

> *A supersaturated solution is a solution that contains more solute than a saturated solution would at the same temperature.*

SUSPENDED SOLIDS

Some mixtures may appear to be solutions, but are not. One such mixture is a **suspension**. When muddy water is allowed to stand for a while, the large particles of mud settle on the bottom. If you look at the water carefully you will see fine particles of mud spread throughout the water: this is known as a "suspension". These suspended solids give water a brown colour, and have to be removed when water is treated before being piped to your home. You may have seen the instruction "shake before use" on bottles of medicine. Many of these medicines are suspensions, so the suspended solids have to be mixed evenly throughout the bottle before use.

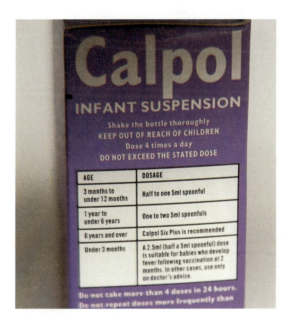

SUMMARY

- A compound is formed when two or more elements are chemically combined.
- A mixture is formed when two or more substances are put together but do not combine chemically.

	Mixture	**Heat**
Composition	Variable	Fixed
Properties	Each substance keeps its properties	New substance has new properties
Separation	Easy	Usually difficult
Heat	No heat necessary to make mixture	Heat usually need to make compound

- A solution is a mixture of a solute and a solvent.
- The solubility of a substance is the mass of the substance that dissolves in 100 g of solvent at a given temperature.
- A concentrated (strong) solution has a lot of solute dissolved in the solvent.
- A dilute (weak) solution has little solute dissolved in the solvent.
- A saturated solution is a solution that has dissolved as much solute as it can hold at that temperature.
- A supersaturated solution is a solution that contains more solute than a saturated solution would at this temperature.

QUESTIONS

Section A

1. A concentrated solution is _____.
2. A compound is formed when _____.
3. Iron and sulphur react to form _____.
4. Fill in the missing information in the table.

	Mixture	Compound
Composition		Fixed
Properties	Each substance keeps its properties	
Separation		

5. Underline the mixtures in the following list: air, water, mercury, sodium chloride, brass.
6. A dilute solution is _____.
7. Crystallisation is _____.
8. Suspended solids are _____.
9. Name a solvent you would use to remove a tar stain from wool. _____
10. Salt is added to warm water until no more salt dissolves. Is the solution now: (*a*) concentrated, (*b*) saturated, (*c*) dissolved, (*d*) evaporated?

Section B

1. Describe a simple experiment to show that iron and sulphur form a compound when heated strongly together. Use the headings Aim, Method, Result and Conclusion. Draw a simple diagram of the apparatus. List the safety precautions you must take in doing this experiment.
2. You want to get a big crystal from a warm solution of a salt. What would you do to the solution to get the biggest possible crystal? (*a*) Cool it quickly, (*b*) heat it quickly, (*c*) heat it slowly, (*d*) cool it slowly. Explain why you think the answer you give will work.
3. Your biro broke and you spilt some ink on your nylon jumper. Which of the following solvents **could** you use to remove the ink? Which solvent **should** you use?

Solvent	Ink	Nylon
A	Insoluble	Insoluble
B	Soluble	Soluble
C	Insoluble	Soluble
D	Soluble	Insoluble

Explain your answers.

4. Describe a simple experiment to show how you would concentrate a solution. Use the headings Aim, Method, Result and Conclusion. Draw a simple diagram of the apparatus. Why might you want to concentrate a solution?
5. When the elements iron and sulphur are heated strongly together they form a compound. Is this compound: (*a*) identical to sulphur, (*b*) identical to iron, (*c*) a mixture of bits of iron and sulphur, (*d*) unlike iron or sulphur? Give reasons for your choice of answer.

SEPARATING MIXTURES   CHAPTER 25

People have always had the problem of separating one substance from another. Wheat grain is separated from straw when harvested, and the wheat kernels are separated from the chaff. When the wheat is milled the flour is often separated from the fibre. In the dairy, cream is separated from milk to make butter. These are all examples of separations used to process our food.

Crude oil is a mixture of different liquid fuels, tar and dissolved gases. They are separated in an oil refinery.

One common job in chemistry is to separate one substance from other substances. The method we use varies from substance to substance but **you must find some physical difference between the substances in order to separate them**.

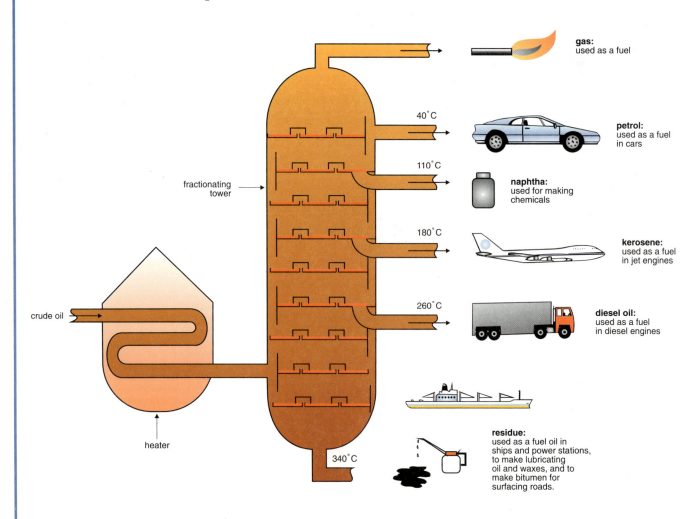

Farmers separate grain from chaff by winnowing: allowing both to fall through a current of air (a good breeze would do!). The lighter chaff is carried further by the air and is separated from the heavier grain.

Cream floats on top of milk. You can use this fact to separate cream from milk. We separate many other substances by similar methods.

SEPARATING ONE SOLID FROM ANOTHER

Sieving

Where one solid has a bigger particle size than the other, you can separate them by sieving. This is used to separate sand from gravel.

Magnetic separation

If one substance is ferrous (contains iron) the mixture is passed under a strong electromagnet. This pulls the ferrous substance away from the other substance. You have already used this process to separate iron from sulphur; it is also used to separate iron from rubbish. This iron is refined and reused: a process called **recycling**.

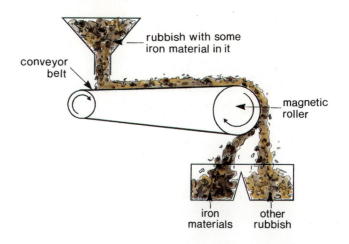

BREAKTHROUGH SCIENCE

Separating one solid from another where one solid dissolves in water.

EXPERIMENT 25.1
AIM: TO SEPARATE SAND FROM ROCK SALT (SODIUM CHLORIDE—NaCl)

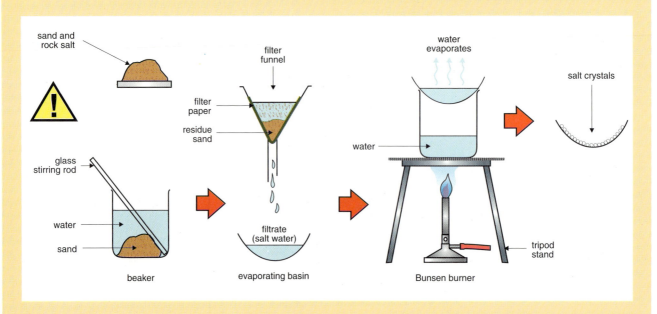

Method

1. Add water to the mixture. Stir to dissolve the salt. Heat the water if it is cold.
2. Fold a disc of filter paper into two and then into four. Open one side of the folded paper, and form the paper into a cone. Place this in a funnel. Place an evaporating basin under the funnel.
3. Pour the solution of salt water and sand through the filter. The sand is retained in the filter. The salt water solution passes through the filter.
4. Transfer the salt water solution to an evaporating dish. Place it on top of a water bath. When most of the water has evaporated, turn off the Bunsen burner and allow the salt to crystallise. This can take some time: you could leave it to happen overnight.

A filter is like a sieve with microscopic openings that stops small particles of solid from passing through it.

CHEMISTRY · SEPARATING MIXTURES

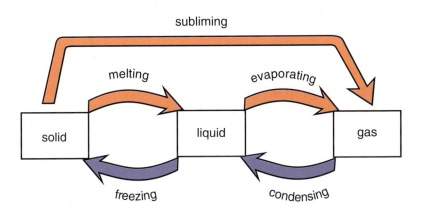

Sublimation
Ammonium chloride has the unusual property of turning from a solid directly to a gas when heated. Iodine and solid carbon dioxide (called "dry ice") are other substances that sublime.

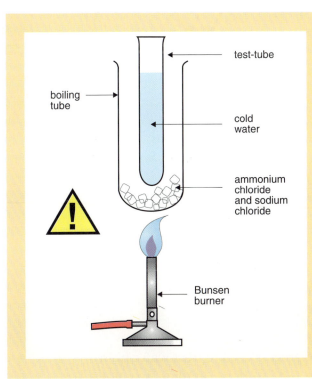

EXPERIMENT 25.2
AIM: TO SEPARATE A MIXTURE OF AMMONIUM CHLORIDE (NH_4Cl) FROM SODIUM CHLORIDE (NaCl)

Method
1. Heat the mixture gently. The ammonium chloride sublimes (changes to the gas state).
2. The ammonium chloride condenses on the cold outside of the test-tube.
3. Remove the test-tube and collect the ammonium chloride. The sodium chloride remains in the boiling tube.

SEPARATING AN INSOLUBLE SOLID FROM A LIQUID

EXPERIMENT 25.3
AIM: TO SEPARATE SAND FROM WATER

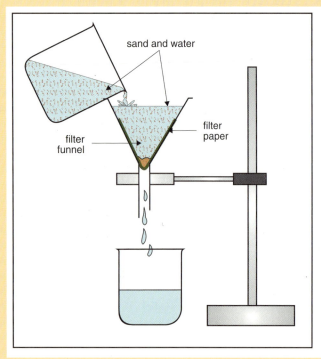
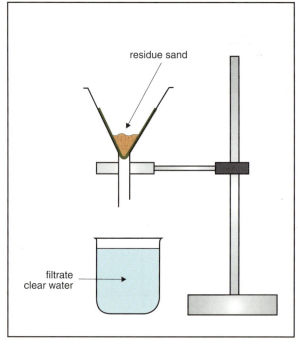

Method
1. Prepare a sheet of filter paper and place it in a funnel.
2. Pour the mixture of water and sand through the filter.
3. The water in the receiving beaker is clear.
4. The sand is retained in the filter. This is part of the process used to treat water from a river or a lake. The water is passed through gravel and sand beds to filter out solid particles in the water.

SEPARATING LIQUIDS

This is quite easy when the liquids do not mix (dissolve in each other).

Liquids that do not mix are immiscible.

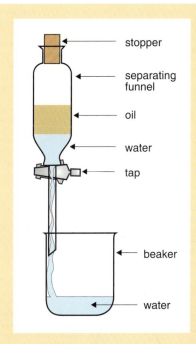

EXPERIMENT 25.4
AIM: TO SEPARATE IMMISCIBLE LIQUIDS (OIL AND WATER)

Method
1. Pour the liquids (oil and water) into a separating funnel. Allow to settle.
2. Open the tap and pour off the denser liquid (the water) into a beaker. Stop when the boundary between the two liquids approaches the tap.
3. Pour the liquid near the boundary into a separate beaker. When the boundary layer has clearly passed, close the tap.
4. Pour the less dense liquid (the oil) into a third beaker.

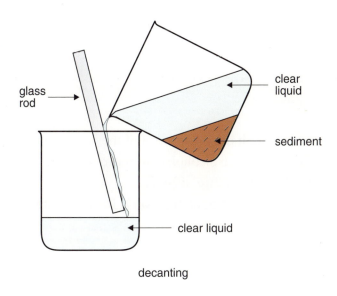

decanting

If the two liquids are in a beaker, we could get the same result by pouring the oil layer carefully from the top of the beaker. This process is **decanting**. Decanting is used to pour off a liquid from a solid where the solid has settled on the bottom. Sediment sometimes forms at the bottom of a bottle of wine. If the wine is poured out carefully—decanted—the solid is not disturbed.

DISTILLATION

Two liquids that mix are **miscible**. We separate miscible liquids by **distillation**.

Distillation is vaporisation followed by condensation.

EXPERIMENT 25.5
AIM: TO SEPARATE ALCOHOL FROM WATER

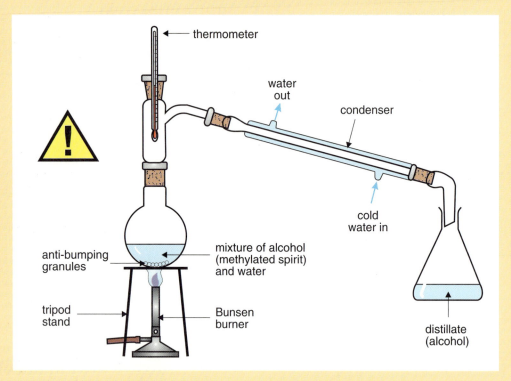

Method
1. Set up the distillation apparatus as shown.
2. Turn on the water tap to get cooling water flowing through the glass jacket of the condenser.
3. Heat the mixture carefully until a steady flow of drops, rich in alcohol, comes through the condenser tube and the thermometer shows a steady temperature of 78 °C.
4. Collect this alcohol in a conical flask. The liquid collected during distillation is the **distillate**. The water remains in the flask.

The process is repeated a number of times to make sure that the liquids collected are pure. Irish whiskey is distilled three times.

CHEMISTRY · SEPARATING MIXTURES

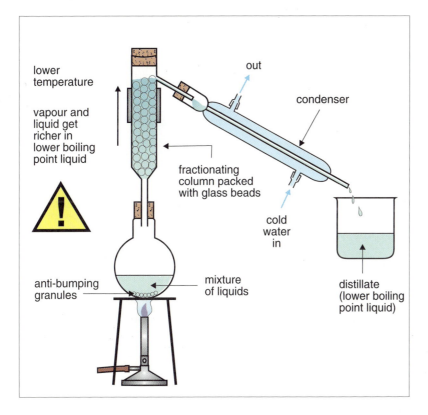

Fractional distillation

When the boiling points of the two liquids are close, we use **fractional distillation** to separate them. A fractionating column is added to the distillation apparatus. This column ensures that when the temperature of the glass beads reaches the boiling point of one of the liquids, only that liquid distils over through the condenser. The vapour of any liquid with a higher boiling point condenses on the beads and falls back down into the flask.

Fractional distillation of crude oil is used in an oil refinery.

Where a liquid has solid impurities dissolved in it, distillation can be used to purify the liquid. This process is used in a **solar still** to get pure drinking water from salt water.

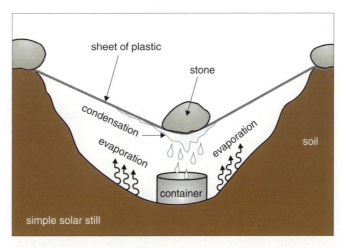

A simple solar still can be made as shown in the diagram. This will work in sunny weather.

CHROMATOGRAPHY

Chromatography is used to separate substances when only small samples of a mixture are available. It is used mainly to identify substances.

Chromatography is a method of separating substances by allowing a solvent to carry them along a strip of filter paper.

BREAKTHROUGH SCIENCE

EXPERIMENT 25.6
AIM: TO SEPARATE A MIXTURE OF INKS USING CHROMATOGRAPHY

Method

1. Place a small amount of water in a shallow glass trough.
2. Put a spot of ink from each of several different-coloured water-soluble felt pens on the end of a rectangular piece of filter paper.
3. Dip the end of the paper into the water. Hold the filter paper vertically.
4. The water rises up the filter paper (by capillary action) and carries the inks with it.

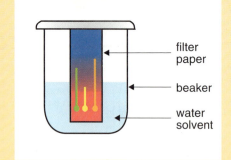

Result: Some inks stick to the filter paper and do not move very much. Other inks do not stick to the paper and so are carried further up the paper by the water. This separates the inks into the various dyes from which they were made. Methylated spirit can be used as a solvent in this experiment if "permanent" felt pens are used.

We can also use chromatography to show the different-coloured substances in plants.

Other methods of chromatography use glass tubes filled with filter materials to separate mixtures of liquids and gases. The mixture is carried through the tube by the flow of a solvent or by a gas. Chromatography is one of the most useful tools in the analysis of chemicals. Chemicals with similar physical properties are separated, and the amounts of chemicals needed for testing are small.

SUMMARY

- You must find some physical difference between substances in order to separate them.
- Sand is separated from sodium chloride by filtration followed by evaporation.
- Ammonium chloride is separated from sodium chloride by sublimation of ammonium chloride.
- Sublimation is a change of state from a solid directly to a gas.
- Distillation is vaporisation followed by condensation.
- Chromatography is a method of separating substances by moving them along a strip of filter paper with a solvent.

QUESTIONS

Section A

1. Sand is insoluble in water. What does the word "insoluble" mean? _____
2. Sugar dissolves in water to form a sugar solution. Water is the solvent. What does the word "solvent" mean? _____
3. A separating funnel is used to separate _____ liquids. Two such liquids are _____ and _____.
4. Name the method you would use to separate: salt and sand _____; salt and ammonium chloride _____.
5. A mixture of alcohol and water can be separated by _____.
6. Vaporisation followed by condensation is _____.
7. Steel cans can be separated from rubbish for recycling by _____.
8. Sand and water can be separated by _____.

Section B

1. Describe, using diagrams, a simple experiment to show the difference between a compound and a mixture.
2. A bucket of sea water was collected at the water's edge on a sandy beach. There was some sand in the water, and this was allowed to settle (see diagram). The water was then separated from the sand by decanting. Finally, the water was purified by distillation. Describe another method of separating the sand and the sea water. Use the headings Aim, Method, Result and Conclusion. Draw a simple diagram of the apparatus.
3. What is meant by decanting? What would you use this method of separation for? Why is this method sometimes used instead of filtration?
4. Describe a simple experiment to show the distillation of sea water. Use the headings Aim, Method, Result and Conclusion. Draw a simple diagram of the apparatus used. What is the difference between **distillation** and **fractional distillation**?
5. In the semi-desert regions of Australia, people can get water from leaves and other plant materials using a simple solar still. Draw a simple diagram of a solar still. Explain how it works.
6. To separate a mixture of iron filings and sawdust, would you use (a) a thermometer, (b) an evaporating dish, (c) a filter funnel, (d) a magnet? Describe, using a simple diagram, how this method could be used to separate steel cans from other rubbish.
7. Fill the blanks below using the words in the following list: distillation, evaporation, filtration.
   Refining oil _____
   Removing dirt from water _____
   Getting salt from sea water _____

THE AIR WE BREATHE — CHAPTER 26

The layer of air that surrounds the earth is the atmosphere. Air is a mixture of gases. We shall prove that air contains oxygen, carbon dioxide and water vapour. Air also contains the unreactive gases nitrogen and argon, and some small amounts of other unreactive gases. The amount of water vapour in air varies with the weather from day to day.

EXPERIMENT 26.1
AIM: TO SHOW THE APPROXIMATE AMOUNT OF OXYGEN IN THE AIR

Method
1. Place a graduated bell jar in a trough of water.
2. Read the volume of air in the bell jar.
3. Place a lighted night-light in the bell jar.
4. Fit the stopper to the bell jar.

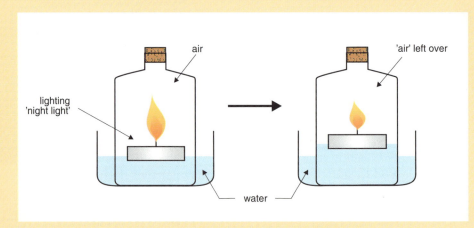

5. When the night-light stops burning, leave the apparatus for a few minutes.
6. Add water to the trough until the level is the same inside and outside the bell jar.

Result: About 1/5 of the air was consumed. This experiment shows that about 20% of the air is a gas that enables substances to burn (supports combustion). This gas is oxygen (O_2).

The gas in air that does not burn is nitrogen. Nitrogen is very unreactive and does not support combustion. In fact, nitrogen gas is pumped into empty oil tankers to prevent explosions of the air and oil vapour mixture in the tanks.

Air contains 78% nitrogen (N_2), 21% oxygen (O_2), 1% argon (Ar), 0·03% carbon dioxide (CO_2).

ALTERNATIVE DEMONSTRATION EXPERIMENT
AIM: TO FIND THE AMOUNT OF OXYGEN IN AIR

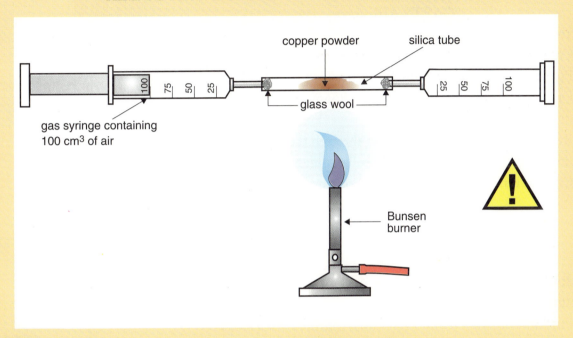

Method
1. Set up the apparatus as shown.
2. Heat the copper powder strongly.
3. Slowly push the plunger of the glass syringe.
4. Continue to push the air from one syringe to the other across the copper powder.
5. Read the volume of air in the gas syringes.
6. Repeat until the volume is constant.

Result:

Volume of air in syringe before heating	100 cm^3
Volume of air in syringe after heating for 2 min	82 cm^3
Volume of air in syringe after heating for 3 min	79 cm^3
Volume of air in syringe after heating for 4 min	79 cm^3

Conclusion: 21% of the air is oxygen.

EXPERIMENT 26.2
AIM: TO SHOW THAT THERE IS CARBON DIOXIDE GAS IN AIR

Method

1. Connect up the apparatus as shown in the diagram.
2. Put limewater into the large test-tube.
3. Draw air through the apparatus as shown in the diagram.

Result: The incoming air bubbles through the limewater. The limewater turns milky after a while.

Conclusion: Air contains carbon dioxide.

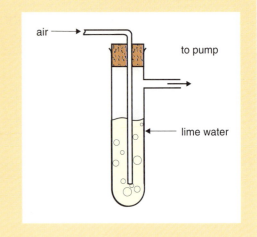

EXPERIMENT 26.3
AIM: TO SHOW THAT THERE IS WATER VAPOUR IN AIR

Method

1. Put some crushed ice and salt into a test-tube. Fit a stopper tightly to the test-tube.
2. After a while, drops of a liquid condense on the outside surface of the test-tube. Add this liquid to anhydrous copper sulphate (white colour).

Result: The bright blue colour proves that the liquid is water.

You could also test for water with blue cobalt chloride paper. This turns pink when water is added to it.

Conclusion: Air contains water vapour.

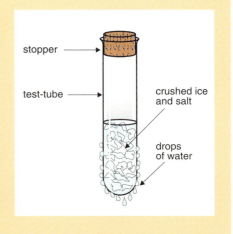

Test for water	Result for water
Melting point	0°C
Boiling point	100°C
Density	1 g/cm^3
Add blue cobalt chloride paper	Pink colour
Add anhydrous copper sulphate	Bright blue colour

CHEMISTRY · THE AIR WE BREATHE

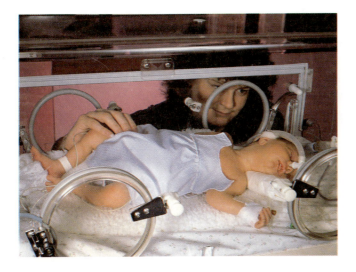

COMMERCIAL USES OF AIR

Many of the gases used in industry are produced from air. The air is first filtered to remove any dust particles. It then goes through a solution of sodium hydroxide to remove the carbon dioxide and is passed through a drying tower to remove water vapour. The gas is liquefied by cooling and compressing. The liquid is then distilled using a fractionating column. Liquid nitrogen, with a lower boiling point, is distilled off first. Liquid oxygen is collected next. Large quantities of both gases are produced in this way.

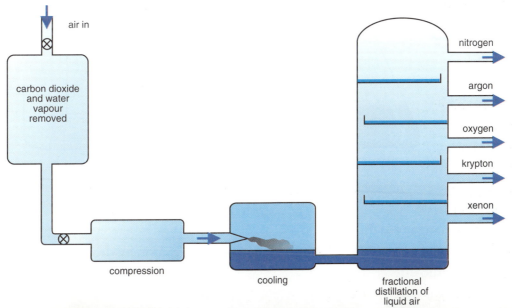

Removing a specimen which had been stored in liquid nitrogen

PREPARATION OF OXYGEN

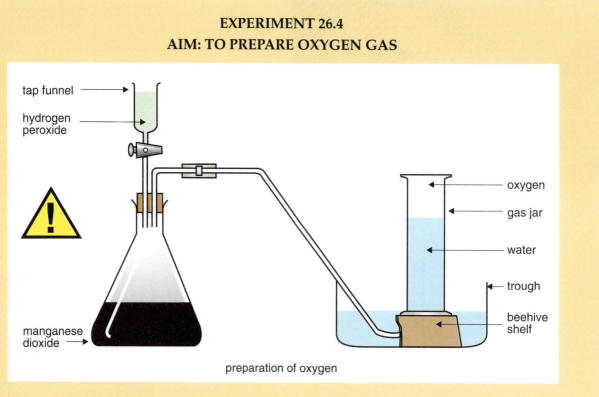

EXPERIMENT 26.4
AIM: TO PREPARE OXYGEN GAS

preparation of oxygen

Method
1. Put some **manganese dioxide** into a flask.
2. Add **hydrogen peroxide** (20 volumes) **solution** slowly, drop by drop, onto the manganese dioxide.
3. Collect the gas over water.

Chemical equation for the preparation of oxygen:

$2 H_2O_2$	$\rightarrow$	O_2	+	$2H_2O$
hydrogen peroxide	manganese dioxide catalyst	oxygen		water

CHEMISTRY · THE AIR WE BREATHE

Physical properties	Chemical properties
A colourless gas	Does not burn
Has no smell	Supports combustion—relights a glowing splint
Denser than air	Does not alter the colour of litmus, so it is neutral
Slightly soluble in water	Reacts with carbon to form carbon dioxide

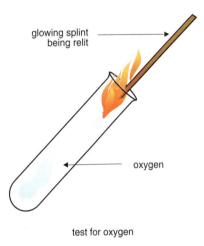

test for oxygen

Test for oxygen: Oxygen relights a glowing splint (a splint of wood that was burning and has just been extinguished).

What is a catalyst?
Hydrogen peroxide breaks down (decomposes) slowly into **oxygen** and water when left in the sunlight or when heated. This is why it is kept in dark bottles in a cool place. When **manganese dioxide** is added, the hydrogen peroxide breaks down faster. The manganese dioxide is not used up in the reaction, but simply speeds up the reaction.

A catalyst is a substance that speeds up a chemical reaction but is not used up in the reaction.

Dropping hydrogen peroxide onto a piece of fresh liver also releases oxygen from the hydrogen peroxide. This shows the action of the *biological catalyst catalase*. Biological catalysts are enzymes.

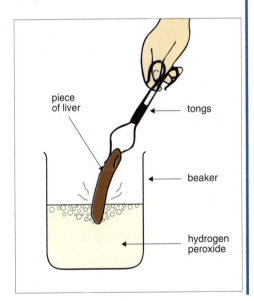

OTHER CHEMICAL PROPERTIES OF OXYGEN
Oxygen is a very reactive substance, and joins with most other elements to make compounds. These compounds of oxygen and another element are called oxides.

An oxide is a compound of oxygen and **one** *other element.*

BREAKTHROUGH SCIENCE

EXPERIMENT 26.5
AIM: TO BURN CARBON IN OXYGEN

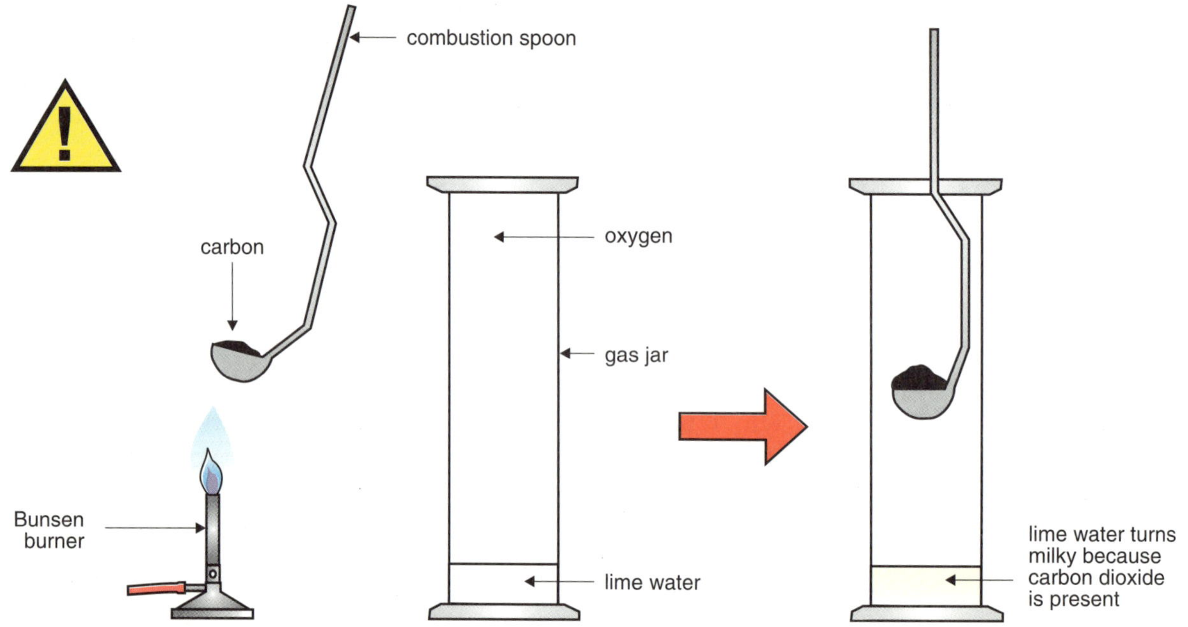

Method

1. Put some carbon (charcoal) into a combustion spoon.
2. Heat the carbon until it is red hot.
3. Plunge the carbon into a gas jar of oxygen. The carbon glows brightly.
4. Put some moist blue litmus paper into the gas jar. The litmus turns red—the gas is acidic.
5. Add some limewater to the gas jar. Shake the jar.
   Result: The limewater turns milky.
   Conclusion: This shows that the gas is carbon dioxide (CO_2): limewater doesn't turn milky with any other gas.

$$C + O_2 \rightarrow CO_2$$
carbon + oxygen → carbon dioxide

The oxides of non-metallic substances like carbon are usually acids.

CHEMISTRY · THE AIR WE BREATHE

The French chemist Antoine Lavoisier (1743–1794) was a founder of modern chemistry. In his text book *Elements of Chemistry*, Lavoisier created a system based on chemical elements and conservation of mass in chemical reactions. Previously everything had been though to be made of earth, wind, fire and water. Lavoisier showed experimentally that oxygen in air is involved in combustion, rusting and respiration. Lavoisier was guillotined during the Reign of Terror in the French Revolution.

DEMONSTRATION EXPERIMENT 26.6 (TEACHER ONLY)
AIM: TO BURN THREE METALS IN OXYGEN

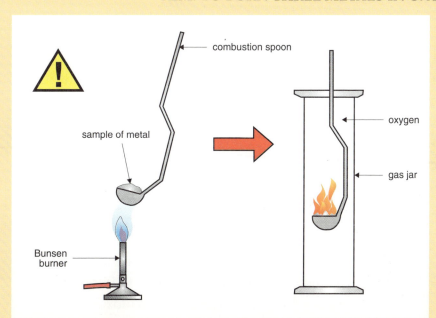

Method
1. Put a small piece of **sodium** into a combustion spoon.
2. Heat the sodium until it ignites.
3. Plunge the sodium into a gas jar of oxygen. The sodium burns with a bright flash.
   Result: A white powder is formed.
4. Add some water to the gas jar and shake. The white powder dissolves in the water.
5. Put some red litmus paper into the gas jar. The litmus turns blue, indicating a base.
   Conclusion: The white powder is sodium oxide.
   Repeat steps 1-5 with **magnesium**. Magnesium burns with an intense flame to give a white powder, magnesium oxide.
6. This powder is only slightly soluble in water. It turns red litmus blue—magnesium oxide is a base.
   Repeat steps 1–5 with **calcium**. Calcium burns with a bright red flame to give a white powder, calcium oxide.
7. This powder is soluble in water. It turns red litmus blue, indicating a base.

Substance	Reaction with oxygen	Compound formed	Soluble in water?	Litmus test
Carbon	Glows brightly, gas formed	Carbon dioxide (CO_2)	Yes, slightly	Acidic
Sodium	Bright light, forms white solid	Sodium oxide (Na_2O)	Yes	Basic
Magnesium	Intense white flame, forms white powder	Magnesium oxide (MgO)	Yes, slightly	Basic
Calcium	Bright red flame, forms white powder	Calcium oxide (CaO)	Yes	Basic

| *The oxides of metals are usually bases.*

SUMMARY

- Air contains 78% nitrogen (N_2), 21% oxygen (O_2), 1% argon (Ar), 0·03% carbon dioxide (CO_2).
- Air also contains a minute quantity of other inert gases.
- Air usually contains water vapour.
- Oxygen is consumed when things burn in air.
- Liquid oxygen and nitrogen are produced by fractional distillation of liquefied air.
- Oxygen is prepared by dropping hydrogen peroxide onto manganese dioxide (a catalyst).

Physical properties	Chemical properties
A colourless gas	Does not burn
Has no smell	Supports combustion—relights a glowing splint
Denser than air	Does not alter the colour of litmus, so it is neutral
Slightly soluble in water	Reacts with carbon to form carbon dioxide

- Catalyst: a substance that speeds up a chemical reaction but is not used up in the reaction.

QUESTIONS

Section A

1. Name the gas that makes up about 20% of air. _____
2. A piece of pink cobalt chloride paper was left in air. It turned blue. What does this tell you about air? _____
3. In the preparation of oxygen, manganese dioxide is used as a _____ to _____ the reaction.

4. The chemical name for rust is _____ _____.
5. What is a catalyst? _____
6. What substance would you use as a catalyst to prepare oxygen? _____ _____
7. Metal oxides are usually _____.
8. Non-metal oxides are usually _____.
9. To prepare the gas oxygen we use the liquid _____ _____ and the solid _____ _____.
10. The symbol for a molecule of oxygen is _____.
11. Why is nitrogen gas pumped into empty oil tankers? _____
12. When air is bubbled through limewater, the limewater turns milky. This shows that _____.
13. Why does the amount of water vapour in air vary? _____
14. A simple test for oxygen is to _____.
15. The fact that fish can live in water shows that water contains dissolved _____.

Section B
1. Air is a <u>mixture</u> of the following <u>elements</u> and <u>compounds</u>: nitrogen, oxygen, carbon dioxide, water vapour, the noble gases.
   (i) Explain the underlined terms.
   (ii) Name the two compounds in the above list.
   (iii) Describe simple experiments, one in each case, to show that the two compounds you have named are present in the air.
   (iv) Describe a simple experiment to prepare oxygen gas.
2. Describe two simple chemical tests to show that an unknown liquid is water. There are also physical tests for water. Name three physical properties of water that could be used to show that an unknown liquid is water.
3. Oxygen is prepared by dropping hydrogen peroxide onto the catalyst manganese dioxide.
   (i) Draw a labelled diagram of the apparatus used to prepare oxygen.
   (ii) Explain what is meant by a catalyst.
   (iii) Oxygen is a very reactive element. Name three substances that react with oxygen and name the compound produced by the reaction in each case.
4. Describe a simple experiment, using a silica tube and gas syringes, you could do to show that air contains 21% oxygen. Draw a diagram and use the headings Aim, Method, Result and Conclusion. What conclusions could you draw from this experiment about the reactivity of copper?

5. Describe a simple experiment you could do to show that air contains water. List the tests you would perform to prove the liquid you collect is in fact water. Draw a diagram and use the headings Aim, Method, Result and Conclusion.
6. Name the two main constituents of air that are elements. Give the percentage of each in the composition of air. Describe a simple experiment you would do to prepare one of these gases.
7. Oxygen reacts with metals and with non-metals to produce oxides. Describe how you would prepare the oxide of one named metal and one named non-metal. What results would you expect to get if you tested each oxide in turn with litmus paper?

CARBON DIOXIDE CHAPTER 27

EXPERIMENT 27.1
AIM: TO PREPARE CARBON DIOXIDE GAS

Method

1. Set up the apparatus as shown in the diagram.
2. Drop the dilute hydrochloric acid onto the marble chips. The marble "fizzes" and carbon dioxide is given off. Continue to add hydrochloric acid to maintain the reaction.
3. The gas can be dried by passing it through fused calcium chloride or silica gel.
4. The gas is collected by the upward displacement of air from a gas jar, because carbon dioxide is denser than air. A lighted splint is put into the top of the gas jar to check that it is full. When the gas jar is full the lighted splint goes out.

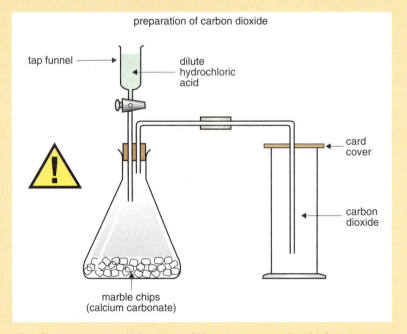

preparation of carbon dioxide

Calcium carbonate	hydrochloric acid		calcium chloride	water	carbon dioxide
$CaCO_3$ +	$2HCl$	→	$CaCl_2$ +	H_2O +	CO_2

Physical properties

Colourless gas
No noticeable smell
Denser than air
Dissolves slightly in water

Chemical properties

Does not burn
Extinguishes a flame
Turns limewater milky when bubbled through it
Carbon dioxide solution turns blue litmus red.
The solution is acidic.

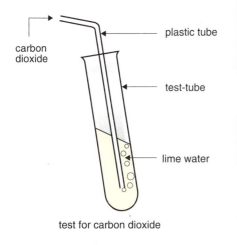

test for carbon dioxide

Test for carbon dioxide: carbon dioxide turns limewater milky when bubbled through it.

FUEL

A fuel is any substance used to produce heat from a chemical or nuclear reaction.

Photosynthesis uses solar energy to turn carbon dioxide and water into carbohydrates in plants. These plants can be burned as fuel (wood and straw). These fuels are **renewable**, because a new crop of plants can be grown on the same land.

Fossil fuels

Fossil fuels come from photosynthesis that happened many millions of years ago. Turf, coal, oil and gas are all made by the accumulation of decaying animals and plants over long periods of time. Turf is the youngest fossil fuel. It is an important fuel

BREAKTHROUGH SCIENCE

in some areas of Ireland for home heating and electricity generation.

Over millions of years, spongy, fibrous turf buried under rock formations is turned into hard coal. Oil and gas are formed similarly. These are hydrocarbons: compounds of hydrogen and carbon. There is much more coal in the world than oil and gas.

EXPERIMENT 27.2
AIM: TO BURN A FUEL AND TEST THE PRODUCTS

Method
1. Set up the apparatus as shown in the diagram.
2. Burn the candle for ten minutes.
3. Test the liquid collected in test-tube A with blue cobalt chloride paper.

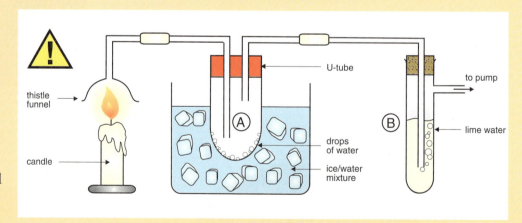

Result: It turns pink. The limewater in test tube B turns milky.
Conclusion: Hydrocarbons burn to give water and carbon dioxide.

Fossil fuels are hydrocarbons (compounds of hydrogen and carbon).

Fossil fuels took many millions of years to form, and cannot be made any faster. Coal, oil and natural gas are **non-renewable** resources.

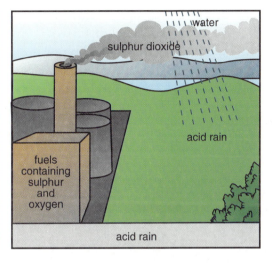

ACID RAIN

Rain is always slightly acidic, because carbon dioxide gas dissolves in rainwater to make weak carbonic acid. Oil and coal and other fuels often contain compounds of sulphur. These burn to form sulphur dioxide. Sulphur dioxide dissolves in rainwater to make a strong acid. 100 million tonnes of sulphur dioxide was put into the atmosphere in 1980 by burning and other industrial processes. Acid rain damages vegetation, dissolves stone in buildings, and injures your lungs.

Stone carvings damaged by acid rain

Oil refineries now remove most of the sulphur compounds and produce low-sulphur fuels. Natural gas from the Kinsale gas field is a clean fuel. It is almost 100% pure methane (CH_4), and contains no sulphur compounds. Coal is the major cause of acid rain because it is not treated to remove sulphur compounds.

Most plants cannot grow in acid conditions. This affects the animals that depend on the plants for their food supply. Acid rain also destroys plant and animal life in streams, rivers and lakes. Many lakes in Scandinavia and the United States are so acidic that no plant or animal lives in them. An acid rain pH level of 1·5—almost as strong as sulphuric acid—was recorded at Wheeling in

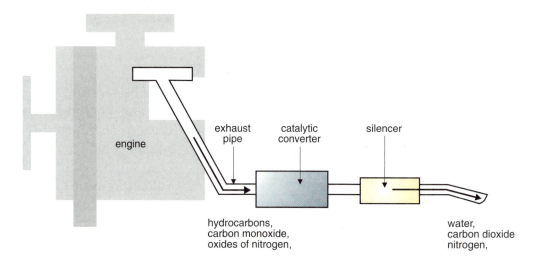

West Virginia.

Acid rain causes enormous damage to the environment. It can be reduced by removing sulphur dioxide from the exhaust gases of coal- and oil-burning industries and by reducing the acidic exhaust gases from cars.

FIRE AND FIRE EXTINGUISHERS
Three conditions are necessary in order to light a fire: fuel, heat and oxygen.

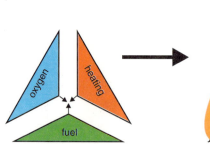

Fire prevention
Most fires occur in ordinary homes. They often cause death and serious injury. Fires don't happen—they are caused! If you think about the three conditions for a fire, you see that fire prevention means making sure that fuel—anything that can burn—is kept from sources of heat. Cigarettes, matches, open fires and electric heaters are all sources of heat.

The kitchen is an especially dangerous spot because of cooking. You should be very careful when cooking anything with a lot of fat in it. Trailing electric flexes and hot water can also cause serious scald burns in the kitchen.

Think safety: prevention is better than cure!

Putting out a fire

Remove the fuel: turn off the gas or electricity. Chip-pan fire: **don't use water**; cover the pan with a lid or with a fire blanket. This cuts off the supply of oxygen. If a child's clothes catch fire, smother the flames with fire blanket or by rolling on the ground. Cool immediately with lots of water.

Firemen extinguish fires by removing at least one of the three conditions: they cut off the supply of fuel, blanket the flame to keep out air (oxygen), and cool the fire with water. Fires are classified according to the material that is on fire.

Type	Materials	Example	Extinguisher type
Class A fire	Paper, wood, and cloth	Rubbish fire	Water or water-based
Class B fire	Flammable liquids, fats and grease	Chip-pan fire	Dry chemicals, carbon dioxide, foam, halon
Class C fire	Electrical equipment	TV fire	Dry chemicals, carbon dioxide, halon
Class D fire	Combustible metals such as magnesium	Chemical plant fire	Special smothering/heat-absorbing agent that does not react with the burning metal

How a carbon dioxide fire estinguisher works

Carbon dioxide is used in many types of fire extinguisher. The water/carbon dioxide extinguisher contains a sealed cartridge of pressurised carbon dioxide and a lot of water in the body of the extinguisher. A lever or plunger on the extinguisher breaks open the gas cartridge. The pressure of the gas forces out a spray of water and carbon dioxide. This smothers (and cools) the fire.

The carbon dioxide gas extinguisher contains pressurised carbon dioxide gas. The gas is sprayed onto the fire. The blanket of dense carbon dioxide prevents air from feeding the fire and smothers the flames.

The foam fire extinguisher contains carbon dioxide gas and a foaming substance. This is sprayed onto the fire; it blankets the fire and smothers it.

USES OF CARBON DIOXIDE

Carbon dioxide gas puts the "fizz" into cola and lemonade. Solid carbon dioxide, called "dry ice" has a temperature of –79°C and is used as a refrigerant (to keep things cold).

When a baker makes bread, yeast is added to the dough. Yeast is a living organism that gives off carbon dioxide as it lives. The bubbles of carbon dioxide cause the dough to rise, and give the bread its open texture.

Cake mixtures usually contain a "raising agent", or baking soda. This is a mixture of sodium hydrogen carbonate and tartaric acid. When these chemicals are wet, they react to produce bubbles of carbon dioxide, water and the harmless salt sodium tartarate. The bubbles of carbon dioxide make the mixture "rise".

Carbon dioxide is produced in large quantities during the brewing of beer and other alcoholic drinks. This is the major commercial source of carbon dioxide.

SUMMARY

- Carbon dioxide is a compound of carbon and oxygen.
- Carbon dioxide is prepared by the action of dilute hydrochloric acid on marble chips (calcium carbonate).
- Properties of carbon dioxide

Physical properties	Chemical properties
Colourless gas	Does not burn
No noticeable smell	Extinguishes a flame
Denser than air	Turns limewater milky when bubbled through it
Dissolves slightly in water	Carbon dioxide solution turns blue litmus red.
	Carbon dioxide solution is acidic.

- Carbon dioxide combines with water to form carbonic acid—a weak acid.
- A fuel is any substance used to produce heat from a chemical or nuclear reaction.
- Three conditions are necessary in order to light a fire: fuel, heat and oxygen.

CHEMISTRY · CARBON DIOXIDE

QUESTIONS

Section A
1. Name two gases that dissolve in water to give acid rain. _____
2. What type of fire extinguisher is suitable for an oil fire? _____ _____
3. Give two uses of carbon dioxide. _____
4. Carbon dioxide combines with water to form _____ _____, which turns _____ litmus _____. This shows that it is _____.
5. To prepare the gas carbon dioxide, we use the liquid _____ and the solid _____ _____.
6. The symbol for a molecule of carbon dioxide is _____.
7. Three conditions are necessary for fire: (1) _____ (2) _____ (3) _____.
8. A fuel is any substance that _____.
9. Waste paper caught fire in a store. This type of fire is a class ____ fire. You would use a _____ extinguisher to put it out.
10. A car engine has caught fire. What type of extinguisher could be used to put it out? _____
11. Hydrocarbon fuels are compounds of _____ and _____.
12. Catalytic converters are fitted to cars to reduce pollutant gases. Name two gases removed by a catalytic converter. _____
13. Fossil fuels are non-renewable. This means that _____.
14. Your TV set has just caught fire. What two things should you do to extinguish the fire? _____
15. Hydrocarbon fuels burn to produce _____ and _____.

Section B
1. (a) (i) What two gases make up the biggest proportion of the air? State the normal percentage (%) of each in air.
   (ii) Briefly describe how you would prepare and collect carbon dioxide in the laboratory. Draw a diagram of the apparatus you would use.
   (b) The graph in the diagram shows how the percentage of carbon dioxide varied above the surface of a grass field at regular intervals over a 24 hour period during calm weather.

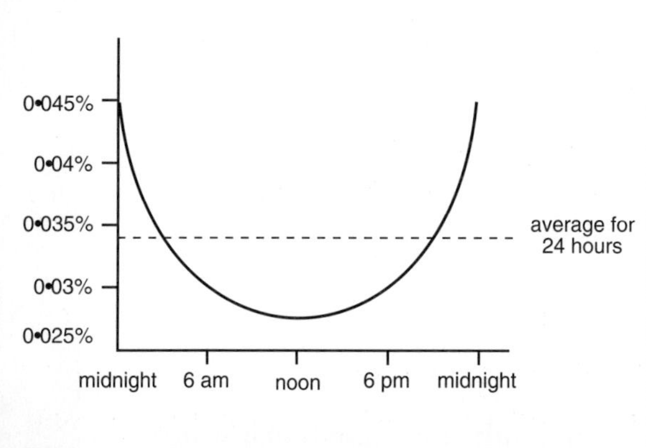

(i) Why is the level of carbon dioxide in the air at its lowest about midday?

(ii) Name the two biological changes that cause these variations in the percentage of carbon dioxide in the air.

2. "Acid rain can severely damage limestone buildings." Describe a simple chemical experiment you could do to show that this statement is true.

3. You have lit a bonfire, and the wind is blowing it towards your garden shed. What do you do to extinguish this fire?

   You are making chips at home (while your parents are out!). The chip-pan catches fire. What do you do?

   Your young sister brushes against an electric fire. Her clothes catch fire and she runs around wildly. What two things would you do to extinguish the flames and prevent serious burns?

4. Describe an experiment you would do to show the substances produced when a fuel is burned in air. Use the headings Aim, Method, Result and Conclusion. Draw a diagram of the apparatus you would use. Name the substances produced. Describe simple tests, one in each case, to identify these substances.

5. Enormous quantities of carbon dioxide are released into the atmosphere during volcanic eruptions. Explain one chemical process that naturally removes this gas from the air. Explain a biological process that removes carbon dioxide from the air.

WATER CHAPTER 28

Water is the most common compound on earth. It covers more than 70% of the earth's surface. All living things consist mostly of water: your body is about two-thirds water. Water is a very good solvent: it dissolves many substances. Pure water is a poor conductor of electricity, but water is a good conductor when it has salts dissolved in it.

Water changes volume with temperature in an unusual way. As warm water cools, it contracts until it reaches its maximum density at 4°C. Cooling it further causes it to expand until the liquid freezes to ice. Can this explain how fish survive in frozen ponds in winter?

We have already shown that water boils at 100°C and freezes at 0°C.

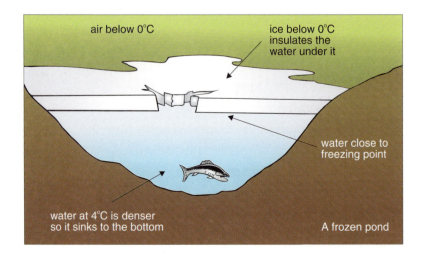

A frozen pond

Water is one of the most widely used chemicals in the world, and is essential in many chemical industries.

It takes 90,000 litres of water to make one tonne of paper.

It takes 75,000 litres of water to make one tonne of steel.

160 litres per day is the domestic/home use of water per person in Ireland.

WATER IS A COMPOUND

A water molecule is made of one oxygen atom joined to two hydrogen atoms.

We cannot break up water molecules by heating. We have to break the bonds joining the oxygen atoms to the hydrogen atoms in a water molecule to free the atoms. One way of doing this is by **electrolysis**. The electrolysis of water is described on page 261.

WATER IS A SOLVENT

Water dissolves many compounds: solids, liquids and gases. Water is a good solvent. Sea water contains many dissolved compounds. These compounds dissolve in the rainwater flowing over rocks and through the soil, and are carried in rivers and streams to the sea.

We shall show by a few simple experiments that tap water contains dissolved gases, dissolved solids and sometimes contains undissolved (suspended) solids.

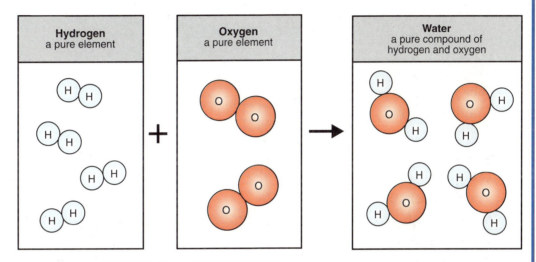

EXPERIMENT 28.1
AIM: TO SHOW THAT WATER CONTAINS DISSOLVED GASES

Method
1. Fill the boiling tube, delivery tube and test-tube with water.
2. Heat the boiling tube gently.
3. Continue to heat until gas stops bubbling into the test-tube.
4. Test the gas in the test-tube with a glowing splint.

Result: The splint relights.

Conclusion: The gas dissolved in water contains oxygen.

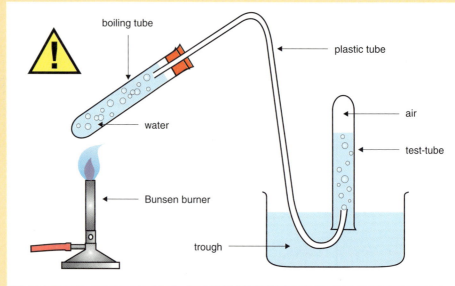

The other main gas in air, nitrogen, also dissolves in water. Oxygen is far more soluble in water than nitrogen. When air dissolves in water, the water is richer in oxygen than the air you breathe.

	Oxygen	Nitrogen
Air dissolved in water	34%	66%
Air in the atmosphere	21%	78%

Fish breathe the oxygen dissolved in water. Air dissolves in water as rain falls through the air, when water splashes over rocks in fast-moving streams, and through the surface of ponds and lakes. Water in fish tanks is aerated by bubbling air (using an air pump) through it.

CHEMISTRY · WATER

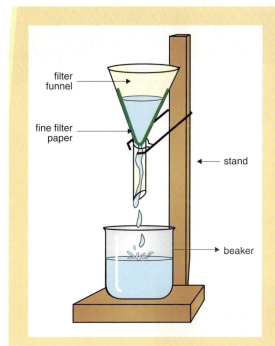

EXPERIMENT 28.2
AIM: TO SHOW THAT WATER CONTAINS UNDISSOLVED (SUSPENDED) SOLIDS

Method
1. Place a fine filter in a funnel.
2. Allow water to flow through the filter for few minutes.
3. Examine the filter carefully.

Result: Some fine solids are left in the filter. These were not dissolved in water, but were just carried along by the water.

Conclusion: Some water supplies contain more suspended solids than others. You will see a lot of suspended solids in streams and rivers after heavy rain.

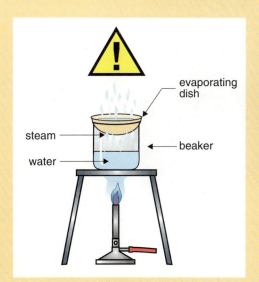

EXPERIMENT 28.3
AIM: TO SHOW THAT WATER CONTAINS DISSOLVED SOLIDS

Methods
1. Fill three-fourths of an evaporating dish with water filtered in Experiment 28.2.
2. Put the dish on a water bath (a beaker half-full of water would do).
3. Heat the dish on the water bath until all the water in the dish has evaporated.
4. Examine the dish carefully.

Result: There are some solids in the bottom of the dish.

Conclusion: This shows that water contains dissolved solids.

Water in some areas contains a lot of dissolved solids. The amount of dissolved solids you find in Experiment 28.3 depends on the nature of your water supply.

BREAKTHROUGH SCIENCE

SURFACE TENSION

Water appears to have a skin on its surface that can produce unusual effects. You can slightly overfill a glass without its spilling. You can also "float" a steel needle on water. Water insects and spiders walk on water, and water snails crawl along the underside of the water surface.

Surface tension is the force that holds the surface of water together.

Surface tension happens because molecules of water have an attraction for each other. The molecules on the surface pull towards each other and form a skin. Surface tension also causes water to form round drops. Adding soap to water weakens the surface tension, and allows it to stretch to hold in a bubble.

CAPILLARITY

Water rises up in narrow tubes. The narrower the tube, the higher the water will rise. Capillarity is due to the attraction of water molecules for glass. This pulls the water up at the sides of a glass tube.

Capillarity also helps water to rise in plants. Absorbent paper and sponges soak up water easily by capillarity. This is because they contain many tiny tubes. If you try out different materials

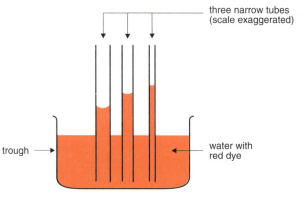

you will find that open-weave cloth soaks up water better than smoother materials. The wax in a candle rises up the wick by capillarity and keeps the candle burning.

| *Capillarity causes water to rise up in narrow tubes.*

OUR WATER SUPPLY

Most of the earth's water is salt water in the seas and oceans. Very little of it is fresh: only about 3%. Much of the fresh water is found in the polar ice-caps, in glaciers and underground in wells. Very little fresh water is in lakes and rivers, ready for water treatment. The water cycle of evaporation and rainfall constantly renews our water supply.

A damp proof course in a house prevents water rising into the walls

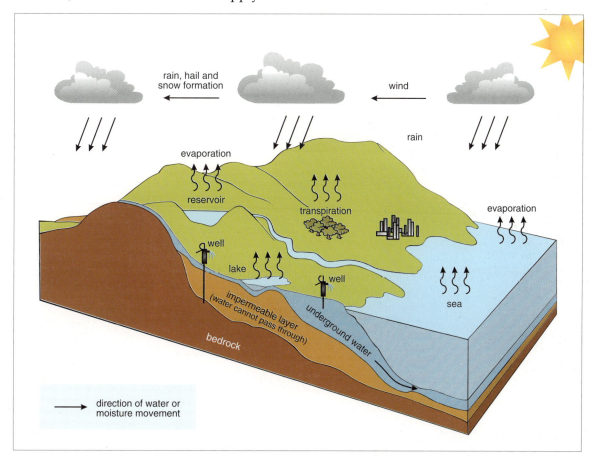

THE WATER CYCLE

Most of our rainfall returns to the atmosphere by evaporation from lakes and rivers and by plant transpiration. This water helps forests, crops, and other vegetation to grow. The remainder of the rainfall enters our water system in rivers and streams. Evaporation from the seas adds to the return of water to the atmosphere. This eventually falls again as rain.

WATER TREATMENT

Water must be treated to remove solids and debris, bacteria (germs) that could cause sickness, traces of compounds such as pesticides, and any substance producing a colour, odd taste or smell. Water is treated by a combination of physical and chemical processes to produce clean drinking-water. Clean water is essential for good health. Bacteria and parasites in unpurified water used for drinking cause millions of deaths in third-world countries.

Screening

Coarse screens remove large floating and suspended debris from the water. The water is then filtered through sand to remove other suspended materials, and left in settling tanks for chemical treatment and further filtering.

Sedimentation/settling

Aluminium sulphate is added to water to get tiny suspended particles to form bigger clumps. This material falls to the bottom of the settling tank, clearing the water of any brown colour.

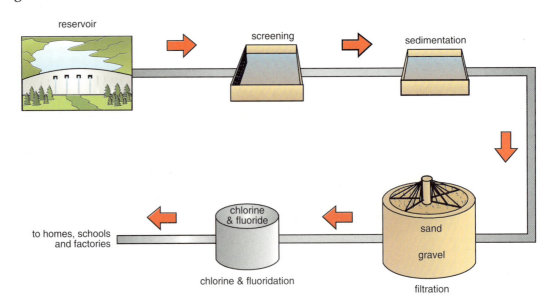

Filtration

The water is then filtered by fine sand.

Chlorination

A carefully controlled amount of chlorine is added to the water. This kills bacteria and other harmful organisms.

Fluoridation

Fluoride is added to the water in many areas to help prevent tooth decay.

If the water is acidic, some lime is added to reduce the acidity.

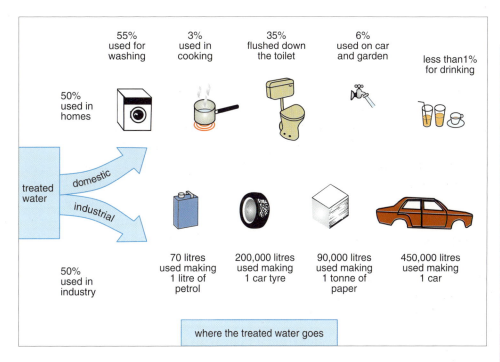

where the treated water goes

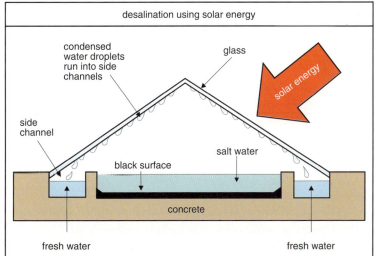

desalination using solar energy

DESALINATION

Salt water is converted to fresh water in countries where freshwater supplies are scarce. There are more than 2000 desalination plants in the world—most of them in the Middle East. These produce a total of 2 billion gallons of fresh water from salt water each day.

SUMMARY

- A water molecule is made of one oxygen atom joined to two hydrogen atoms.
- Water is an excellent solvent, and dissolves many substances.
- Pure water is a poor conductor of electricity.
- Water boils at 100 °C and freezes at 0 °C.
- Surface tension is the force that holds the surface of water together.
- Capillarity causes water to rise in narrow tubes.

QUESTIONS

Section A
1. What property of water allows an insect to walk on it? _____ _____ _____
2. Name two of the steps in the treatment of water for use in the home. _____ and _____
3. Builders use a damp-proof course to prevent dampness rising by _____.
4. The curved surface on water in a graduated cylinder is called a _____.
5. Fish breathe in _____ dissolved in water through their _____.
6. A water molecule is made of _____ oxygen atom joined to _____ _____ atoms.
7. The salt in sea water comes from _____ in rocks dissolved in _____ and carried to the sea in rivers.
8. A razor blade can be placed on the surface of water without sinking. This demonstrates _____ in water.
9. When you mop up water from a floor with a cloth you are using _____ action in the cloth.
10. The stages of the water cycle are: evaporation, _____ and _____.
11. _____ is added to treated water to kill bacteria.
12. A meniscus on water in a test-tube is caused by _____.
13. Fluoride is added to water to _____.
14. Wax rises in the wick in a candle by _____.

Section B
1. Water for domestic use has to be purified. Why? There are several stages in water purification. What happens during the settling stage? What is removed during the filtration stage? Why is the water chlorinated?
2. Sea water and the water in lakes and rivers contains dissolved oxygen which is used by fish and other aquatic organisms for respiration. Describe two ways in which the dissolved oxygen gets into the water.
3. Describe, with a labelled diagram, a simple experiment to show that water contains dissolved gases. How could you show that one of these gases is oxygen?
4. Describe, with a labelled diagram, a simple experiment to show that water contains dissolved solids. Use the headings Aim, Method, Result and Conclusion.
5. Water treatment is an expensive process. Outline the main steps in the treatment of water for use in the home. List three ways in which water is used wastefully. Suggest two ways in which the waste of water could be reduced.
6. Name two ways in which water gets into the atmosphere. Water falls as rain, hail and snow. Is there any other way that water can reach the ground? Water passes through rocks and soil before forming streams and rivers. This makes water treatment necessary to produce clean water. What changes take place that make treatment of river water necessary? Describe one way that countries such as Saudi Arabia could produce fresh water from sea water.

CHAPTER 29 · WHAT ARE ATOMS MADE OF?

Up to the end of the nineteenth century, scientists believed that all matter consisted of atoms that could not be divided. They thought that nothing smaller than an atom existed.

In 1897, J.J. Thomson discovered a particle smaller than an atom—the **electron**. This particle is found in all atoms. The electron has a negative electric charge, and a mass much smaller than that of the lightest atom.

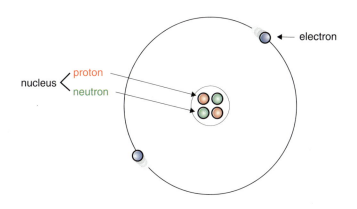

Early in the twentieth century, two other particles smaller than an atom were discovered. These particles are the **proton** and the **neutron**. The proton has a positive electric charge and a mass slightly less than the lightest atom—the hydrogen atom. The neutron has no electric charge and a mass almost identical to that of the proton.

These three particles—the proton, neutron and electron—make up all atoms. Scientists also discovered that the protons and neutrons make up the dense centre or solid core of an atom—the nucleus. The electrons are tiny but take up most of the space occupied by an atom by whizzing round the nucleus. Electrons move round the nucleus in definite orbits.

As the masses of the proton, neutron and electron are very small, we measure them on a scale that uses a unit called the **atomic mass unit (amu)**. This scale is based on an atom of carbon, which is taken as having a mass of 12 atomic mass units. All other substances are measured using this scale.

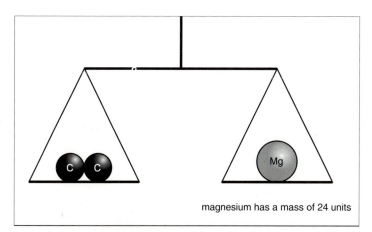

magnesium has a mass of 24 units

	Mass	Electric charge	Location
Proton	1 amu	+1	In the nucleus
Neutron	1 amu	0	In the nucleus
Electron	1/1840 amu	−1	Orbiting around the nucleus

BREAKTHROUGH SCIENCE

Because atoms are electrically neutral, the number of positive protons in an atom equals the number of negative electrons.

The position of an element on the table of elements, numbering from the lightest element, hydrogen, is called the **atomic number** of the element. Early in the twentieth century it was discovered that the atomic number is the same as the number of positive protons in the nucleus of an atom.

The atomic number is the number of protons in the nucleus.

HOW MANY NEUTRONS ARE THERE IN AN ATOM?

Since the mass of the electron is just 1/1840 amu, we ignore it when calculating masses. The mass of an atom can be taken as **mass of protons + mass of neutrons**. Protons and neutrons each have a mass of 1 amu, so that:

Mass number = number of protons + number of neutrons

Number of neutrons = mass number – number of protons

Example: the atomic number of sodium is 11, and its mass number is 23.

Atomic number: 11   11 protons, 11 electrons
Mass number: 23   23 – 11 = 12 neutrons

THE ARRANGEMENT OF ELECTRONS IN ATOMS

We already know that the electrons in an atom are found in orbits whizzing around the nucleus. The arrangement and number of electrons in these orbits are very important to chemists, as they are used to explain how and why chemical reactions take place. We shall use a simplified arrangement of orbits.

The electrons move in orbits at various distances from the nucleus. The number of electrons that a particular orbit can hold is strictly limited. The first orbit or electron shell can have at most two electrons. The second electron orbit can have a maximum of eight electrons.

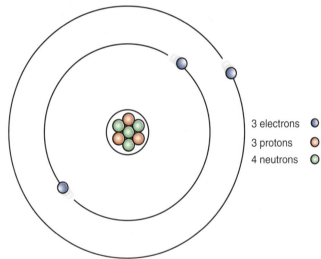

an atom of lithium

New Zealander Ernest Rutherford (1871–1937) discovered two basic forms of radioactivity, and in 1908 received the Nobel Prize for chemistry. He also discovered the structure of the atom as we now know it.

NUMBER OF ELECTRONS IN AN ORBIT

The arrangement of the electrons in an atom is called the **electronic configuration**.

Atomic number	Element	Symbol	First orbit n = 1	Second orbit n = 2	Third orbit n = 3	Fourth orbit n = 4
1	Hydrogen	H	1	-	-	-
2	Helium	He	2	-	-	-
3	Lithium	Li	2	1	-	-
4	Beryllium	Be	2	2	-	-
5	Boron	B	2	3	-	-
6	Carbon	C	2	4	-	-
7	Nitrogen	N	2	5	-	-
8	Oxygen	O	2	6	-	-
9	Fluorine	F	2	7	-	-
10	Neon	Ne	2	8	-	-
11	Sodium	Na	2	8	1	-
12	Magnesium	Mg	2	8	2	-
13	Aluminium	Al	2	8	3	-
14	Silicon	Si	2	8	4	-
15	Phosphorus	P	2	8	5	-
16	Sulphur	S	2	8	6	-
17	Chlorine	Cl	2	8	7	-
18	Argon	Ar	2	8	8	-
19	Potassium	K	2	8	8	1
20	Calcium	Ca	2	8	8	2

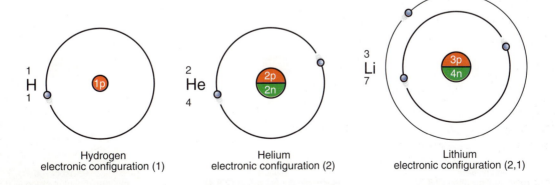

Hydrogen electronic configuration (1)

Helium electronic configuration (2)

Lithium electronic configuration (2,1)

Breakthrough Science

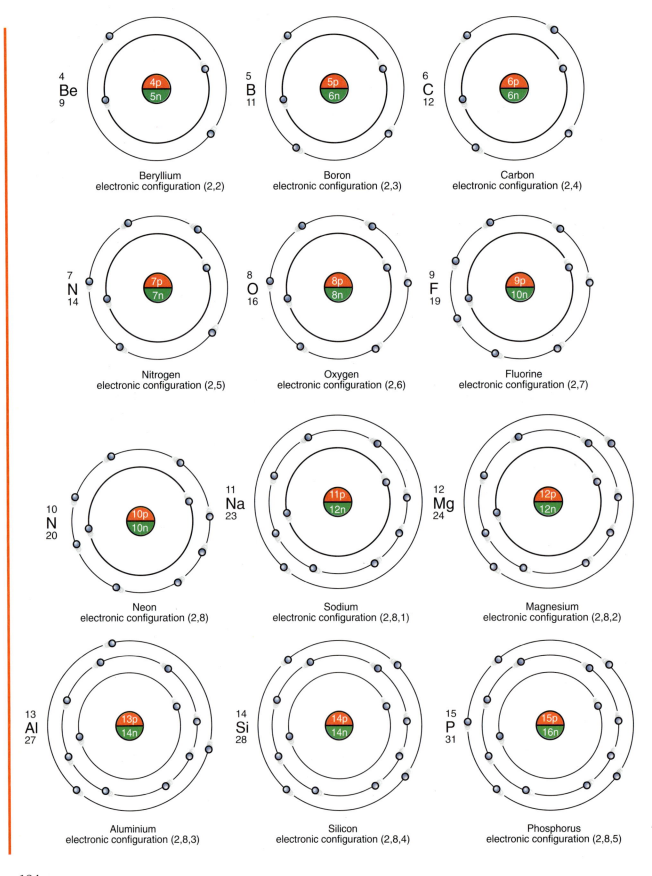

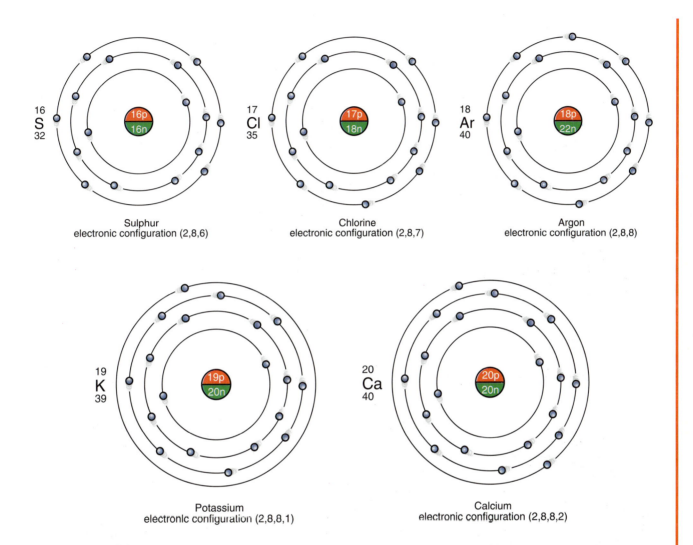

SUMMARY

	Mass	Electric charge	Location
Proton	1 amu	+1	In nucleus
Neutron	1 amu	0	In nucleus
Electron	1/1840 amu	−1	Orbiting around the nucleus

- Atomic number is the number of protons in the nucleus.
- Mass number is the number of protons plus the number of neutrons in the nucleus.
- The arrangement of the electrons in an atom is called the electronic configuration.

QUESTIONS

Section A

1. The three particles that make up an atom are the _____, _____ and _____.
2. The masses of atoms are measured using _____ units.
3. The atomic number of an element is _____.
4. The number of neutrons in the nucleus is calculated by _____.
5. An atom of nitrogen ($_7N^{14}$) has _____ protons, _____ electrons and _____ neutrons.
6. An element has an atomic number of 17. Name the element. _____
7. An atom of sodium ($_{11}Na^{23}$) has _____ protons, _____ electrons and _____ neutrons.
8. The electronic configuration of an atom of an element is 2, 8, 1. Name the element. _____
9. An atom of phosphorus ($_{15}P^{31}$) has _____ protons, _____ electrons and _____ neutrons.
10. The electronic configuration of an atom of an element is 2, 5. Name the element. _____
11. Draw a simple atomic diagram (Bohr diagram) of an atom of helium ($_2He^4$).
12. The electronic configuration of an atom of an element is 2, 8, 3. Name the element.
    _____
13. Draw a simple atomic diagram (Bohr diagram) of an atom of carbon ($_6C^{12}$).
14. Draw a simple atomic diagram (Bohr diagram) of an atom of nitrogen ($_7N^{14}$).

Section B

1. Look up the first 20 elements of the periodic table, and name the elements that fit these descriptions. A: the lightest element; B: the element with atomic number 5; C: the element with six electrons in its atoms; D: the element with seven neutrons in its nucleus; E: any element with six outer-shell electrons.
2. Complete the following table.

	Mass	Electric charge	Location
Proton	1 amu		
Neutron			In nucleus
Electron		−1	

3. Answer the following questions about the element $_5B^{11}$.
   Name this element.
   What is its atomic number?
   What is its mass number?
   How many protons does it have?
   How many neutrons does it have?
   How many electrons does it have?
   Draw a diagram of an atom of this element.

CHEMISTRY · ELEMENTS CAN BE GROUPED TOGETHER

CHAPTER 30: ELEMENTS CAN BE GROUPED TOGETHER

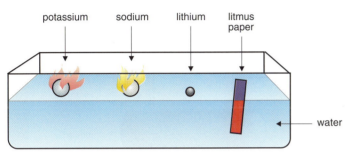

We often find that a number of elements have similar chemical and physical properties. The most reactive group of metal elements is the alkali metals—lithium, sodium, potassium, rubidium and caesium. The group is called the alkali metals because they all react with water to form alkaline (basic) solutions.

Lithium is the least reactive of the alkali metals; reactivity increases with atomic number.

Atomic number	Name	Symbol	Electronic configuration	Melting point	Reactivity
3	Lithium	Li	2, 1	179°C	Increases as we move down the group
11	Sodium	Na	2, 8, 1	97·9°C	
19	Potassium	K	2, 8, 8, 1	63·5°C	

The alkali metals are called Group 1 because they each have one electron in their outer electron shell.

EXPERIMENT 30.1
DEMONSTRATION EXPERIMENT ONLY. AIM: TO EXAMINE THE PROPERTIES OF LITHIUM, SODIUM AND POTASSIUM

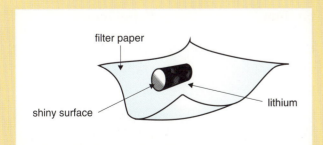

WARNING: Alkali metals are very reactive. Always handle with tongs. Potassium reacts explosively with water. Safety spectacles and a safety screen should be used when doing these tests.

Method
1. Place a small piece of lithium on a piece of filter paper. Cut the lithium with a pen knife. A metallic sheen (shiny surface) is seen. After a few minutes this dulls over and turns black.
2. Try to shape the lithium with the knife. It is difficult to shape.
3. Connect up the test circuit as shown. Push the two probes into the lithium. The bulb lights. This shows that lithium is a conductor of electricity.
4. Put a small piece of lithium into water. The lithium floats on the water (it is less dense than water) and fizzes about. The reaction gives off a gas.
5. Put a lighted splint to this gas. The gas lights with a quiet "pop". This gas is hydrogen.
6. Test the water with red litmus paper. The litmus turns blue, showing that the solution is a base. Feel the water with your fingers—the water has a soapy feeling. Repeat these tests with a small piece of sodium and a tiny piece of potassium. When potassium reacts, the hydrogen gas usually catches fire because of the violence of the reaction. Potassium burns with a lilac flame.

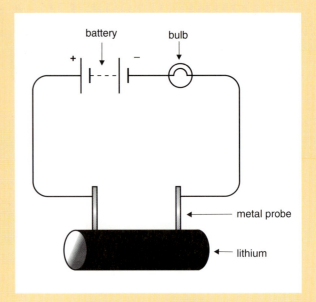

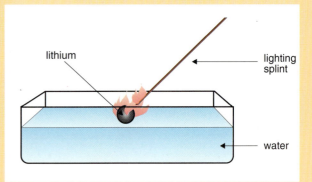

Physical property	Lithium	Sodium	Potassium
Appearance	Black oxide	Cream/silver oxide	Green/grey oxide
Storage	Under oil	Under oil	Under oil
Ease of shaping	Quite difficult	Easy	Very easy
Conduction of electricity	Conductor	Conductor	Conductor
Colour of flame	Red flame	Yellow flame	Lilac flame

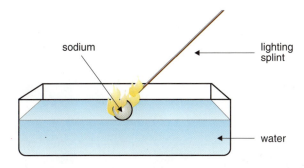

CHEMICAL REACTIONS

With water
Sodium reacts with water extremely vigorously.

$2Na + 2H_2O \rightarrow 2NaOH + H_2$
sodium　water　　sodium hydroxide　hydrogen

With oxygen
Alkali metals react with oxygen when heated to give metal oxides. The reaction is more vigorous with potassium than with sodium, and with sodium than with lithium.

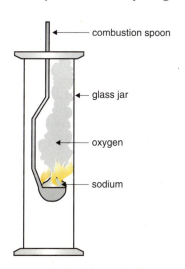

EXPERIMENT 30.2

AIM: TO REACT SODIUM WITH OXYGEN

1. Heat some sodium metal on a combustion spoon.
2. Put this into a gas jar of oxygen in a fume cupboard. The sodium reacts with the oxygen to give sodium oxide.

$4Na + O_2 \rightarrow 2Na_2O$
sodium　　oxygen　　sodium oxide

With acids
Alkali metals react in an extremely explosive manner with acids, forming the metal salt of the acid and giving off hydrogen gas.

WARNING: This reaction is very dangerous and should not be done.

$2Li + 2HCl \rightarrow 2LiCl + H_2$
lithium　hydrochloric acid　lithium chloride　hydrogen

USES

Sodium is used in yellow street light bulbs. Sodium hydroxide is used in the manufacture of soap and industrial cleaners. It is also used to remove carbon dioxide (CO_2) gas from chemical reactions. Sodium carbonate is used as a raw material in the production of glass and detergents. Sodium chloride (common salt, NaCl) and potassium chloride (KCl) are also widely used.

ALKALINE EARTH METALS

The elements magnesium and calcium belong to the alkaline earth metal group. Compounds of these elements are found in many common rocks. Chalk, marble and limestone are three common rocks that contain calcium carbonate.

The alkaline earth elements are less reactive than the alkali metals. Calcium is more reactive than magnesium.

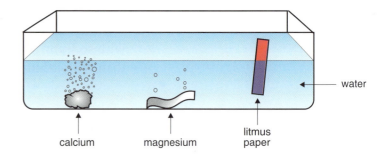

Atomic number	Name	Symbol	Electronic configuration	Reactivity
12	Magnesium	Mg	2, 8, 2	Calcium is more reactive than magnesium
20	Calcium	Ca	2, 8, 8, 2	

The alkaline earth metals are called Group 2 because they each have two electrons in the outer electron shell.

Physical property	Magnesium	Calcium
Appearance	Silver-coloured metal, white oxide coat	Silver-coloured metal, white oxide coat
Stored	In a desiccator (dry container)	In a desiccator
Ease of shaping	Difficult	Difficult: brittle
Conduction of electricity	Conductor	Conductor

The oxide coatings on magnesium and calcium can be removed using emery paper.

CHEMISTRY · ELEMENTS CAN BE GROUPED TOGETHER

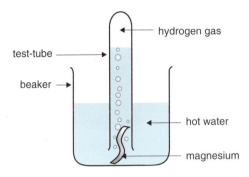

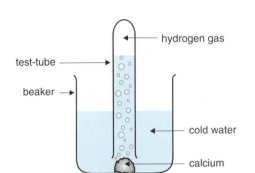

CHEMICAL REACTIONS

With water

Magnesium reacts very slowly with cold water. The reaction can be speeded up by passing steam over the piece of magnesium (it will also react with hot water). A gas is given off which lights with a "pop"—hydrogen gas.

The solution produced turns red litmus blue—it is a base.

$$Mg + H_2O \rightarrow MgO + H_2$$
magnesium water magnesium oxide hydrogen

Calcium reacts vigorously with cold water. The reaction gives off a gas. The solution produced turns red litmus blue, and so it is alkaline. The gas lights with a "pop" and burns gently—hydrogen gas.

$$Ca + 2H_2O \rightarrow Ca(OH)_2 + H_2$$
calcium water calcium hydroxide hydrogen

USES

Alloys of magnesium are used where lightweight metals are needed in engines and aircraft. Calcium compounds are a necessary part of our food for proper bone development. Calcium compounds are also widely used in buildings: limestone and marble are both forms of calcium carbonate, and gypsum is a form of calcium sulphate.

BREAKTHROUGH SCIENCE

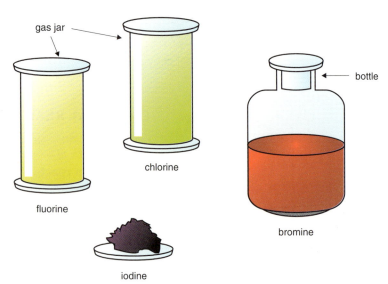

THE HALOGENS

The halogens are the elements fluorine, chlorine, bromine, iodine and astatine. The halogens are the most reactive **non-metallic** elements. The most reactive halogen is fluorine, and their reactivity decreases with increasing atomic number. Iodine is considerably less reactive than Fluorine. All the halogens are poisonous.

Atomic number	Name	Symbol	State	Colour	Electronic configuration
9	Fluorine	F	Gas	Pale yellow	2, 7
17	Chlorine	Cl	Gas	Yellow-green	2, 8, 7
35	Bromine	Br	Liquid	Dark red	7 electrons in outer shell
53	Iodine	I	Solid	Shiny dark grey/black	7 electrons in outer shell

> *The halogens are called Group 7 because they each have seven electrons in the outer electron shell.*

Properties of chlorine

Physical properties
Green-yellow gas
Much denser than air
Poisonous choking gas, affects the respiratory system
Dissolves in water

Chemical properties
Bleaches moist litmus, ink, grass, etc.
Reacts with hot sodium to form sodium chloride

The halogen you are most likely to use is probably **iodine**. This is sometimes used in an alcohol solution as an antiseptic on cuts and wounds. Iodine solution is used to test for starch. Iodine is also an essential element in your food, because your body uses it to make the hormone thyroxin.

Iodine

CHEMISTRY · ELEMENTS CAN BE GROUPED TOGETHER

USES OF HALOGENS

Halogens are used in the chemical industry to make a wide variety of substances used in everyday life.

Chlorine is used to make many household bleaches and kitchen and bathroom cleaners. These contain sodium chlorate [NaClO]. These cleaners should not be used with other types of cleaner, because sodium chlorate reacts with even weak acids to release dangerous chlorine gas.

Chlorine is used in chemicals to kill bacteria in drinking water and swimming pools, and also in antiseptics, plastics, insecticides and solvents for dry cleaning. Fluorine compounds are used in toothpaste and in drinking water to help prevent tooth decay.

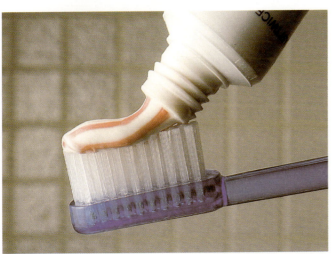

SUMMARY

- The elements in Group 1 of the periodic table of the elements are known as the alkali metals.
- The alkali metals (Group 1) each have one electron in the outer electron shell.

Alkali metals

Physical properties	Chemical properties
Soft and easily cut	One electron in outer shell
Less dense than water	React with water to give hydrogen and metal hydroxide
Good conductors of heat and electricity	
Low melting and boiling points	React with oxygen to give metal oxide
	React with acids to form the salt of the acid and give off hydrogen

- The alkaline earth metals (Group 2) each have two electrons in the outer electron shell.
- The elements of Group 7 of the periodic table of the elements are the halogens.
- The halogens have seven electrons in the outer electron shell.

QUESTIONS

Section A
1. The alkali metals each have _____ electron in the outer shell.
2. Name the most reactive alkali metal. _____
3. Name the most reactive halogen. _____
4. Name the least reactive alkali metal. _____
5. The alkaline earth metals are called **Group**_____ because they each have _____.
6. Sodium is stored in oil because _____.
7. Why is fluoride added to drinking water and to toothpaste? _____
8. Sodium and potassium are in the same group because they have the same number of _____.
9. Give two reasons why lithium, sodium and potassium are placed in the same group of the periodic table. _____ and _____
10. Why is chlorine added to drinking water? _____
11. Sodium reacts with cold water to release the gas _____.
12. When a piece of sodium is held in a Bunsen flame, a _____ colour is seen.
13. Magnesium and calcium are called alkaline earth metals because compounds of these elements are found in _____.
14. Bleaches often contain the halogen _____.

Section B
1. Lithium, sodium and potassium are three alkali metals. Write down the electronic configuration of each metal.
   Why are these elements placed in the same group?
   Describe what happens when each metal is put into water.
   With what colour flame does each burn?
   Each metal liberates a gas when added to water. Describe a simple test you would use to establish that this gas is hydrogen.
2. Why are the elements chlorine and fluorine in the same group?
   Name the group to which both elements belong.
   Draw a simple diagram of the arrangement of electrons in fluorine.
   Name another element in the same group.
   Why is chlorine added to water during water treatment?
3. Alkaline earth metals are so called because many rocks contain these elements. Name two alkaline earth metals.
   Give two reasons why you would place both these metals in the same group.
   Describe a simple experiment to show the reaction of one of these metals with water.
   Name the substance produced in this reaction.
   Name two uses of alkaline earth metals.

CHEMISTRY · ORGANISING THE ELEMENTS

CHAPTER 31 — ORGANISING THE ELEMENTS

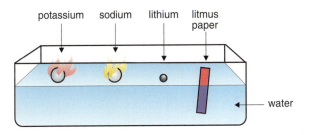

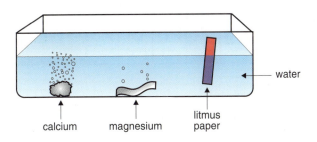

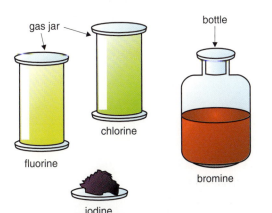

You would find it very difficult to deal with each of the 92 natural elements separately. You would have to learn long lists of properties and many different reactions.

Early in the last century, chemists noticed that some elements were very alike in the way they reacted. The German chemist Döbereiner noticed that the three elements chlorine (Cl), bromine (Br) and iodine (I) were alike in many ways. He also noticed other arrangements of three similar elements, which he called **triads**. The English chemist Newlands found when he arranged the elements in order of their mass numbers that every eighth element seemed to be similar. (He didn't know about the inert gases, as they hadn't been discovered.) Newlands called these **octaves**.

We saw these patterns of similar properties in our experiments with the alkali metals, the alkaline earth metals and the halogens.

THE PERIODIC TABLE

The first of our present type of table of elements was devised by the Russian chemist Dmitri Mendeleev in 1869. Mendeleev arranged the elements in order of increasing mass number. Only 57 elements were known at the time. He placed elements with similar chemical and physical properties underneath each other in columns. When an element did not fit, he left a blank space in his table and predicted that the space would be filled with an element yet to be discovered. Mendeleev even predicted the properties of the undiscovered elements. Fluorine (F) was discovered in 1886, and fitted into the gap he had left in the table above chlorine, bromine and iodine.

Dmitri Mendeleev (1834–1907), a Russian chemist, developed the periodic table. He published his discovery of the periodic law of elements in 1869.

205

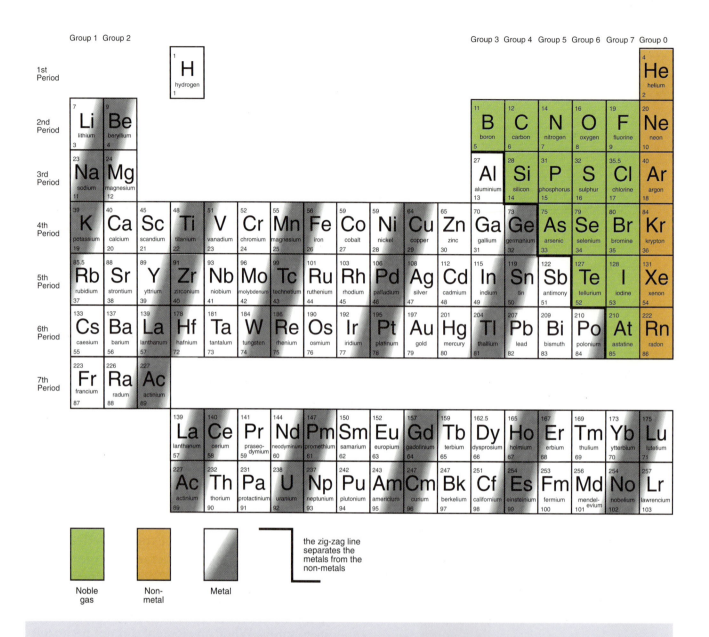

YOU CAN GET A LOT OF INFORMATION FROM THE PERIODIC TABLE OF THE ELEMENTS

1. The symbol of an element—you have to know the name yourself!
2. The atomic number of an element.
3. The mass number of an element.
4. Elements in the same group or vertical column have similar chemical and physical properties.
5. The group number of the element tells you the number of electrons in the outer shell of its atoms.
6. The period number (number of the horizontal row) tells you the number of electron orbits in its atoms.
7. Elements on the left and centre of the table are generally metals.
8. Elements on the right of the table are generally non-metals.

Example: $^{23}_{11}Na$

- Na is sodium.
- Atomic number is 11 (11 protons and 11 electrons).
- Mass number is 23. It has 12 (i.e. 23 − 11) neutrons.
- It is in group 1, and so has one electron in its outer electron orbit.
- It is in the same group as lithium and potassium, and so has similar physical and chemical properties to these elements.
- It is in Period 3 (third row), and so has a total of three electron orbits.
- It is on the left of the table, and so is a metal.

Example: $^{10.8}_{5}B$

- B is boron.
- Atomic number is 5 (5 protons and 5 electrons).
- Mass number is 10·8. We round this up to 11. It has 6 (i.e. 11 − 5) neutrons.
- It is in group 3, and so has three electrons in its outer electron orbit. It is in the same group as aluminium, and so has similar physical and chemical properties.
- It is in Period 2 (second row), and so has a total of two electron orbits.
- It is on the left of the table, and so is a metal.

SUMMARY

- The atomic number gives the number of protons in the nucleus.
- In a neutral atom, the number of protons = the number of electrons.
- Mass number is the total number of protons and neutrons in the nucleus of the atom.
- The number of neutrons = mass number − atomic number.
- The electronic configuration is the arrangement of electrons in the electron shells of an atom.
- Elements in the same group have similar chemical and physical properties.
- The group number tells you the number of electrons in an element's outer shell.
- Elements on the left of the table are generally metals.
- Elements on the right of the table are generally non-metals.

BREAKTHROUGH SCIENCE

> **QUESTIONS**
>
> 1. How many electrons are there in the outer shell of magnesium? _____
> 2. How many electrons are there in the outer shell of argon? _____
> 3. How many electrons are there in the outer shell of sulphur? _____
> 4. How many electrons are there in the outer shell of carbon? _____
> 5. How many electron shells are there in an atom of oxygen? _____
> 6. How many electron shells are there in an atom of chlorine? _____
> 7. How many electron shells are there in an atom of potassium? _____
> 8. Groups 1 and 7 contain very reactive elements. What groups would you expect to contain the next most reactive sets of elements? _____ and _____
> 9. Write down the electronic configuration of aluminium. _____
> 10. Write down the electronic configuration of nitrogen. _____

HOW DO ELEMENTS FORM COMPOUNDS?

CHAPTER 32

Why do elements combine to produce new substances? We have discovered that many elements are chemically reactive. Some elements are so reactive that they must be stored in a special way to prevent them from reacting. Phosphorus must be stored under water to prevent it from reacting with the oxygen in the air and bursting into flames. Sodium and potassium are so reactive with water or moisture in the air that they must be stored under liquid paraffin. Other elements such as gold, platinum and most precious metals do not react easily. Some elements (the noble gases) do not react at all.

Phosphorus is very reactive with air

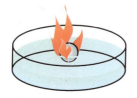

Potassium is very reactive with water

Gold is very unreactive

CHEMISTRY · HOW DO ELEMENTS FORM COMPOUNDS?

Neon lights

Halogen bulb (left) and conventional bulb containing an inert mixture of nitrogen and argon

Helium-filled balloons

THE NOBLE (INERT) GASES

The elements helium (He), neon (Ne) and argon (Ar) are in Group 0—the **noble gases**. The noble gases are **inert**, i.e. they do not normally react with other elements. These elements are all gases found in the atmosphere. They are monatomic—they exist as single atoms.

Helium is the least dense gas other than the highly flammable hydrogen. It is used to fill airships (weather balloons are still filled with hydrogen). Liquid helium at a temperature of −270°C is used to keep things very cold. The noble gases argon and neon are used in light bulbs.

The noble gases do not react and have eight electrons in their outer electron shell (helium, with two electrons, is the exception). This tells us having eight electrons in the outer shell is a very stable state. The elements in all the other groups do not have eight electrons in their outer shells. Chemists explain that these elements react in order to get a full outer shell and in order to become stable like the noble gases. Elements can do this by combining with other substances to get eight electrons in their outer electron shell, i.e. the electronic configuration of the nearest noble gas. A full outer shell usually has eight electrons.

Niels Bohr (1885–1962) was a Danish physicist who investigated atomic structure and devised the idea of electron shells or electronic configuration that we now use to explain chemical bonding.

The noble gases do not react, and have eight electrons in their outer electron shells.
Elements react to gain eight electrons in their outer shells.

Hydrogen and lithium are two exceptions to this rule. They react with other elements to get two electrons in their outer shell. When hydrogen and lithium have two electrons in their shell they have the stable electronic configuration of the noble gas helium.

IONIC BONDING

When two atoms join together, we say that they "bond". One way that atoms join is by forming an **ionic bond**.

Sodium (Group 1) and chlorine (Group 7) react to form an **ionic bond**.

	Electronic configuration	Electronic configuration of nearest noble gas
Sodium	2, 8, 1	2, 8
Chlorine	2, 8, 7	2, 8, 8

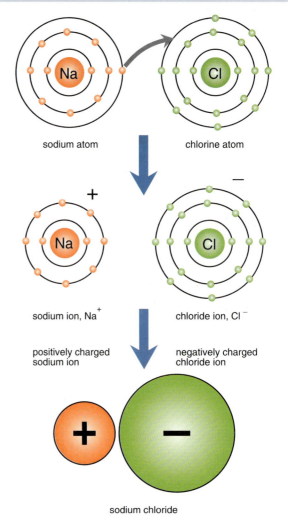

sodium atom · chlorine atom

sodium ion, Na^+ · chloride ion, Cl^-

positively charged sodium ion · negatively charged chloride ion

sodium chloride

Sodium reacts to give the electron in its outer shell to chlorine. Sodium then has eight electrons in its outer shell and, with one electron fewer, has a **positive** electrical charge. The sodium atom has become a positive sodium **ion**.

The chlorine atom has the extra electron it got from sodium. It now has a eight electrons in its outer shell. Chlorine, with one electron more, has a **negative** electrical charge. The chlorine atom has become a negative chloride **ion**.

An ionic bond is formed by the transfer of electrons between two atoms.

A positive ion is an atom that has lost an electron; a negative ion is an atom that has gained an electron.

An ion is an electrically charged atom or group of atoms formed by the loss or gain of electrons.

The positive sodium ion and the negative chloride ion attract each other (opposite electrical charges attract!). The attraction between the ions is the **ionic bond**. It is a very strong type of bond.

An ionic bond is formed by the force of attraction between a positive ion and a negative ion.

CHEMISTRY · HOW DO ELEMENTS FORM COMPOUNDS?

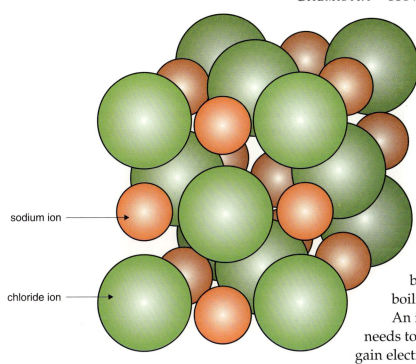

sodium ion

chloride ion

Ionic crystal
Electric charges attract in all directions, so each sodium ion attracts all the chloride ions around it. Likewise, the chloride ion attracts the sodium ions around it. In this way a large number of sodium ions and chloride ions attract each other and form a solid crystal of salt. The crystal is held together by strong ionic bonds, and so has high melting and boiling points.

An **ionic bond** is formed when one element needs to lose electrons and the other needs to gain electrons in order for each to have eight electrons in its outer shell.

Ionic bonds are usually formed between the metallic elements (in Groups 1, 2 and 3) and the elements in Groups 6 and 7 of the periodic table of the elements.

Ionic bonds are formed between metallic and non-metallic elements.

In some ionic compounds, one of the ions is a group of atoms that reacts as a single unit.

COMMON IONS AND IONIC SUBSTANCES

Positive ion		Negative ion		Ionic compound
Sodium	(Na^+)	Chloride	(Cl^-)	Sodium chloride ($NaCl$)
Calcium	(Ca^{2+})	Oxide	(O^{2-})	Calcium oxide (CaO)
Potassium	(K^+)	Hydroxyl	(OH^-)	Potassium hydroxide ($NaOH$)
Magnesium	(Mg^{2+})	Carbonate	(CO_3^{2-})	Magnesium carbonate ($MgCO_3$)
Copper (II)	(Cu^{2+})	Sulphate	(SO_4^{2-})	Copper (II) sulphate ($CuSO_4$)

Ionic compounds can be easily identified. They are crystalline solids with high melting and boiling points. They generally dissolve in water and conduct electricity when dissolved (or in the molten state). Ionic reactions in solution are very fast.

Breakthrough Science

> **DEMONSTRATION EXPERIMENT 32.1**
> **AIM: TO SHOW THE FAST REACTION OF IONIC COMPOUNDS**
>
> **Method**
> 1. Put some lead nitrate solution into a beaker.
> 2. Add a few drops of potassium iodide solution.
>   **Result:** A yellow colour is immediately seen.
>   **Conclusion:** This is due to the formation of lead iodide.
>   Lead nitrate + potassium iodide = lead iodide (yellow) + potassium nitrate

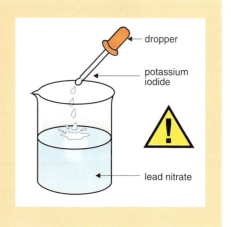

HOW TO GET THE FORMULA OF AN IONIC COMPOUND

If you know that a substance is ionic, the correct formula for the compound is got by balancing the electric charges, i.e. by arranging the numbers of each ion so that the total electric charge of the positive ions equals the total electric charge of the negative ions.

Examples

1. **Potassium chloride** contains two kinds of ion: potassium ion with a single positive electric charge K^+; chloride ion with a single negative electric charge Cl^-.
2. As a single positive charge balances a single negative charge, the correct formula for potassium chloride is K^+Cl^-.
3. Write the formula without the charges: **KCl**.

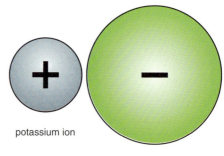

1. **Magnesium chloride** contains two kinds of ion: magnesium ion with a double positive electric charge Mg^{++}; chloride ion with a single negative electric charge Cl^-.
2. As a double positive charge needs to be balanced by two single negative charges, the correct formula for magnesium chloride is $Mg^{2+}(Cl^-)_2$.
3. Write the formula: **MgCl$_2$**.

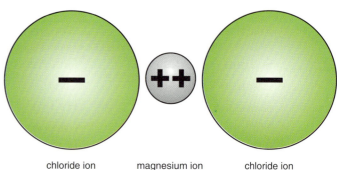

1. **Calcium hydroxide** contains two kinds of ion: calcium ion with a double positive electric charge Ca^{2+}; hydroxyl ion with a single negative electric charge OH^-.
2. To balance the two positive electric charges of calcium requires two hydroxyl ions.
3. Formula: **Ca(OH)$_2$**.

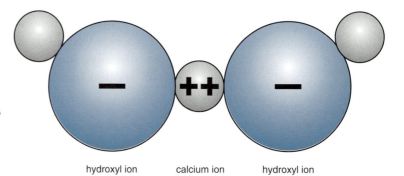

> *The first part of the chemical formula of an ionic compound is always a positive ion. The positive ion is a metal or hydrogen ion (H^+) or ammonium (NH_4^+) ion. The negative ion forms the second part of the formula.*

Other examples

Substance	Positive ion	Negative ion	Balanced formula	Written as
Sodium chloride	Na^+	Cl^-	Na^+Cl^-	NaCl
Calcium chloride	Ca^{2+}	Cl^- (×2)	$Ca^{2+}Cl^-_2$	$CaCl_2$
Sodium sulphate	Na^+ (×2)	SO_4^{2-}	$Na^+_2SO_4^{2-}$	Na_2SO_4
Aluminium sulphate	Al^{3+} (×2)	SO_4^{2-} (×3)	$Al^{3+}_2(SO_4)^{2-}_3$	$Al_2(SO_4)_3$
Potassium carbonate	K^+	CO_3^{2-}	$K^+_2CO_3^{2-}$	K_2CO_3

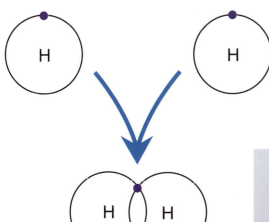

COVALENT BONDING

Not all substances are ionic. Gases or liquids, with low melting and boiling points and that do not conduct electricity, are clearly not ionic.

Hydrogen gas (H_2) is a diatomic molecule—each molecule is made up of two atoms of hydrogen. This is an example of a molecule held together by covalent bonds.

	Electronic configuration	Electronic configuration of nearest noble gas
Hydrogen	1	2

A hydrogen atom has **one** electron in its electron shell. Hydrogen needs **two** electrons to have a full electron shell. To get this stable state, two hydrogen atoms join by each atom sharing its electron with the other atom. The two electrons now orbit around both atoms and hold the molecule together. The shared pair of electrons is a covalent bond.

> *A covalent bond is formed by the sharing of electrons between two atoms.*

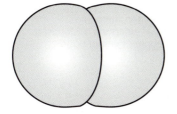

hydrogen molecule
Two hydrogen atoms get full outer electron shells by sharing electrons.

	Electronic configuration	Electronic configuration of nearest noble gas
Oxygen	2, 6	2, 8

Oxygen gas is another diatomic molecule. Oxygen (O_2) molecules are held together by covalent bonds.

An oxygen atom has **six** electrons in its electron shell. Oxygen needs **eight** electrons to have a stable electronic configuration. To get this stable state, two oxygen atoms join by each atom sharing two of its electrons with the other atom. These two pairs of electrons now orbit around both atoms and hold the molecule together. Each shared pair of electrons is a covalent bond, so oxygen is held together by two covalent bonds.

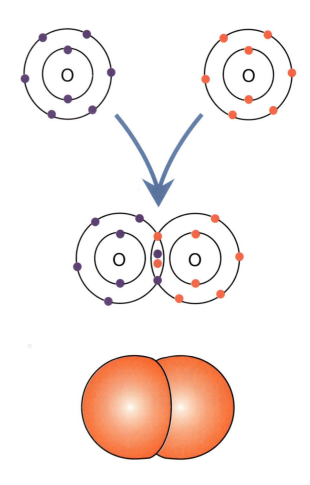

oxygen molecule
Two pairs of electrons are shared.
A double covalent bond.

Water (H_2O) is another example of a molecule held together by covalent bonds. Hydrogen needs to gain **one** electron to have a full electron shell; oxygen needs to gain **two** electrons to have eight electrons in its outer shell. So two hydrogen atoms share two pairs of electrons with one oxygen atom. Each shared pair contains one electron from hydrogen and one electron from oxygen. Each shared pair of electrons is a covalent bond.

	Electronic configuration	Electronic configuration of nearest noble gas
Hydrogen	1	2
Oxygen	2, 6	2, 8

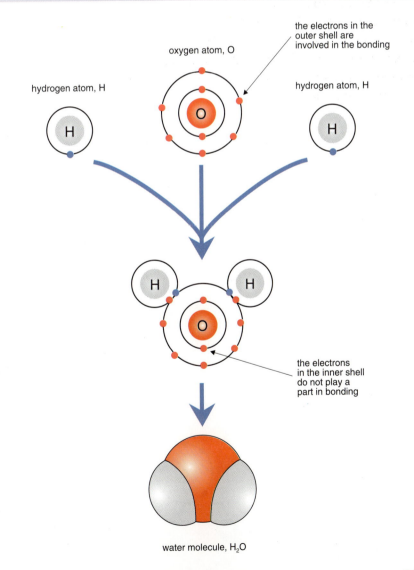

water molecule, H_2O

Other examples of covalent molecules are: fluorine (F_2), chlorine (Cl_2), nitrogen (N_2), ammonia (NH_3), methane (CH_4), carbon dioxide (CO_2).

COVALENT SUBSTANCES
Covalent bonds link atoms only to the atoms they share electrons with. They do not normally form strong crystals. Covalent compounds are usually gases, liquids or soft solids. Covalent compounds have low melting and boiling points, and react slowly.

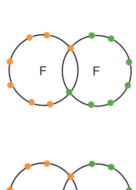
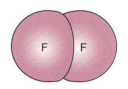

fluorine molecule, F_2

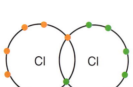

chlorine molecule, Cl_2
one pair of electrons is shared

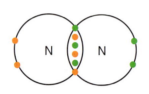
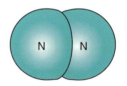

nitrogen molecule, Cl_2
three pairs of electrons are shared so the bond is a triple bond.

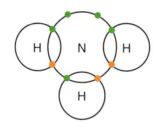
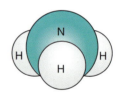

ammonia molecule, NH_3

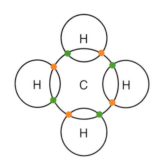
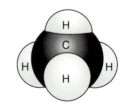

methane molecule, CH_4

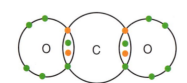
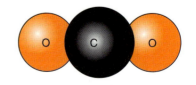

carbon dioxide molecule, CO_2

CHEMISTRY · HOW DO ELEMENTS FORM COMPOUNDS?

Covalent substances are gases, liquids or soft solids

Properties of covalent compounds	Properties of ionic compounds
Usually gases or liquids, but can be soft solids	Solid crystalline substances
Low melting and boiling points	High melting and boiling points
Generally do not dissolve in water; do not conduct electricity	Usually dissolve in water: the solution conducts electricity
React slowly, as one covalent bond must be broken before another bond can be formed	Reactions between ionic substances in solutions are very fast

SUMMARY

- Noble gases do not react. They have eight electrons in the outer electron shell.
- Elements react to get the electronic configuration of the nearest noble gas.
- An ion is an electrically charged atom or group of atoms.
- A negative ion is an atom that has gained an extra electron.
- A positive ion is an atom that has lost an electron.
- An ionic bond is the force of attraction between a positive ion and a negative ion.
- Ionic substances are solid crystalline substances which have high melting and boiling points, dissolve in water, and conduct electricity in solution or in the molten state.
- A covalent bond is formed by two atoms sharing a pair of electrons.
- Covalent substances are usually gases or liquids, do not dissolve in water, and do not conduct electricity.

QUESTIONS

Section A
1. Noble gases do not react because _____.
2. Why do most elements react? _____
3. What is an ion? _____
4. A negative ion is an atom that has _____ electrons.
5. A positive ion is an atom that has _____ electrons.
6. A covalent bond is formed by two atoms _____ electrons.
7. The atoms in a hydrogen molecule are joined by a _____ bond.
8. Magnesium carbonate is a substance joined by _____ bonds.
9. The atoms in a carbon dioxide molecule are joined by a _____ bond.
10. Calcium oxide is a substance joined by _____ bonds.
11. The atoms in a methane molecule are joined by a _____ bond.
12. Potassium hydroxide is a substance joined by _____ bonds.
13. A shared pair of electrons form a _____ bond.
14. In a negatively charged ion, the number of _____ is always greater than the number of _____.
15. An ionic bond is formed by the _____ of electrons between two atoms.

Section B
1. Name three noble (inert) gases. Choose any one of these and explain, in terms of electrons, why it does not normally undergo chemical reactions.
   What type of bond holds lithium fluoride together? Describe with the aid of simple diagrams how the bond is formed.
2. The main salt in sea water is sodium chloride. Name the type of chemical bond found in sodium chloride. Use simple diagrams to explain briefly, in terms of electrons, how the bond is formed. Sodium chloride is normally found in crystals. Explain how these crystals are formed.
3. Complete the following table.

	Covalent substances	Ionic substances
State		
Dissolve in water?		
Conducts electricity?		
Melting and boiling points		

4. One of the gases found in air is oxygen. Name the type of chemical bond found in oxygen. Use simple diagrams to explain briefly, in terms of electrons, how the bond is formed.
5. Carbon dioxide is a gas found in air. What type of bond holds the molecule of carbon dioxide together? Describe with the help of simple diagrams how a molecule carbon dioxide is formed. How would you test a gas to find out whether or not it is carbon dioxide?

CHAPTER 33
VALENCY, EQUATIONS AND ENERGY

We use the idea of **valency** to help us work out the chemical formula of a compound. Valency is a number we give to each element. Valency is the number of electrons an atom of an element can give, take or share when forming chemical bonds. This means that the valency tells us the number of covalent bonds an element can make. Valency also tells us the electric charge of the ion if a substance forms ionic bonds.

Valency of an element is the number of electrons an atom of that element can give, take or share when forming chemical bonds.

We already know the valency of many elements. We know that sodium and potassium can give one electron when forming ionic bonds, and that hydrogen and chlorine can take (or share) one electron when reacting.

In general, we get the valency of an element from its group number.

Group	Number of electrons		Valency
1	Give or share	1	1
2	Give or share	2	2
3	Give or share	3	3
4	Share	4	4
5	Take or share	3	3
6	Take or share	2	2
7	Take or share	1	1

The valency of an ion is the same as the electric charge on the ion.

	Formula	Valency
Hydroxyl	OH^-	1
Carbonate	CO_3^{2-}	2
Sulphate	SO_4^{2-}	2
Ammonium	NH_4	1

BREAKTHROUGH SCIENCE

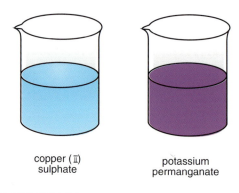

transition metal ions

copper (II) sulphate

potassium permanganate

VARIABLE VALENCIES

In some reactions copper gives away one electron and is called copper (I); in other reactions it gives two electrons and is called copper (II). Iron can have a valency of two (iron (II)) or three (iron (III)). Elements with variable valencies are found in the centre part of the periodic table of the elements, and are called **transition metals**. As well as having variable valencies, they form coloured ions like the blue copper (II) ion you saw in copper (II) sulphate or the purple manganate (VII) ion in potassium permanganate.

HOW TO USE VALENCY TO GET THE FORMULA OF A COMPOUND

The correct formula for a covalent compound is got by balancing the valencies of the two parts of the compound.

Example 1
1. **Hydrogen** molecules are covalent and are made of two hydrogen atoms. Write down the valencies. The valency of each hydrogen atom is 1.
2. Balance the valencies of the two atoms—the correct formula for the hydrogen molecule is H_2.
3. Write the formula H_2.
   This is also the case for most other elements found as gases (except the inert gases, which exist as single atoms). The correct formulas for these covalent molecules are H_2, N_2, O_2, F_2, Cl_2.

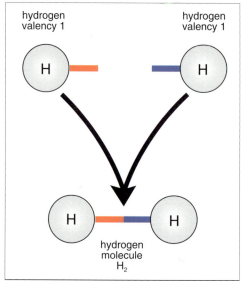

Example 2
1. **Water** is a covalent molecule of hydrogen and oxygen. Hydrogen has a valency of 1; oxygen has a valency of 2.
2. To balance oxygen's valency of 2 takes two hydrogen atoms, each with a valency of 1. The molecule is made up of two hydrogen atoms bonded to an oxygen atom.
3. The correct formula is H_2O.

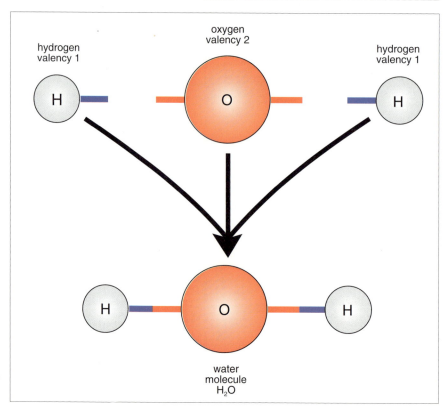

CHEMISTRY · VALENCY, EQUATIONS AND ENERGY

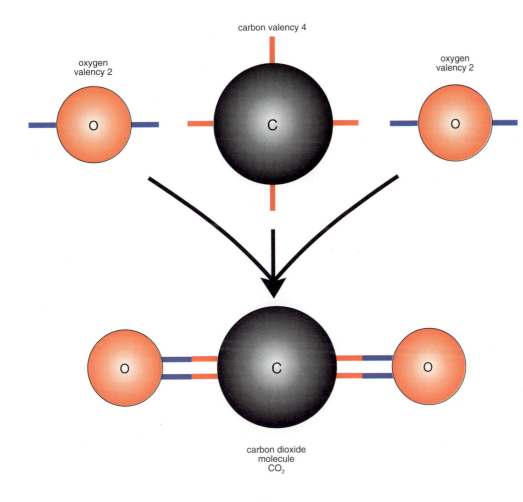

carbon dioxide molecule CO_2

Example 3
1. **Carbon dioxide** is a covalent molecule of carbon and oxygen. Carbon has a valency of 4; oxygen has a valency of 2.
2. To balance carbon's valency of 4 takes two oxygen atoms, each with a valency of 2. The molecule is made up of one carbon atom bonded to two oxygen atoms.
3. The correct formula is CO_2.
We can also use this method with ionic compounds.

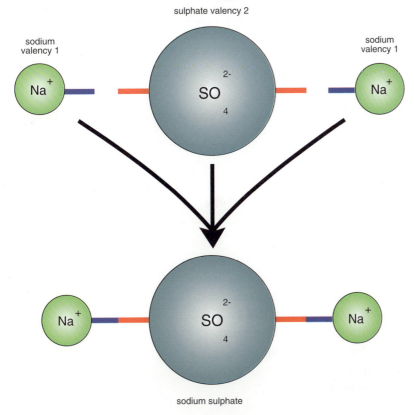

sodium sulphate

Example 4
1. **Sodium sulphate** is a compound of sodium and the sulphate group ion. Sodium has a valency of 1; sulphate has a valency of 2.
2. The simplest way that these valencies can be balanced is to balance two sodium ions against one sulphate.
3. The correct formula for sodium sulphate is Na_2SO_4.

HOW TO BALANCE AN EQUATION

The two rules for balancing equations are as follows.
1. **Correct formulas**: The correct formula must be used for each of the substances (molecules) taking part in the reaction. When the reaction is with water you must use the formula H_2O. You can't change this to H_2O_2, because that is the formula for hydrogen peroxide and not for water.
2. **Number of atoms**: The number of each of the molecules taking part in the reaction must be adjusted so that the number of atoms of each element on the left side of the equation equals the number of atoms of the same element on the right side of the equation.

Equations are balanced as follows.
1. Write the reactions in words.
2. Write the reaction in symbols, using the chemical formula for each substance.
3. Balance the equation.

Example
1. Reaction in words: **Magnesium** reacts with **oxygen** to form **magnesium oxide**.
2. Reaction in symbols

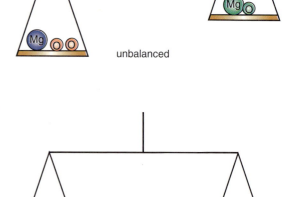

Mg   + O_2 →   MgO (not balanced)
1 atom Mg   2 atoms O   1 atom Mg and 1 atom O

To balance this reaction, use two atoms of magnesium on the left side. This would give two molecules of magnesium oxide on the right.
3. Balancing

2Mg   + O_2 →   2MgO (balanced)
2 atoms Mg   2 atoms O   2 atoms Mg and 2 atoms O

Remember: change the numbers in front of the formula but don't change the formula. Use only whole numbers.

CHEMICAL ENERGY AND HEAT

How do you heat a room in winter? You burn coal or gas or some other fuel. The reaction between the fuel and the oxygen in the air gives off a lot of heat. In our chemistry experiments, when we burn magnesium or when sodium reacts with water, heat is given off by the reaction. Heat is also given off when sulphuric acid dissolves in water. We call reactions that give off heat **exothermic**.

CHEMISTRY · VALENCY, EQUATIONS AND ENERGY

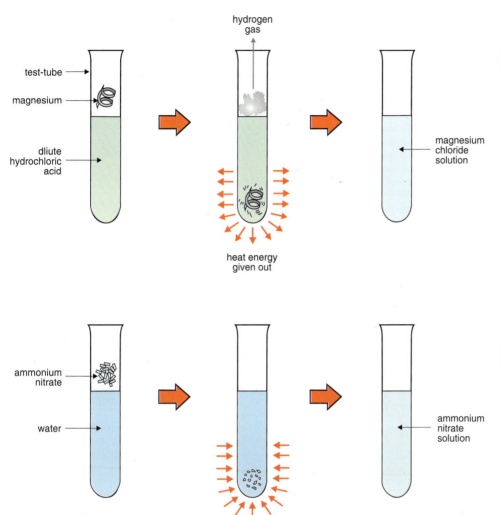

An exothermic reaction gives off heat.

In some other reactions we must put in heat to get the reactants to react. When we want iron to react with sulphur we must heat the mixture to get the substances to form a compound—iron (II) sulphide. Heat is also taken in when ammonium nitrate dissolves in water. We say that these reactions are **endothermic**.

An endothermic reaction takes in heat.

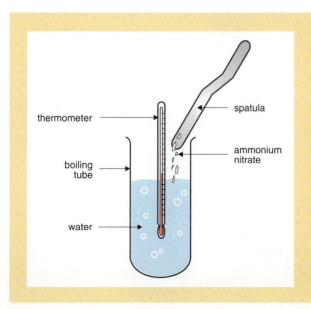

EXPERIMENT 33.1
AIM: TO SHOW THAT DISSOLVING AMMONIUM NITRATE IN WATER IS ENDOTHERMIC

Method
1. Put 20 cm³ of water into a boiling tube.
2. Put a thermometer into the boiling tube. Read the temperature.
3. Add some ammonium nitrate to the water. Stir gently to dissolve. Note the temperature change when the ammonium nitrate dissolves.

223

Repeat this procedure with anhydrous copper sulphate. What do you find? Note the highest temperature reached when anhydrous copper sulphate dissolves in water.

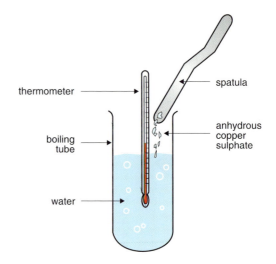

SUMMARY

- The valency of an element is the number of electrons it gives, takes or shares when forming a compound.
- When you are balancing chemical equations, the correct formula for a substance must not be altered.
- In a balanced equation, the number of atoms of each element on the left side of the equation must equal the number of atoms of the same element on the right.
- An exothermic reaction gives off heat.
- An endothermic reaction takes in heat.

QUESTIONS

Section A

1. What is the valency of: magnesium? ___ oxygen? ___
2. Chemical reactions that give off heat are called _____ reactions.
3. The formula for glucose is $C_6H_{12}O_6$. How many atoms are there in a molecule of glucose?
4. The electronic configuration of an element is 2, 7. Give the name and valency of this element. Name: _____ Valency: ___
5. Chemical reactions that take in heat are called _____ reactions.
6. The valency of an atom is defined as _____.
7. The formula for copper sulphate is $CuSO_4$. The valency of copper in this compound is ___.
8. Coloured ions are a feature of elements with _____.
9. What compound is formed when magnesium burns in air? Give the formula for this compound.
10. Complete and balance the following equation: Mg + HCl = _____.

Section B
1. Complete and balance the following equations.
   (a) $Zn + HCl =$
   (b) $NaOH + HCl =$
   (c) $CuO + H_2 =$
   (d) $Na + H_2O = NaOH +$
   (e) $H_2O_2 = O_2 + H_2O$
   (f) $CaCO_3 + HCl = CaCl_2 + H_2O + CO_2$
2. Describe, using a simple diagram, an experiment to show that dissolving ammonium chloride in water is an endothermic reaction. Use the headings Aim, Method, Result and Conclusion.
3. The following chemical equation is balanced.
   $Na_2SO_3 + 2HCl \rightarrow 2\ NaCl + H_2O + X$
   Write down the formula of substance X.
4. Write down the correct formula for each of the following.
   (a) Potassium sulphate
   (b) Magnesium hydroxide
   (c) Sodium carbonate
   (d) Aluminium sulphate
   (e) Sodium sulphate
5. The chemical formula for a hydrocarbon fuel is $C_{11}H_{24}$. How many atoms are there in this molecule? Write down a balanced equation to describe the complete burning of this fuel in oxygen to give carbon dioxide and water. Which type of reaction is this, endothermic or exothermic?

ACIDS AND BASES — CHAPTER 34

We already know quite a lot about acids and bases. We have used them in many chemical reactions, and have made some of them.
Let's summarise what we know about acids and bases.

ACIDS

1. Acids are corrosive—they break down substances. If you spill some strong acid on your clothes it will "burn" a hole through the cloth.
2. Most acids are solutions of oxides of non-metallic elements such as sulphur and nitrogen.
3. Acids react with zinc and many other metals to give hydrogen gas (and the metal salt of the acid). All acids contain hydrogen.
4. Acids react with calcium carbonate (limestone or marble chips) to give carbon dioxide.
5. Acids turn blue litmus red.

An acid turns blue litmus red.

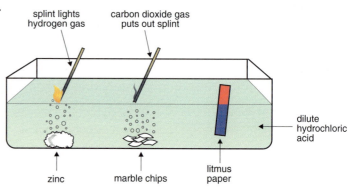

BASES

1. Bases are also corrosive.
2. Bases are usually oxides or hydroxides of metals.
3. Bases neutralise acids (make them harmless), forming salts and water.
4. Bases turn red litmus blue.
   An **alkali** is a base that dissolves in water. Sodium hydroxide (NaOH) is an alkali.

A base turns red litmus blue.

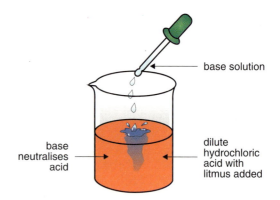

INDICATORS

We used litmus to find out which substances are acids and which are bases. Litmus is an **indicator**. An indicator has two coloured states—one colour when you put it into an acid solution, and the other when you put it into a basic solution. The two coloured states of litmus are red in acid and blue in base. Litmus is made from an extract of lichens. An indicator can also be extracted from red cabbage.

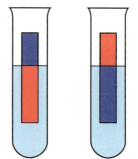

litmus indicator

red in acid   blue in base

EXPERIMENT 34.1
AIM: TO PREPARE AN INDICATOR SOLUTION FROM RED CABBAGE

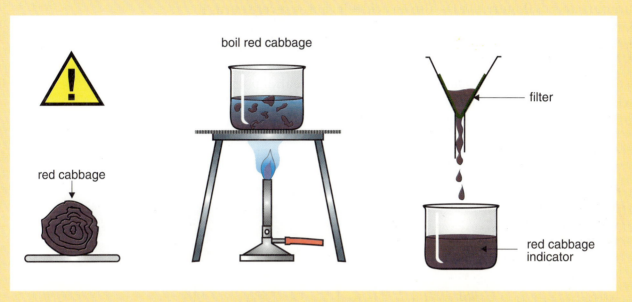

Method
1. Chop up some red cabbage leaves and add to a small quantity of water.
2. Bring to the boil and boil for about ten minutes. Pulp the leaves.
3. Leave to stand for a few minutes. Filter the mixture.

Result: The coloured liquid (blue or purple) is an indicator. Add a few drops of the indicator to hydrochloric acid: deep red colour. Add to sodium hydroxide: yellow colour. The two coloured states of the red cabbage indicator are a deep red in acid and a yellow in base.

An indicator is a substance we use to tell us by a colour change whether a solution is an acid or a base.
The following are common indicators.

	Colour in strong acid solution	Colour in strong basic solution
Litmus	Red	Blue
Phenolphthalein	Colourless	Red
Methyl orange	Red	Yellow

THE PH SCALE: A SCALE FOR MEASURING THE STRENGTH OF ACIDS AND BASES

Lemon juice is an acid. We can drink lemon juice without any harmful effect. Sulphuric acid is also an acid—it causes severe burns to skin and can kill. Clearly, we need some sort of scale to measure the strength or power of acids and bases. We call this scale the **pH scale**. The pH scale goes from 0 to 14, and numbers acids and bases according to their strength. The mid-point of the scale, 7, is the neutral point. Acids have a pH between 0 and 7; bases have a pH between 7 and 14.

| *pH measures the strength of an acid or a base.*

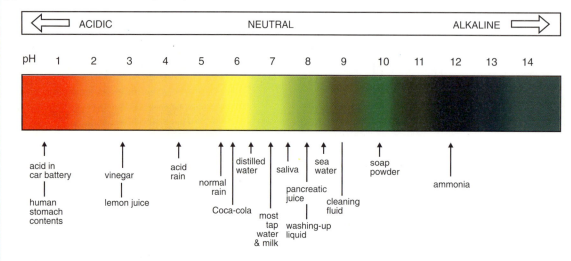

Each solution has the same concentration.

EXPERIMENTS WITH ACIDS

EXPERIMENT 34.2
AIM: TO MEASURE THE pH OF SOME ACIDS

Collect samples of each of the following: lemon juice, vinegar, sour milk, soda water, dilute sulphuric acid, dilute hydrochloric acid.

Method

Add 2–3 drops of universal indicator solution to each sample in a test-tube (universal indicator is a mixture of indicators). Compare with the colour chart to determine its pH (or use a pH meter).

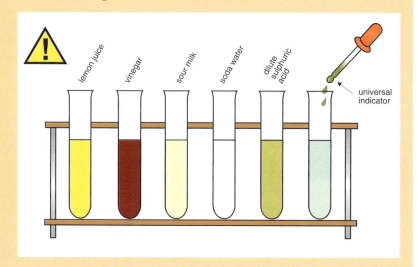

Use the results of these tests to complete the table.

Substance	pH	Substance	pH
Water	7	Soda water	
Lemon juice		Dilute sulphuric acid	
Vinegar		Dilute hydrochloric acid	
Sour milk			

DEMONSTRATION EXPERIMENT 34.3
AIM: TO TEST THE REACTIONS OF A NUMBER OF SUBSTANCES WITH HYDROCHLORIC ACID

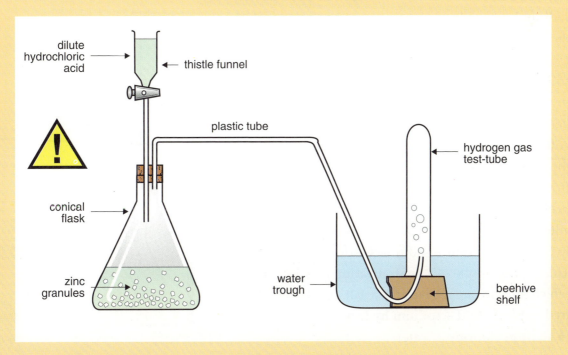

(a) Zinc

Method

1. Add hydrochloric acid to some zinc granules.
2. Collect the gas produced in a test-tube. A lighted splint gets the gas to burn with a quiet "pop" and a blue flame—hydrogen gas.

Result:

hydrochloric acid + zinc → zinc chloride + hydrogen

Breakthrough Science

(b) Calcium carbonate
Method
1. Add some hydrochloric acid to some calcium carbonate (marble chips).
2. Collect the gas produced in a test-tube. The gas extinguishes a lighted splint—carbon dioxide gas.
3. Bubble some gas through limewater. The limewater turns milky—carbon dioxide gas.

Result:

hydrochloric acid + calcium carbonate →
calcium chloride + water + carbon dioxide

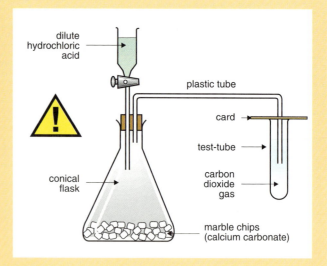

(c) Sodium hydroxide
Method
1. Use a pipette to measure 10 cm^3 of a solution of sodium hydroxide into a conical flask. Add a few drops of methyl orange indicator solution. A yellow colour is seen.
2. Add hydrochloric acid solution from a burette, drop by drop, until the indicator changes colour from yellow to red. The sodium hydroxide has been neutralised by the hydrochloric acid.

Result: Hydrochloric acid reacts with sodium hydroxide to produce sodium chloride and water. This is an example of a **neutralisation** reaction, in which an **acid** and a **base** give a **salt** and **water**.

hydrochloric acid + sodium hydroxide
→ sodium chloride + water

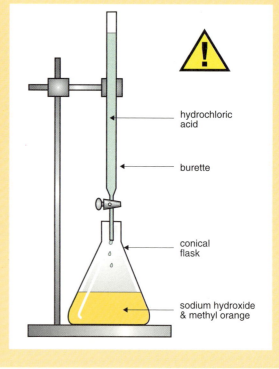

NEUTRALISATION

Sometimes your stomach can produce too much acid. This causes a pain—indigestion. The medicine your doctor gives you for this contains a mild base which neutralises the acid and stops the pain. The medicine is a suspension of magnesium hydroxide or aluminium hydroxide.

Soil can also become too acidic. Most plants do not grow properly in acidic soils. Farmers add lime—a base—to the land to neutralise the acidity in the soil.

EXPERIMENT 34.4
AIM: TO TEST THE REACTIONS OF A NUMBER OF SUBSTANCES WITH SULPHURIC ACID

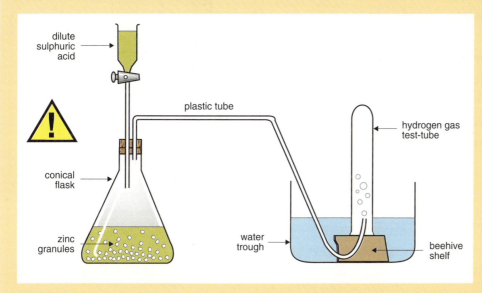

(a) Zinc
Method
1. Add sulphuric acid to some zinc granules.
2. Collect the gas produced in a test-tube. A lighted splint gets the gas to burn with a blue flame —hydrogen gas.

Result:

H_2SO_4	+	Zn	→	$ZnSO_4$	+	H_2
sulphuric acid		zinc		zinc sulphate		hydrogen

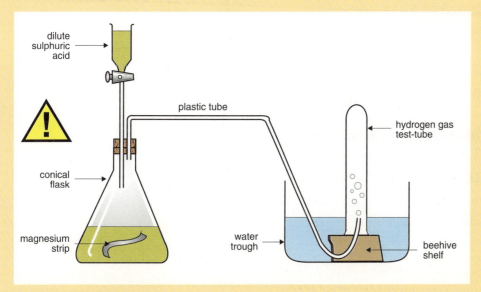

(b) Magnesium
Method
1. Add some sulphuric acid to a short strip of magnesium (clean first with emery paper).
2. Collect the gas produced in a test-tube. A lighted splint gets the gas to burn with a blue flame— hydrogen gas.

Result:

H_2SO_4	+	Mg	→	$MgSO_4$	+	H_2
sulphuric acid		magnesium		magnesium sulphate		hydrogen

(c) **Sodium carbonate**

Method
1. Add some sulphuric acid to some sodium carbonate.
2. Collect the gas produced in a test-tube. The gas extinguishes a lighted splint—carbon dioxide gas.
3. Bubble some gas through limewater. The limewater turns milky—carbon dioxide gas.

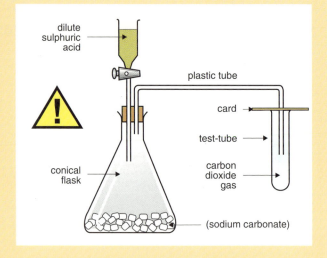

Result:
$$H_2SO_4 + Na_2CO_3 \rightarrow Na_2SO_4 + H_2O + CO_2$$
sulphuric acid + sodium carbonate → sodium sulphate + water + carbon dioxide

(d) **Sodium hydroxide**

Method
1. Measure 10 cm^3 of a solution of sodium hydroxide into an evaporating dish. Add a few drops of methyl orange indicator solution. A yellow colour is seen.
2. Add sulphuric acid solution from a burette, drop by drop, until the indicator changes colour from yellow to red.

Result: The sodium hydroxide has been neutralised by the sulphuric acid.

$$H_2SO_4 + 2NaOH \rightarrow Na_2SO_4 + 2H_2O$$
sulphuric acid + sodium hydroxide → sodium sulphate + water

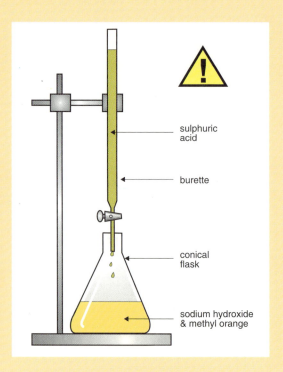

ORGANIC ACIDS

Organic acids are found in nature and are usually compounds of carbon, hydrogen and oxygen. They are much weaker than sulphuric acid or hydrochloric acid. Organic acids are present in lemon juice (**citric acid**), sour milk (**lactic acid**) and vinegar (**ethanoic acid**).

EXPERIMENTS WITH BASES

EXPERIMENT 34.5
AIM: TO FIND OUT THE pH OF A NUMBER OF SUBSTANCES

Method

1. Collect samples of solutions of each of the following: soap, milk of magnesia, toothpaste, washing soda, oven cleaner, window cleaner, sodium hydroxide.
2. Add a drop of universal indicator solution to each sample. Compare with the colour chart to determine its pH. Alternatively, use a pH meter. Enter your results in the table on the next page.

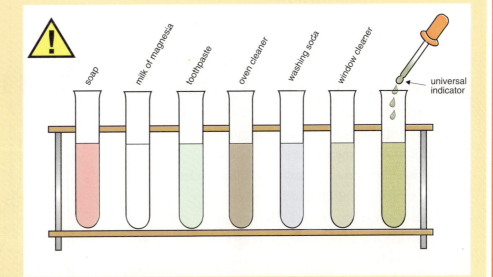

Substance	pH

SALTS

The salt sodium chloride is formed when hydrochloric acid reacts with sodium hydroxide.

HCl	+	NaOH	→	NaCl	+	H₂O
hydrochloric acid		sodium hydroxide		sodium chloride		water
Acid	**+**	**Base**	**→**	**Salt**	**+**	**Water**

An acid reacts with a base to neutralise it, forming a salt and water only.

EXPERIMENT 34.6
AIM: TO PRODUCE THE SALT SODIUM CHLORIDE BY TITRATION

Method

1. Use a pipette to measure 25 cm³ of a solution of sodium hydroxide into a conical flask. Add a few drops of methyl orange indicator solution. A yellow colour is seen.
2. Add hydrochloric acid solution from a burette, drop by drop, until the indicator changes colour from yellow to red. The sodium hydroxide has been neutralised by the hydrochloric acid. Note the volume of hydrochloric acid used.

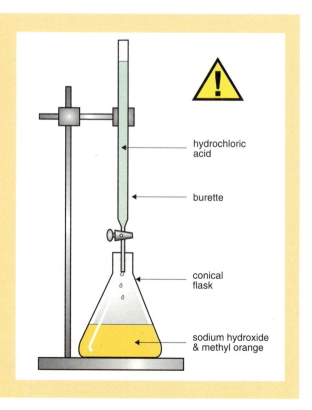

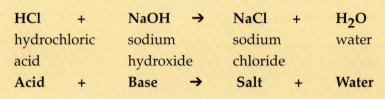

HCl	+	NaOH	→	NaCl	+	H_2O
hydrochloric acid		sodium hydroxide		sodium chloride		water
Acid	+	**Base**	→	**Salt**	+	**Water**

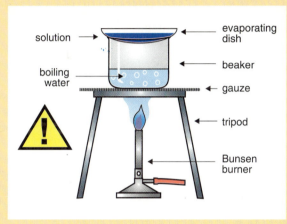

3. Empty the conical flask. Rinse it. Put 25 cm^3 of the same sodium hydroxide solution into the conical flask. **Do not use any indicator this time.**
4. Add exactly the same amount of hydrochloric acid solution as before.
5. Pour the contents of the flask into an evaporating dish and put the evaporating dish on a water bath. Evaporate the water until the dry salt is obtained.

SUMMARY

- An acid turns blue litmus red and has a pH less than 7.
- A base turns red litmus blue and has a pH greater than 7.
- An indicator is a substance we use to tell us by a colour change whether a solution is acidic or alkaline.
- The pH scale measures how acidic or how basic the solution is.
- An acid reacts with a base to neutralise it, forming a salt and water only.

QUESTIONS

Section A
1. What acid is used in car batteries? _____ What is the formula for this acid? _____
2. You could use vinegar to treat a wasp sting because _____.
3. Underline the acids in the following list: toothpaste, sour milk, vinegar, tea, milk of magnesia.
4. If you accidentally spilled some acid, what common household substance could you use to neutralise it? _____

5. Which of the following is the likely pH of a strong solution of
   (a) hydrochloric acid,
   (b) sodium hydroxide:
   1, 5, 7, 9, 14?
6. You could use ammonia to treat a bee sting because _____.
7. Indigestion tablets contain a base such as magnesium hydroxide because
   _____.
8. Name an indicator. _____ What is the colour of this indicator in basic solution? _____
9. Write down an equation for the reaction between hydrochloric acid and calcium carbonate.
10. Write down an equation for the reaction between sulphuric acid and magnesium.

Section B
1. An indicator is a chemical that changes colour when it is put into an acid or a base.
   (i) Name an indicator you have used in the laboratory.
   (ii) Describe a simple experiment to obtain an indicator from red cabbage.
2. The pH scale is used to compare the strength of acids and bases. Here is a table of liquids with their pH values.

Liquid	A	B	C	D	E
pH	1	5	7	9	12

   (i) Which liquid is the strongest acid?
   (ii) Which liquid is the strongest base?
   (iii) Which liquid is neutral?
   (iv) What would you use to measure the pH of a liquid?
3. The following is a brief description of a neutralisation reaction. "25 cm^3 of sodium hydroxide solution was placed in a flask and a few drops of litmus were added. Dilute hydrochloric acid was added until the indicator changed colour. It was found that 21 cm^3 of acid had been used."
   (i) Draw a labelled diagram of titration apparatus for this neutralisation.
   (ii) What piece of apparatus should be used to measure out accurately 25 cm^3 of sodium hydroxide solution?
   (iii) What colour was the solution in the flask at the start of the titration?
   (iv) What colour did it turn when the alkali had been neutralised?
   (v) Was the acid more concentrated or less concentrated than the alkali?
   (vi) Name the salt formed in this neutralisation.
   (vii) How would you obtain pure crystals of the salt produced by this experiment?

4. Which acids are represented by the following: HCl, H$_2$SO$_4$?
   What element is common to all acids?

5. Study the diagram.
   (i) Which metal reacts most easily?
   (ii) What gas is released when this metal reacts?
   (iii) How would you identify this gas?
   (iv) Give one property of this gas.

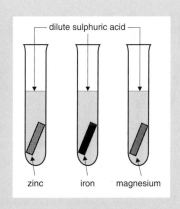

6. A solution was made by dissolving 1 g of acid in 10 cm^3 of water.
   (i) How would you show that the solution is acidic?
   (ii) How many grams of acid are contained in 50 cm^3 of the solution?
   (iii) If 50 cm^3 of the solution neutralises 20 cm^3 of an alkali (base solution). How many grams of acid will be needed to neutralise 80 cm^3 of the alkali?
   (iv) Explain the term *neutralisation*.

7. The following is a list of some common solutions: iodine, hydrogen chloride, sulphur dioxide, sodium hydroxide, calcium carbonate. From the list, name
   (i) a solution with a pH greater than 7;
   (ii) a solution with a pH less than 7.

8. The labels on five solutions have been lost. The solutions include sodium chloride, sulphuric acid, window cleaner (ammonia solution), sodium hydroxide and vinegar (ethanoic acid). The solutions were marked A to E, and were tested with universal indicator solution to find their pH. The results are shown in the diagram.
   Fill in the following table.

	pH	Colour
Sodium chloride		
Sulphuric acid		
Ammonia solution		
Sodium hydroxide		
Ethanoic acid		

Which solution is:
(i) neutral?
(ii) strongly acidic?
(iii) weakly acidic?
(iv) strongly alkaline?

Now identify each of the solutions A to E.

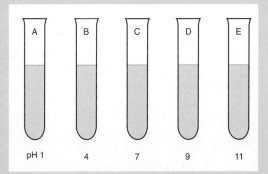

HARD AND SOFT WATER   CHAPTER 35

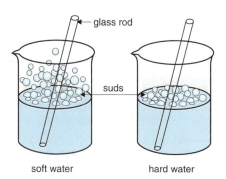

We know from our experiments in chapter 28 that water has substances dissolved in it. The amount and the type of substances dissolved in water change the nature of the water. Water in some areas makes suds (lather) very easily; in other areas it takes a lot of soap to make suds. Water in some areas also causes a white stony substance to form on the inside of kettles and pipes.

Water that makes suds easily with natural soap is *soft water*.
Water that takes a lot of soap to make suds is *hard water*.

EXPERIMENT 35.1
AIM: TO TEST FOR WATER HARDNESS
Method

1. Collect a number of samples of water from different sources, such as a spring (or use bottled spring water), a stream, rainwater and sea water. Another sample can be made as follows: (*a*) add calcium sulphate to water; (*b*) shake to dissolve; (*c*) filter any undissolved calcium sulphate.
2. Make up a solution of soap using natural soap such as Lux flakes.
3. Place 10 cm^3 of each sample of water in a test-tube.
4. Add soap solution to it, 1 cm^3 at a time. Shake the test-tube.
5. Continue to add soap solution until you get a good amount of suds on the water for one minute. Note the amount of soap solution each sample needed to go "sudsy". Record your results in a table similar to this one.

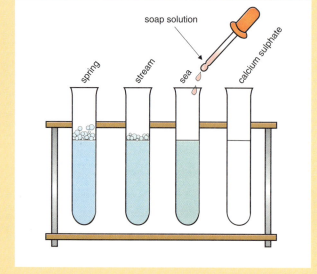

Sample	Volume of soap solution needed to give suds	Type of water
Spring water		
Stream water		
Rainwater	1	Soft
Sea water	12	Hard
Calcium sulphate solution	13	Hard

CHEMISTRY · HARD AND SOFT WATER

This shows that sea water and calcium sulphate solution are hard and rainwater is soft. Spring water and stream water can be hard or soft, depending on the types of rocks and soil in your area. *Hard water is caused mainly by the soluble salts of calcium and magnesium.*

These are usually calcium hydrogen carbonate, magnesium hydrogen carbonate, calcium sulphate and magnesium sulphate.

TEMPORARY HARDNESS IN WATER

EXPERIMENT 35.2
AIM: TO INVESTIGATE TEMPORARY HARDNESS IN WATER

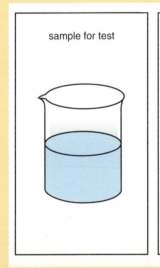
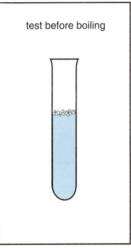
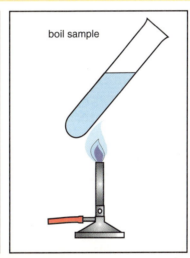
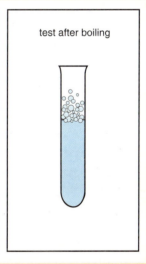

Method
1. Make up solutions of: (*a*) calcium hydrogen carbonate in water; (*b*) magnesium hydrogen carbonate in water. Filter each solution before use.
2. Test 10 cm³ of each solution for hardness, as described in the previous experiment.
3. Boil each solution. Filter after boiling.
   The filter contains some solid material. This must have come from the solution as a result of boiling it.
4. Test 10 cm³ of each solution for hardness after boiling. Enter your results in a table like the one at the top of the next page.

Solution	Volume of soap solution needed to give suds	
	Before boiling	After boiling
Calcium hydrogen carbonate	14	2
Magnesium hydrogen carbonate	12	1

Conclusion: Calcium hydrogen carbonate and magnesium hydrogen carbonate make water hard, but the hardness is removed when the water is boiled. This type of hardness is **temporary hardness**.

Temporary hardness in water is hardness that can be removed by boiling the water.

Temporary hardness in water is caused by calcium hydrogen carbonate and magnesium hydrogen carbonate.

Calcium hydrogen carbonate and magnesium hydrogen carbonate turn into **insoluble** calcium carbonate and magnesium carbonate when heated.

calcium hydrogen carbonate (soluble) ➜ calcium carbonate (insoluble) + carbon dioxide + water

LIME SCALE OR "FUR"

Temporary hardness is a serious problem in some areas. When water in these areas is heated, the soluble calcium hydrogen carbonate and magnesium hydrogen carbonate turn into insoluble calcium carbonate and magnesium carbonate. These substances form a solid stony cement-like coating on the inside of kettles, pipes and boilers, making it difficult to heat water. This lime scale or "fur" can also block central heating pipes.

Limescale on kettle element

DESCALING KETTLES

Lime scale in kettles is removed by leaving a solution of a weak acid such as citric acid in the kettle. The acid reacts with the solid carbonate to produce carbon dioxide, water, calcium citrate and magnesium citrate. The kettle must be rinsed out thoroughly before use.

PERMANENT HARDNESS IN WATER

Repeat experiment 35.2 with solutions of calcium sulphate and magnesium sulphate. The hardness produced by these salts is not removed by boiling the water. We call this type of hardness **permanent hardness**.

Permanent hardness in water is hardness that cannot be removed by boiling.

Permanent hardness in water is usually caused by calcium sulphate and magnesium sulphate.

WHAT MAKES WATER HARD?

We know that the soluble salts of metals (except sodium and potassium) make water hard. But how do these salts get into the water?

Rainwater is soft, but the water in many streams and rivers is hard. Rainwater is acidic because it dissolves carbon dioxide (and in polluted areas sulphur dioxide gas). The rainwater flows over rocks and through the soil into streams and rivers. When it flows over rocks such as limestone (calcium carbonate), chalk (calcium carbonate), gypsum (calcium sulphate) or dolomite (magnesium carbonate and calcium carbonate), the rocks react and dissolve in the acidic water. This makes the water hard. Clearly, whether water is hard or soft depends on the types of rock and soil through which it flows.

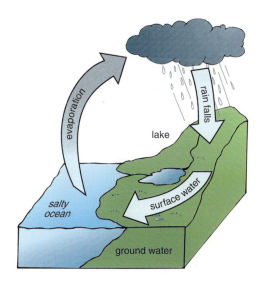

WATER SOFTENING

Water softening means removing the hardness from the water.

One of the simplest ways of treating hard water is to add sodium carbonate (washing soda). This makes the water soft. The soft water needs much less soap to make suds. Washing soda was added to water for washing before soapless detergents were developed.

When sodium carbonate is added to hard water containing soluble calcium sulphate it forms insoluble calcium carbonate, which leaves the water free of calcium ions, and as a result the water is soft. The soluble sodium sulphate that is also formed in the reaction does not make the water hard.

Bath salts are a coloured and scented form of sodium carbonate. This softens the water and allows the soap to make a lot of suds.

Permanent and temporary hardness can be removed by adding sodium carbonate to the water.

BREAKTHROUGH SCIENCE

WATER SOFTENERS

All the methods to soften temporary and permanent hard water mentioned so far produce an insoluble salt of calcium or some other metal ion. This produces a scale or fur on the inside of boilers and pipes.

Water used in commercial boilers is softened by passing it through a water softener. The softener is a container of **sodium alumino silicate**, which removes calcium ions and other metal ions and replaces them with sodium ions. This leaves the water soft, and does not produce any insoluble salts which produce scale.

When the sodium alumino silicate is used up, the process is reversed by adding sodium chloride (common salt) to the container. This regenerates the sodium alumino silicate in the water softener, and it can be used again.

DEIONISED WATER

Water softened in a water softener still contains ions. In the example given above it contains sodium sulphate. We sometimes need water that is free of all ions for chemical experiments and for topping up car batteries. This water is called deionised water. It is of course soft water, as it contains no metal ions.

Deionised water is produced by passing water through specially prepared plastic beads called **ion exchange resins**. One type of bead replaces positive metal ions with positive hydrogen ions (H^+). The other type of bead replaces negative ions with negative hydroxyl ions (OH^-). Most of the hydrogen ions (H^+) join with hydroxyl (OH^-) ions to form water (H_2O).

The advantages and disadvantages of hard water are as follows.

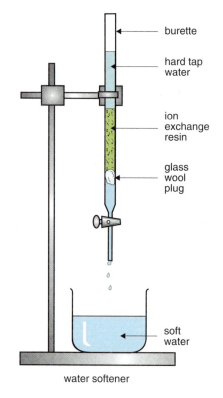

water softener

Advantages	Disadvantages
Provides some of the calcium needed for healthy teeth and bones	Blocks pipes and valves
	Leaves scale in boilers and kettles
Tastes better than soft water	Uses a lot more soap than soft water
Makes better-tasting beer	With soap, produces a scum

The pH of water—a measure of acidity and alkalinity—is another important water-quality factor. Stream waters usually range from pH 6·5 (slightly acid) to pH 8·5 (somewhat alkaline). Rainwater is slightly acidic (pH 5·6).

SUMMARY

- Soft water is water that makes suds (lather) easily with soap.
- Hard water is water that does not easily form suds with soap.
- The soluble salts of metals (except sodium and potassium) make water hard.
- Temporary hardness in water can be removed by boiling the water.
- Temporary hardness is caused by calcium hydrogen carbonate and magnesium hydrogen carbonate.
- Permanent hardness in water cannot be removed by boiling.
- Permanent hardness is mainly caused by calcium sulphate and magnesium sulphate.
- Permanent hardness can be removed by adding sodium carbonate (washing soda) or by passing the water through an ion exchanger.
- Hard water can be softened by distillation.

QUESTIONS

Section A

1. Hard water is water that _____.
2. The two types of hardness of water are _____ and _____.
3. Pure water will not conduct electricity because _____.
4. One advantage of soft water is _____.
5. One advantage of hard water is _____.
6. The "fur" in kettles is caused by _____.
7. Limestone does not dissolve in pure water. Why does it dissolve in rainwater?
   _____
8. Soft water is water that _____.
9. Name two gases that dissolve in water to form acid rain. _____ and _____
10. Temporary hardness is caused by _____.
11. Temporary hardness can be removed by _____.
12. Lime scale is _____ caused by _____.
13. Water softeners replace the ions that make water hard with _____.
14. Hardness is caused by the soluble salts of _____.

Section B

1. What is hard water? Is water obtained as a result of distilling sea water hard or soft? Explain your answer.
2. A = distilled water, B = tap water, C = calcium hydroxide solution, D = sodium chloride solution, E = unknown solution. Ten drops of soap solution were added to 10 cm^3 of each of these liquids. After shaking for a minute, the heights of lather shown in the diagram were obtained.

   (a) Does the tap water come from a hard water or a soft water area? Explain your answer.
   (b) Does sodium chloride form a precipitate with the soap solution? Explain your answer.
   (c) Does calcium hydroxide cause hardness? How can you tell from the experiment?
   (d) What can you say about liquid E, the unknown solution?

   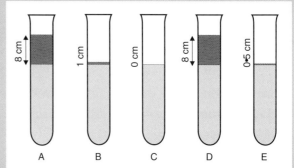

3. You are given a sample of hard water containing two dissolved compounds, one causing temporary hardness in the sample and the other causing permanent hardness.
   (i) Name a substance that could have caused temporary hardness.
   (ii) Name a substance that could have caused permanent hardness.
   (iii) How would you remove the temporary hardness from the sample?
   (iv) Describe how you would soften the water by ion exchange.
4. (i) State the cause of hardness in water.
   (ii) State the difference between temporary hardness and permanent hardness.
   (iii) Explain how water samples may be tested for hardness.
   (iv) Give two advantages and two disadvantages of hard water.
   (v) Describe how you would remove temporary hardness from water.
5. Hardness in water is caused by substances dissolved in it. Would you expect spring water in the Burren to be hard or soft? Describe a method of removing permanent hardness from a sample of water.
6. Three samples of tap water, A, B and C, were collected from different sources. 20 cm^3 of sample A was placed in a test-tube, soap solution was added in 1 cm^3 portions and the mixture was shaken. This procedure was repeated until a permanent lather was obtained. The experiment was repeated with a boiled portion of A and also with a portion of A that had been passed through an ion exchange resin.
   The experiment was repeated with samples B and C. The volumes of soap solution required to give a permanent lather in each case are given.

Water sample	Volume required (cm³) Before treatment	After boiling	After passing through ion-exchange resin
A	25	1	1
B	25	25	1
C	25	10	1

State, giving reasons for your answer in each case, which of the samples contain compounds that cause: (i) temporary hardness only; (ii) permanent hardness only; (iii) both temporary hardness and permanent hardness.

CHAPTER 36 METALS

Most elements are metals. Not all metals, however, are elements. This is because metal elements form mixtures with one another. **Alloys** are mixtures of metals.
- Bronze is an alloy of copper and tin.
- Brass is an alloy of copper and zinc.
- Steel is an alloy of iron and carbon.
- Stainless steel is an alloy of iron, chromium and nickel.

| *An alloy is a mixture of metals.*

All metals have a number of things in common.
Metals
- are **solids**: mercury—a liquid at room temperature—is an exception
- are **shiny** (metallic lustre): all metals, even the very reactive alkali metals, are shiny when freshly cut or polished—many lose their shine when exposed to the air (tarnish)
- are **ductile** (can be stretched): hot steel can be drawn into wires
- are **malleable** can be hammered and shaped): metals can be rolled into long sheets and hammered to form other shapes
- **conduct electricity**: electrical energy is sent through metal wires —copper is usually used, but the less dense aluminium is used in overhead ESB cables
- are **strong** (except the alkali metals): alloys of iron and tungsten are used in the cutting surfaces of high-speed drills and saw blades

BREAKTHROUGH SCIENCE

- **conduct heat**: copper is a good conductor of heat
- are **dense**: most metals are denser than water
- have **high melting points**: most metals have high melting points—iron has a melting point of 1540°C
- **form oxides**: most metals form oxides on their surface in air—iron forms rust.

iron + air + water ⟶ rust

CONDUCTION OF HEAT

You have already done a simple experiment to compare how well different metals conduct heat. You can also try out rods of wood and glass as examples of non-metals.

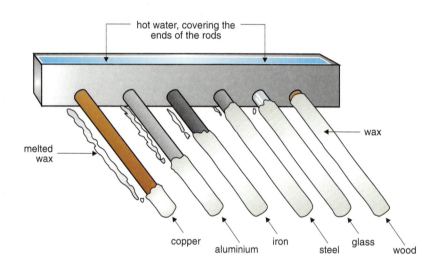

CONDUCTION OF ELECTRICITY

Set up the test circuit as shown. Connect the crocodile clips to the material being tested. If the substance is a conductor, the bulb lights up.

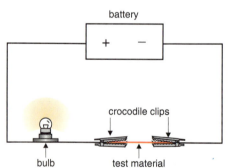

HARDNESS

An easy way to compare the hardness of metals is to see if a metal will scratch a piece of another metal. Could you scratch your name on a piece of steel with a piece of lead? Why not? It's because the lead is softer than steel.

If you wanted to inscribe designs on steel you would use a tungsten-steel tipped scribing tool.

Using a scribing tool

CHEMISTRY · METALS

RUSTING

When you burned magnesium with oxygen you made magnesium oxide. This reaction with oxygen is called **oxidation**. When you burn coal and other fuels, the fuel combines with oxygen—**burning** is also an oxidation reaction. When iron rusts, no great amount of heat is given off and no flames are seen, but the iron still reacts with the oxygen. When a metal is oxidised by combining with oxygen in the air we call it **corrosion**. **Rusting** is iron corrosion.

| *Oxidation of a metal by oxygen in the air is corrosion.*

EXPERIMENT 36.1
AIM: TO INVESTIGATE THE CONDITIONS NECESSARY FOR IRON TO RUST

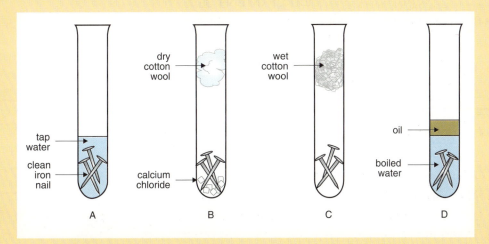

Method

1. Set up four test-tubes, each containing three identical iron nails as shown in the diagram. Label the test-tubes as shown.
2. Leave the test-tubes to stand for a few days.

Result: The nails in test-tube C show the greatest amount of rusting. The nails in test-tube A also show some rust. The nails in test-tubes B and D show no signs of rust.

Conclusion: Water and air (or more exactly, the oxygen in the air) are necessary for rusting to take place. Test tubes A and C have both of these. Rusting takes place faster when salt water is present in the air, and also at higher temperatures.

| *Water and oxygen are necessary for rusting.*

Breakthrough Science

Rust costs millions of pounds

Rusting of iron (and iron alloys such as steel) is a costly chemical reaction because it weakens anything made of iron. When iron rusts in very dry air it forms a thin oxide layer that protects it. However, when water is present iron forms a flaky iron oxide and the rusting spreads. Replacing rusted machinery and protecting buildings against corrosion costs many millions of pounds each year.

Corrosion of other metals is often confined to the surface. Aluminium pots and pans do not rust even though they are heated in air with plenty of water present. This is because aluminium corrodes to form a coating of aluminium oxide on its surface. This is a tough colourless barrier that stops oxygen reaching any more aluminium.

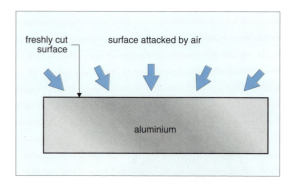

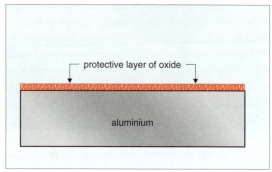

EXPERIMENT 36.2
AIM: TO COMPARE HOW EASILY VARIOUS METALS, IMPURE METALS AND ALLOYS CORRODE

1. Set up a number of different metal strips as shown in the diagram.
2. Leave the metal strips to stand for a few days. Examine the metals for signs of corrosion.
3. Try this experiment using salt water instead of tap water.

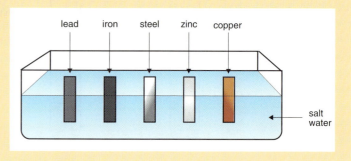

EXPERIMENT 36.3
AIM: TO SHOW THAT AIR IS USED IN THE RUSTING OF IRON

Method

1. Dampen some iron wool and place in a boiling tube as shown.
2. Invert the boiling tube over a trough of water. Mark the initial water level with an elastic band.
3. Leave for a week.
   Result: The iron wool shows rust and the water level has risen. Mark the new water level.
   Conclusion: 20% of the air has been used up—the oxygen in the air reacted with the iron to form rust. This shows that air is used in the rusting of iron.

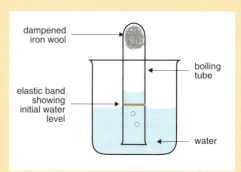

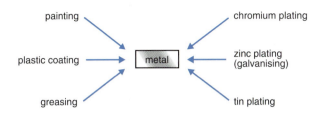

ways of preventing corrosion

RUST PREVENTION

Some samples of a metal rust more easily than others. This is often due to impurities in the metal. Mild steel rusts faster than pure iron, but mild steel is more flexible than iron and can be used to make car bodies.

Steel and other metals are protected against corrosion by

- painting them: this prevents the air and water from coming into contact with the metal surface
- covering them with a layer of grease, wax or plastic material
- coating them with a layer of another metal: this is used in **galvanised** steel, where the steel is covered with a layer of **zinc** by dipping the steel in hot zinc or by electrolysis.

STAINLESS STEEL

Although some metal impurities in iron can accelerate rusting, other metals added to iron can prevent it. Alloys of iron can be made that are corrosion-resistant. Stainless steel is an alloy of iron, chromium and nickel. The most common stainless steel is the alloy known as stainless steel 18-8, which is iron with 18% chromium, 8% nickel and 0·15% carbon.

SUMMARY

- Metals conduct electricity.
- Metals conduct heat.
- Metals usually have high melting points.
- Metals form oxides.
- An alloy is a mixture of metals.
- Oxidation of a metal by oxygen in the air is **corrosion**.
- Water and air (oxygen) are necessary for iron to rust.
- Alloys of metals can be made corrosion-resistant.
- Galvanised steel is covered with a layer of zinc.

BREAKTHROUGH SCIENCE

QUESTIONS

Section A

1. Metals are good conductors of _____ and _____.
2. A mixture of metals is called _____.
3. Metals are malleable: this means that _____.
4. Steel coated with a layer of zinc is _____.
5. Metals are ductile: this means that _____.
6. Stainless steel is an example of _____.
7. Why is aluminium foil considered suitable for wrapping food? _____.
8. A student set up an experiment as shown. After one week the iron nails had rusted in test-tube B only. What does this tell you about the conditions necessary for rusting?
   _____
9. Why is iron often coated with zinc?
   _____
10. What is an alloy? _____ Name one alloy. _____

Section B

1. (a) Name two everyday metals. Give one use for each of the metals you have named.
   (b) Describe a simple experiment to show that a metal is a good conductor of heat.
2. (a) Describe, using diagrams, a simple experiment to show that oxygen and water are necessary for iron to rust.
   (b) Describe, using diagrams, a simple experiment to compare how easily different metals corrode.
   (c) Name three things you might do to prevent parts of your bicycle from rusting.
3. What is meant by the term "corrosion"? In order to find out whether or not paint is better than grease at preventing rusting, you are given the following materials: iron nails, test-tubes, paint, paintbrush, grease, water.
   Describe how you would carry out the experiment.
   Galvanising is another method of rust prevention. What is meant by the term "galvanising"?

4. (i) Metals are good conductors of heat and electricity. Give three other properties of metals.
(ii) Metals are often mixed to produce metals with different properties. What are mixtures of metals called?
(iii) Name two such mixtures.
(iv) Give two uses for these mixtures.

CHAPTER 37 HOW REACTIVE ARE METALS?

One of the gold and silver treasures of the National Museum

We know from our experiments that the alkali metals and alkaline earth metals are very reactive. Potassium is the most reactive and sodium is the next most reactive, followed by calcium and magnesium. You would hardly expect to find lumps of sodium and potassium in the ground. Reactive metals like these are found only as metal compounds or ores. In fact most metals are reactive and are found as ores, often mixed with other metal ores in rock formations.

Gold is so unreactive that it is found free in nature in the form of nuggets. Silver and copper are sometimes found free in nature. These metals are very unreactive and do not form compounds easily. Because gold, silver and copper are unreactive, they are often used to make coins and jewellery. Gold jewellery many thousands of years old can be seen in the National Museum. Items of iron more than a few hundred years old are rarely found, as most of them have rusted away.

Reactive metals are never found free in nature; unreactive metals are often found free in nature.

We see from the following experiment why copper was one of the first metals used by humans.

251

BREAKTHROUGH SCIENCE

DEMONSTRATION EXPERIMENT 37.1
AIM: TO MAKE COPPER FROM COPPER ORE AND CHARCOAL

Method
1. Mix some copper (II) oxide (this is a copper ore) with charcoal and put it into a test-tube.
2. Heat the test-tube strongly. Bubble any gases formed through fresh limewater.

Result: A coppery-coloured substance is formed in the test-tube, and the limewater turns milky.

Conclusion: Copper (II) oxide reacts with charcoal when heated to produce copper metal and carbon dioxide.

$$2CuO + C \rightarrow 2Cu + CO_2$$

| copper (II) oxide | charcoal (carbon) | copper | carbon dioxide |

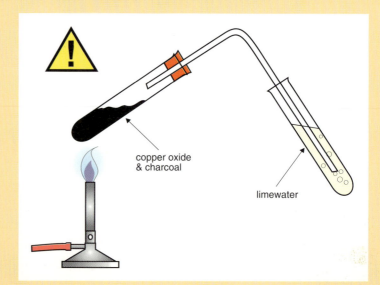

The Bunsen burner does not produce a very hot flame, so copper can be made from copper ore with ordinary fires. Many other metals are produced by a similar reaction in a furnace.

ACTIVITY SERIES OF THE METALS

You already know the order of reactivity of potassium, sodium, calcium and magnesium from your study of the alkali metals and the alkaline earth metals. We shall now compare the elements copper, iron, silver and zinc to place them in order of their reactivity.

Molten steel glowing inside a furnace

CHEMISTRY · HOW REACTIVE ARE METALS?

DEMONSTRATION EXPERIMENT 37.2
AIM: TO DETERMINE THE ORDER OF ACTIVITY OF IRON AND ZINC

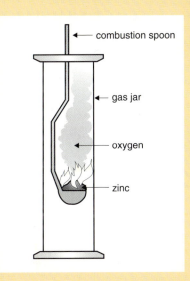

Method
1. Prepare a few gas jars of oxygen.
2. Put some zinc dust into a combustion spoon. Heat the zinc dust.
3. Plunge the combustion spoon into a jar of oxygen. The zinc dust burns with a brilliant flame.
4. Add dilute hydrochloric acid to a small sample of zinc dust.
   Result: The zinc reacts vigorously with the acid to give hydrogen gas and zinc chloride.
5. Repeat steps 2–4 with iron dust. Iron glows red in oxygen. It reacts slowly with the acid.
   Conclusion: Zinc is more reactive than iron. Similar tests with copper and silver show these to be less reactive than zinc or iron with copper the more reactive of the two.

We can now write down our table of metal elements showing their activity:

Most reactive	Potassium
	Sodium
	Calcium
	Magnesium
	Zinc
	Iron
	Copper
Least reactive	Silver

Displacement of a metal from a solution by a more active metal

EXPERIMENT 37.3
AIM: TO SHOW THAT A MORE ACTIVE METAL DISPLACES A LESS ACTIVE METAL FROM A SOLUTION OF ITS SALT

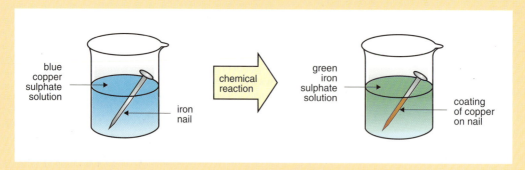

Method
1. Prepare small test pieces of iron and copper. Clean the metal surface with fine sandpaper and wash with methylated spirits and distilled water.
2. Make up solutions of copper (II) sulphate and iron (II) sulphate.
3. Place the piece of iron in a test-tube of copper (II) sulphate solution.
   Result: Copper forms in the test-tube.
4. Place the piece of copper in the test-tube of iron (II) sulphate. Nothing happens.
   Conclusion: The more active metal, iron, displaces the ions of the less active metal, copper, from the solution of copper (II) sulphate. Iron enters the solution as iron ions, and copper ions leave the solution as copper metal.
   copper sulphate + iron → iron sulphate + copper

You could use experiments similar to this to establish the order of reactivity of metals by testing a sample of one metal in the salt solution of another.

A more active metal displaces a less active metal from a solution of its salt.

CHEMISTRY · HOW REACTIVE ARE METALS?

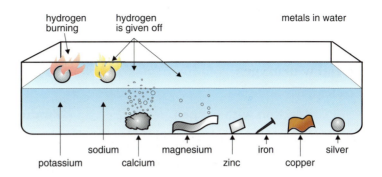

Reaction of metal with water

Metal	Reaction with water	Products
Potassium	Violent with cold water	Hydrogen and potassium hydroxide
Sodium	Violent with cold water	Hydrogen and sodium hydroxide
Calcium	Vigorous with cold water	Hydrogen and calcium hydroxide
Magnesium	Vigorous with hot water	Hydrogen and magnesium oxide
Zinc	Steam reacts with hot zinc	Hydrogen and zinc oxide
Iron	Steam reacts with hot iron	Hydrogen and iron hydroxide
Copper	No reaction	
Silver	No reaction	

Reaction of metal with hydrochloric acid

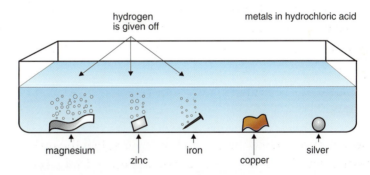

Metal	Reaction with hydrochloric acid	Products
Potassium	**WARNING: explosive!!**	Hydrogen and potassium chloride
Sodium	**WARNING: explosive!!**	Hydrogen and sodium chloride
Calcium	**WARNING: explosive!!**	Hydrogen and calcium chloride
Magnesium	Very vigorous	Hydrogen and magnesium chloride
Zinc	Very vigorous	Hydrogen and zinc chloride
Iron	Reacts with acid	Hydrogen and iron chloride
Copper	No reaction	
Silver	No reaction	

Reaction of metal with oxygen

Metal	Reaction with oxygen	Products	pH of solution
Potassium	Burns with a crimson flame	Potassium oxide	14
Sodium	Burns with a yellow flame	Sodium oxide	13
Calcium	Burns with a red flame	Calcium oxide	10
Magnesium	Burns with a bright white flame	Magnesium oxide	8
Zinc	Burns with a bright flame	Zinc oxide	7
Iron	Glows brightly, forms a black powder	Iron oxide (insoluble)	
Copper	Does not burn, becomes coated with black powder	Copper oxide (insoluble)	
Silver	No reaction		

SUMMARY

- Potassium and sodium are found only in compounds because they are **reactive**. They are **never** found free in nature.
- Gold, silver and copper are found free in nature because they are **unreactive**. They are used in coins and jewellery.
- A more active metal displaces the ions of a less active metal from a solution of its salt.

QUESTIONS

Section A

1. The _____ metals are never found free in nature.
2. Gold, silver and copper are used to make coins because _____.
3. Calcium reacts with water to give _____ and _____.
4. Magnesium burns in oxygen to give _____.
5. Underline the element in the following list that is never found free in nature, and give a reason for your choice. Gold, sodium, silver, copper.
6. Zinc reacts with steam to give _____ and _____.
7. Name a metal that can displace copper from copper sulphate solution.
8. Iron reacts with hydrochloric acid to give _____ and _____.
9. Calcium burns in oxygen to give _____.
10. Sodium reacts with water to give _____ and _____.

Section B

1. The following is a list of metals in decreasing order of their reactivity: sodium, calcium, magnesium, zinc, iron, copper.
   (a) (i) Name the metal that will react most vigorously with cold water. Name one product of the reaction.
   (ii) Which metal will not react with dilute hydrochloric acid?
   (b) One of these metals is soft and silvery, and was stored under oil. When dropped onto cold water it reacted vigorously, and a gas was given off which burned with a "pop".
   (i) Name the metal.
   (ii) Name the gas given off.

2. The table summarises some properties of three metals, X, Y and Z.

	Metal X	Metal Y	Metal Z
Occurrence	Never found free in nature	Rarely found free in nature	Often found free in nature
Reaction with water	Very vigorous, gas A produced	When heated, reacts with steam to produce gas A	No reaction
Reaction with dilute acid	Explosive, gas A produced	Gas A produced	No reaction

   The three metals are copper, zinc and sodium, not necessarily in that order.
   (a) Identify the three metals X, Y and Z.
   (b) Name the gas A. Explain why metal X is never found free in nature.

3. List the following metals in order of **increasing** reactivity: zinc, copper, potassium, calcium.
   (i) Give the chemical equation for the reaction of calcium with water.
   (ii) Give the chemical equation for the reaction of copper with oxygen.
   (iii) What is observed when dilute sulphuric acid is added to zinc?

4. When a piece of zinc is added to a solution of copper sulphate, some brown copper falls to the bottom of the beaker.
   (i) Which is the more reactive, copper or zinc?
   (ii) What would happen if you added a piece of magnesium to a solution of zinc sulphate?
   (iii) Describe a simple experiment to establish the order of reactivity of iron, copper, magnesium and zinc.

5. The following observations were recorded during an experiment.

Element	Behaviour	Reaction with cold water	Reaction with dilute hydrochloric acid
A	Sinks	Fast reaction, gas released, white solid formed	Dissolves rapidly, gas released
B	Floats	Vigorous reaction, violet flame	Not attempted
C	Sinks	None observed	Dissolves rapidly, gas released
D	Floats	Fast reaction, hissing sound, yellow sparks	Not attempted

(i) Given that the four elements are magnesium, potassium, sodium and calcium, identify A, B, C and D and place them in order in an activity series.

(ii) Write a balanced equation for the reaction of A with cold water.

(iii) Write a balanced equation for the reaction of C with dilute hydrochloric acid.

6. (i) Describe a simple experiment to show that calcium reacts with water. Use the headings Aim, Method, Result, Conclusion, and draw a diagram of the apparatus.

(ii) Write a balanced equation for the reaction of calcium with water.

(iii) List the following elements in order of increasing reactivity:

copper, calcium, iron, gold, sodium.

ELECTRICITY FROM CHEMICALS CHAPTER 38

When you buy a 1·5 volt battery for your personal stereo, you are really buying an electric cell. A battery is made up of a number of these cells connected to give you 3 volts, 4·5 volts or 6 volts.

An electric cell produces the voltage that pushes the electrons round a circuit.

EXPERIMENT 38.1
AIM: TO MAKE A SIMPLE ELECTRIC CELL

Method
1. Connect up the circuit as shown.
2. Place the strip of copper and the strip of zinc in a beaker of dilute hydrochloric acid (or copper sulphate solution). Read the voltage from the voltmeter. What do you notice happening to the voltage after a few seconds?

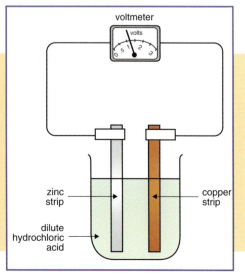

EXPERIMENT 38.2
AIM: TO COMPARE THE VOLTAGES PRODUCED BY DIFFERENT COMBINATIONS OF METALS IN A SIMPLE CELL

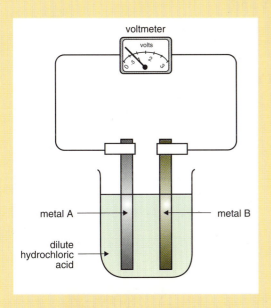

Method
1. Connect up the circuit as shown.
2. Place a piece of metal A and a piece of metal B (see the table below) in the beaker of dilute hydrochloric acid. Read the voltage from the voltmeter.
3. Repeat with the other combinations of metals listed. Enter your results in the table.

Metal A	Metal B	Voltage produced
Copper	Zinc	
Copper	Iron	
Copper	Lead	
Zinc	Iron	
Zinc	Lead	
Iron	Lead	

What do you notice?

Conclusion: The further apart the two metals are on the activity series of the metals, the greater is the voltage produced. The metal pieces are called electrodes. The ionic solution between them is the electrolyte.

An electrolyte is a solution that conducts electricity by the movement of ions.

Thomas Alva Edison (1847–1931) was an American inventor. His many inventions included the first practical electric light bulb, the record player, the film projector and the alkaline storage battery.

BREAKTHROUGH SCIENCE

Would you use potassium and silver electrodes to get a very big voltage? Of course not! Potassium is much too dangerous and silver is too expensive. Your choice of electrodes depends on factors other than the voltage produced between them. Scientists are working to make new types of cell to give more electricity and be lighter and cheaper. Scientists must also develop safe and "environment friendly" batteries.

HOW DOES THE SIMPLE CELL PRODUCE ELECTRICITY?

You already know that zinc is more reactive than copper. When zinc and copper are both placed in sulphuric acid (electrolyte), some zinc atoms (at the negative electrode) each give up two electrons and go into the electrolyte as zinc ions (Zn^{2+}). The zinc electrode has a surplus of electrons on it, and is negatively charged.

Hydrogen ions (H^+) in the electrolyte each gain one electron from the copper electrode. These hydrogen atoms leave the solution by sticking to the positive electrode. The copper electrode is now positively charged.

When the positive copper electrode is connected to the negative zinc electrode, the electrons flow through the connecting wire. This flow of electrons is electricity.

The more reactive metal is always the negative electrode of the cell.

MAKING A BATTERY: THE DRY CELL

Many millions of **dry cells** are used in Ireland every year. You probably use some yourself in a radio or personal stereo or torch. These cells are sometimes called zinc–carbon cells, because these two electrodes are used to make them.

The mixture of manganese dioxide and charcoal that surrounds the carbon rod prevents the voltage of the cell from dropping after a few minutes of use. The dry cell has a voltage of 1·5 volts. A number of cells—called a battery—are joined in series to give bigger voltages—3 V, 4·5 V, 6 V, etc. Each unit of the battery is a cell. Dry cells last well in radios and low-power devices. They are used up quickly in personal stereos and equipment with motors, because the motor takes a lot of energy to work it.

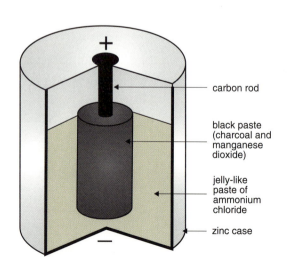

ELECTROLYSIS: USING ELECTRICITY TO BRING ABOUT A CHEMICAL CHANGE

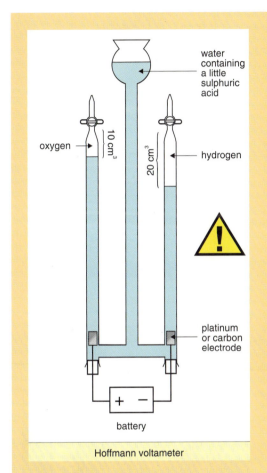

Hoffmann voltameter

EXPERIMENT 38.3
AIM: TO BREAK DOWN WATER INTO ITS ELEMENTS (ELECTROLYSIS OF WATER)

Method
1. Set up a Hoffmann voltameter. Add a few drops of sulphuric acid to the water. Connect to a battery or d.c. supply.
2. Pass an electric current through the voltameter for about 30 minutes.

Result: Collect the gas from the positive electrode. The gas relights a glowing splint—it is oxygen. Collect the gas from the negative electrode. The gas burns with a blue flame and a slight "pop"—it is hydrogen.

Conclusion: This experiment shows that water is a compound of hydrogen and oxygen.

> *Electrolysis is the chemical decomposition of an electrolyte by the passing of an electric current through it.*

The volume of hydrogen produced is twice that of oxygen. We conclude that water contains two atoms of hydrogen to one atom of oxygen.

BREAKTHROUGH SCIENCE

Electroplating with copper

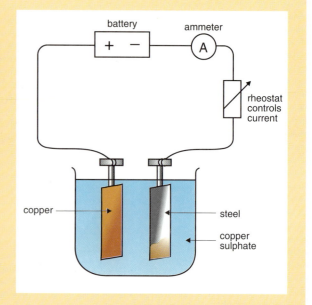

EXPERIMENT 38.4
AIM: TO COPPER-PLATE A PIECE OF STEEL
Method
1. Prepare the steel by rubbing it with emery paper and then cleaning with methylated spirits.
2. Place the steel strip and the copper strip in a beaker of copper sulphate. Connect the steel strip to the negative pole of the battery, and the copper strip to the positive pole.
3. Switch on the current and allow it to flow for about 30 minutes.

Result: The steel strip is now covered with a layer of copper—we say it is copper-plated. It is important to keep the current flowing small (about 0·5 A) to make sure that the copper plating sticks to the steel.

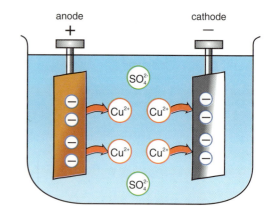

Copper sulphate is an ionic substance. When it dissolves in water the crystal breaks down into positive copper ions (Cu^{2+}) and negative sulphate ions (SO_4^{2-}).

At the negative electrode (cathode): Copper ions gain two electrons each and form a layer of copper on the cathode.

Cu^{2+}	+	$2e^-$	→	Cu
copper ions		2 electrons		copper atoms

At the positive electrode (anode): Copper atoms lose electrons and enter the solution as copper ions.

Cu	→	Cu^{2+}	+	$2\,e^-$
copper atoms		copper ions		2 electrons

Energy is needed to transfer the copper from one electrode to the other. This energy is supplied by the battery.

CHEMISTRY · ELECTRICITY FROM CHEMICALS

Uses of electroplating

Electroplating is used to coat a metal surface with a thin layer of another metal by electrolysis. It is done to improve the appearance of the metal or to protect it against corrosion. The surface of the item is plated with silver, nickel, chromium, zinc, gold or copper. If you see EPNS written on knives and forks it means **E**lectro **P**lated **N**ickel **S**ilver.

The metal being plated must be carefully prepared and the size of the current controlled in order to make good electroplated coatings.

SUMMARY

- A simple cell is made of any two metals and an electrolyte.
- Copper and zinc in copper sulphate solution is a simple cell.
- An electrolyte is a solution that conducts electricity by the movement of ions.
- The more reactive metal is always the negative pole of the cell.
- Electrolysis is the chemical decomposition of an electrolyte by the passing of an electric current through it.
- Electroplating is the coating of a metal surface with a thin layer of another metal by electrolysis.

QUESTIONS

Section A

1. What is an electrolyte? _____
2. A simple cell is made from _____.
3. What is electroplating? _____
4. The gases produced in the electrolysis of water are _____ and _____.
5. A spoon has the letters EPNS stamped on it. What does this mean? _____
6. The negative pole of a cell is always the _____ _____ metal.
7. In an electrolysis of water, the volume of hydrogen is 40 cm^3. The volume of oxygen is _____.
8. When a piece of metal is being copper-plated, a solution of _____ is used.
9. The positive role of a dry cell is made of _____.
10. Electrolysis is _____.

Section B

1. Describe a simple experiment to show how you could coat a nail with copper. Draw a diagram of the apparatus used in the experiment.
   (i) Is the nail at the anode or the cathode?
   (ii) What liquid would you use?
   (iii) Draw a diagram of the apparatus.
   (iv) What does the term "electroplating" mean?

2. The diagram shows an electric current being passed through acidulated water.
   (i) What does this experiment show about the composition of water?
   (ii) What gas collects in tube A? Describe the test you would do to identify this gas.
   (iii) Why is acid added to the water?
   (iv) In which tube, A or B, is the negative electrode?

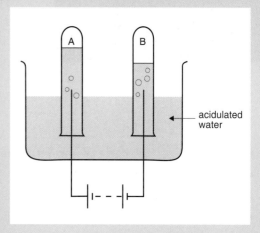

3. The diagram shows a simple cell.
   (i) What is the function of a simple cell?
   (ii) Name two metals that could be used in the simple cell.
   (iii) What term is used for the dilute sulphuric acid in the simple cell?

4. (*a*) A dry cell converts chemical energy into electrical energy. Draw a diagram of the dry cell. Name the substances used in the following parts of the cell:
   (i) the positive pole
   (ii) the negative pole.
   (*b*) Dry cells are connected in series to form batteries. How many dry cells would you combine to get a 6 V battery?

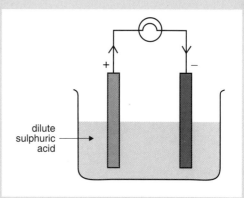

CHAPTER 39 OXIDATION AND REDUCTION

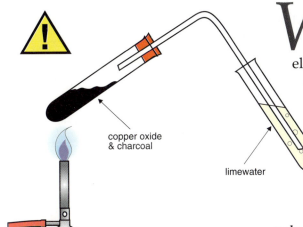
copper oxide & charcoal
limewater

When oxygen combines with an element to form an oxide, the reaction is called **oxidation**. All reactions of oxygen with metal and non-metal elements are oxidation reactions. You have done many experiments that involved burning things in air or in oxygen—all these are oxidation reactions.

When copper (II) oxide is heated in a stream of hydrogen gas, copper metal and water are produced. This process of changing a metal compound (metal ore) to the pure metal is called **reduction**. When you made copper metal by heating copper oxide with charcoal, you were reducing copper oxide to copper metal.

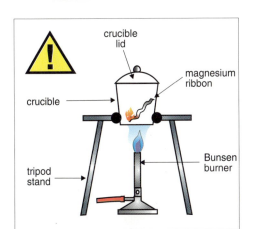
crucible lid
magnesium ribbon
crucible
Bunsen burner
tripod stand

OXIDATION
Magnesium combines with oxygen: magnesium is oxidised.
This happens in two stages:
(a) a magnesium atom loses two electrons to an oxygen atom
$$2Mg + O_2 \rightarrow 2Mg^{2+} + 2O^{2-}$$

(b) magnesium ions and oxygen ions combine
$$2Mg^{2+} + 2O^{2-} \rightarrow 2MgO$$

Magnesium—electrons lost—oxidised.
Oxygen—electrons gained—reduced.

A similar reaction is that of magnesium and chlorine.
$$Mg + Cl_2 \rightarrow Mg^{2+} + 2Cl^-$$
$$Mg^{2+} + 2Cl^- \rightarrow MgCl_2$$

Magnesium—electrons lost—oxidised.
Chlorine—electrons gained—reduced.

Any substance that loses electrons *in any reaction* is oxidised. We don't restrict oxidation to reactions involving oxygen.

| *Oxidation means a loss of electrons.*

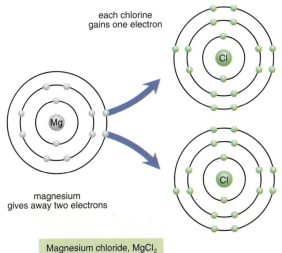
each chlorine gains one electron
magnesium gives away two electrons
Magnesium chloride, MgCl$_2$

BREAKTHROUGH SCIENCE

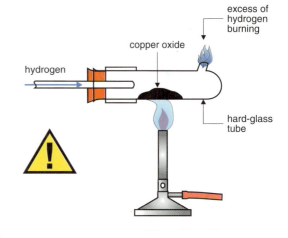

REDUCTION

Copper (II) oxide reacts with hydrogen to produce copper and water.

$Cu^{2+}O^{2-}$ + H_2 → Cu + H_2O
copper (II) oxide   hydrogen   copper   water

Hydrogen—electrons lost—oxidised.
Copper—electrons gained—reduced.

The copper ion in copper (II) oxide is a double positive ion. It takes a gain of two electrons to turn the dipositive copper ion into neutral copper atoms.

When you put a piece of zinc into copper (II) sulphate solution, the dipositive copper (II) ions are displaced as neutral copper atoms. The neutral zinc atoms enter the solution as dipositive zinc ions. In this reaction, copper is reduced even though the reaction does not involve the removal of oxygen.

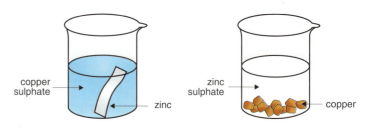

$Cu^{2+}SO_4^{2-}$ + Zn → Cu + $Zn^{2+}SO_4^{2-}$
copper (II) sulphate   zinc   copper   zinc sulphate

Zinc—electrons lost—oxidised.
Copper—electrons gained—reduced.

Any substance that gains electrons *in any reaction* is reduced.

Reduction means a gain of electrons.

Oxidation	Reduction
Is	Is
Loss of electrons	Gain of electrons

Whenever an element is oxidised, another element in the same reaction is reduced. This must be the case, as one element cannot lose electrons unless another substance gains those electrons. In each of the examples given, one element is oxidised and another is reduced.

Oxidation and reduction—you can't have one without the other!

SUMMARY

- Oxidation means loss of electrons.
- Reduction means gain of electrons.
- Oxidation   Reduction
  Is   Is
  Loss of electrons Gain of electrons

QUESTIONS

Section A
1. When a substance combines with oxygen it is _____.
2. Reduction means a _____ of electrons.
3. When magnesium burns in oxygen, magnesium is _____.
4. Copper oxide reacts with hydrogen to give copper and water. In this reaction, copper is _____ and hydrogen is _____.
5. Calcium reacts with oxygen to form calcium oxide

    $2Ca \quad + \quad O_2 \quad \rightarrow \quad 2CaO$

    In this reaction, calcium is _____.
6. Oxidation is the _____ of electrons.
7. Copper oxide reacts with charcoal to give copper and carbon dioxide. In this reaction, copper is _____ and charcoal is _____.
8. Magnesium reacts with hydrochloric acid to release hydrogen gas and form magnesium chloride. In this reaction, magnesium is _____.
9. When calcium reacts with chlorine to form calcium chloride (an ionic bond), calcium is _____ and chlorine is _____.
10. When magnesium is added to a solution of zinc sulphate the zinc falls out of solution and is replaced by magnesium ions. In this reaction, _____ is oxidised and _____ is reduced.

Section B
1. When zinc is added to a solution of copper sulphate, the copper falls out of solution and the zinc goes into solution as zinc sulphate.
    (i) Explain why this reaction takes place.
    (ii) Write a balanced equation for this reaction.
    (iii) Which substance undergoes (a) oxidation, (b) reduction in this reaction? Explain your answers in terms of electron transfer.
2. What is meant by (i) oxidation, (ii) reduction? Sodium reacts with chlorine to form sodium chloride. State the substance oxidised and the substance reduced in the following chemical reaction.

    $2Na \quad + \quad Cl_2 \quad \rightarrow \quad 2NaCl$

    Explain your answer in terms of electron transfer.

3. (i) Describe a simple experiment to show the reduction of a metal ore to the pure metal. Use the headings Aim, Method, Result, Conclusion and a draw a diagram of the apparatus used.

   (ii) When one substance is oxidised, another substance is always reduced. Use the experiment in (i) to explain why this is true.

4. When silver nitrate is added to salt solution (sodium chloride), a white substance (silver chloride) and a solution of sodium nitrate are formed. The following equation describes the reaction.

   $Ag^+Cl^- \quad + \quad Na^+NO_3^- \quad \rightarrow \quad Ag^+Cl^- \quad + \quad Na^+NO_3^-$

   Is this reaction an oxidation-reduction one? Explain your answer in terms of electron transfer.

5. Given that the relative position of these metals in the activity series is as follows

   Na Ca Zn Fe Cu Ag

   state what you would expect to observe when a piece of copper metal is placed in a solution of silver nitrate. Give an equation for the reaction that takes places. State which substance undergoes (i) oxidation, (ii) reduction, in this reaction.

SECTION 3

BIOLOGY

STARTING BIOLOGY CHAPTER 40

Biology is the study of living things (organisms). In other words, it is the study of plants and animals. You know what plants are, but do you know what animals are? Is there an animal in the room with you at the moment? What do you mean, "don't be silly"? A fly is an animal, a bee is an animal, a human being is an animal. What you must remember is that there are many different kinds of animals. Elephants, people, birds and bees are all animals.

CHARACTERISTICS OF LIVING THINGS

Look around you. How many living things can you see? I'll give you a start. Can you see other people? A fly? A tree? How many non-living things can you see? The plastic in your biro, the glass in the window, the metal door-handle; these are all non-living things. (Non-living is not the same as dead: non-living things were never alive.) But how can we tell the difference between living and non-living things? Is it always this easy?

Scientists say that living things have certain features which make them different from non-living things. These are known as the characteristics of living things.

1. **Feeding**

    All living things need food so that they can get energy and grow. Plants can make their own food; animals get food by eating plants or by eating animals that have already eaten plants.

2. **Respiration**

    The chemical energy contained in food has to be released. Respiration is the way that living things release energy from food.

3. **Movement**

    Animals move from place to place by walking, running, flying, etc. Plants cannot move from place to place; they can only move parts of their bodies. For example, the petals of a flower open during the day and close up at night.

4. **Sensitivity**

    All living things are aware of changes taking place around them. Animals respond quickly to sound, light, etc. Plants also respond to light, but they do it more slowly.

Feeding

Respiration

Movement

BIOLOGY · STARTING BIOLOGY

Sensitivity

Excretion

5. **Growth**

   All living things grow. Grass grows quickly, trees grow slowly; but of course trees grow to be much taller than grass.

6. **Reproduction**

   All living things produce others like themselves. If they could not do this they would die out.

7. **Excretion**

   All living things produce waste and must get rid of it from their bodies. This is called excretion.

Growth   Reproduction

THE IMPORTANCE OF ANIMALS

Agriculture

In agriculture, animals are raised as a source of food and clothing.

Animals can also do harm. Rabbits, pigeons and locusts attack crops. Some animals damage other animals. Liver fluke and warble fly damage cattle.

Medicine

Animals such as monkeys, rabbits and rats are used to test new drugs. Vaccines can also be extracted from animals.

On the other hand, some animals help spread disease. Mosquitoes carry malaria, cows carry brucellosis, and the badger is being blamed for the spread of bovine TB, although some people dispute this.

Leisure

Many people keep cats and dogs as pets. Children love them and they provide great companionship for those that live alone. Some people like to visit zoos and wildlife parks. Others like horse racing or a trip to the dog track.

Mariculture

Mariculture means using and cultivating the natural resources of the sea. The waters around our coast are rich in mackerel, herring, whiting, etc., but due to overfishing the EU has had to impose quotas on each country in order to conserve stocks. On the other hand there has been a great increase in shellfish farming, particularly along our west coast.

IMPORTANCE OF PLANTS

Agriculture

A great variety of plants, such as wheat (flour), vegetables and fruit are grown for food. Grass provides food for animals, which in turn provide food for us.

Medicine

Plants have been used to heal the sick for thousands of years. Many useful drugs such as painkillers and antibiotics are extracted from

SUMMARY

- The characteristics of living things are feeding, respiration, movement, sensitivity, growth, reproduction and excretion.
- Animals are useful to humans for food, clothing, medicine, sports and leisure.
- Plants are useful to humans for food, clothing, medicine, paper and timber.

plants. Morphine, one of the most widely used of all painkillers, is extracted from the white poppy plant. The antibiotic penicillin is extracted from the penicillium fungus.

People benefit from plants in other ways. It is very pleasant to sit in a garden full of flowers. Streets lined with trees are very pleasant to walk on.

Carolus Linnaeus (1707–1778), a Swedish botanist, developed a system of names for classifying plants, animals and micro-organisms. He used the name *Homo sapiens* to classify humans.

QUESTIONS

Section A

1. Give two characteristics of living things. _____, _____
2. If someone sticks a pin in you, you jump. What characteristic of living things are you showing? _____
3. Underline the non-living things in this list: a cat, a rosebush, a dead bird, a glass, a stone.
4. Underline all the animals in this list: tree, cat, frog, stone, elephant, wasp.
5. Underline three animal products in this list: cotton, wool, silk, mahogany, leather, plastic.
6. Apart from food, give two ways in which animals are useful to humans. _____, _____
7. Name two animals that attack crops. _____, _____
8. Name two animals that harm other animals. _____, _____
9. The drug _____ is extracted from the _____ plant.
10. Underline the plant products in this list: wool, paper, glass, linen, ivory.
11. What industry do you associate with (a) barley, _____ (b) flax? _____
12. What does the word "mariculture" mean ? _____

Section B

1. Did you ever hear the word "respiration" before? What did you think it meant? Do you know what it means now?
2. What does "sensitivity" mean? You can read this question because you are sensitive to what? Name four other things you are sensitive to.
3. Why is it necessary for animals to be able to move around from one place to another? Why do plants not need to move around in this way?
4. List the seven characteristics of living things. A car releases chemical energy from petrol, produces waste gases and moves. Does this mean that a car is alive? Explain your answer.

CHAPTER 41 CELLS

The Dutch biologist Anton van Leeuwenhoek (1632–1723) made simple microscopes as a hobby. Leeuwenhoek made over 400 microscopes, many of which still exist. The most powerful of these instruments can magnify objects about 275 times. Leeuwenhoek was the first person to observe single-celled animals (protozoa), red blood cells and sperm cells with a microscope.

All living things are made up of cells. Some living things, such as bacteria, consist of a single cell. Your body contains millions of cells. Plant cells and animal cells are not quite the same, as we shall see if we examine them under a microscope.

USING A MICROSCOPE

1. Clean the lenses with a lens tissue.
2. Switch on the microscope light.
3. Place a drop of water on a clean glass slide.
4. Place a very small piece of newspaper on top of the drop of water.
5. Lower a coverslip gently over the paper so as not to trap any air bubbles.
6. Clip the slide into position on the stage so that the paper is in the centre of the opening.
7. Position the low-power objective lens over the slide and use the coarse adjustment to bring the lens as close to the slide as possible without touching it (careful!).
8. Looking through the eyepiece, use the coarse adjustment to move the lens slowly away from the slide until the image is clearly focused.
9. You can now position the medium- and high-power lenses over the slide and use the fine adjustment to focus. If you have time, examine thread, human hair, etc. to get used to using a microscope.

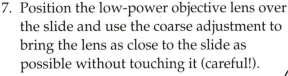

LOOKING AT PLANT CELLS

1. Peel off a tiny piece from one of the inner layers of an onion's skin.
2. Spread the onion skin out flat on a slide and add a drop of iodine. (The iodine stains the nuclei of the cells orange, making them easier to see.)
3. Carefully lower a coverslip onto the stain so as not to trap air bubbles.

BREAKTHROUGH SCIENCE

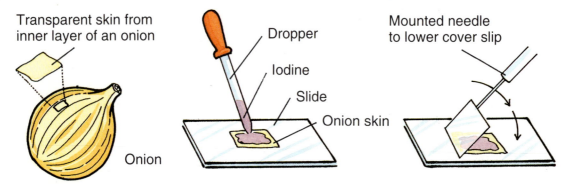

4. Soak up excess stain with filter paper.
5. Prepare a second slide as a control using water instead of iodine.
6. Examine the slides under the microscope and draw a sketch of what you see.

LOOKING AT ANIMAL CELLS

1. Scrape your index finger along the inside of your cheek.
2. Smear the liquid you get onto a clean glass slide.
3. Add one or two drops of methylene blue stain to the smear, wait 5 minutes and rinse off the stain with water.
4. Wait for the slide to dry and add a coverslip carefully before examining it under a microscope.
5. Prepare a second slide as a control using water instead of methylene blue.

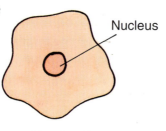

COMPARISON OF PLANT AND ANIMAL CELLS

These are typical plant and animal cells. Can you spot any differences?

Have you ever wondered why most plants, such as grass and trees, are green? A very important difference between plant cells and animal cells is that plant cells contain **chloroplasts**. The chloroplasts contain a green chemical called **chlorophyll** which plants use to make food. Animals cannot make food because their cells do not contain chloroplasts.

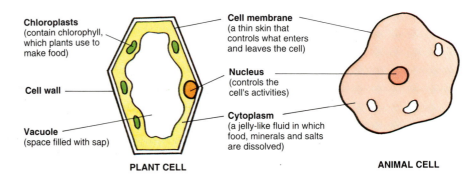

276

Plant cells	Animal cells
1. Cellulose walls	1. No cellulose walls
2. Large vacuoles	2. Small vacuoles
3. Chloroplasts	3. No chloroplasts

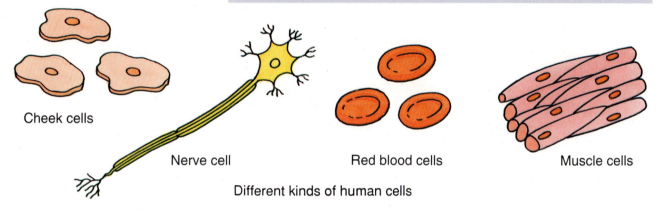

Different kinds of human cells

TISSUES

Your body contains many different kinds of cells which do different jobs, e.g. blood cells, skin cells, nerve cells and muscle cells.

Muscle tissue is composed of muscle cells which can help your body to move, whereas nerve tissue is made up of nerve cells which can carry messages from one part of your body to another.

A group of similar cells with a special function is called a tissue.

ORGANS

Your body has many organs. Your nose, eyes and heart are all organs. Each organ has a special job. What job has the nose? Your nose contains bone tissue, skin tissue and blood tissue.

An organ consists of a group of tissues working together to perform a special function.

SYSTEMS

Your body is made up of a group of different systems, e.g. your digestive system, your nervous system and your blood system. Each system consists of a group of organs working together. For example, your digestive system consists of your stomach, your intestines, etc.

A group of organs working together forms a system.

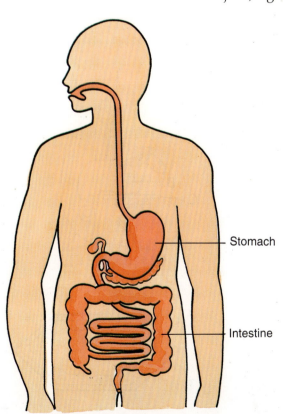

The digestive system is made up of the stomach, intestines, etc.

SUMMARY

- All living things are made of cells.
- Cell membrane: a thin skin that controls what enters and leaves the cell.
- Cytoplasm: a jelly-like fluid in which food, minerals and salts are dissolved.
- Nucleus: controls the activities of the cell.
- Vacuoles: spaces filled with sap, i.e. sugar and salts dissolved in water.

Plant cells	Animal cells
1. Cellulose walls	1. No cellulose walls
2. Large vacuoles	2. Small vacuoles
3. Chloroplasts	3. No chloroplasts

- Chloroplasts contain a green chemical called chlorophyll which plants use to make food.
- A tissue is a group of similar cells with a special function.
- An organ consists of a group of tissues working together to perform a special function.
- A group of organs acting together forms a system.

QUESTIONS

Section A

1. The top diagram shows a microscope. Name the parts marked A, B and C. _____
2. What does 10x stamped on the objective lens of a microscope mean? _____
3. State three ways in which a plant cell differs from an animal cell. _____
4. Why is iodine added to the slide of an onion cell?
5. Animal cells have no _____ and no _____
6. In the plant cell in the bottom diagram, name A and B. _____
7. Name the green substance in plant cells that enables them to make food. _____
8. Apart from skin tissue, name two types of tissue in the human body. _____
9. Name three organs in the human body.

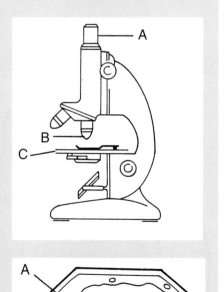

10. Groups of similar cells form _____
11. Which part of the cell controls the activities of the cell?
    _____
12. What is the function of the cell membrane?
    _____

Section B
1. Describe how you would prepare plant cells for examination under the microscope. Why would you use iodine stain?
2. What effect does methylene blue have on a cheek cell?
3. A person with poor eyesight wants to read the small print in a newspaper. Which would be better, a magnifying glass or a microscope? Explain your answer.
4. Draw a diagram of a plant cell, and label (*a*) the cell wall, (*b*) the nucleus, (*c*) a vacuole, (*d*) cytoplasm.
5. Give the function of each of the following parts of a cell: membrane, nucleus, vacuole, cytoplasm.
6. Do plant cells have cell walls? Why, or why not? Do animal cells have cell walls? Why, or why not?
7. Underline the parts in the following list that all cells have in common: cell wall, membrane, nucleus, cytoplasm, chloroplasts. Explain the function of each part you have underlined.
8. Match each item in column A with an item in column B.

Column A	Column B
Membrane	Controls the activities of the cell
Tissue	A very thin skin
Organ	A space in the cytoplasm filled with sap
Nucleus	Jelly-like fluid
Cytoplasm	Building blocks of life
Vacuole	Made of many cells
Cells	A group of tissues
System	A group of similar cells
Multicellular	A group of organs acting together

FOOD CHAPTER 42

Do you remember the characteristics of living things from Chapter 40? If you do, you know that all living things need food. For example, our bodies need food for heat, movement, growth, repair and protection against disease. Plants can make their own food because their cells contain chlorophyll. Animals, such as humans, get food by eating plants or by eating other animals.

Humans eat a great variety of food, but the five major constituents of the food we eat are carbohydrates, proteins, fats, vitamins and minerals. These are called **nutrients**. The main sources of these nutrients, and their functions, are given in the following table:

Food type	Source	Function
Carbohydrate	Bread, potatoes, sugar	Provides energy quickly
Protein	Lean meat, fish, cheese, milk	Growth and repair
Fat	Butter, margarine	Provides energy slowly
Vitamin C	Oranges, potatoes	Healthy skin and gums
Calcium (a mineral)	Milk, cheese, yoghurt	Strong bones and teeth
Fibre	Wholemeal wheat, bran	Prevents constipation

Fats contain twice as much energy as other foods, but the rate of release of energy from fat is lower. Fat also provides the body with insulation, while vitamin C is taken to avoid colds.

A BALANCED DIET

It is not good for us to eat too much of one kind of food. We should have a balanced diet. A balanced diet contains the correct amount of carbohydrates (sugar and starch), proteins, fats, vitamins, minerals, water and fibre (roughage). But what is the correct amount? The amount you need depends on your age, sex, size, lifestyle and general health. Would a pregnant woman need the same diet as a bodybuilder? Of course not!

DEFICIENCY

A person that does not eat a balanced diet could suffer from a deficiency disease. For example, in 1993 a girl in Northern Ireland was found to be suffering from scurvy, which is very rare in this

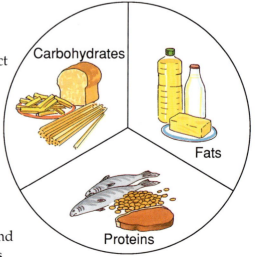

NUTRITION INFORMATION			
		Typical value per 100g	Per 30g Serving with 125ml of Semi-Skimmed Milk
ENERGY	kJ	1550	700 *
	kcal	370	170
PROTEIN	g	8	7
CARBOHYDRATE	g	83	31
(of which sugars)	g	(8)	(9)
(starch)	g	(75)	(22)
FAT	g	0.7	2.5 *
(of which saturates)	g	(0.2)	(1.5)
FIBRE	g	1.0	0.3
SODIUM	g	1.1	0.4
VITAMINS:		(%RDA)	(%RDA)
VITAMIN D	µg	2.8 (55)	0.9 (17)
THIAMIN (B_1)	mg	1.2 (85)	0.4 (30)
RIBOFLAVIN (B_2)	mg	1.3 (85)	0.6 (40)
NIACIN	mg	15 (85)	4.6 (25)
VITAMIN B_6	mg	1.7 (85)	0.6 (30)
FOLIC ACID	µg	333 (165)	110 (55)
VITAMIN B_{12}	µg	0.8 (85)	0.8 (75)
IRON	mg	7.9 (55)	2.4 (17)

part of the world. The reason was a deficiency in vitamin C. Her condition would have been worse but for the fact that there was a certain amount of potato in her diet. Potato is the main source of vitamin C in the Irish diet, though other countries rely on citrus fruit such as oranges.

ENERGY

One of the reasons we eat food is to get energy. Most food packets give details of the energy content. For example, 100 g of cornflakes provides 1550 kilojoules of energy. The energy in food is measured in kilocalories (kcal) or kilojoules (kJ).

How much energy do you need per day? To give you an idea, the energy required per day by a 15-year-old is roughly as follows:

Boy	12,600 kJ
Girl	9,600 kJ

SUMMARY

- Our bodies need food for energy, growth and repair and to fight disease.
- A balanced diet consists of the correct amounts of carbohydrates, proteins, fats, fibre, vitamins, minerals and water.

QUESTIONS

Section A
1. Name four different food types. _____
2. State two reasons why we need to eat protein. _____
3. Name two foods rich in fat. _____
4. Name two foods containing starch. _____
5. Name two good sources of fibre. _____
6. Why do we need fibre in our diet? _____
7. What is meant by a balanced diet? _____
8. What food type does sugar belong to? _____
9. For strong bones and teeth we need the mineral _____
10. Babies drink a lot of milk because _____
11. We drink orange juice because we need vitamin _____
12. Meat, fish, butter, eggs, cheese. From this list name a food that is (a) not a good source of protein, _____ (b) a good source of calcium. _____

13. Underline the food in the following list that would give a rapid supply of energy: milk, glucose, cheese, beef, fruit juice.
14. Apart from being a food type, what use has fat in the body? _____
15. Name one deficiency disease. _____

Section B
1. Choose one food from the drawing that has a large amount of (*a*) carbohydrate, (*b*) calcium. What is the function of fat in the human body? What is the function of protein in the human body?

2. What vitamin is contained in orange juice? What deficiency disease is caused by a lack of this vitamin? What is the main source of this vitamin in the Irish diet?
3. Examine the labels of three food packets used regularly in your home. Write down the names of the three foods, and say which contains the most (*a*) carbohydrates, (*b*) protein, (*c*) fat. Which provides the greatest amount of energy per 100 g?

4.

Food	% Carbohydrate	% Protein	% Fat	Vitamins	Energy (J/kg)
Bread	52·6	9·3	1·3	B1	11·1
Butter	—	1	85	A, D	33·3
Sugar	100	—	—	—	17·2
Milk	4·5	3·3	4	A, B, D	3
Bacon	—	9·9	67·4	B1	2·7
Egg	—	13·2	12	A, B1, B2, D	7·1

From the above table of breakfast foods, answer the following questions.
(*a*) What vitamin is not provided by any of these foods? What might you add to the above list to provide this vitamin?
(*b*) One of the foods in the table is often called a 'complete' food. Which one do you think it is? Would you agree with the description?
(*c*) Which food is not suitable for a small baby? Why?
(*d*) Which of these foods would you leave out if you were trying to lose weight? How would this affect the amount of energy in your breakfast? What could you do about this?

(e) Which of the foods would give you energy quickly?

(f) Which has the greatest percentage of energy? Is it easy for your body to release this energy from the food?

5. Fish is a very good source of protein, yet we in Ireland eat very little of it even though we are an island. Why do you think this is? Discuss this with your classmates.

6. Do some research to find out what the recommended daily intake of calcium is for *(a)* children, *(b)* adults. Explain the difference.

7. What does the body use calcium for? What does the body use iron for? Do some research and draw up a table to compare the calcium and iron requirements of a woman who is not pregnant and a woman who is. Explain the difference.

8. Keep a strict record of **everything** you eat for a single day. (This will involve weighing each item on a kitchen scale, so it is best done on a Saturday or Sunday.) Do some research to find out the energy value of 100 g of each food you have eaten or drunk, and calculate your total energy intake for the day. How does this compare with the average recommended for someone your age? If you take in more energy than you use, what do you think happens to it?

CHAPTER 43 THE DIGESTIVE SYSTEM

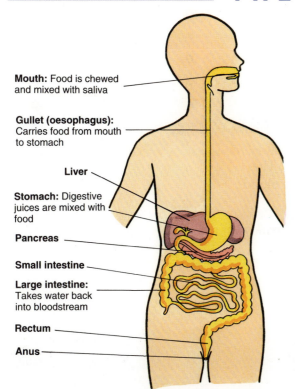

Mouth: Food is chewed and mixed with saliva

Gullet (oesophagus): Carries food from mouth to stomach

Liver

Stomach: Digestive juices are mixed with food

Pancreas

Small intestine

Large intestine: Takes water back into bloodstream

Rectum

Anus

Before we can use the food we eat, it must be chewed into small pieces and broken down into soluble form so that the nutrients can be absorbed into the bloodstream and carried round to the cells. This is the job of the digestive system.

DIGESTION

1. The breakdown of carbohydrates (starches and sugar) begins in the mouth, when the food is mixed with saliva and chewed.

2. The food is pushed from the mouth to the stomach by a muscular action called **peristalsis**.

3. The breakdown of protein begins in the stomach, when digestive juices are added and the food is churned round for several hours.

4. The food passes into the small intestine, where the fat is broken down and the nutrients are absorbed into the bloodstream and carried round to all the cells of the body.

5. Anything that has not been digested or absorbed passes into the large intestine, where it is dried out and stored in the rectum before being passed out of the body through the anus.

NUTRITION

Ingestion means taking food into the mouth.

Digestion means food is broken down by teeth in the mouth and by chemicals in the stomach and small intestine.

Absorption means the digested food passes out through the walls of the small intestine into the bloodstream.

Assimilation means the food is used by the body cells for energy, growth and repair.

Egestion means the undigested and unabsorbed material is passed out of the body.

LIVER AND PANCREAS

Although food does not pass through the liver or pancreas, they both play a part in digestion. The **liver** produces **bile**, which is used to break down fats. The **pancreas** produces **digestive enzymes**, which are chemicals that break down food.

DIGESTIVE ENZYMES

There are lots of chemical reactions going on inside your body. Many of these reactions would be much too slow were it not for the presence of catalysts. (A catalyst is a substance that changes the speed of a chemical reaction but is not used up in the reaction.) The catalysts in your body are called enzymes. We are concerned here only with digestive enzymes. A digestive enzyme is a protein that acts as a catalyst in breaking down food. Each digestive enzyme can break down only one kind of food. The enzyme **amylase** breaks down starch into a sugar called **maltose**. The enzyme **maltase** breaks this down further into a simple sugar called **glucose**. The glucose molecules are small enough to be absorbed into the blood stream and carried round to the body cells.

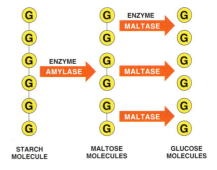

A digestive enzyme is a protein that acts as a catalyst in breaking down food.

The end products of digestion are given in the following table.

Food type	End product
Carbohydrate	Simple sugars
Protein	Amino acids
Fat	Fatty acids and glycerol

BIOLOGY · THE DIGESTIVE SYSTEM

EXPERIMENT 43.1
AIM: TO SHOW THE ACTION OF A DIGESTIVE ENZYME (AMYLASE)

(Note: Saliva contains the enzyme amylase)

Method
1. Pour starch solution to a depth of 2 cm into each of two test-tubes, A and B.
2. Add 2 cm of saliva to each tube, and shake to mix the contents.
3. Place both tubes in the water for 15 minutes.
4. Add a sample from test tube A to iodine solution on a dropping tile.
   Result: If starch were present it would turn black. It isn't.
5. Add 2 cm of Benedict's solution to test-tube B and bring the water to 90°C. The colour changes to red/orange, showing the presence of glucose.
   Conclusion: The enzyme amylase breaks down starch to sugar.

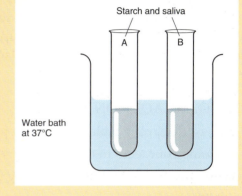

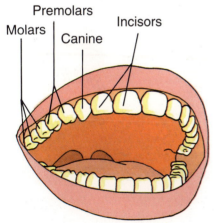

TEETH

An adult has 32 teeth, 16 in the upper jaw and 16 in the lower jaw. There are four different types of teeth.

1. **Incisors:** These are sharp teeth used for cutting food.
2. **Canines:** These are pointed teeth used for tearing food.
3. **Premolars:** These are fairly flat teeth used for grinding food.
4. **Molars:** These are larger than premolars and are also used for grinding food.

Structure

Although there are four different types of teeth, they all have the same basic structure, as shown in the diagram.

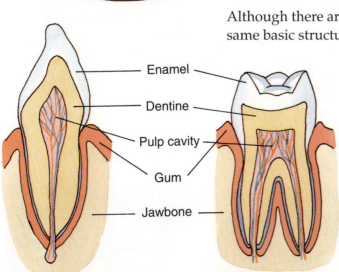

1. **Enamel:** A hard non-living substance which protects the tooth.
2. **Dentine:** A bone-like substance which is softer than enamel.
3. **Pulp cavity:** This contains living cells, blood vessels and nerves.
4. **Cement:** This holds the root of the tooth firmly in place.

Canine for tearing food   **Molar for grinding food**

BREAKTHROUGH SCIENCE

PLAQUE

Tooth decay is caused by a substance called plaque forming on your teeth. Plaque contains sugar and bacteria. If you eat sugary food such as sweets and chocolate, it sticks to your teeth. The bacteria in the plaque change this food to acid which eats through the enamel into the dentine and the pulp cavity. This allows bacteria to enter and cause infection. Plaque can also cause gum disease.

TO SHOW PLAQUE ON TEETH

1. Chew a disclosing tablet to stain the plaque on your teeth.
2. Examine your teeth in the mirror to see the amount of plaque.
3. Brush your teeth, and time how long it takes to remove the plaque.

CARING FOR TEETH

The simplest and most effective way to care for your teeth is to watch your diet. Try to avoid sugary food such as sweets and chocolate, which sticks to your teeth and forms plaque. Use a fluoride toothpaste (fluoride hardens the enamel). Brush each tooth thoroughly as shown in the diagram. If this fails to remove plaque from between the teeth use dental floss. The use of an anti-plaque mouthwash is also helpful. Fluorine is added to the public water supply to help prevent tooth decay.

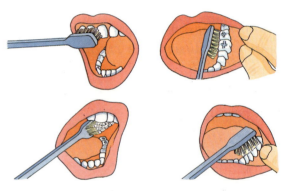

Make sure you brush all surfaces of your teeth thoroughly

SUMMARY

- Digestion means breaking down food so that it can be absorbed into the bloodstream.
- Peristalsis is a muscular squeezing action which keeps the food moving through the system.
- A digestive enzyme is a protein that acts as a catalyst in breaking down food.
- The enzyme amylase changes starch to sugar.

Food type	End products of digestion
Carbohydrate	Simple sugars
Protein	Amino acids
Fat	Fatty acids and glycerol

- The stages of nutrition are ingestion, digestion, absorption, assimilation and egestion.
- An adult should have 32 teeth.
- Plaque contains saliva and bacteria.

QUESTIONS

Section A

1. List the following in the order in which they occur in the human body: assimilation, digestion, absorbtion. _____
2. Where does the digestion of lean meat begin?
3. If you chew bread for a long time it begins to taste sweet. Why?
4. The liver produces _____, which is used to break down _____
5. The pancreas produces _____ and _____, which are chemicals that break down food.
6. A chemical that breaks down food in the human body is called an _____
7. Name the enzyme found in saliva that breaks down starch. _____
8. At the end of the digestive process, protein has been broken down into _____
9. At what temperature do human digestive enzymes work best? _____
10. Can you feel the enamel in your teeth decaying? _____
    Why? _____
11. What is plaque? _____
12. Underline which of the following is the normal number of pemanent teeth in an adult human: 20, 26, 28, 32.
13. A layer of bacteria on teeth is called _____
14. In the diagram, name the type of teeth labelled L, P, Q. 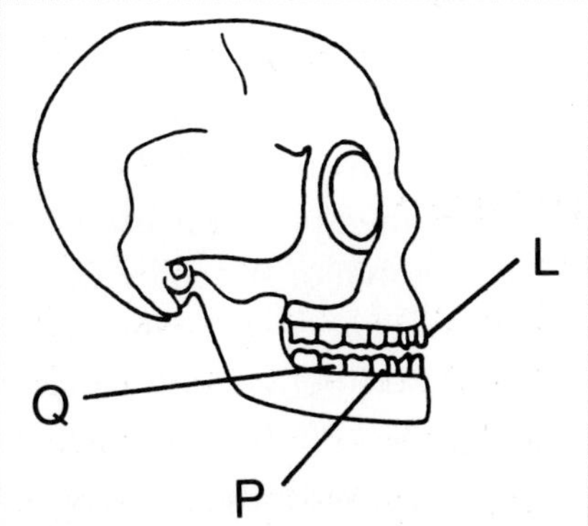
    _____
    _____
    _____
15. What chemical is added to the domestic water supply to protect teeth from decay? _____

Section B

1. Human nutrition occurs in the following five stages, but not in this order. **Assimilation, digestion, egestion, ingestion, absorption**. Give these stages in the order in which they occur. Name two of the parts labeled A, B, C, D in the diagram. Using the letters in the diagram, state where (i) ingestion, (ii) absorption occur. Explain what happens at the assimilation stage.

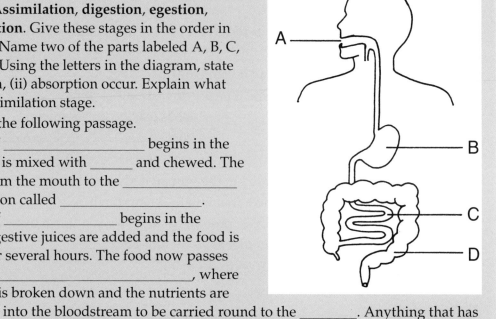

2. Fill in the gaps in the following passage.
   The breakdown of _____ begins in the mouth when food is mixed with _____ and chewed. The food is pushed from the mouth to the _____ by a muscular action called _____.
   The breakdown of _____ begins in the stomach when digestive juices are added and the food is churned round for several hours. The food now passes into the _____, where the _____ is broken down and the nutrients are _____ into the bloodstream to be carried round to the _____. Anything that has not been _____ or _____ passes into the _____, where it is dried out to be stored in the _____. It is then passed out of the body through the _____.

3. What is the purpose of digestion? Give the function of each of the parts labeled B, C, D in question 1.

4. If you take carbohydrates into your mouth, some of the carbohydrates will eventually reach the cells in your toes. Describe their journey in detail, using terms such as ingestion, enzyme, peristalsis, digestion, absorbtion and assimilation.

5. What is an enzyme? The diagram shows the apparatus used to show the breakdown of starch by the salivary enzyme amylase. Write an account of the experiment, using the headings Aim, Method, Result and Conclusion.

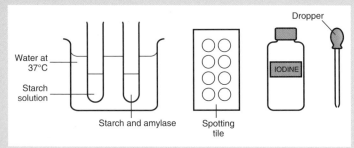

6. Describe, **with the aid of a diagram**, how starch is digested in the human body. Refer in your answer to two digestive enzymes and two sugars.

7. What are digestive enzymes? What happens to starch when it is digested? Glucose molecules don't have to be digested. Why? Do you understand now why athletes who need energy quickly take glucose drinks?

8. What enzyme is contained in saliva? In an experiment to show the action of this enzyme, why is the water bath at 37°C? What is the iodine used for? What is Benedict's solution used for?

9. Devise a simple experiment to show the effects of soft drinks, such as cola, on teeth. Use a control. Use the headings Aim, Method, Result and Conclusion to describe the experiment.

CHAPTER 44 RESPIRATION AND BREATHING

When food is absorbed into the bloodstream it is carried round to the cells, where energy is released from the food in a process called respiration.

Respiration is the release of energy from food.

The equation for respiration, in words, is

glucose + oxygen ⟶ carbon dioxide + water + energy

The balanced chemical equation for respiration is

$$C_6H_{12}O_6 + 6O_2 \longrightarrow 6CO_2 + 6H_2O + \text{energy}$$

RESPIRATION AND BREATHING

It is important to realise that respiration and breathing are not the same thing, although they are closely connected. Respiration is the release of energy from food. Like all living things, we must carry out respiration or we will die. As you can see from the equation for respiration, oxygen is necessary and waste carbon dioxide gas is produced, so we must have some means of taking in oxygen and getting rid of carbon dioxide. This is the job of the breathing system.

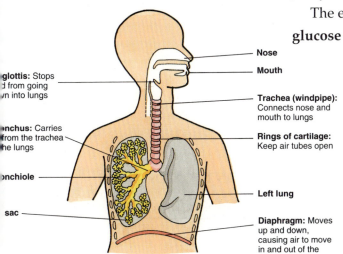

- **Nose**
- **Mouth**
- **Trachea (windpipe):** Connects nose and mouth to lungs
- **Rings of cartilage:** Keep air tubes open
- **Left lung**
- **Diaphragm:** Moves up and down, causing air to move in and out of the lungs
- **glottis:** Stops [food] from going [dow]n into lungs
- **[bro]nchus:** Carries [air] from the trachea [to t]he lungs
- **[bro]nchiole**
- **[air] sac**

THE BREATHING SYSTEM

The breathing system works like this. A muscular sheet called the **diaphragm** moves down, causing air to be drawn in through the mouth and nose, down through the **trachea**, into the **bronchus**, along a number of smaller tubes called **bronchioles** and into air sacs called **alveoli**. (This can be demonstrated using the model of the breathing system shown in the diagram.) When the rubber sheet is pulled down, air is drawn into the balloons.)

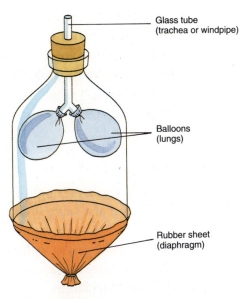

- Glass tube (trachea or windpipe)
- Balloons (lungs)
- Rubber sheet (diaphragm)

Model of the breathing system

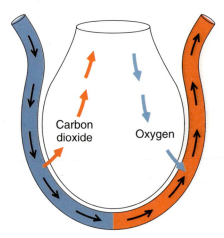

Gas exchange in the alveoli

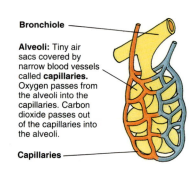

The alveoli are covered by a network of narrow blood vessels called **capillaries**. Oxygen passes from the alveoli into the capillaries and is carried round into the body cells. Waste carbon dioxide is carried back from the body cells into the capillaries, from which it passes into the alveoli. When the diaphragm moves up, air containing the waste carbon dioxide is forced out of the lungs. (Once again, this can be demonstrated using the model of the breathing system. When the rubber sheet moves up, air is forced out of the balloons.)

EXPERIMENT 44.1
AIM: TO SHOW THAT EXPIRED AIR CONTAINS MORE CARBON DIOXIDE THAN INSPIRED AIR

Note: Carbon dioxide gas turns limewater milky.

Method
1. Blow in through tube A and time how long it takes for the limewater to turn milky.
2. Suck air out through tube B and time how long it takes for the limewater to turn milky.

Result: It takes much longer for the limewater in tube B to turn milky.

Conclusion: The air we breathe out contains more carbon dioxide than the air we breathe

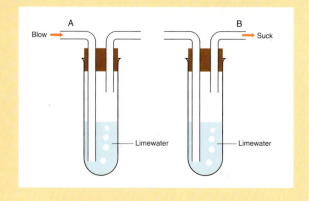

SMOKING

We know that smokers cough and that smoking can lead to lung cancer, but it can also have several other bad effects.

1. Carbon monoxide in cigarette smoke means that the blood is not able to carry oxygen as well as it should. This means that the smoker must breathe faster, and the heart must beat faster which can lead to heart strain.
2. Smoking during pregnancy reduces the amount of oxygen available to the baby.

BIOLOGY · RESPIRATION AND BREATHING

EXPERIMENT 44.2

AIM: TO SHOW HOW SMOKING AFFECTS YOUR LUNGS

Method

1. Set up the apparatus as shown in the diagram.
2. Light the cigarette and let the filter pump draw air through it.

   Result: Tar is trapped in the glass wool.

   Conclusion: Smoking causes tar to build up on your lungs.

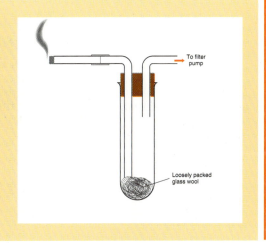

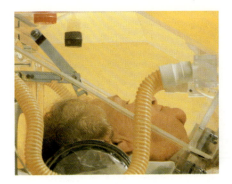

BREATHING RATE

This is a simple test you can do at home or in school.

1. Sit down and count the number of times you breathe in and out in a minute.
2. Do ten push-ups or run on the spot for two minutes. Now count the number of times you breathe in and out in a minute. Are you breathing faster? Of course you are.

When you were doing the push-ups you were using more energy, so you needed more oxygen to release the energy from your food. You had to breathe faster. You breathe faster when you exercise. There is nothing wrong with this: exercise is good for you. In fact, more and more people are attending aerobics classes. Aerobic exercise makes us breathe more deeply and make our lungs more efficient.

Some people breathe quickly even when they are doing light exercise or no exercise at all. There could be a number of reasons for this.

1. They are unfit. They do not play games or take regular exercise, and so their lungs are not very efficient.
2. They are suffering from a lung disease such as bronchitis or emphysema.
3. They have damaged their lungs by smoking.

OTHER ORGANISMS

Up to now we have only considered respiration in humans, but all living things respire and so all living things produce carbon dioxide.

EXPERIMENT 44.3
AIM: TO SHOW THAT LIVING ORGANISMS PRODUCE CARBON DIOXIDE

Method
Set up the apparatus as shown in the diagram. The limewater will turn milky after a while.

Result: Why do you think it takes so long for the limewater to turn milky?

Conclusion: Living organisms (e.g. woodlice) produce carbon dioxide.

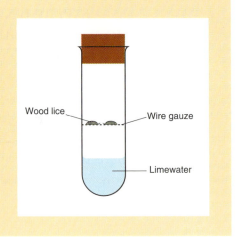

EXPERIMENT 44.4
AIM: TO SHOW THAT HEAT IS PRODUCED DURING RESPIRATION

Method
1. Set up the apparatus as shown in the diagram.
2. Take the temperature in each flask once a day for four days.

Result: The temperature rises in flask A but not in flask B.

Conclusion: Heat is produced during respiration.

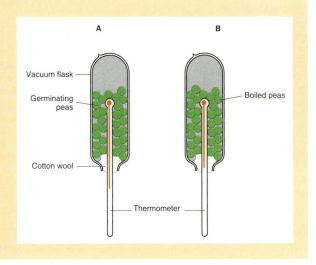

OTHER SYSTEMS OF GAS EXCHANGE

Not all animals have lungs. Fish breathe through gills. Insects breathe through a system of holes called spiracles. Earthworms breathe through holes in their skin.

Have you ever noticed that, after a very heavy shower, earthworms crawl out of the soil onto concrete paths or any other relatively dry surface they can find? Why do you think they do this?

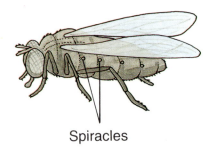

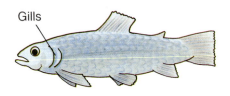

BIOLOGY · RESPIRATION AND BREATHING

SUMMARY

- Respiration is the release of energy from food.
- Glucose + oxygen → carbon dioxide + water + energy
- $C_6H_{12}O_6 + 6O_2 \rightarrow 6CO_2 + 6H_2O$ + energy
- Breathing is a means of taking in the oxygen necessary for respiration and getting rid of the carbon dioxide produced during respiration.
- In the lungs, exchange of gases takes place in the alveoli (air sacs).

QUESTIONS

Section A

1. What does "respiration" mean? _____
2. Complete the word equation for respiration.
   glucose + _____ = _____ + water + energy
3. Complete the chemical equation
   $C_6H_{12}O_6$ + ____ = $6CO_2$ + ____ + energy
4. What gases are exchanged in the alveoli? _____
5. Where **exactly** in the human body does gas exchange take place? _____
6. The lungs are the organs of gas exchange in humans. Give one alternative system of gas exchange and name an animal using that system. _____
7. What are alveoli? _____
8. What is the function of the alveoli?
   _____
9. Give another name for the trachea.
   _____
10. What chemical is used to test for carbon dioxide in expired air? _____

Section B

1. The diagram shows the human breathing system. Name the parts labelled A, B, C, D. What is the function of the rings of cartilage in A? Name the gas absorbed from the air in the lungs. Name the gas excreted into the air by the lungs. What form of pollution is particularly damaging to the lungs?

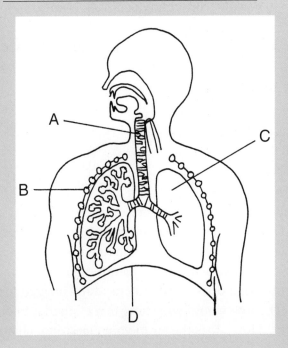

2. The diagram on the right shows a model of the human breathing system. What do the balloons represent? What does the rubber represent? What happens when the rubber sheet is pulled down?
3. What is the purpose of experiment X in the diagram below? What is the purpose of experiment Y? (Soda lime removes carbon dioxide from the air.)

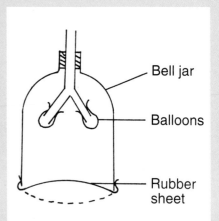

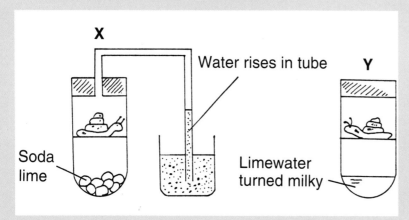

4. The diagram on the right shows an experiment on respiration. A student breathes in and out through the mouthpiece by opening and closing the clips A and B. What is the purpose of the experiment? Name the liquid in flask X. Through which flask is air drawn in? What is the result of the experiment?

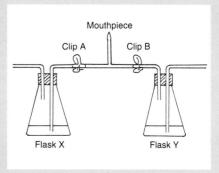

5.

	Air expired Fresh air	Air expired by by a person at rest	a person taking exercise
Nitrogen	78%	78%	78%
Oxygen	21%	17%	12%
Carbon dioxide	0·03%	4%	9%
Water vapour	Variable	Saturated	Saturated

From this table, answer the following questions.
(a) Do we use all the oxygen we take in?
(b) Why does the person taking exercise use more oxygen?
(c) Is exercise good for your lungs?
(d) What does your body need the oxygen for?
(e) How would you show that expired air contains water vapour?
(f) How does the air in your classroom change from the time you enter it?

6. While doing experiment 44.1, a student blew into tube A and it took the limewater 12 seconds to turn milky. She then sucked tube B, and it took the limewater 30 seconds to turn milky. Why is there a difference in these times? What difference would it have made if she had carried out the experiment
   (a) in the open air?
   (b) in the lab first thing in the morning?
   (c) in the lab after one hour?
   (d) after taking a rest?
   (e) after playing a game of basketball?
7. In experiment 44.2 we saw that smoking causes tar to build up on the lungs. How exactly does this affect the lungs? What diseases do smokers suffer from? What other parts of the body are affected by smoking? (Look at the photograph in Chapter 49 and the diagram in chapter 57.) Do you know what passive smoking is? Is it dangerous? Why?

CHAPTER 45 — THE CIRCULATORY SYSTEM

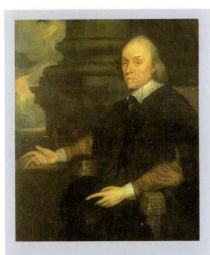

The English physician William Harvey (1578–1657) proved that blood circulates in the human body. Harvey also measured the amount of blood in the circulatory system. His experiments were published in *On the Motions of the Heart* in 1628.

Nutrients and oxygen must be transported round to all the cells in your body, and waste carbon dioxide and water must be carried back to the lungs. All these things are carried in your blood.

BLOOD

Blood consists of a straw-coloured liquid called **plasma** which has **red corpuscles**, **white cells** and **platelets** suspended in it. The body of an adult contains about 5 litres of blood on average.

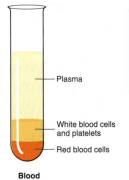

Blood

RED CELLS

Most cells have a nucleus, but there are some exceptions. Red blood cells have no nucleus, and are called red corpuscles. They contain a substance called haemoglobin which can absorb oxygen. Oxygen from the lungs passes into the blood, where it combines with the **haemoglobin** to form **oxyhaemoglobin**, which is bright red. The oxygen is later released to the body cells.

Red blood cells

WHITE CELLS

White cells protect us from disease. Some white cells surround germs and digest them. Other white cells produce chemicals called **antibodies** which kill germs. Each type of antibody can kill only one type of germ.

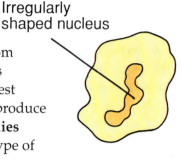

White blood cell

PLATELETS

These are pieces of cells that clot the blood and stop your bleeding. The platelets produce tiny fibres which combine with red cells to form a blood clot. This hardens into a scab which keeps out dirt and bacteria until new skin grows over the wound. The dry scab then falls off.

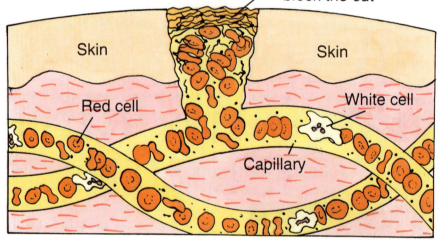

Red corpuscles	White cells	Platelets	Plasma
Carry oxygen	Kill bacteria, produce antibodies to fight disease	Clot blood	Carries nutrients to the cells and carbon dioxide back from the cells

You should now examine a commercially prepared blood slide under the microscope.

All the blood cells you see are produced in the bone marrow. Note that there are a large number of red corpuscles, but only a few white cells.

The functions of the blood are
- to carry food and oxygen to each cell
- to remove waste from the cells
- to protect the body from infection
- to help maintain body temperature
- to transport hormones round the body.

BIOLOGY · THE CIRCULATORY SYSTEM

THE CIRCULATORY SYSTEM

Have you ever studied a map of a city's transport system? Well, diagram (a) is a **very simplified** map of the body's transport system. There are two circuits: one from the heart to the lungs and back (this is called the **pulmonary** circuit) and one from the heart to the rest of the body and back. Now look at diagram (b) and note the following.

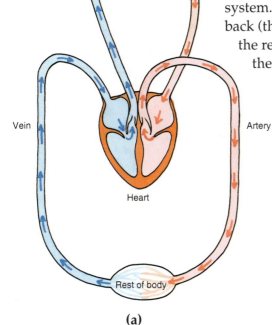

(a)

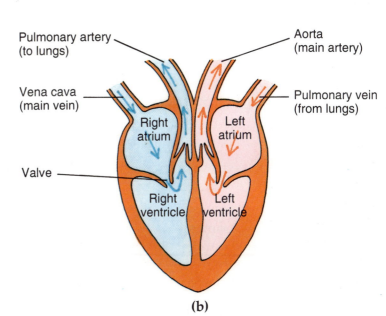

(b)

1. The heart is divided down the centre by a wall so that blood cannot flow directly from one side to the other.
2. The **right atrium** has thin walls.
3. The **right ventricle** has thicker walls because it has to pump blood up to the lungs.
4. The **left atrium** has thin walls.
5. The **left ventricle** has very thick walls because it has to pump blood all round the body.

THE BLOOD'S JOURNEY

Oxygenated blood is pumped from the left ventricle out through the **arteries**. The arteries have thick walls so that they can withstand the pressure.

From the arteries, the blood flows into a network of tiny tubes called **capillaries**. The walls of the capillaries are so thin that oxygen can pass through them into the body cells, and carbon dioxide can pass through them from the body cells into the blood.

Artery

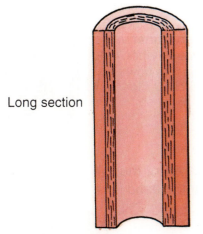

Long section

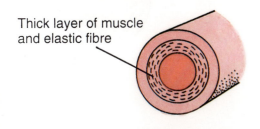

Thick layer of muscle and elastic fibre

Cross-section

297

This **deoxygenated** blood flows into tubes called **veins**. At this stage the pressure is fairly low, so veins do not need thick walls: in fact they need valves to stop the blood from flowing backwards.

The veins carry the blood into the right atrium. From there it passes through a valve into the right ventricle, and is pumped out through the pulmonary artery to the lungs to get rid of the carbon dioxide and collect oxygen. The oxygenated blood passes back through the pulmonary vein to the left atrium and on into the left ventricle, where the cycle begins all over again.

At this stage it would be a good idea to dissect a sheep's heart or to study a model of the heart.

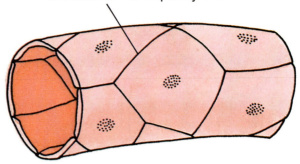

Capillary

Veins	Arteries	Capillaries
Carry blood into the heart	Carry blood out of the heart	Connect arteries to veins
Carry blood at low pressure	Carry blood at high pressure	Food and oxygen pass through the walls of the capillaries into the cells
Have thin walls	Have thick walls	
Have valves	Have no valves (except for the aorta and the pulmonary arteries)	Waste from the cells passes through the walls of the capillaries into the blood

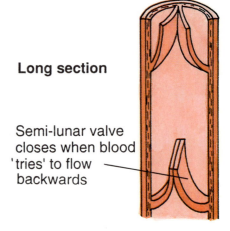

Vein

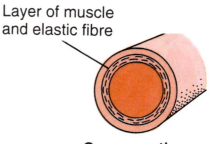

BIOLOGY · THE CIRCULATORY SYSTEM

REVISION DIAGRAM OF THE CIRCULATORY SYSTEM
(follow the captions in order from 1 to 5)

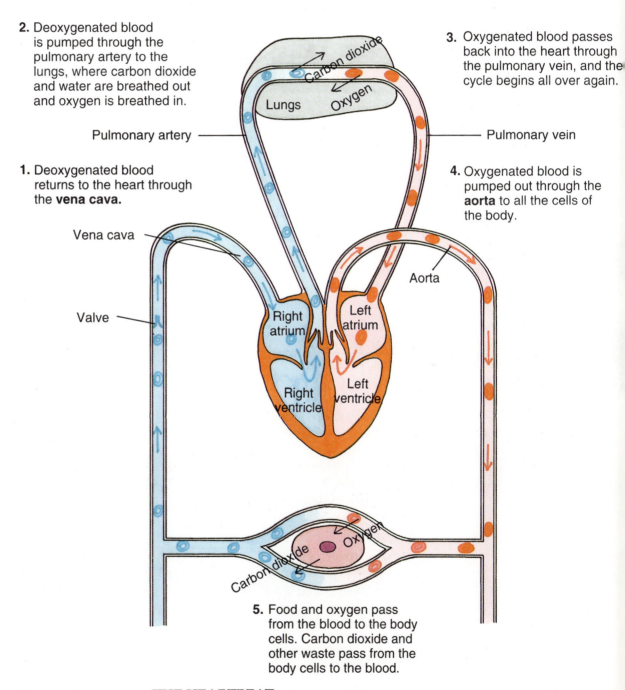

2. Deoxygenated blood is pumped through the pulmonary artery to the lungs, where carbon dioxide and water are breathed out and oxygen is breathed in.

Pulmonary artery

1. Deoxygenated blood returns to the heart through the **vena cava.**

Vena cava

Valve

3. Oxygenated blood passes back into the heart through the pulmonary vein, and the cycle begins all over again.

Pulmonary vein

4. Oxygenated blood is pumped out through the **aorta** to all the cells of the body.

Aorta

5. Food and oxygen pass from the blood to the body cells. Carbon dioxide and other waste pass from the body cells to the blood.

THE HEARTBEAT

The walls of your heart are made of cardiac muscles which keep working day and night all through your life. The adult heart beats an average of 72 times per minute, the teenage heart beats faster.

To study the effect of exercise on the heartbeat, take hold of your friend's wrist as shown in the diagram and find the pulse. Count the number of beats per minute. Now ask your friend to do five push-ups or to run on the spot for two minutes. Count the number of beats per minute again. Can you explain the difference?

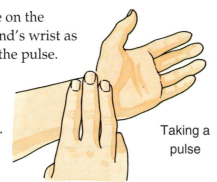
Taking a pulse

HEART DISEASE

In Ireland, thousands of people die each year of heart disease. Although the heart pumps blood all round the body, the heart itself needs a constant supply of food and oxygen, which reaches it through the coronary arteries. A fatty substance called cholesterol can stick to the walls of these arteries and make them very narrow. A blockage in the coronary arteries causes a heart attack. You can reduce the risk of heart disease by following some simple rules.
- Take regular exercise.
- Don't smoke.
- Eat less fat, red meat, butter, cream.
- Eat more fish, chicken, vegetables, fruit.

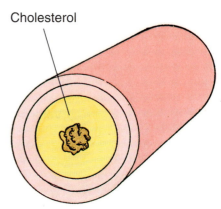
Artery almost completely blocked by cholesterol

SUMMARY

Blood contains:

Red corpuscles	White cells	Platelets	Plasma
Carry oxygen	Kill bacteria, produce antibodies to fight disease	Clot blood	Carries nutrients to the cells and carbon dioxide back from the cells

Veins	Arteries	Capillaries
Carry blood into the heart	Carry blood out of the heart	Connect arteries to veins
Carry blood at low pressure	Carry blood at high pressure	Food and oxygen pass through the walls of the capillaries into the cells
Have thin walls	Have thick walls	
Have valves	Have no valves (except for the aorta and the pulmonary arteries)	Waste from the cells passes through the walls of the capillaries into the blood

- Veins carry deoxygenated blood, except for the pulmonary vein.
- Arteries carry oxygenated blood, except for the pulmonary artery.

BIOLOGY · THE CIRCULATORY SYSTEM

QUESTIONS

Section A

1. What is the function of the heart? _____
2. Name two substances carried by the blood to the body cells. _____
3. State two functions of the blood. _____
4. What is the function of white cells in the blood? _____
5. What is oxygenated blood? _____
6. What substance in the blood enables the red cells to absorb oxygen? _____
7. Name the blood cells in the diagram on the right. _____
8. Give one function of each of the blood cells in question 7. _____
9. Name the parts of the heart marked X and Y in the diagram on the right. _____
10. Which of the diagrams below represents a section through an artery? _____ Give a reason for your answer. _____
11. In the diagram on the right, what are the tiny blood vessels marked Y called? _____
12. The pulmonary artery carries blood from the _____ to the _____ .
13. Give two ways in which arteries differ from veins. _____
14. After leaving the left atrium blood travels through the other chambers of the heart in which of the following orders?
    (a) Right atrium, right ventricle, left ventricle.
    (b) Left ventricle, right ventricle, right atrium.
    (c) Left ventricle, right atrium, right ventricle.
    (d) Right ventricle, left ventricle, right atrium.

Breakthrough Science

Section B

1. Give four functions of the blood. Which part of the blood (*a*) carries oxygen, (*b*) carries digested food, (*c*) fights infection?

2. State the function of the heart.
   _____
   Name the parts labelled A, B, C in the diagram of the heart.
   _____
   Give three means by which heart disease may be prevented.
   _____

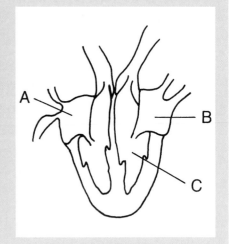

3. The diagram (below right) shows part of the circulatory system. What type of blood vessels are marked A and D? Name the parts labelled B and X. What type of blood is carried in the blood vessel F?

4. Fill in the gaps in the following table.

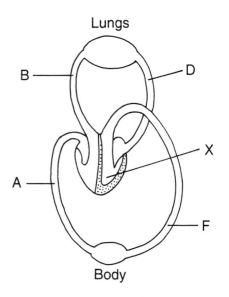

Blood vessel	Blood flow From	To	Type of blood
Aorta	Heart	Body	
	Heart	Lungs	
		Heart	Oxygenated
	Body	Heart	

5. If you scraped your knuckles against the wall you would burst some capillaries and bleed a little. What are capillaries? Why would you bleed only a little? Explain, using the word "platelets", what would happen next. Would it be more serious if you cut an artery? Why? Why do you think the arteries are as deep as possible in the body, while capillaries and veins are near the surface?

6. Name the four main tubes entering and leaving the heart. Name a vein that carries oxygenated blood. Name an artery that carries deoxygenated blood.

7. Name the four chambers of the heart. Give the order in which the blood passes through these four chambers, starting with the right atrium. Which chamber has the thickest wall? Why? With the aid of simple diagrams, show the **structural differences** between a vein and an artery, and explain the reason for these differences.

8. What are the main causes of heart disease, and how can they be avoided? Write to the Irish Heart Foundation for information.

9. When a British pop singer and actor was asked why he had to have a heart bypass operation, he replied "My mother's frying pan". What did he mean by this?

CHAPTER 46 EXCRETION

Chemical reactions that take place in your body cells produce waste substances such as **carbon dioxide** and **water**. Also, if there is excess protein in the diet it is broken down to amino acids, and a waste substance called urea is formed. Your body must get rid of this waste, because (except for water) it is poisonous. This is the job of the excretory system.

Excretion means getting rid of the waste produced by chemical reactions in the cells.

EXCRETORY ORGANS

The excretory system consists mainly of the **kidneys, lungs** and **skin**.

Kidneys

The kidneys filter the blood and remove urea and other waste in the form of a solution called **urine**.

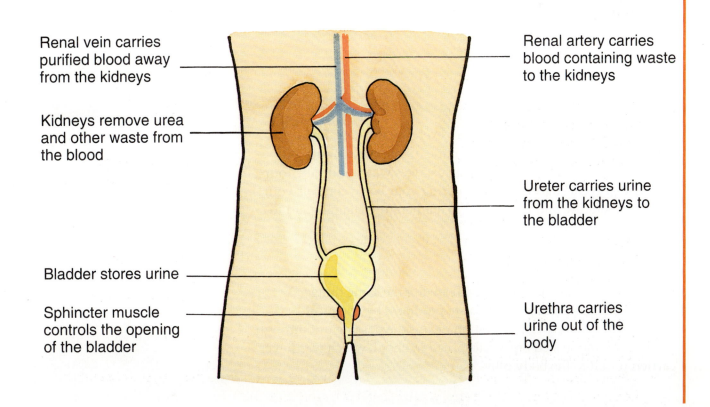

Renal vein carries purified blood away from the kidneys

Kidneys remove urea and other waste from the blood

Bladder stores urine

Sphincter muscle controls the opening of the bladder

Renal artery carries blood containing waste to the kidneys

Ureter carries urine from the kidneys to the bladder

Urethra carries urine out of the body

Lungs

We have already done an experiment to show that the lungs breathe out carbon dioxide. The following experiment shows that they also breathe out water vapour.

> **EXPERIMENT 46.1**
>
> **AIM: TO SHOW THAT EXPIRED AIR CONTAINS WATER VAPOUR**
>
> **Note:** Blue cobalt chloride paper turns pink if water is present.
>
> **Method**
> 1. Hold a small mirror in front of your mouth and nose, and breathe on it.
> 2. You will see small droplets of clear liquid forming on the mirror, but how do you know that this is water?
> 3. Place a piece of blue cobalt chloride paper, which has been dried in an oven, on top of the liquid and see what happens. (Water turns blue cobalt chloride paper pink.)
>
> **Result:** The cobalt chloride paper turns pink.
>
> **Conclusion:** Expired air contains water vapour.

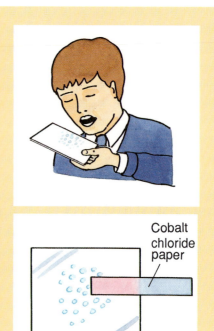

Skin

The next time you are sweating after exercise, rub your fingertip across your forehead and place it on the tip of your tongue. You should recognise the taste. Your skin excretes water and salt. Your skin also prevents bacteria from entering your body, controls your body temperature and is sensitive to temperature changes and to pain.

EXCRETION AND EGESTION

Excretion and egestion (Chapter 43) are not the same thing. Excretion means getting rid of waste produced by chemical reactions in the body cells. Egestion means the passing out through the anus of undigested and unabsorbed material, which was never carried round to the body cells.

SUMMARY

- Excretion means getting rid of waste produced by chemical reactions in the body cells.

Excretory organ	What it excretes
Kidneys	Urea
Lungs	Carbon dioxide and water
Skin	Salt and water

QUESTIONS

Section A

1. What does excretion mean? _____
2. The diagram shows one of the excretory systems in humans. Name the parts marked A and B.

   _____
3. Name two substances excreted by the skin

   _____
4. Name one substance excreted by the kidneys. _____
5. Name the tubes that connect the kidneys to the bladder. _____
6. Name the tube that carries urine out of the body. _____
7. Name three organs of excretion in the human body. _____
8. Which of the following is not excreted by the human body: urea, water, salt, oxygen?
9. Urea results from an excess of _____ in the diet.

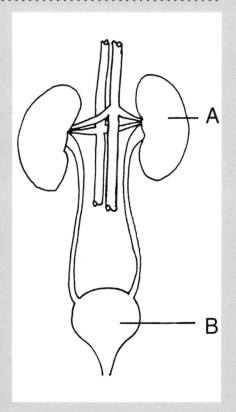

Section B

1. Draw a labelled diagram of the human urinary system, giving the name and function of each part.
2. Describe a simple experiment to show that your lungs excrete carbon dioxide.
3. A student who was given a piece of cobalt chloride paper in the lab complained that it was already pink. What colour should it have been? How do you think it became pink? What experiment do you think the student was preparing for? What should be done with the cobalt chloride paper before starting the experiment?
4. Say, giving reasons, whether each of the following would increase or decrease the amount of urine produced by the kidneys: (*a*) exercise, (*b*) hot weather, (*c*) cold weather.

BREAKTHROUGH SCIENCE

SUPPORT, PROTECTION AND MOVEMENT

CHAPTER 47

Plant cells have strong cellulose walls so that plants can support themselves. Animal cells do not have walls, so an animal needs a frame or skeleton to support its body and give it shape. Some animals, like the crab, have a skeleton on the outside of the body, called an exoskeleton. Others, like humans, have an inside skeleton, called an endoskeleton. Your skeleton has a number of functions.

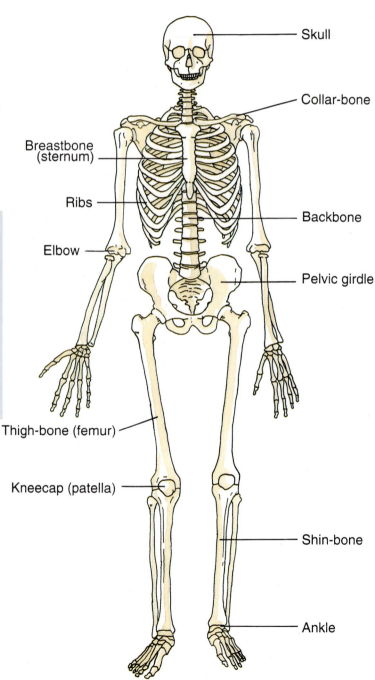

FUNCTIONS OF THE SKELETON

1. It protects delicate organs (e.g. the skull protects the brain).
2. It supports the body.
3. Along with the muscles, it enables us to move.
4. It gives the body shape.
5. Blood cells are made in the bone marrow.

Look at the leg bones in the diagram of the skeleton. These bones support your entire body, which is amazing considering they are hollow! It was probably this that gave people the idea for tubular steel legs for desks and hollow metal poles for scaffolding. What makes bones so strong? Try the following simple experiment.

BIOLOGY · SUPPORT, PROTECTION AND MOVEMENT

EXPERIMENT 47.1
AIM: TO FIND OUT WHAT MAKES BONES STRONG

Warning: Acid is corrosive.

Method
1. Set up the apparatus as shown in the diagram, and let it stand for 24 hours.
2. Remove the bones and wash them under a tap.
3. Now test them to see if they are as strong and as rigid as before.

Result: The bone in the acid lost its strength and rigidity.

Conclusion: Calcium makes bones strong. Acid attacks the calcium and weakens the bone.

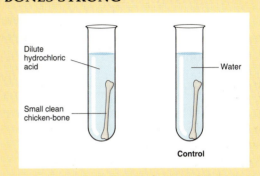

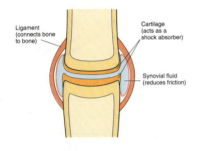

JOINTS

A joint is a point in the skeleton where bones meet.

As you can see from the diagram, ligaments connect bone to bone.

Ligaments connect bone to bone.

If your arm was all one bone, you would not be able to bend it. We can move the way we do because our skeleton has joints. Your body has many movable joints. These are called synovial joints.

Ball and socket joint
This type of joint is found at the shoulder and at the hip. It allows movement in all directions.

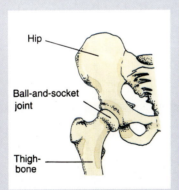

Pivot joint
This is found at the base of your skull. It allows you to turn your head around.

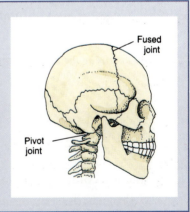

Hinge joint
This type of joint is found at the knee and elbow. It can bend in one direction only, just like the hinge on a door.

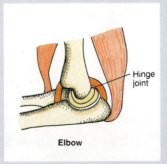

Gliding joint
This is found at the wrist and ankle. It allows the bones to glide over one another.

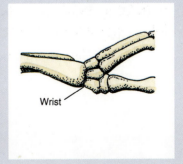

Another type of joint is the fused joint. This type of joint does not allow any movement. An example can be found between the bony plates in the skull.

Your backbone is also jointed. If it were a single bone you would not be able to bend your back. In fact it contains 24 separate bones, cushioned from each other by cartilage discs. It also contains nine other bones in two fused sections. The backbone is hollow and protects the spinal cord. In all, there are over 200 bones in your skeleton. The femur (thigh bone) is the largest bone in the human body.

MUSCLES

Bones can move only when they are pulled by muscles. As you can see from the diagram, tendons connect muscles to bones.

When your arm is hanging down by your side, the biceps muscle in your arm is relaxed. It cannot get any longer than it is now. It can, however, contract (get shorter) and bend your arm. Another muscle, called the triceps, must contract if you want to straighten your arm.

When one muscle bends a joint and another straightens it, they are called antagonistic muscles.

Muscles like these are voluntary muscles. They are controlled by the brain and get tired very fast. The muscles in your stomach are involuntary: they work away on their own.

Tendons connect muscles to bones.

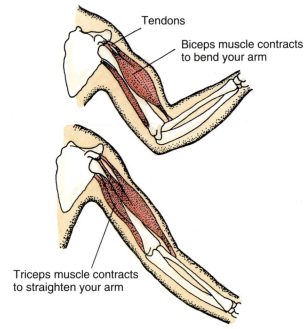
Tendons
Biceps muscle contracts to bend your arm
Triceps muscle contracts to straighten your arm

SUMMARY

- Functions of the skeleton
  1. Protects delicate organs
  2. Supports the body
  3. Enables the body to move
  4. Gives the body shape
  5. Blood cells are made in the bone marrow.

- Types of joint
  Ball and socket: hip, shoulder
  Hinge: elbow, knee
  Pivot: base of skull
  Fused: in skull
  Gliding: wrist and ankle
- Ligaments connect bone to bone.
- Antagonistic muscles: two muscles; one bends a joint and the other straightens it.
- Tendons connect muscles to bones.

QUESTIONS

Section A
1. Give two functions of the skeleton. _____
2. Give an example of a hinge joint. _____
3. Give an example of a ball and socket joint. _____
4. Name one mineral that makes bones hard. _____
5. _____ join bones to bones.
6. _____ join muscles to bones.
7. In the diagram of a movable joint (right), A is _____ and B is _____
8. What is the longest bone in the human body? _____
9. When you raise your forearm, the _____ muscle contracts.
10. What are antagonistic muscles? Give an example. _____

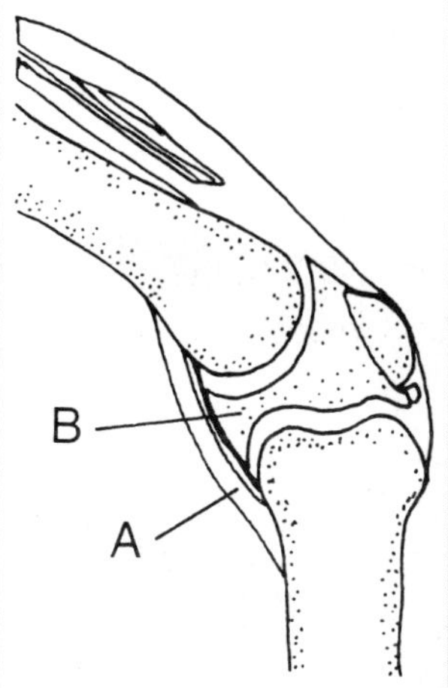

Section B
1. Describe a simple experiment to show what happens to a bone when calcium is removed.
2. What type of joint is shown at G in the diagram? What type of movement does this type of joint allow? Name two places in the body where this type of joint is found. Name one other type of joint found in the human body.
3. Make a list of the types of joint found in the skeleton, and give an example of each.
4. Complete the following. The skull protects the _____, the ribs protect the _____ and _____, the _____ protects the spinal cord.
5. If you look at the top of a young baby's head, you can see a pulse beating beneath the skin. What has this got to do with the skeleton? Do some research and find out.
6. Name the parts A, B, C, D, E, F, in the diagram. _____ When muscle D contracts, what happens to C and to the forearm? _____ 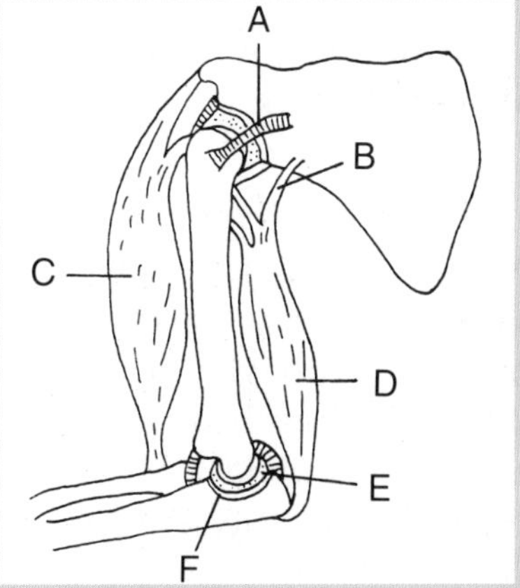
7. Complete the following table by writing 'Voluntary' or 'Involuntary' in each box.

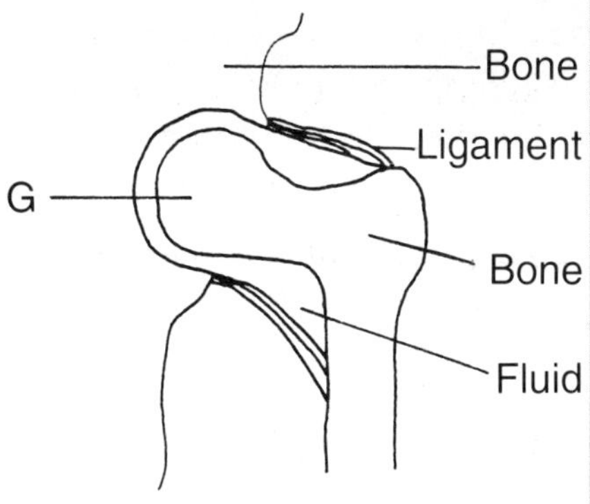

Action / Type of muscle	Walking	Chewing	Digesting	Blinking	Heartbeat

SENSITIVITY AND CO-ORDINATION

CHAPTER 48

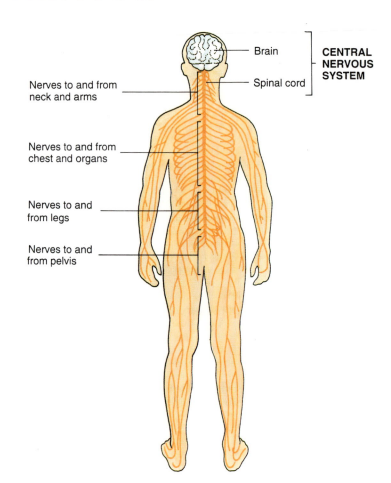

I bet the first thing you noticed when you opened this page was the juicy burger. Your eyes sensed it and sent a message to your brain. Your body responded. Saliva poured into your mouth. You were all ready to gobble it up. But of course there was no burger, just a photograph. I know that's mean, but at least it may help you to understand how your senses work.

SENSE ORGANS

You have five senses: sight, hearing, smell, taste and touch. If one of your sense organs picks up what scientists call a "stimulus", your body responds. The system in your body that deals with the stimulus is called your nervous system.

THE NERVOUS SYSTEM

The nervous system consists of the **brain** and the **spinal cord** (called the central nervous system) and the nerves that connect the spinal cord to all parts of the body (called the peripheral nervous system).

Nerves are made up of bundles of nerve cells called **neurons**. The system works like this:

(a) a sense organ, such as the ear, receives a stimulus

(b) the message passes along a **sensory** nerve to the brain

(c) the brain decides what to do

BIOLOGY · SENSITIVITY AND CO-ORDINATION

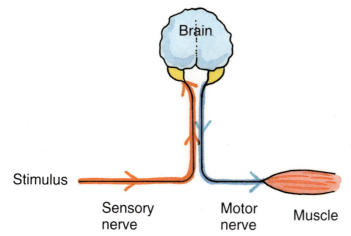

(*d*) The brain sends a message along a **motor** nerve to the muscles, telling them what to do. This is called the response.

| *Sensory nerves carry messages to the brain.*

| *Motor nerves carry messages from the brain.*

The eye is another sense organ that can receive a stimulus (light) and pass a message along a sensory nerve to the brain.

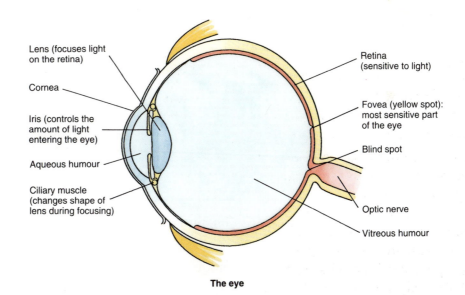

The eye

HOW THE EYE WORKS

Light from an object enters the eye and is focused by the lens, so that an upside-down image is formed on the retina, which is sensitive to light. The light is converted to electrical impulses, which travel along the optic nerve to the brain, where the image is turned right-way-up.

At this stage you should dissect a sheep's eye or examine a model of the eye.

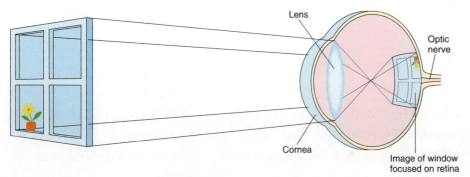

THE ENDOCRINE SYSTEM

The nervous system controls your body by sending electrical impulses along nerves to all parts of the body. The endocrine system controls certain organs in your body by sending chemical messengers through the bloodstream to these organs. These chemical messengers are called **hormones**. They are produced by glands in various parts of your body and released directly into the bloodstream. This system is much slower than the nervous system, but the effects last much longer.

One example is a gland called the **pancreas** which produces a hormone called **insulin** that controls the level of sugar in the blood. People who suffer from **diabetes** produce little or no insulin, so they must take insulin each day and follow a special diet.

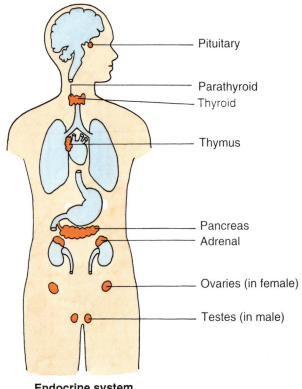

Endocrine system

SUMMARY

- The nervous system consists of the brain, the spinal cord and nerves connected to all parts of the body.
- Sensory nerves carry messages to the brain.
- Motor nerves carry messages from the brain.
- A hormone is a chemical messenger released by a gland into the bloodstream.

QUESTIONS

Section A
1. Name your five senses. _____
2. _____ nerves carry messages to the brain.
3. _____ nerves carry messages from the brain.
4. Messages are carried from the eye to the brain by the _____ nerve.
5. In the eye the _____ focuses light on the _____.
6. The _____ controls the amount of light entering the eye.

7. Which organ in your body is responsible for your sense of touch?
8. What is a hormone?
9. Name one hormone and the gland that produces it.
10. The endocrine system is much _____ than the nervous system but the effects
_____

Section B

1. You see a five-pound-note on the ground and you pick it up. What was the stimulus? Which of your sense organs was stimulated? What type of nerve carried the message to the brain? Name this nerve. What type of nerve carried a message back from the brain? What was your response?

2. What is the central nervous system? What is the peripheral nervous system? What are neurons?

3. Name the parts labelled A, B, C in the diagram of the human eye. State the function of each of these parts. Name two other human sense organs.

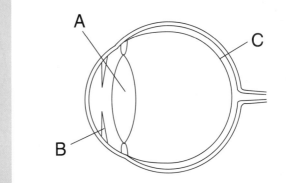

4. Match the items in Box A with descriptions in box B.

Box A	Box B
Iris	Focuses light on the retina
Retina	Carries messages to the brain
Lens	Change the shape of the lens
Ciliary muscles	The most sensitive part of the retina
Optic nerve	Controls the amount of light entering the eye
Yellow spot	Sensitive to light

5. Where in the body is insulin produced? Give one function of this hormone. What illness is caused by a deficiency in this hormone?

REPRODUCTION CHAPTER 49

Remember what we said about hormones in the last chapter? Well, when a person is somewhere between the ages of 10 and 15 years old, sex hormones cause the body to change. In a boy, the voice deepens, hair grows around the testes (pubic hair), chest, face and armpits, and **the testes begin to produce sperm.**

In a girl, the hips widen, breasts develop, pubic and underarm hair grows and **the ovary begins to release eggs** (ova).

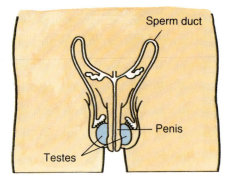

Male reproductive organs

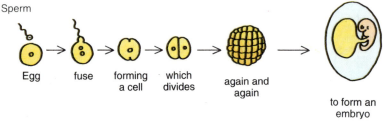

SEXUAL REPRODUCTION

Most animals reproduce sexually. The two sexes, male and female, each produce sex cells called gametes. (In humans, the sperm is the male **gamete** and the egg contains the female gamete.) For reproduction to take place, the male gamete must fuse (join) with the female gamete to produce a new cell called a zygote which will grow by cell division to form a new individual.

But how is this achieved in humans?

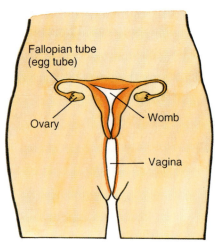

Female reproductive organs

PROCESS OF REPRODUCTION

The process takes place in several stages, beginning with **ovulation**.

The release of an egg from the ovary is called ovulation.

If a man and a woman want to have a baby, they must have sexual intercourse. This means that the man's penis must be inserted into the woman's vagina, and sperm released (this is called **insemination**).

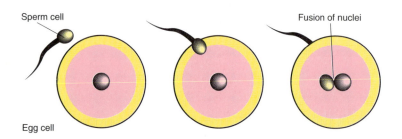

BIOLOGY · REPRODUCTION

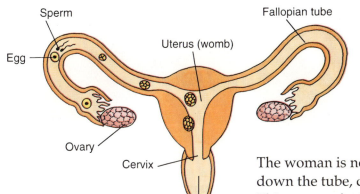

The sperm then travels up through the womb into the **fallopian tubes**. If an egg is present in one of the tubes, then a sperm may fuse (join) with it. This is called **fertilisation** (**conception**).

Fertilisation means that the sperm fuses with the egg.

The woman is now **pregnant**. The fertilised egg moves back down the tube, dividing into more and more cells as it goes. This group of cells, called an **embryo**, becomes attached to the soft lining of the womb. This is called implantation.

Implantation means that the embryo becomes attached to the lining of the womb.

During the next 40 weeks or so the embryo develops in the womb, protected by a bag of fluid which acts as a cushion. A tube called

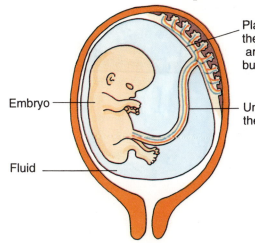

the **umbilical cord** connects the embryo to the **placenta**. During pregnancy the mother breathes, digests food and excretes waste for the embryo.

In the placenta the mother's blood and the baby's blood do not mix, but food and oxygen can pass from the mother's blood to the baby's. Carbon dioxide and other waste pass from the baby's blood to the mother's. Can you understand from this why pregnant women should avoid smoking and alcohol?

BIRTH

When the bag of fluid bursts, this is a sign that the baby is about to be born. The baby usually comes out head first. The umbilical cord is then clamped and cut. Shortly afterwards the placenta (afterbirth) also comes out.

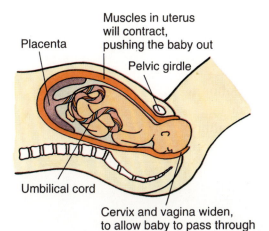

THE MENSTRUAL CYCLE

Every 28 days or so a woman's body goes through a series of changes known as the menstrual cycle. The reason for this is that her body is preparing for pregnancy. A spongy lining full of blood vessels builds up on the wall of the womb. An egg is released from the ovary. If it is fertilised it should become attached to the wall of the womb, and pregnancy should proceed as normal.

If it is not fertilised, the lining is not needed so it breaks down and flows out of the body, with blood, through the vagina. This is known as **menstruation** or having a **period**. After this the whole cycle begins all over again.

Somewhere between the ages of 50 and 55 a woman's menstrual cycles stop. This is known as menopause or the "change of life".

FAMILY PLANNING

If parents want to control the number of children they have, they must control the number of times that fertilisation takes place. There are a number of methods for doing this.

Natural method

When an egg is released from the ovary, a woman's temperature rises slightly. If she keeps a chart of her temperature every morning for a few months, she should be able to tell when her next egg will be released. If intercourse is avoided around this time, then fertilisation should not take place. This is called "the rhythm method".

Artificial methods

The "pill" is a small dose of hormones that prevents ovulation. This means that even if intercourse takes place, there is no egg present to be fertilised. The contraceptive pill can cause health problems for some women. A condom can be used to prevent the sperm from reaching the egg.

SUMMARY

- Sexual reproduction involves two sex cells, a male gamete and a female gamete, which join to form a zygote.
- The sperm is the male sex cell (male gamete).
- Sperm is produced in the testes.
- The egg contains the female sex cell (female gamete).
- The egg is produced in the ovary.
- Ovulation is the release of an egg from the ovary.
- Fertilisation means that the sperm fuses with the egg.
- Fertilisation usually takes place in the fallopian tube.
- Implantation means that the embryo becomes attached to the lining of the womb.
- The umbilical cord connects the embryo to the placenta.
- Menstruation means that the lining of the womb breaks down and flows, with blood, out of the vagina.

QUESTIONS

Section A

1. Sperm are produced in the _____.
2. Ovulation means the release of _____ from the _____.
3. Fertilisation means that the _____.
4. In the diagram of the female reproductive system, name the parts labelled S and T. _____
5. In the diagram in question 4, mark with an X the place where fertilisation usually occurs. Mark with a Y the place where the foetus is carried.
6. Food and _____ pass from the mother to the foetus through the _____.
7. One substance that passes from the baby to the mother is _____.
8. During menstruation the lining of the _____ breaks down.
9. What is the function of the placenta? _____
10. Name three methods of family planning. _____

Section B

1. Fill in the spaces using the correct word from this list: ovulation, uterus, fertilisation, nine.

   During menstruation, the lining of the _____ breaks down.

   About 14 days from the end of the cycle, _____ occurs.

   If _____ occurs, menstruation will not occur.

   In humans a normal pregnancy lasts about _____ months.

2. Name the parts labelled A, B, C, D, E, F in the diagram. State (i) where fertilisation takes place, (ii) where the embryo develops. What name is given to the gametes produced by (i) the ovary (ii) the testis? State three differences between the male gamete and the female gamete.

3. Write a brief account of what happens from the time that fertilisation takes place until the baby is born, assuming that everything proceeds normally.

4. Complete the following sentences.

   (a) The female sex cell is called the _____.

   (b) The female sex cell is made in the _____.

   (c) The male sex cell is called the _____.

   (d) The male sex cell is made in the _____.

   (e) The _____ carries the sperm from the testes to the penis.

   (f) The egg passes from the ovary into the _____.

   (g) Fertilisation usually takes place in the _____.

   (h) Food and oxygen are carried to the baby by the _____.

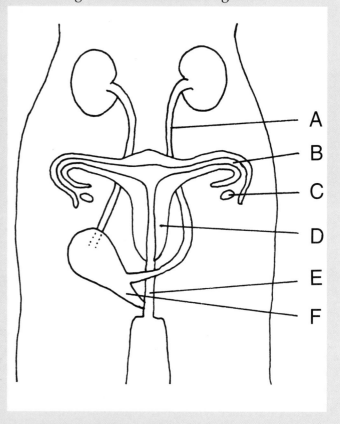

5. Name the organ where the mother's blood runs alongside the baby's blood. Name the tube connecting the baby to this organ. Name two substances that pass from the mother to the baby. Name two substances that pass from the baby to the mother.

6. Do some research to find out (a) how smoking during pregnancy can affect the baby; (b) whether the mother can pass the AIDS virus to the baby, and if so, how; (c) what causes German measles, and how German measles in the mother can affect the unborn baby. Can anything be done about the danger of German measles?

7. Do some research to find out if there are forms of family planning other than those mentioned in this chapter. Is any method perfect?

CHAPTER 50 GENETICS (THE STUDY OF HEREDITY)

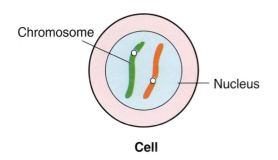
Cell

So, the sperm has fertilised the egg and the baby has developed in the womb for about 9 months. Now for the big moment: the baby is about to be born! Will it have brown eyes or blue eyes, dark hair or fair hair, dark skin or light skin? To find the answers to these questions, we have to look inside the sperm and the egg. The nucleus of each sex cell contains 23 thread-shaped parts called chromosomes.

> *Chromosomes are thread-shaped structures found in the nucleus of a cell.*

These chromosomes carry a series of chemicals called genes, which contain the instructions for forming the new baby. We get half our genes from our father and half from our mother.

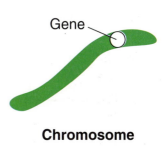
Chromosome

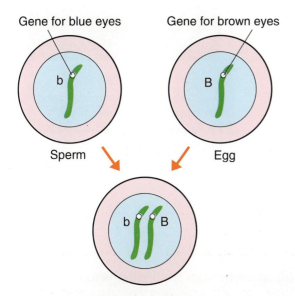

> *Genes are chemicals that are responsible for passing on characteristics from parents to children.*

One gene is responsible for hair colour, another for eye colour, another for height, and so on. But wait a minute! Suppose the gene from your mother's egg says brown eyes, and the gene from your father's sperm says blue eyes! What happens then?

The answer is that some genes are stronger than others. The gene for brown eyes is **dominant** over the gene for blue eyes, which is recessive. Capital letters are used to represent dominant genes, and small letters are used to represent recessive genes.

Make out a bar chart of all the students in your class who

(*a*) have dark hair/fair hair

(*b*) have brown eyes/blue eyes

(*c*) have attached/free ear lobes

(*d*) can/cannot roll their tongues

These are all **inherited characteristics**.

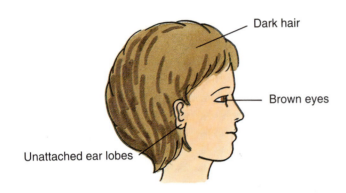

Dominant characteristics

The British biologist Charles Darwin (1809–1882) proposed a theory of evolution. This theory was presented in *On the Origin of Species* in 1859 and *The Descent of Man* in 1871.

Gregor Mendel (1822–1884), an Austrian monk, discovered the basic principles of heredity through experiments with peas he grew in the monastery garden.

Wouldn't it be wonderful if you were born knowing how to drive a car and speak French, just because your mother can drive a car and speak French? Unfortunately, this does not happen. Certain things are not inherited; you have to learn them. Scientists call these **non-inheritable characteristics**.

SUMMARY

- Chromosomes are thread-shaped structures found in the nucleus of a cell.
- Genes are chemicals that are responsible for passing on characteristics from parents to children.

QUESTIONS

Section A

1. What are chromosomes? _____
2. Where in a cell are chromosomes found? _____
3. How many chromosomes are there in a human sperm cell? _____
4. What are genes? _____
5. The gene for _____ is dominant; the gene for _____ is recessive.
6. _____ _____, who was born in 1822, discovered by doing experiments with pea plants that genetic characteristics are inherited.
7. Which of the following can be inherited: eye colour, ability to roll tongue, accent, ability to speak French. _____
8. Underline the non-inheritable characteristics in this list: hair style, hair colour, swimming ability, eye colour.

Section B

1. Name (*a*) three inherited characteristics, (*b*) three non-inheritable characteristics, (*c*) three dominant genes, (*d*) three recessive genes.
2. The gene for tallness is dominant over the gene for shortness. Does this mean that, if a girl is born with a gene for tallness, she will be tall no matter what sort of diet she eats or what kind of environment she lives in? Research this question in the library.

CHAPTER 51 PLANTS

Many people are interested in animals. They keep cats and dogs as pets, they bring their children to the zoo and they join organisations like the Society for the Prevention of Cruelty to Animals. Not many people, except gardeners, are interested in plants. This is because they don't know how important plants are. You already know that plants are living things and that they have the seven characteristics of living things. (Can you remember what these are?)

You also know that they supply us with food, timber, medicine, and even the paper on which this book is printed. Of course there are many different kinds of plants. Most of them have roots, stems, leaves and flowers. Each part of the plant is there for a purpose, as we can see from the diagram at the top of the next page.

Examine some flowering plants and see if you can identify the parts.

For humans, the most important thing about plants is that they can make food. Do you know what happened in Ireland from 1845 to 1847 when the potato crop failed? What happens in third-world countries when crops fail? Now can you see how important plants are?

THE PLANT AS A FACTORY

Have you ever visited a factory? I have. It was a factory for making school desks. The raw materials (tubular steel and timber) were taken in at one end of the factory and wheeled along a corridor to the workshops. There, using electrical energy from the ESB, the tubular steel and timber were cut, shaped and put together to make desks. The desks were then stored in various parts of the factory until they were needed.

A plant is like a factory. The raw materials (minerals and water) are taken in by the roots. They pass up through the stem to the leaves. There, with carbon dioxide and the help of sunlight energy, they are made into food which is then stored in the leaves, stem or roots, depending on the plant.

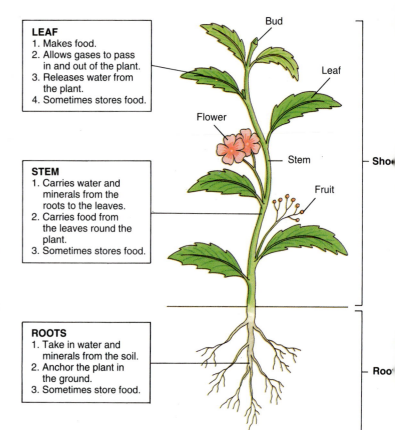

LEAF
1. Makes food.
2. Allows gases to pass in and out of the plant.
3. Releases water from the plant.
4. Sometimes stores food.

STEM
1. Carries water and minerals from the roots to the leaves.
2. Carries food from the leaves round the plant.
3. Sometimes stores food.

ROOTS
1. Take in water and minerals from the soil.
2. Anchor the plant in the ground.
3. Sometimes store food.

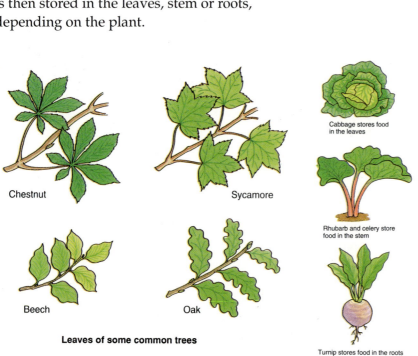

Chestnut   Sycamore

Beech   Oak

Leaves of some common trees

Cabbage stores food in the leaves

Rhubarb and celery store food in the stem

Turnip stores food in the roots

SUMMARY

- Plants are useful because they supply the materials for timber, food, medicine, clothing and paper.
- The functions of the various parts of a plant are given in the table below.

Root	Stem	Leaf
To take in water and minerals	To carry water and minerals from the roots to the leaves	To make food
To anchor the plant in the ground	To store food	To release water from the plant
Sometimes, to store food	To carry food from the leaves to the rest of the plant	To allow gases to pass in and out of the plant
	To hold the leaves up high so that they can get sunlight	Sometimes, to store food

QUESTIONS

Section A

1. What are the seven characteristics of living things?
2. Roots _____ plants in the soil and absorb _____ and _____ from the soil.
3. Give two functions of the stem. _____
4. Give two functions of the leaf. _____
5. Name one plant that is useful in medicine and the drug that can be got from it. _____
6. The plant uses sunlight and the gas _____ _____ to make food.
7. Name a plant that stores food in its roots. _____
8. Name a plant that stores food in its stem. _____
9. Name a plant that stores food in its leaves. _____
10. What plant do the leaves in the diagram come from? _____

PHOTOSYNTHESIS CHAPTER 52

One of the biggest differences between plants and animals is that plants can make food and animals cannot. Plants can make food because they have a green chemical called chlorophyll which can trap light energy and use it, with carbon dioxide and water, to make a simple sugar called glucose. The way in which it does this is called "photosynthesis".

Photosynthesis is the way in which plants make food.

The word equation for photosynthesis is

$$\text{carbon dioxide} + \text{water} \xrightarrow[\text{chlorophyll}]{\text{sunlight}} \text{glucose} + \text{oxygen}$$

The balanced chemical equation for photosynthesis is

$$6CO_2 + 6H_2O \xrightarrow[\text{chlorophyll}]{\text{sunlight}} C_6H_{12}O_6 + 6O_2$$

So photosynthesis is vital because
- it provides plants and all other living things with food.
- it renews the earth's supply of oxygen.

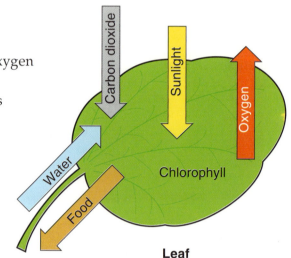

Leaf

THE USE OF GLUCOSE BY PLANTS

The glucose produced in photosynthesis is used in a number of ways.
1. It is used as a source of energy for the plant.
2. It is stored as starch in the roots, stem and leaves (starch can be changed back to glucose when needed).
3. It is used to make cellulose cell walls.
4. It is combined with minerals to make protein.

THE STRUCTURE OF LEAVES

Leaves are specially designed for photosynthesis.
1. They have broad flat surfaces to absorb as much light as possible.
2. They are thin so that carbon dioxide can easily reach the cells.
3. They have stomata (tiny holes) on their underside to let water out and gases in and out.

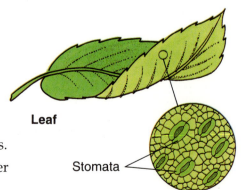

Leaf

Stomata

EXPERIMENT 52.1
AIM: TO TEST A LEAF FOR STARCH

Warning: Alcohol is highly inflammable. When boiling a leaf in alcohol, it is best to turn off the Bunsen burner before placing the test-tube containing the alcohol into the beaker of water. Since alcohol boils at 78°C, it will boil under these circumstances.

Method
1. Take a leaf from a plant that has been standing in sunlight.
2. Boil the leaf in water to kill it and break down the cell membranes, so that iodine can get into the cells.
3. Boil the leaf in alcohol to remove the chlorophyll.
4. Dip the leaf in boiling water to soften it.
5. Spread the leaf out on a white tile and drop iodine solution on it.

    Result: If starch is present it will turn black. A yellow/brown colour indicates no starch.

EXPERIMENT 52.2
AIM: TO SHOW THAT LIGHT IS NECESSARY FOR PHOTOSYNTHESIS

Method
1. Store a pot plant in darkness for 48 hours to destarch the leaves.
2. Cover part of one leaf and leave the plant in good light for a day.
3. Remove the chlorophyll from the leaf as in the last experiment.
4. Use iodine solution to test the leaf for starch.

    Result: Only the part of the leaf that was exposed
    to light turns black.

    Conclusion: Light is necessary for photosynthesis.

EXPERIMENT 52.3
AIM: TO SHOW THAT CARBON DIOXIDE IS NECESSARY FOR PHOTOSYNTHESIS.

Method
1. Store two pot plants in the dark for 48 hours to destarch the leaves.
2. Set up the apparatus as shown in the diagram and leave in good light for about 8 hours.
3. Take a leaf from each plant and test them for starch.

    Result: The leaf from plant A does not turn black, but the leaf from plant B does.

    Conclusion: Carbon dioxide is necessary for photosynthesis.

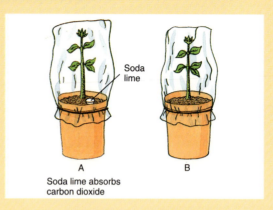

EXPERIMENT 52.4
AIM: TO SHOW THAT CHLOROPHYLL IS NECESSARY FOR PHOTOSYNTHESIS

Method
1. Take a variegated leaf from a plant.
2. Remove the chlorophyll as shown above.
3. Use iodine to test the leaf for starch.

    Result: Only the part of the leaf that contained chlorophyll turns black.

    Conclusion: Chlorophyll is necessary for photosynthesis.

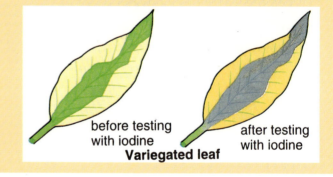

EXPERIMENT 52.5
AIM: TO SHOW THAT OXYGEN IS GIVEN OFF DURING PHOTOSYNTHESIS

Method
1. Set up the apparatus as shown in the diagram.
2. Leave it for a few days to allow enough gas to collect in the test-tube.
3. Test the gas with a glowing splint.

    Result: The glowing splint ignites.

    Conclusion: The gas given off during photosynthesis is oxygen.

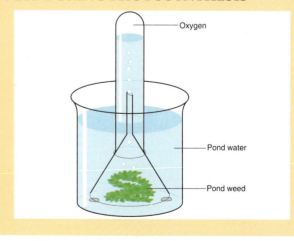

PHOTOSYNTHESIS AND RESPIRATION

Respiration: glucose + oxygen $\longrightarrow$ carbon dioxide + water + energy

Photosynthesis: carbon dioxide + water $\xrightarrow[\text{chlorophyll}]{\text{sunlight}}$ glucose + oxygen

If you compare the equation for respiration with the equation for photosynthesis, you will see that they are the reverse of each other. This is because respiration uses up food and oxygen. If this process were not reversed by photosynthesis, the earth's supply of food and oxygen would eventually run out.

Respiration	Photosynthesis
All living things	Green plants only
All the time	During daylight only
Oxygen in	Oxygen out
Carbon dioxide out	Carbon dioxide in
No chlorophyll	Chlorophyll needed

MINERAL NUTRITION IN PLANTS

Plants make food by photosynthesis, but they also need minerals, particularly nitrogen, potassium and phosphorus. Nitrogen is needed for healthy leaves, potassium for healthy flowers and fruit, and phosphorus for healthy roots.

In fact it is possible to grow plants in solutions containing minerals.

EXPERIMENT 52.6

AIM: TO SHOW THE EFFECTS OF MINERAL DEFICIENCY IN PLANTS

Method
1. Set up four jars as shown in the diagram.
2. Examine the seedlings each day for a fortnight.

Result: As shown in the diagram.

Conclusion: Certain minerals are essential for healthy plant growth.

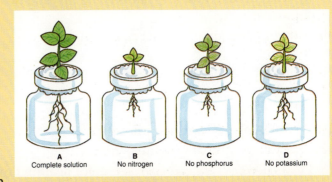

Deficiency	Result
Nitrogen	Yellow leaves, weak stem
Phosphorus	Poor root growth
Potassium	Poor flowers and fruit

SUMMARY

- Photosynthesis is the way in which plants make food.
- The word equation for photosynthesis is

$$\text{carbon dioxide} + \text{water} \xrightarrow[\text{chlorophyll}]{\text{sunlight}} \text{glucose} + \text{oxygen}$$

The balanced chemical equation for photosynthesis is

$$6CO_2 + 6H_2O \xrightarrow[\text{chlorophyll}]{\text{sunlight}} C_6H_{12}O_6 + 6O_2$$

Respiration	Photosynthesis
All living things	Green plants only
All the time	During daylight only
Oxygen in	Oxygen out
Carbon dioxide out	Carbon dioxide in
No chlorophyll	Chlorophyll needed

QUESTIONS

Section A

1. What is photosynthesis? _____
2. Name the green chemical necessary for photosynthesis. _____
3. Name the gas produced by photosynthesis. _____
4. Give the word equation for photosynthesis. _____
5. Plants are placed in an aquarium to provide _____ for the fish by _____

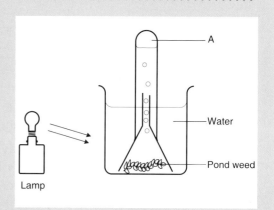

6. What gas is produced by the pondweed in the diagram on the right? _____
7. The apparatus shown in the diagram (below right) was kept in the dark for two days and then exposed to sunlight for 8 hours. What is being investigated in this experiment? _____
8. Name a common plant that cannot make its own food. _____
9. Write out the chemical equation for photosynthesis. _____

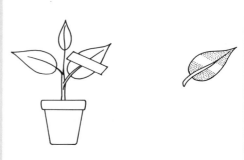

10. What chemical is used to test for starch? _____
11. When do plants carry out photosynthesis? _____
12. When do plants carry out respiration? _____
13. Plants need _____ for healthy flowers and fruit.
14. Plants need nitrogen for healthy _____.
15. Plants need _____ for healthy roots.

Section B

1. In question 6, section A, what gas is being produced? Describe a simple test for this gas. Name the biological process responsible for producing the gas.
2. Write an account of an experiment to show that light is necessary for photosynthesis. Use the headings Aim, Method, Result, Conclusion.
3. Photosynthesis can be represented by the following equation:
   A + water, B and chlorophyll = glucose + C.
   Identify A, B and C.
4. Two healthy potted destarched plants were set up as shown in the diagram and left in a sunny position for a number of hours. The experiment is an investigation of the conditions necessary for photosynthesis.

   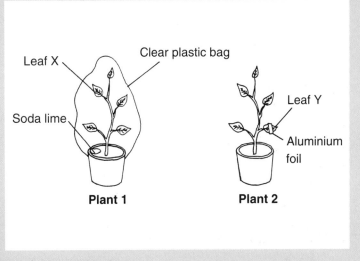

   (a) Which condition is being investigated using (i) plant 1, (ii) plant 2?
   (b) Outline the procedure you would use to test leaves X and Y for starch at the end of the experiment. What result would you expect from (i) leaf X, (ii) leaf Y? Describe how you would destarch the plant leaves before starting the investigation. Why is it necessary to destarch the leaves?

5. (a) What is photosynthesis?
   (b) Where does it occur?
   (c) What is the green substance in leaves called?
   (d) What does a plant use glucose for?
   (e) How is the glucose stored?
   (f) What else is produced during photosynthesis?

6. The four tubes in the diagram were set up in a well-lit position.

   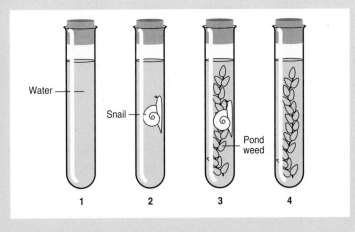

   (a) In which tubes does photosynthesis take place?
   (b) In which tubes is oxygen produced?
   (c) In which tubes does respiration take place?
   (d) In which tube will the snail live longer? Why?

7. In an experiment, the apparatus was set up as shown in the diagram below. The bubbles produced in three minutes with the lamp placed at a distance of 10 cm from the water plant were counted. This was repeated with the lamp at distances of 20 cm and 30 cm from the plant.

The results were as follows:

Distance from lamp to plant (cm)	Number of gas bubbles produced in three minutes
10	189
20	45
30	21

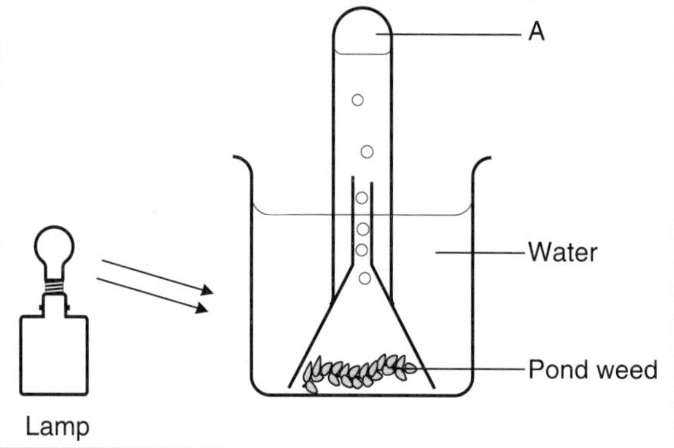

(a) Name gas A. Describe a simple test for it.
(b) Name the biological process taking place in the water plant that is responsible for producing this gas.
(c) Calculate the rate of gas production in bubbles per minute for each of the three distances.
(d) Suggest why the rate of gas production decreases as the distance increases.

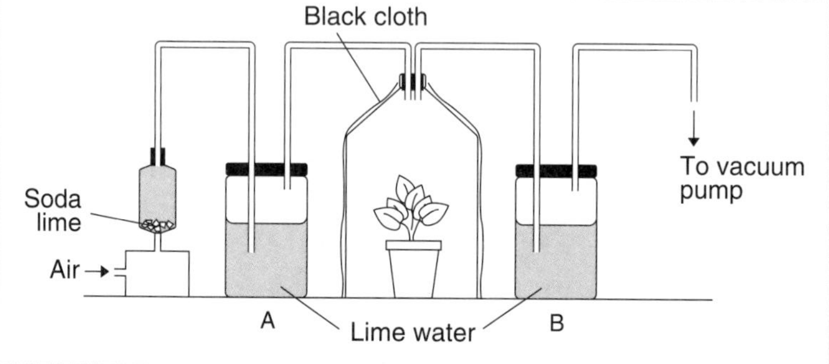

8. In the experiment shown in the diagram, the soda lime absorbs carbon dioxide from the air.
   (a) What would you expect to happen to the limewater in jar A? Why?
   The bell jar containing the plant is covered with a black cloth so that the plant is in darkness.
   (b) Can photosynthesis take place? Why/why not?
   (c) What process do you think is being carried out by the plant in the bell jar?
   (d) What should be happening to the limewater in jar B? Why?
   (e) What do you learn about respiration from this experiment?
   (f) What gas do plants take in when they respire?
   (g) What gas do plants give off when they respire?
   (h) Where in the plant does respiration take place?
   (j) What difference would it make if the black cloth were removed?
   (k) Where in a plant does photosynthesis take place?
   (l) What gas does a plant take in during photosynthesis?

9. Make a chart like the one below, comparing photosynthesis and respiration, and fill it in.

	Photosynthesis	Respiration
Gas taken in		
Gas given off		
Conditions necessary		
Product		
When does it take place?		
Where does it take place?		

CHAPTER 53 TRANSPORT IN PLANTS

Most living things contain thousands, if not millions, of cells. This means they must have a transport system to carry food and oxygen to the cells and to carry waste away from the cells. In your body, there is a continuous flow of blood from the heart to the cells and back again.

In plants there is a flow of water (and minerals) from the roots to the leaves. Also, food made in the leaves is carried all round the plant.

WATER TRANSPORT

The flow of water from the roots to the leaves is called the transpiration stream.

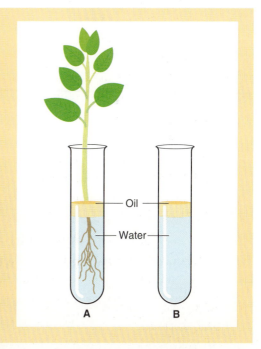

EXPERIMENT 53.1
AIM: TO SHOW THE ABSORPTION OF WATER BY THE ROOTS

Method
1. Set up the apparatus as shown in the diagram.
2. Mark the water level in each tube and let them stand for about a week.
3. Note the water level in each tube.

Result: the water level in tube A has fallen but the level in tube B has not.

Conclusion: The roots of the plant in tube A have absorbed water.

EXPERIMENT 53.2
AIM: TO SHOW THE MOVEMENT OF WATER IN A PLANT

Method
1. Set up the apparatus as shown in the diagram and leave it for 3 or 4 days.
2. Remove the celery from the coloured water and rinse it under a tap.

3. Examine the veins in the leaves.
4. Use a sharp blade or scissors to cut across the stem and leaves, and examine each with a magnifying glass.

   Result: You will see that water has travelled up through the stem into the leaves.

The movement of water in a plant can also be shown like this

EXPERIMENT 53.3
AIM: TO SHOW TRANSPIRATION IN A PLANT

Method

Note: Water causes cobalt chloride to change colour from blue to pink.

1. Set up the apparatus as shown in the diagram and leave it in a warm position for several hours.
2. Test the droplets of colourless liquid on the inside of the bag with cobalt chloride paper.

   Result: The liquid causes the paper to change colour from blue to pink.

   Conclusion: Water is lost from the leaves of a plant.

Transpiration

BIOLOGY · TRANSPORT IN PLANTS

A cactus reduces water loss by having spines instead of leaves.

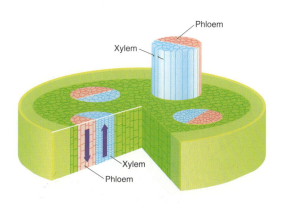

TRANSPIRATION

We have already mentioned the transpiration stream. Obviously, if water keeps flowing up from the roots to the leaves then it must escape from the leaves into the air.

The loss of water from the surface of a plant is called transpiration.

In fact, about 95% of the water is lost through the leaves.

Functions of transpiration

The functions of transpiration are
1. to carry water to the leaves for photosynthesis
2. to carry minerals up from the roots
3. to cool the plant.

Factors affecting transpiration

Would you hang clothes out to dry on a wet day? On a cold day? On a calm day? They certainly would not dry on a wet day. On a calm day it would depend on the temperature. The best time to hang them out would be on a warm, dry, windy day. Can you say why?

The factors that affect the drying of clothes are basically the same as those that affect transpiration in plants: temperature, humidity (the amount of water in the air) and wind. Another factor, of course, is the amount of water available to the plant in the soil. The plant tries to control the amount of water loss: in a drought, a plant tries to reduce its water loss to ensure its survival.

FOOD TRANSPORT

If you look at the cross-section of the stem in the diagram below you will see two kinds of tubes: **xylem** tubes, which carry minerals and water from the roots to the leaves, and **phloem** tubes, which carry food from the leaves all round the plant. Scientists refer to these tubes as **transport tissue**.

SUMMARY

- Transpiration is the loss of water from the surface of a plant.
- The transpiration stream is the flow of water from the roots to the leaves.
- Xylem tubes carry water and minerals up from the roots.
- Phloem tubes carry food from the leaves round the plant.
- Water is lost through tiny holes called stomata in the underside of the leaf.
- The factors affecting the rate of transpiration are temperature, humidity, wind and the amount of water in the soil.

QUESTIONS

Section A
1. What is the transpiration stream? _____
2. Name the type of cell that carries water from the roots to the leaves. _____
3. What is transpiration? _____
4. The loss of water takes place through tiny holes called _____ on the underside of the leaves.
5. Give three functions of transpiration. _____
6. State two atmospheric (weather) conditions that increase the rate of transpiration.
   _____ _____
7. How would you test the colourless liquid given out by the leaves of a plant to show that it is water?
8. What is the function of the phloem cells in a plant? _____
9. The _____ cells carry water and minerals up the plant.
10. The _____ cells carry food around the plant.

Section B
1. Describe a simple experiment to show the movement of water in a plant.
2. Choose your answer to the following questions from this list: respiration, transpiration, photosynthesis.
   (i)   Which of the above is most affected by humidity?
   (ii) Which stops completely in darkness?
   (iii) Which occurs in every living cell?
   (iv) Which is responsible for pulling water up a plant?
   (v)   Which causes carbon dioxide to be given off?
   (vi) Which causes oxygen to be given off?
3. What is meant by transpiration?
   The apparatus shown in the diagram was weighed at noon, 6 p.m. and midnight.
   The masses were as follows:

Time	12 noon	6 p.m.	12 midnight
Mass	491 g	482 g	479 g

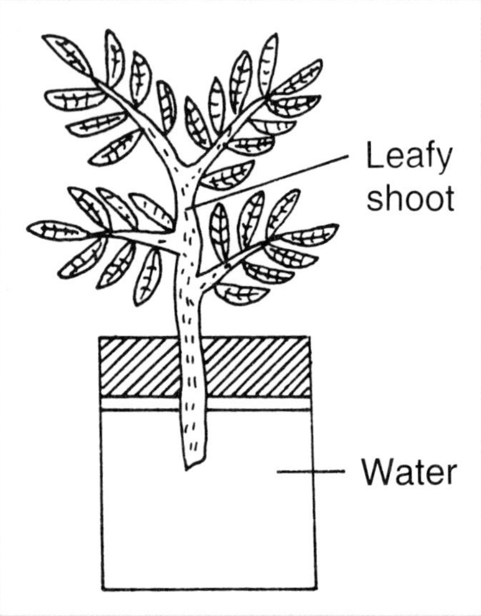

   Assuming the changes in mass are due entirely to transpiration, calculate the average transpiration rate in grams per hour for the period from (i) noon to 6 p.m., (ii) 6 p.m. to midnight. Suggest a reason for the difference between the two rates.

4. (a) What is the transpiration stream?
   (b) Give three functions of transpiration.
   (c) When is the transpiration stream at its fastest?
   (d) What happens if water is lost from the leaves faster than it is taken in through the roots? How is (i) holly, (ii) cactus adapted to prevent this?
   (e) Do some research to find out what guard cells are and what they have to do with transpiration.

5. (a) What is the purpose of the experiment in the diagram?
   (b) Why was a plastic cover put on the pots?
   (c) What are the drops of liquid in jar A?
   (d) How would you prove this?
   (e) Why were the leaves removed from the plant in jar B?
   (f) In an experiment, what is B called?

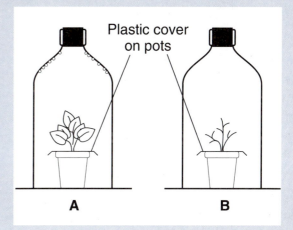

CHAPTER 54 RESPONSIVENESS IN PLANTS

Do you remember the characteristics of living things? One of them is sensitivity. This means that living things are aware of changes in their environment and respond to them. Of course plants respond much more slowly than animals, but they still respond, as you will see in this chapter.

Light and water are very important to plants. Without them, they cannot make food. For this reason, plants are sensitive to light and water. They are also sensitive to gravity because the roots must grow down into the soil to get water and minerals and the shoot must grow up to get light. Plants respond to light, water and gravity by growing (slowly) in certain directions. These responses are called tropisms.

| *A tropism is a growth response of a plant to a stimulus.*

PHOTOTROPISM

Plants respond to light by growing towards it and by turning their leaves, so as to get as much light as possible for photosynthesis.

Breakthrough Science

The growth response of a plant to light is called phototropism. Either of the simple experiments in the diagram on the right can be used to demonstrate phototropism.

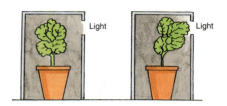

GEOTROPISM

Can you imagine what it would be like if a farmer had to examine each seed to make sure it was the right way up before putting it into the soil? How long do you think it would take to sow a field of corn? Fortunately, no matter how you plant a seed, the roots will always respond to gravity by growing down. This is called positive geotropism. The stem responds to gravity by growing up. This is called negative geotropism.

The growth response of a plant to gravity is called geotropism.

Geotropism can be demonstrated by setting up the simple experiment in the diagram.

Put some bean seeds between a roll of blotting paper and the inside of a glass jar. Pour a little water into the jar and put it in a sunny place. After a few days roots begin to grow down towards the water and shoots will grow up to the light.

Plants can move too. They move when they grow towards the light; also their stems grow upwards and their roots grow down. You can see this by taking a pot plant and putting it on its side. Its horizontal stem will bend and gradually turn upwards.

SUMMARY

- A tropism is the growth response of a plant to a stimulus.
- Phototropism is the growth response of a plant to light.
- Geotropism is the growth response of a plant to gravity.

QUESTIONS

Section A
1. What is a tropism? _____
2. _____ is the growth response of a plant to light.
3. The growth response of a plant to gravity is called _____
4. The _____ responds positively to gravity, the _____ responds negatively to gravity.
5. A plant is placed on a windowsill. What is the stimulus? What will the response be? _____
6. Of what advantage is phototropism to a plant? _____

336

Section B

1. Describe a simple experiment to show geotropism in plants.
2. Movement and sensitivity are two of the characteristics of living things. Explain how plant movement and sensitivity differ from animal movement and sensitivity.
3. Match each item in Box A with the corresponding item in Box B.

Box A	Box B
Positive phototropism	The response of a plant shoot to gravity
Negative phototropism	The response of a plant root to light
Positive geotropism	The response of a plant shoot to light
Negative geotropism	The response of a plant root to gravity

4. (a) What advantage does phototropism have for plants?
   (b) What advantage does geotropism have for plants?
   (c) If plants did not have phototropism or geotropism, what difference would it make to farmers in spring? What difference would it make at harvest time? How would it affect world food supplies?

CHAPTER 55 PLANT REPRODUCTION

Do you remember reading about human reproduction earlier in this book? The male nucleus fuses with the female nucleus. This is called sexual reproduction.

Sexual reproduction occurs when a male nucleus fuses with a female nucleus.

Scientists say that a male gamete fuses with a female gamete to form a zygote.

A gamete is a sex cell.

Sexual reproduction also occurs in plants. A plant's sex organs are in the flower. Different plants have flowers that look very different, but they all contain the same basic parts.

Petal: attracts insects by its colour and scent

Stamen: produces pollen grains which contain the male sex cells

Carpel: produces the female sex cells

Sepal: protects the flower when it is in bud

At this stage, it would be a good idea to get a number of different flowers and take them apart. Stick the parts onto a page of your copy.

STAGES IN PLANT REPRODUCTION (SEXUAL)

Pollination

The first step towards reproduction is pollination.

> *Pollination is the transfer of pollen from the anther to the stigma.*

There are two main types of pollination: insect pollination and wind pollination.

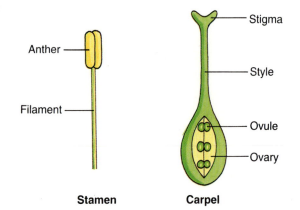

Insect pollination

This means that bees and other insects carry the pollen from one flower to another.

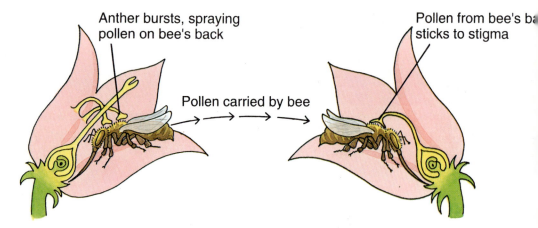

Wind pollination

In this case, the wind blows the pollen from one flower to another.

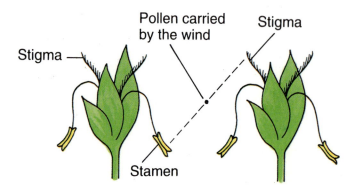

Insect-pollinated flowers and wind-pollinated flowers look quite different. Insect-pollinated flowers are generally colourful and scented, with the anther and stigma inside the flower.

Wind-pollinated flowers have no smell, and are often green. Indeed, some people would not recognise them as flowers at all (e.g. in grasses).

BIOLOGY · PLANT REPRODUCTION

The ovary becomes a fruit

	Insect-pollinated	Wind-pollinated
Petals	Coloured, scented	Often green
Stamens	Inside flower	Hang outside flower
Stigma	Small, inside flower	Outside flower, feathery to catch pollen
	Nectar	No nectar

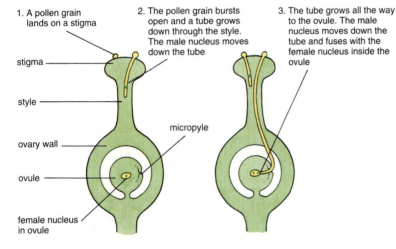

1. A pollen grain lands on a stigma
2. The pollen grain bursts open and a tube grows down through the style. The male nucleus moves down the tube
3. The tube grows all the way to the ovule. The male nucleus moves down the tube and fuses with the female nucleus inside the ovule

Labels: stigma, style, ovary wall, ovule, female nucleus in ovule, micropyle

Fertilisation

Pollination is followed by fertilisation, as shown in the diagram.

Fertilisation occurs when the male nucleus fuses with the female nucleus.

The fertilised ovule becomes the seed. The petals and the stamen now die and fall off. The ovary swells and becomes a fruit.

Seed dispersal

If too many plants are growing in a small space, there will not be enough water, minerals or light for all of them.
To avoid this, the seeds of a plant must be spread as far away from the parent plant as possible. This happens in a number of different ways.

Dandelion Sycamore

Wind dispersal

Dandelion, thistle, sycamore and ash seeds are scattered by the wind.

Ash Thistle

Burdock Strawberry Blackberry

Animal dispersal

Blackberries and strawberries are eaten by animals and the seeds are passed out later. Burdock seeds stick to animals and are knocked off later.

Self-dispersal

The seed pods of the furze, pea and lupin burst open when they are dry and scatter the seeds.

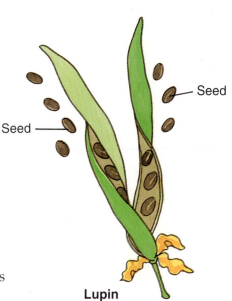
Lupin

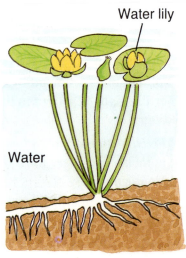

Water dispersal

Many plants, such as the water lily, produce seeds that can float for long distances away from the parent plant.

Germination

Seeds are usually dispersed in the autumn and lie dormant for the winter because it is too cold. In the spring they germinate.

Germination means that the seed begins to grow into a new plant.

We have already seen how a seed germinates, in the simple experiment in the last chapter to demonstrate geotropism.

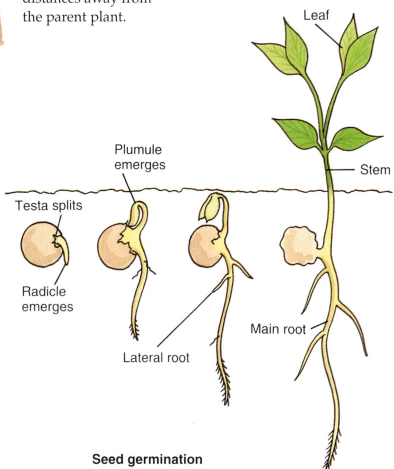

Seed germination

EXPERIMENT 55.1
AIM: TO SHOW THAT WATER, OXYGEN AND HEAT ARE NECESSARY FOR GERMINATION

Method:
1. Set up four test-tubes as shown in the diagram.
2. Put test-tubes 1, 2 and 3 in a warm spot and place test-tube 4 in a fridge.
3. Leave them for a few days, making sure that the cotton wool in 3 and 4 does not dry out.

Result: Only the seeds in 1 germinate.

Conclusion: Water, oxygen and heat are all necessary for germination.

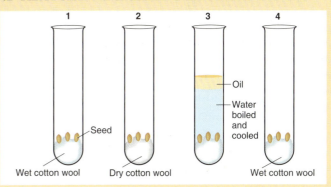

LIFE CYCLE

The diagram shows the life cycle of a flowering plant. Why not buy a packet of sweet pea seeds and sow them in your garden so that you can follow their life cycle for yourself?

ASEXUAL REPRODUCTION

Some plants reproduce asexually. This does not involve male gametes, female gametes or fertilisation. In fact, only one parent is involved.

Asexual reproduction involves only one parent.

A common example of asexual reproduction is reproduction by means of spores. Remove the stalk from a ripe mushroom and place the head on a piece of paper. Cover it with a glass cover. Examine it over the next few days: you will see spores on the paper.

Another example of asexual reproduction is reproduction by means of bulbs, e.g. in daffodils.

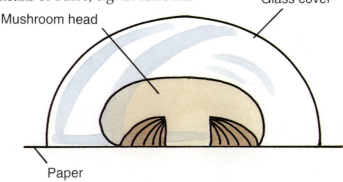

SUMMARY

- Sexual reproduction means that the male gamete fuses with the female gamete to form a zygote.
- Pollination is the transfer of pollen from the anther to the stigma.
- Fertilisation means that the male nucleus fuses with the female nucleus.
- Germination means that the seed begins to grow into a new plant.
- In order to germinate, a seed needs water, oxygen and heat.
- Asexual reproduction involves only one parent.

QUESTIONS

Section A

1. Sexual reproduction involves the union of a _____ and a _____.
2. One function of the petal in a flower is _____ .
3. One function of the sepal in a flower is _____ .
4. What is meant by pollination? _____ .
5. Fertilisation occurs when the _____ fuses with the _____ .
6. Name three methods of seed dispersal. _____
7. How are the seeds of the plants shown in the diagram scattered? _____ .
8. Why is it necessary that seeds be dispersed? _____
9. What is meant by germination? _____
10. State two conditions necessary for seed germination. _____
11. Which of the following is essential for germination: soil, light, water, carbon dioxide, darkness?
12. What gas is released by germinating seeds?
13. Put a mark beside the correct order of events in plant reproduction.
    (a) pollination, germination, fertilisation, dispersal
    (b) pollination, fertilisation, germination, dispersal
    (c) pollination, dispersal, fertilisation, germination
    (d) pollination, fertilisation, dispersal, germination.
14. What is asexual reproduction? _____
15. Give an example of asexual reproduction. _____

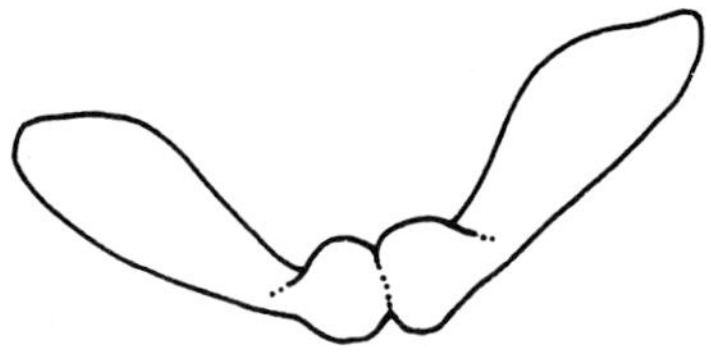

Sycamore

Burdock

Blackberry

Section B

1. Name four of the parts A, B, C, D, E shown in the diagram. Explain the term 'pollination', referring in your answer to the labelled parts of the flower in the diagram.
2. Give four methods of seed dispersal, and name a plant to go with each method.
3. Using the headings Aim, Method, Result and Conclusion, describe an experiment to show the factors necessary for germination.
4. Design a simple experiment to see whether light is necessary for seed germination. Use a control and the headings Aim, Method, Result, Conclusion.
5. Match each item in Box X with the corresponding item in Box Y.

Box X	Box Y
Ovary	Attracts insects
Petal	Produces eggs
Sepal	Protects young flower
Anther	Produces male cells

6. Why do seeds need to be dispersed? Complete the following table.

Plant	Type of dispersal	How fruit or seeds are suited to this
Thistle	Wind	Parachute on seeds
Strawberry		
Sycamore		
Lupin		
Poppy		
Burdock		

7. What part of the flower becomes a seed? What part becomes a fruit? How would the seeds shown in the diagram be dispersed? Plants with seeds like (C) and (D) produce much more seeds than plants with seeds like (A) and (B). Why do you think this is?

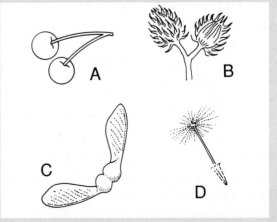

8. What do you think the terms "self-pollination" and "cross-pollination" mean? Go to the library and look for diagrams of the flowers of self-pollinating plants. How do they differ from the flowers of cross-pollinating plants?

ECOLOGY   CHAPTER 56

"To waste and destroy our natural resources instead of increasing their usefulness will undermine the very prosperity which we are obliged to hand down to our children."

When do you think that was said? Last week? Last year? Wrong! It was said by President Theodore Roosevelt to the US Congress in 1907. Unfortunately, they were not listening. It took a long time for people to realise that plants and animals, including humans, do not live in isolation. They are affected by each other and by their surroundings or environment.

Ecology is the study of the relationship between plants and animals and their environment.

HABITATS

The place where a plant or animal lives is called its habitat. There are many different habitats: woodland, grassland, hedgerow, pond, rocky seashore, etc. A habitat may contain a smaller habitat, known as a micro-habitat. For example, a woodland may contain a fallen tree trunk which is the habitat of many insects.

A pond is the habitat of a frog, but the frog is not the only organism (living thing) in the pond. All the organisms in a habitat form a community of plants and animals.

A community and its environment are called an ecosystem.

A very compact food chain
Plant → greenfly → ladybird

TRANSFER OF ENERGY (Energy Flow)

All our energy comes from the sun, and all living things need energy, so there must be a way to transfer energy from the sun to all living things. There is, and, put simply, it works like this.

Plants trap sunlight energy and convert it into food (chemical energy). Plants that can make food are called **producers**. A food chain always begins with a producer.

A producer is a plant that can make food.

These plants are eaten by a group of animals called herbivores.

BIOLOGY · ECOLOGY

| *A herbivore is an animal that eats plants only.*

The herbivores are then eaten by carnivores.

| *A carnivore is an animal that eats animals only.*

These carnivores are eaten in turn by other carnivores in what is called a **food chain**. A food chain is a feeding relationship between plants and animals in a habitat. So how long is a food chain? Usually no more than four or five links. This is because at each stage a great deal of energy is used up in movement, growth, etc., while more energy is lost to the environment in heat energy, excretion, etc. As a result, by the time the fourth or fifth link is reached there isn't enough energy left to support another link.

CONSUMERS

Just as the plants at the start of the food chain can make food, and are therefore called producers, the animals in the food chain cannot make food, and are therefore called **consumers**.

| *A consumer is a living thing that cannot make food.*

So the producer is eaten by a primary consumer, which is eaten by a secondary consumer, which is eaten by a tertiary consumer, and so on.

| *The position of an organism in the food chain is called its trophic level.*

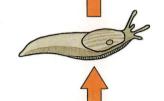

FOOD WEB

A single food chain gives a simple idea of feeding relationships in a habitat, but the situation is much more complex than that. For example, a rabbit

345

Food web

might eat a variety of different plants and be eaten by a variety of different animals, such as the stoat, fox, etc.

A number of interconnected food chains form a food web.

You might think at this stage that all animals are either herbivores or carnivores, but this is not true. Take humans, for example. When you eat a burger and chips, are you a herbivore or a carnivore? In fact you are an omnivore.

An omnivore is an animal that eats both plants and animals.

Badgers, blackbirds and rats are all omnivores.

DECOMPOSERS

What happens if an animal in the food chain dies? Is the chemical energy stored in its body lost forever? No, this is where decomposers come in. Decomposers such as bacteria and fungi break down dead plant and animal matter into mineral salts such as nitrates, which can be absorbed from the soil by plants. In this way the chemical energy stored in the dead organism is recycled back into the food chain.

BIOLOGY · ECOLOGY

Decomposers break down dead plant and animal matter.

Bacteria and fungi are decomposers.

PYRAMID OF NUMBERS

Primary consumers use up a great deal of energy in movement and respiration. This leaves only enough energy to support a smaller number of secondary consumers, and so on. Secondary consumers are also larger, and a number of secondary consumers will eat a greater number of primary consumers.

Pyramid of numbers

COMPETITION

We already know that plants disperse their seeds to avoid competition for water, minerals and light. Animals compete for shelter and for food, as you saw in the food web. They also compete for territory, mates, etc.

ADAPTATION

Living things have to adapt to their surroundings in order to survive. The rabbit has large ears to enable it to hear the fox approaching, and powerful hind legs for burrowing.

The squirrel's eyes are large and to the side of its head to give it a wide field of view. Many insects, such as the caterpillar, are the same colour as the plants they feed on. Why do you think this is?

Adaptation means that plants and animals have developed special features that enable them to survive in their environment.

The leaf insect is the same colour as leaves. It has leaf-like markings on its wings, which help camouflage it from predators

INTERDEPENDENCE

Interdependence means that plants and animals depend on each other for survival.

How many examples of plants depending on animals can you count from the diagram? How many examples of animals depending on plants?

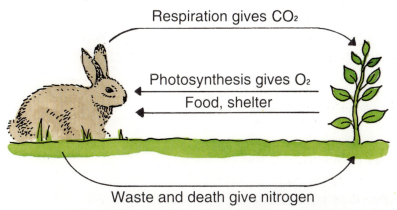

Interdependence of plants and animals

THE BALANCE OF NATURE

Traditionally, farmers have regarded the fox as their enemy. This is understandable. Given a chance the fox will steal hens, ducks or whatever else it can get. At one stage, foxes were doing so much damage that the government placed a bounty on them. A certain amount of money was paid out for every fox killed. Between this and modern farming methods which destroyed much of the fox's natural habitat, the number of foxes declined dramatically. The short-term aim had been achieved. But what if the fox had been wiped out entirely? The rabbit population would have grown out of control. More and more rabbits would eat more and more plants. The effects on the ecosystem could be disastrous! Fortunately it didn't happen that way, but it could have. The point is that feeding relationships in an ecosystem are very complex and very finely balanced. If this balance is upset—if, for example, one organism is removed from the food web—the ecosystem can be destroyed. Before we interfere with nature, as we often do by cutting down forests, draining marshes, declaring war on foxes, etc., we must consider not just the short-term aim but also the long-term effects.

SUMMARY

- Habitat: the place where a plant or animal lives.
- Producer: a plant that can make food.
- Consumer: any living thing that cannot make food.
- Herbivore: an animal that eats plants only.
- Carnivore: an animal that eats animals only.
- Omnivore: an animal that eats both plants and animals.
- Trophic level: the position of an organism in the food chain.
- Decomposers break down dead plant and animal matter.
- Adaptation: plants and animals have developed special features that enable them to survive in their environment.
- Interdependence: plants and animals depend on each other for survival.

QUESTIONS

Section A

1. Give an example of a food chain. _____
2. Animals get food by eating _____ or by eating _____.
3. Every food chain begins with a _____.
4. Name two consumers shown in the diagram. _____

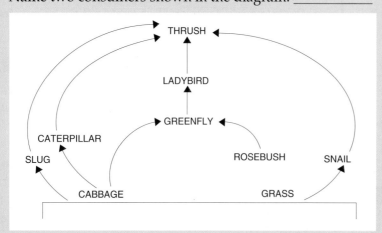

5. What is a producer? Name one from the diagram. _____

6. Give an example of how plants and animals help each other.
7. Name one way in which animals compete with each other.
8. In the pyramid of numbers shown in the diagram, name one animal that might be found at level X and one that might be found at level Y.
   _____

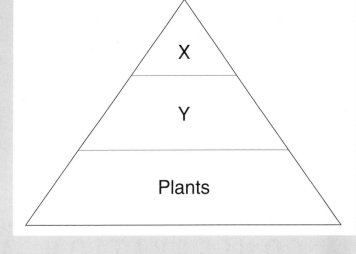

9. What does the word "ecology" mean? _____
10. Explain the term "trophic level".
    _____
11. What is a decomposer? Name one.
    _____
12. What is meant by 'adaptation'?
    _____
13. Name two carnivores. _____
14. Give an example of adaptation by naming an animal and saying how it has adapted.
    _____
15. What does interdependence mean? _____

Section B
1. From the diagram in section A, question 4, name a producer, a primary consumer, a secondary consumer, a herbivore and a carnivore. Extract a food chain from the food web. How many trophic levels are present in this food web?
2. Explain the terms producer, primary consumer, secondary consumer. A gardener discovered that greenfly were attacking his rose bushes, so he brought ladybirds into the garden to eat the greenfly. Construct a simple food chain to illustrate this, and match each member of the food chain with one of the terms given above. If all the rosebushes died, what effect would this be likely to have on the population of (a) greenfly, (b) ladybirds?
3. What is meant by the term "energy flow in the food chain"? Is any energy lost along the chain? How is it lost? Why are there usually no more than four or five links in a food chain? If an organism in the food chain dies, is all the energy contained in that organism lost or wasted? Explain.
4. If a group of schoolchildren on a visit to a habitat in spring take all the frogspawn back to the science lab to see how it will develop, how will this affect the plants, insects, spiders and hawks in the habitat?

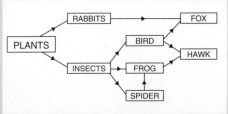

5. Explain each of the following words, giving an example: producer, consumer, herbivore, carnivore, omnivore, decomposer.

Breakthrough Science

6. Here is a very good class experiment.
   (a) Make 100 caterpillars out of dough.
   (b) Paint 20 green, 20 red, 20 yellow, 20 white and 20 brown.
   (c) When the forecast is for dry weather, place them on green plants outdoors.
   (d) Check them each day for seven days.
   (e) Draw a bar chart to show how many of each colour are left.
   (f) Which colour did the birds eat most of? Why do you think this was?
   (g) How did the green caterpillars do? Why do you think this was?
   (h) Would birds find it easy to see the red caterpillars? How did the red caterpillars do? One of the students who did this experiment used the phrase "warning colouration". What did she mean by this?

CONSERVATION AND POLLUTION — CHAPTER 57

CONSERVATION

Conservation is the protection, preservation and careful management of our natural resources.

Would you say that we use the environment wisely? Look at the photographs on this page. Would you say that the land in these photographs is useless? That nothing lives there? Wrong! Lots of

plants and animals live there. Some of these plants and animals are quite rare, and becoming even rarer. People cause enormous damage to wildlife by destroying habitats.

Woodlands containing oak, ash, beech, etc. are cut down to make way for crops or commercial forestry. Natural grasslands containing wild grasses, flowers and insects are ploughed up to make way for crops. Ponds and marshes, which are the natural habitat of birds, frogs and butterflies, are drained for building land. Almost 90% of our bogs have already been destroyed.

In Brazil, tropical rainforests are being cut down at an alarming rate **(deforestation)**. Think of the effect this will have on the amounts of oxygen and carbon dioxide in the air.

Desertification

The cutting down of forests, particularly in warm climates, results in large areas of dry dusty soil. This, combined with soil erosion, causes the spread of deserts. One solution to this problem is to plant young trees to replace those cut down. Another solution is to create nature reserves such as the national parks at Killarney, Connemara and the Burren.

Fishing

World stocks of fish are under threat because of the use of nets with very small mesh which catch young fish as well as adults. Have you heard of people being arrested for using monofilament nets?

POLLUTION

Pollution is any undesirable change in the environment caused by human activity.

There are two main types of pollution: water pollution and air pollution.

Water pollution

Fertilisers are washed into rivers and lakes, causing an increase in the amount of algae. This in turn leads to an increase in the number of bacteria, which use up all the oxygen in the water. As a result, fish and other organisms in the water die. Sewage and slurry in water have much the same effect.

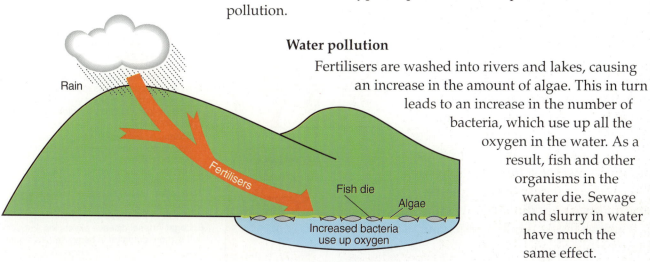

Breakthrough Science

Oil spills destroy beaches and coat the feathers of birds, making them unable to fly. As a result, the birds starve. Some countries will not allow oil tankers within a certain distance of their coastline.

Air pollution

Air pollution is caused by the burning of fossil fuels such as coal and oil. Sulphur dioxide and nitrogen oxides combine with water in the air to form acids. The result is acid rain, which attacks buildings, plants, etc., and can destroy all life in lakes and rivers.

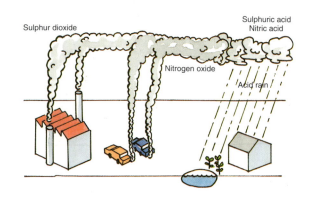

EXPERIMENT 57.1
AIM: TO STUDY THE EFFECTS OF ACID RAIN ON PLANTS

Method

1. Using a forceps, soak a piece of cotton wool in a solution of sodium metabisulphite and place it in a dish.
2. Soak a piece of cotton wool in water and place it in a dish.
3. Now set up the apparatus as shown in the diagram, and observe the results over a few days.

Smoke

Smoke is another form of air pollution. It can cause bronchitis, and severe difficulty for those who already suffer from respiratory problems.

It is bad enough to have to live in an environment where the air is polluted: imagine, then, the damage that smokers are doing to themselves and to others around them. Smoking can cause lung cancer, heart disease and emphysema, and that's just for starters. (Emphysema is a disease that destroys lung tissue and makes it very difficult to breathe. It is a serious illness. Look it up in your library.)

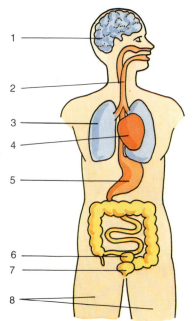

1. **Head:** Nicotine is a drug that causes dependence. This makes smoking a difficult habit to break.
2. **Larynx:** Chemicals in the smoke irritate the air passages.
3. **Lungs:** Coughs, bronchitis and respiratory infections are more common among smokers. Lung tissue damage leads to breathlessness and sometimes to cancer.
4. **Heart:** Nicotine raises the blood pressure and constricts the blood vessels. Heart disease kills many more smokers than non-smokers.
5. **Stomach:** Excessive smoking irritates the stomach and prevents healing of stomach ulcers.
6. **Uterus:** Heavy smoking in pregnancy increases the risk of miscarriage.
7. **Bladder:** Heavy cigarette smokers are more likely than non-smokers to have cancer of the bladder.
8. **Legs:** Arterial disease affects the limbs. It is more common in heavy smokers.

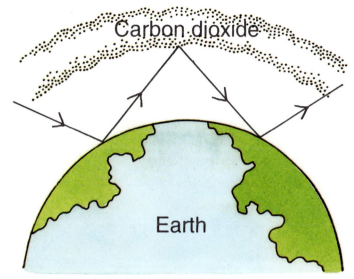

The greenhouse effect

When we burn fossil fuels, large amounts of carbon dioxide are produced. Some of this is used by plants in photosynthesis, some is dissolved in the seas, but a great deal remains in the atmosphere and leads to what is called 'the greenhouse effect'. In a greenhouse, heat from the sun is trapped by the glass, causing the temperature in the greenhouse to rise. In much the same way, heat from the sun is trapped by the carbon dioxide in the atmosphere, causing the temperature of the earth to rise.

The ozone layer

Much of the sun's radiation is harmful to us. Fortunately, this radiation is absorbed by a layer of gas surrounding the earth, known as the ozone layer. This layer is being damaged by gases called chlorofluorocarbons. (Phew! No wonder they call them **CFCs** for short.) These are used mainly as propellants in aerosol cans and as coolants in refrigerators. **CFCs** have now been banned in many countries.

Air pollution can be reduced by using smokeless fuels and lead-free petrol, and by controlling emissions from factory chimneys.

SUMMARY

- Conservation: protection, preservation and careful management of our natural resources.
- Pollution: any undesirable change in the environment caused by human activity.

QUESTIONS

Section A
1. What is meant by "conservation"? _____
2. What is meant by "pollution"? _____
3. Name two causes of water pollution. _____
4. Name two gases that cause acid rain. _____
5. What gas causes the greenhouse effect? _____
6. Name two diseases caused by smoking. _____

Section B

1. Name two major causes of water pollution in Ireland. Explain how one of these pollutants affects aquatic life.
2. An increased amount of carbon dioxide in the atmosphere is leading to what is called "the greenhouse effect". What is the greenhouse effect? Why is the amount of carbon dioxide in the atmosphere increasing? What are the possible consequences of the greenhouse effect? What measures should be taken to reduce the level of carbon dioxide in the air?
3. What are lichens? How can they help in the study of pollution?
4. Explain the terms 'deforestation' and 'desertification'. What effects can they have?
5. What causes air pollution? How can it be reduced? How can it affect your health?

HABITAT—FIELD STUDY CHAPTER 58

If you want to make a study of a habitat, remember a habitat can be anything from a forest to a woodland to a grassland to a rocky seashore to a pond.

FACTORS AFFECTING LIFE IN A HABITAT

1. The type of soil determines what plants and consequently what animals live in the habitat.
2. Climate, i.e. rainfall, wind, temperature, light and humidity.
3. Physical features: whether the ground is flat or sloped, sheltered or exposed; the direction in which it faces and its height above sea level.
4. Biotic features: these are caused by competition and interdependence. For example, plants compete for light and water, while animals compete for food and shelter. Also, if there are no plants there will be no herbivores and therefore no carnivores.

INVESTIGATING A HABITAT

1. Make a simple map of the habitat, showing the compass directions and the main features.

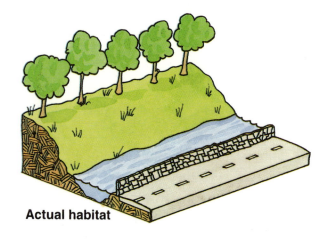
Actual habitat

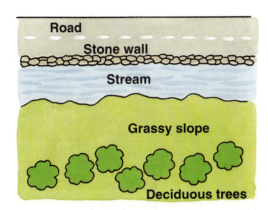
Simple diagram

2. Take some slides or photographs on each visit (autumn, winter, spring) and compare them to see what changes have taken place in the habitat.

3. Bring a thermometer to measure the temperature of air and water, and a soil thermometer to measure soil temperature. A light meter can be used to measure light intensity.

Using a pooter

COLLECTING ANIMALS

Small animals such as woodlice, beetles and spiders can be found inside rotting bark, under rocks and in the soil. A sweeping net can be used to catch insects in long grass, a butterfly net to catch butterflies and moths, and a plankton net to catch plankton in ponds and streams.

Equipment	Used to collect
Pooter	Insects etc.
Pitfall trap	Small animals that walk along the ground
Sweeping net	Insects from tall grass
Butterfly net	Butterflies
Plankton net	Plankton from water
Beating tray	Insects and small animals from trees
Tullgren funnel	Small animals from soil and leaf litter

Other useful items are a trowel, a sieve, a first-aid kit, polythene bags and screw-top jars.

Using an umbrella as a beating tray

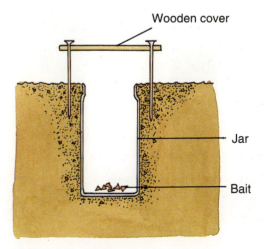
Pitfall trap

BREAKTHROUGH SCIENCE

PLANTS

Bring along a plant and wildflower guide to help you identify plants.

A quadrat can be used to estimate the percentage frequency of a plant in a habitat. A quadrat is a square frame. If you find a 1 m square too big use a 0.5 m square. Throw the quadrat 20 times at random in the habitat. The names of the plants inside the quadrat should be recorded each time in a chart such as the one shown below. (For example, if dandelions are found inside the quadrat in 5 of the 20 throws, the frequency of dandelions in the habitat is taken as 5 out of 20, or 25%.) Use this information to draw a bar chart.

Did you notice that, in the last column of the chart, the plants were placed on the DAFOR scale? This scale can be troublesome, as the terms are not exact. When, for example, does abundant become frequent or occasional become rare?

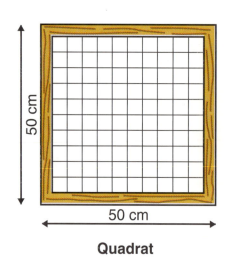

Quadrat

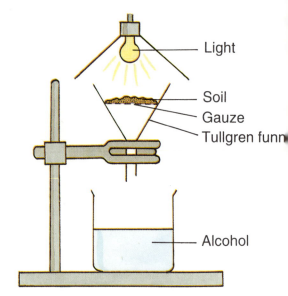

PLANT	QUADRAT NUMBER											TOTAL	PERCENTAGE FREQUENCY	
	1	2	3	4	5	6	7	8	9	10			
CLOVER	✓	o	✓	✓	o	✓	✓	✓	✓	✓	8	$\frac{8}{10}$ = 80%	A
THISTLE	o	o	o	✓	✓	o	o	o	o	o	2	$\frac{2}{10}$ = 20%	O
GRASS	✓	✓	✓	✓	✓	✓	✓	✓	✓	✓	10	$\frac{10}{10}$ = 100%	D
BUTTERCUP	✓	✓	o	✓	✓	o	o	✓	o	o	5	$\frac{5}{10}$ = 50%	F
DAISY	o	o	✓	o	o	✓	o	o	o	o	2	$\frac{2}{10}$ = 20%	O

Dafor Scale: D, dominant; A, abundant; F, frequent; O, occasional; R, rare.

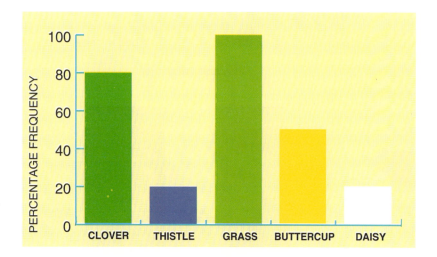

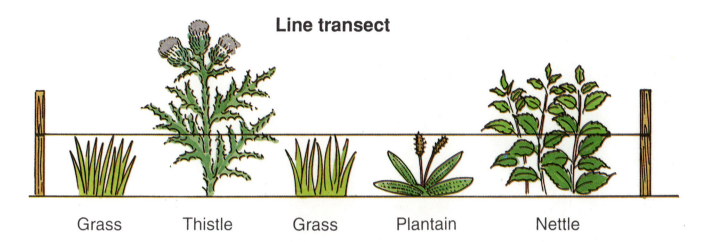

A LINE TRANSECT

This is used to study plant variation from one section of a habitat to another, e.g. from a shaded area to an open area or from one zone of a seashore to another. A line transect consists of a rope knotted at intervals of one metre and stretched across the habitat. The name and height of each plant touching a knot on the rope is recorded. This information can be used to draw a profile of the habitat. You can then discuss why certain plants grow in certain places and why the plants are taller in one place than in another.

HABITAT DETAILS

Woodland

A woodland can be divided into 4 zones:

1. tree or canopy layer
2. shrub layer
3. field or herb layer
4. ground layer.

The trees in a woodland keep the water content and the temperature more stable than elsewhere. The type of plants found in a woodland depends on the pH of the soil.

Layer	Alkaline or mildly acid	Acid
Tree	Ash, elm	Oak
Shrub	Hawthorn, hazel	Holly
Field	Fern, primrose	Wood sorrel
Ground	Mosses, leaf litter	Mosses, leaf litter

Breakthrough Science

Producers	Consumers
Ash	Herbivores: rabbit, butterfly, caterpillar,
Elm	moth, slug, snail
Hazel	
Hawthorn	Carnivores: spider, hedgehog, fox
Primrose	
Bluebell	Omnivores: blackbird, badger

ADAPTATION

Some plants, such as ferns, are adapted to live in the shade. Others, such as the bluebell, grow, flower and produce seed before the leaves come on the trees and block out the light. Caterpillars are camouflaged.

Grassland

In Ireland, many schools have a grassland habitat nearby. Also, many of the plants and animals will be familiar to students.

BIOLOGY · HABITAT—FIELD STUDY

Plants (producers)	Animals (consumers)	
Grasses	Herbivores:	Rabbit, slug, snail, earthworm, butterfly, fieldmouse
Nettle		
Thistle		
Daisy	Carnivore:	Fox, spider
Buttercup		
Plantain	Omnivore:	Rat

Rocky seashore

The main difference between a seashore habitat and other habitats is the great change that takes place with the coming in and going out of the tide. The seashore can be divided into four zones.

1. The splash zone: this region may be sprayed by water.
2. The upper shore: this region is above the average high tide but is covered by spring tides.
3. The middle shore: this is between the average high tide and the average low tide.
4. The lower shore: this region is uncovered by spring tides only.

Zone	Plants	Animals (consumers)	
Splash	Sea pink, lichens	Herbivores:	limpets, periwinkles, sea urchins
Upper	Sea holly, channel weed		
Middle	Bladder wrack, serrated wrack, sea lettuce	Carnivores:	sea anemone, jellyfish, starfish
Lower	Oarweed, laminaria, carragheen moss		
		Omnivores:	mussels, crabs

Two other common habitats are the hedgerow and the pond, although both are disappearing fast, especially the hedgerow.

SUMMARY

- Adaptation: organisms have developed certain features that enable them to survive in the habitat.
- Interdependence: plants and animals depend on each other to survive.
- A quadrat is used to estimate the percentage frequency of a plant in a habitat.
- A pooter is used to catch small animals (insects).
- A tullgren funnel is used to extract small animals from leaf litter.
- A pitfall trap is used to collect small animals that walk along the ground.

QUESTIONS

Section A

1. Name two environmental factors that affect life in a habitat.
   _____

2. Name the instrument shown in the diagram on the right and state its use.
   _____

3. Name a herbivore and a decomposer from a habitat you have studied.

4. The object in the diagram below is called a _____. You would use it in a land habitat to _____.

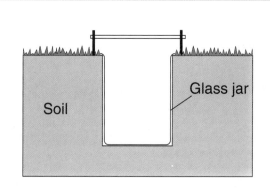

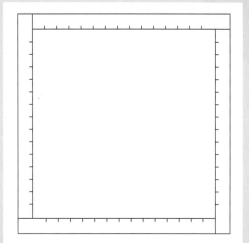

5. What is the tullgren funnel in the diagram below used for? _____

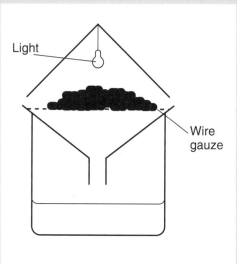

6. 50 pea seeds were sown near a wall, and 50 more in an open plot. When all the plants were fully grown, those near the wall were taller. Suggest a reason for this. _____

7. Give an example of competition from a habitat you have studied. _____
8. Say how a named animal is adapted to live in a habitat you have studied.
   _____
9. Say how a named plant is adapted to live in a habitat you have studied.
   _____
10. Give one example, other than feeding, of how animals depend on plants in a habitat.
    _____

Section B
1. Name a habitat you have studied. Name an animal you found in the habitat and explain how the animal is adapted. Give an example of competition between plants in the habitat. For what purpose would you use a pooter? Choose three living things from the habitat and show how they form a food chain.
2. The diagram shows a bar chart of the percentage frequency of plants growing in a habitat.

   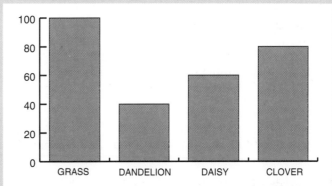

   (a) What type of habitat do you think this was?
   (b) What instrument was used to carry out the survey?
   (c) Assuming 20 throws, give the table that was drawn up before the bar chart was drawn.
   (d) What is the dominant plant in the habitat?
3. Fill in the following table.

Instrument	Used for
Sweeping net	
Plankton net	
Beating tray	
Pooter	
Pitfall trap	
Tullgren funnel	

4. Describe a habitat you have visited, under the following headings.
   (a) A map or diagram of the habitat.
   (b) Three plants and three animals found in the habitat.
   (c) Three food chains involving some of the above plants and animals.
   (d) Three examples, other than food, of the interdependence of plants and animals.
   (e) An example of (i) competition, (ii) adaptation.

BREAKTHROUGH SCIENCE

SOIL CHAPTER 59

Animals depend on plants, and plants depend on soil, so ultimately all life on land depends on soil. Soil particles are formed by the weathering of rocks over thousands of years. Rock is broken down into different-sized particles: gravel (largest), sand, silt, and clay (smallest).

EXPERIMENT 59.1
AIM: TO FIND THE TYPE OF PARTICLES (MINERAL CONTENTS) IN A SOIL SAMPLE

Method
1. Quarter-fill a large measuring cylinder with soil.
2. Add water until it is three-quarters full.
3. Cover with your hand and shake vigorously.
4. Allow to stand overnight.

   Result: The result should be as in the diagram.

HUMUS

Humus is the decayed remains of dead plants and animals.

1. It is rich in minerals.
2. It contains nitrogen-fixing bacteria.
3. It provides food for the organisms living in the soil.

GOOD SOIL

Good soil should contain

1. humus
2. water
3. mineral salts
4. air
5. soil particles
6. living organisms.

The best mixture of soil particles contains sand, silt and clay, and is called **loam**.

EXPERIMENT 59.2
AIM: TO FIND THE PERCENTAGE OF WATER IN THE SOIL

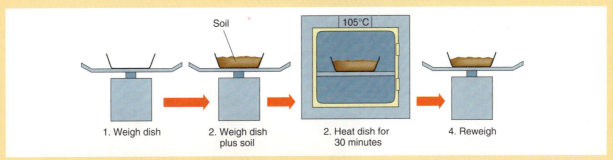

1. Weigh dish
2. Weigh dish plus soil
2. Heat dish for 30 minutes
4. Reweigh

Method
1. Weigh a clean dish.
2. Break some fresh soil into the dish and weigh it again.
3. Place the dish in an oven at 105°C for half an hour.
4. Remove the dish, cool and reweigh.
5. Repeat steps 3 and 4 until there is no further change in weight, showing that all the water has been evaporated.

$$\text{Percentage water} = \frac{\text{loss in weight of soil}}{\text{original weight of soil}} \times 100$$

EXPERIMENT 59.3
AIM: TO FIND THE PERCENTAGE OF HUMUS IN THE SOIL

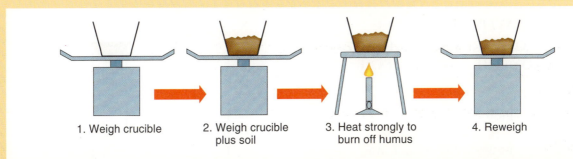

1. Weigh crucible
2. Weigh crucible plus soil
3. Heat strongly to burn off humus
4. Reweigh

Method
1. Weigh an empty crucible.
2. Add some dry soil and reweigh.
3. Heat the crucible strongly to burn off the humus, then allow the crucible to cool and weigh it.
4. Keep doing this until there is no further change in weight.

$$\text{Percentage humus} = \frac{\text{loss in weight of soil}}{\text{original weight of soil}} \times 100$$

EXPERIMENT 59.4
AIM: TO FIND THE PERCENTAGE OF AIR IN THE SOIL

Method

1. Find the volume of a small can by filling it with water and emptying it into a measuring cylinder.
2. Punch a hole in the bottom of the can and fill it with soil, as shown in the diagram.
3. Level the surface of the soil in the can.
4. Block the hole and measure the amount of water that must be poured slowly into the soil to fill the air spaces. This gives the volume of air in the soil.

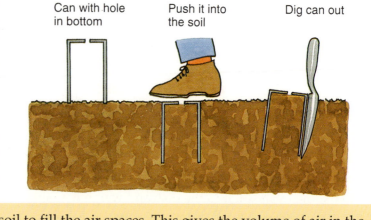

$$\text{Percentage air} = \frac{\text{volume of water added}}{\text{volume of can}} \times 100$$

DRAINAGE

If soil has poor drainage, it holds too much water and becomes waterlogged. If the drainage is too good, it leads to a problem called leaching. This means that the nutrients are washed out of the soil.

EXPERIMENT 59.5
AIM: TO COMPARE THE DRAINAGE OF SANDY SOIL AND CLAY SOIL

Method

1. Set up the apparatus as shown in the diagram.
2. Pour 50 cm of water into each funnel.
3. Wait for five minutes. Which soil has the better drainage?

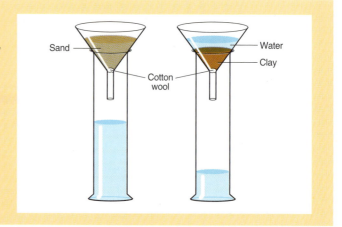

SOIL POLLUTION

Soil pollution is mainly caused by the same substances that cause water pollution. Organic fertilisers such as manure; artificial fertilisers such as nitrates, phosphates and potassium; insecticides and acid rain can cause soil pollution, and later water pollution if they are leached into rivers and lakes.

ACID SOIL

Most garden plants do well in mildly acid soil (pH 6·5 to 7) but not in strongly acid soil. Farmers add lime to the soil to reduce its acidity. Bogland has a pH of 4–5, therefore only certain plants can grow there, e.g. heather.

EXPERIMENT 59.6
AIM: TO TEST THE PH OF A SOIL SAMPLE

Method
1. Put a small amount of soil into a test-tube.
2. Half fill the test-tube with distilled water and shake it vigorously.
3. Filter the mixture into a second test-tube, and test it with universal indicator paper.

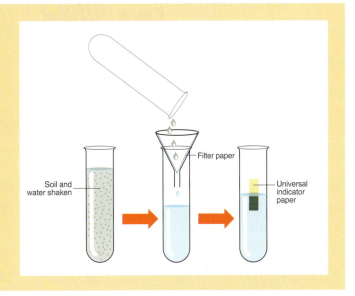

ORGANISMS

Many small organisms such as earthworms, centipedes and millipedes live in the soil. The soil also contains decomposers (bacteria and fungi) which break down humus and release nitrogen into the soil. These organisms need oxygen for respiration. In this regard the earthworm is very helpful because it

1. aerates the soil (burrows into the soil and lets air in)
2. drags leaves down into the soil
3. rotates the soil
4. increases the amount of humus by excreting and by dying and decaying
5. improves drainage.

MICRO-ORGANISMS

Bacteria multiply rapidly if they have food, moisture, warmth and oxygen. A suitable food for laboratory use is nutrient agar.

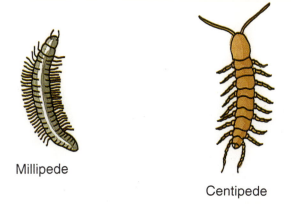
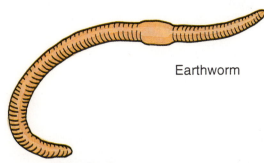

Organisms in the soil

BREAKTHROUGH SCIENCE

EXPERIMENT 59.7
AIM: TO SHOW THAT THERE ARE MICRO-ORGANISMS IN SOIL

Method
1. Prepare two sterile petri dishes containing nutrient agar.
2. Using a sterile spatula, scatter a little fresh soil on the surface of the agar in one dish, then replace the lid and seal it. Seal the other dish.
3. Place the dishes upside down in an oven at 37°C for 48 hours.
4. Examine the surface of the agar for shiny patches: these are colonies of bacteria. (Do not remove the lid.)

Result: Bacteria will be found in the dish with the soil, but not in the other dish.

Conclusion: There are micro-organisms in the soil.

SUMMARY

- Weathering means that rock is broken down to small particles.
- Humus is the decayed remains of plants and animals.
- Leaching means that materials are washed out of the soil.
- Decomposers in the soil break down humus to provide nitrogen for plants.
- Good soil should contain humus, water, mineral salts, air, soil particles and living organisms.

QUESTIONS

Section A
1. What does "weathering" mean? _____
2. What is humus? _____
3. Name two living organisms you might find in soil. _____
4. In the diagram, which part is humus? _____ Which is clay? _____ Which is sand? _____
5. Strong heating removes water and X from soil. What is X, and why is it important in soil? _____
6. Name two types of decomposer found in soil. _____
7. What is meant by "leaching"? _____
8. What is meant by "mildly acid soil"? _____

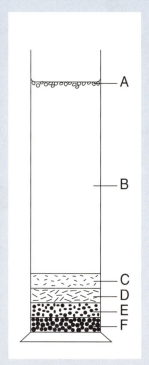

9. Name two ways in which earthworms improve the soil. _____
10. Name two artificial fertilisers commonly added to soil. _____

Section B

1. Describe an experiment to find the percentage of air in a soil sample. Why is air important in soil? Name an organism that helps to aerate the soil.

2. Name two living organisms found in soil, and two non-living parts of soil. Name two decomposers found in soil. What part of the soil do they break down? What element is released into the soil as a result? Why is this element important to plants?

3. What type of soil has poor drainage? What problems can this lead to? What could be added to this soil to improve drainage? What problems would arise if too much of this substance were added to the soil?

4. Name three living things and three non-living things you might find in the soil. Why is it good to have earthworms in the soil?

5. Farmers sometimes take sand from beaches and add it to their soil. Why do they do this? What else might they take from beaches to spread on their land? Why?

6. A student decided to find out the percentage of air in a soil sample, as follows:

    (a) She placed some soil in a graduated cylinder as shown in the diagram on the left.
    (b) She added 50 cm³ of water to the cylinder and shook it vigorously.
    (c) She then allowed the mixture to settle as shown in the diagram on the right.

    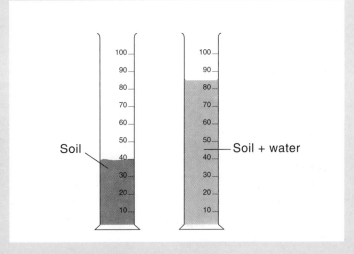

    (i) What was the volume of soil in the cylinder?
    (ii) What volume of water was added?
    (iii) What would she get if she added (i) and (ii)?
    (iv) What was the volume of the mixture according to the second diagram?
    (v) What volume of air was in the soil?
    (vi) What percentage volume of air was in the soil?

BREAKTHROUGH SCIENCE

MICRO-ORGANISMS CHAPTER 60

All living things are known as 'organisms'. Viruses, bacteria and some fungi are so small that they are called 'micro-organisms'.

VIRUSES

Parasites live on or inside a living host. Viruses are parasites that attack animals, plants and even bacteria. They can multiply only inside living cells. They use material from the cell they are attacking to make copies of themselves. They continue to do this until the cell bursts, releasing a large number of viruses.

In humans, viruses cause influenza, measles, the common cold, mumps, polio and AIDS. In other animals, viruses cause foot-and-mouth disease and rabies. Dutch elm disease is caused by a virus.

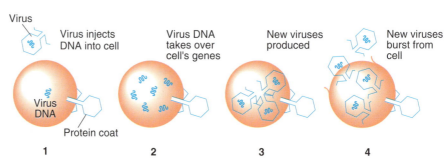

BACTERIA

Bacteria consist of a single cell. Even so, they are bigger than viruses. Given a suitable temperature and food supply, they can reproduce every 20 minutes. This is particularly true inside the human body, but they are also found inside the bodies of other animals, in plants, and in the air, soil and water.

Many bacteria are useful to humans, but some are harmful and cause diseases such as pneumonia, tuberculosis, boils and scarlet fever.

Beneficial bacteria	Harmful bacteria
1. cause dead matter to decay	1. cause disease
2. provide nitrogen in soil	2. spoil food
3. help make yoghurt and vinegar	3. cause milk to turn sour
4. produce antibiotics.	4. cause tooth decay.

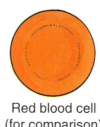

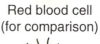
Red blood cell (for comparison)

Tuberculosis

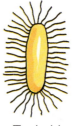
Typhoid

Diphtheria

Different types of bacteria cause different diseases

The green mould on an orange is actually a fungus

FUNGI

Many fungi are small, although some, like mushrooms, are not micro-organisms. Fungi have no chlorophyll, so they cannot make food. Some are parasites that live on living matter, e.g. ringworm, athlete's foot and potato blight. Like bacteria, some fungi are useful to humans.

Beneficial fungi	Harmful fungi
1. produce antibiotics (penicillin)	1. cause disease in humans, e.g. athlete's foot
2. are used in baking (yeast)	2. cause plant disease
3. are used in brewing (yeast)	3. spoil food (fruit mould)
4. are used as food (mushrooms).	4. can be poisonous.

INFECTION

Micro-organisms cause infection. In the last century and for much of this century, the focus of science was on bacteria. In recent times the focus has moved to viruses, although diseases caused by bacteria are still with us.

DISEASES CAUSED BY BACTERIA

Tuberculosis (TB) is caused by bacteria that usually enter the body through the nose or mouth and affect the lungs. It is usually associated with poverty. It took many lives in Ireland in the 1940s and 1950s, but a rise in living standards and the development of antibiotics caused the disease to decline. (Antibiotics are chemicals, produced by some bacteria and fungi, that can be used to fight disease.)

Salmonella poisoning is another illness caused by bacteria. It can be contracted by drinking unpasteurised milk or meat that is not thoroughly cooked, or by handling raw meat and then handling and eating cooked meat, thus transferring bacteria from one to the other.

The German bacteriologist Robert Koch (1843–1910) discovered the bacillus responsible for tuberculosis.

DISEASES CAUSED BY VIRUSES

The common cold and influenza (flu) are caused by viruses. There is no cure.

Antibiotics have no effect on viruses, although they can be used to treat bacterial infection which often follows influenza.

Jonas Salk (1914–   ), an American micriobiologist, developed the first vaccine against polio.

AIDS (acquired immunodeficiency syndrome) is caused by a virus which attacks the white cells in the blood and weakens the body's defences. As a result the person may die of an illness such as pneumonia which he or she would normally be able to fight. AIDS is transmitted by blood contact and those most at risk are drug addicts and practising homosexuals, but it can also be contracted by having sexual intercourse with a member of the opposite sex who has the disease. A woman who carried the AIDS virus can pass it on to her baby either through the placenta or, later, through her milk while breastfeeding.

FIGHTING INFECTION

Your skin is your first line of defence against infection. It prevents bacteria from entering your body. Your white blood cells are the next line of defence. After that your body may need some help, in the form of antibiotics or vaccination. (For example, a small dose of German measles virus is injected into your bloodstream. Your body produces antibodies to kill the virus. If you ever come into contact with German measles again, the antibodies are there in your body ready to fight it.)

Of course the best way to deal with infection is to avoid it in the first place. If you cut yourself, treat the cut immediately. Insist on the highest standards of hygiene, especially where food is concerned. Avoid drugs unless they are prescribed by a doctor and, when the time comes, behave responsibly with regard to sex. There is a great deal of truth in the saying, "A gram of prevention is better than a kilogram of cure".

The English physician Edward Jenner (1749–1823) introduced vaccination against smallpox. Jenner noticed that dairy workers who caught the mild disease cowpox did not later contract smallpox. In 1796 he innoculated an 8-year-old boy with cowpox. Several weeks later Jenner innoculated the boy with smallpox, and the disease failed to develop.

Sir Alexander Fleming (1881–1955) discovered the antibiotic penicillin in 1928, while examining a culture of bacteria contaminated with mould. He shared the 1945 Nobel Prize for Physiology and Medicine with H. Florey and E. B. Chain, who purified and tested the drug.

SUMMARY

- Micro-organisms are bacteria, viruses and small fungi.
- Viruses cause cold, flu, measles, mumps and AIDS.
- Bacteria cause tuberculosis, food poisoning and tooth decay.
- Fungi cause athlete's foot, ringworm and potato blight.

BIOLOGY · MICRO-ORGANISMS

Joseph Lister (1827–1912) revolutionised surgery by introducing antiseptic practices. Lister believed that micro-organisms caused infection, and used carbolic acid to kill them. He also introduced sterile dressings.

Louis Pasteur (1822–1895) was born in France, near the Swiss border, and qualified as a teacher. He made a study to find out why wine and beer turned sour, and discovered that this was caused by microbes. He also discovered that microbes could be killed by heating the wine and then cooling it again. This process, called pasteurisation, is used today mainly for milk and milk products such as yoghurt. Pasteur also discovered vaccines against anthrax and rabies.

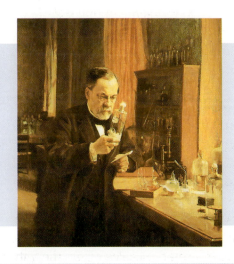

QUESTIONS

Section A

1. Name three types of micro-organism. _____
2. Mention one way in which any of them is useful to humans. _____
3. Which type of micro-organism is responsible for AIDS? _____
4. Name a micro-organism used in the food industry. _____
5. Name a fungus we can eat. _____
6. Name a disease caused by a virus. _____
7. To reproduce, bacteria need a suitable _____ and _____.
8. Which of the following illnesses can be caused by bacteria: Influenza, measles, mumps, tuberculosis?
9. What substance besides soil would you use in an experiment to show that there are micro-organisms in soil? _____
10. What is a parasite? _____
11. Name one way in which fungi are harmful to man. _____
12. Antibiotics have no effect on _____

Section B
1. Bacteria cause tooth decay. How exactly do they do this?
2. What is meant by 'vaccination'? Parents are encouraged to get the "three in one" vaccination for their babies. Against which three illnesses does this protect the baby? Research MMR.
3. Bacteria can be beneficial or harmful. Give three examples of each.
4. Fungi can be beneficial or harmful. Give three examples of each.
5. Is AIDS caused by bacteria or by a virus? What groups of people are most at risk from AIDS? Does this mean that people who are not in these groups have nothing to worry about? Can AIDS be transmitted from a mother to her baby? How?
6. Why should the following rules be obeyed in a food shop?
    (a) Do not keep cooked and uncooked food close together.
    (b) Handle food with a pair of tongs.
    (c) As far as possible, keep food covered.
7. Why are the following scientists famous: (a) Louis Pasteur, (b) Alexander Fleming, (c) Joseph Lister, (d) Edward Jenner?

SECTION 4

EARTH SCIENCE
(Chapters 61–62)

FOOD
(Chapters 63–65)

ELECTRONICS
(Chapters 66–68)

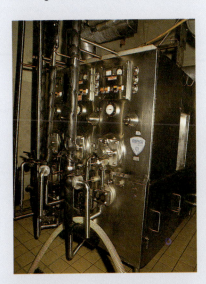
ENERGY CONVERSION
(Chapters 69–70)

HORTICULTURE
(Chapters 71–72)

MATERIALS
(Chapters 73–74)

APPLIED SCIENCE

THE UNIVERSE   CHAPTER 61

The earth is our home in the universe. The universe is made up of all the matter that exists. It contains hundreds of thousands of millions of galaxies. When you look at the sky on a clear night, you see thousands of stars. Our sun is a medium-sized star. It is much closer to us than any other star, and heats and lights the earth. The next star is Proxima Centauri—4 light-years away.

Astronomers measure the enormous distances between stars by using the light year as their unit. A light-year is the distance travelled by light in one year (9.5×10^{15} m: almost 10,000 billion kilometres).

The Milky Way

A light-year is the distance travelled by light in one year.

GALAXIES

We live in a galaxy—the Milky Way—made up of around 150 billion stars. Our sun is just one of these stars. The Milky Way has a diameter of 100,000 light-years. Most of the stars in the Milky Way cannot be seen from the earth, because interstellar dust obscures them. The Milky Way is a spiral galaxy. It is moving through space.

Another large galaxy, the **Andromeda galaxy**, is more than 2 million light-years away from our galaxy.

The gravitational pull between galaxies is enormous. The earth is 150 million km from the sun, and moves in orbit at 30 km/s. The sun is a million million times this distance from centre of the Milky Way, and orbits about the galactic centre at 250 km/s.

A galaxy is a group of stars held together by gravitational pull.

Albert Einstein (1879–1955) was a German physicist who revolutionised modern physics. In 1895 Einstein failed an entrance exam for electrical engineering at the Zurich polytechnic, but succeeded in 1896 and graduated in 1900 as a secondary school teacher of mathematics and physics. Over the next 15 years he put forward many theories on light, gravity, the universe, time and many other subjects. Experimental work, including the splitting of the atom, later proved many of his theories.

APPLIED SCIENCE · THE UNIVERSE

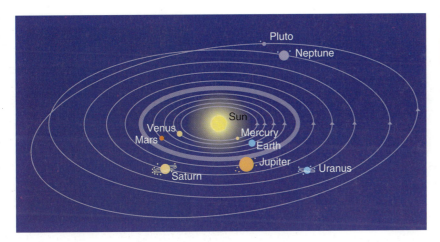

THE SOLAR SYSTEM
The solar system is the sun and the nine planets orbiting around it.

The sun
The sun contains most of the mass (99·9%) of the solar system. The enormous gravitational pull of the sun keeps the planets moving around it. All planets in the solar system (except Pluto) move in the same plane.

The planets
The four inner planets are Mercury, Venus, Earth and Mars. Mercury, Venus and Mars all have rocky crusts and metallic cores like the Earth. The Earth is the largest of the inner planets.

The five outer planets are Jupiter, Saturn, Uranus, Neptune and Pluto. Most of these are giant balls of gas with very small solid cores.

Pluto has a very elliptical orbit that takes it inside the orbit of Neptune. Pluto will be inside Neptune's orbit until 1999.

Asteroids
Asteroids are lumps of rock orbiting the sun. Most of the asteroids are found in the gap between Mars and Jupiter. This is the Asteroid Belt. Some large asteroids have radii of a few hundred kilometres, but most are much smaller.

ARE THERE OTHER SOLAR SYSTEMS?
A few nearby stars are now known to have planet-sized objects moving around them. It is possible that there are solar systems in other parts of the universe.

	Earth	Moon	Mars
Diameter (km)	12,750	3,500	6,750
Distance to sun (million km)	150	150	228
Gravity (newtons/kg)	9·8	1·6	3·6
Average temperature	15°C	−60°C	−50°C
Atmosphere	78% nitrogen, 21% oxygen, 1% argon	None	95% carbon dioxide, 3% nitrogen, 2% argon
Satellites	One—the moon	None	Two—Deimos and Phobos

Breakthrough Science

Galileo Galilei (1564–1642) was an Italian who made a great number of discoveries in physics. He proved that the speed of a falling object was not proportional to weight; he worked out an explanation of the tides; he proved that machines do not create energy, but simply transform it; and he showed that the time for one swing of a pendulum is constant. Galileo made a 20-power telescope that enabled him to see lunar mountains, the stars of the Milky Way, and the moons of Jupiter. His main achievement was in showing that experimental evidence is essential to establish scientific truth.

EARTH—THE BLUE PLANET

The earth looks blue from space. This is because three quarters of the earth's surface is water. Water is necessary to support life.

Earth is at a distance from the sun that provides it with enough energy to keep the earth's surface warm. The average temperature of the earth is 15°C. This keeps most of the water in the liquid state. The sun also provides enough light energy for photosynthesis.

The atmosphere of the earth contains a large amount of oxygen (21%). Oxygen is necessary for respiration in plants and animals.

Earth is the only planet in the solar system with water, oxygen and a temperature range capable of supporting life.

A living planet

The plants and animals on earth affect its temperature range and supply of water and oxygen. Plants recycle oxygen from carbon dioxide. The amount of carbon dioxide in the earth's atmosphere as a result of the burning of fossil fuels has a significant effect on the temperature range (greenhouse effect). Volcanoes also add enormous amounts of carbon dioxide to the atmosphere. Scientists do not fully understand what effect changes in carbon dioxide levels will have on our climate.

A STAR IS BORN

Our sun is a medium-sized star. Stars are formed from clouds of gas in space called nebulae. Gravity pulls the particles of gas towards each other and over millions of years enormous amounts of gas collect into a giant ball of gas. As the gas collects, its energy is released and its mass heats up. Eventually it gets so hot that nuclear reactions begin. The nuclear reactions give out heat and light from the star. Stars vary in size from smaller than the sun to a hundred times the mass of the sun.

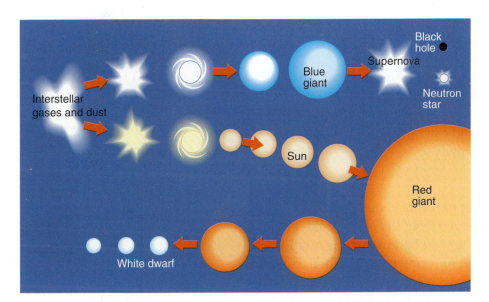

Life cycle of the Sun
Our sun is a typical star. Enough hydrogen and helium gas collected in the formation of the sun to start a nuclear reaction in the hydrogen that produces all the heat and light of the sun. In about 5,000 million years the sun will have used up all the hydrogen near its core. The reaction will then spread outwards, releasing more heat.

Nuclear reactions will also start in the helium core. This heat will cause the sun to expand outwards, shining 1000 times as brightly as now. At this stage it will be a **red giant**. It will get so big that it will spread out as far as the present orbit of Mars. The outer layers will use up their hydrogen fuel, and cool. These layers will drift away into space, leaving a small hot core called a **white dwarf**—a star about the size of the earth but with a mass almost as great as that of the present sun. The white dwarf will give out only a fraction of the present sun's heat and light. It will slowly die out.

Giant stars
Stars about ten times the mass of the sun start like the sun, but with a bigger size and more fuel they get brighter and hotter. They are **blue giants**—they give out a blue-white light. They go to the red stage faster than the sun, and form **red supergiants**. Because they are so hot and dense, the nuclear reactions in red supergiants produce a wide range of elements. They also get even hotter, until the reactions start producing iron. This nuclear reaction absorbs heat and quickly lowers the temperature of the core. The star collapses inwards on the core, causing a massive nuclear explosion which blows the star to bits. This is a **supernova**. A supernova is so bright that it can be seen light-years away even in daylight.

The core of the star remains as a **neutron star**. The pull of gravity in the core and the force of the explosion compress the core into a sphere 20 km in diameter containing the mass of two suns. A teaspoon of neutron star would weigh a billion tonnes.

Breakthrough Science

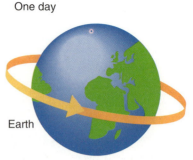

Earth turned once on its axis

Red dwarf stars
These are stars about one tenth the size of the sun. The core of the red dwarf does not get hot enough to start nuclear reactions. It gives out a lot of heat (infra-red radiation), but not much visible light. Red dwarfs are very dull and can only be seen with telescopes. They burn out very slowly.

DAYS, YEARS AND SEASONS

A day is the time the earth takes to rotate about its axis.

Earth once around the sun

It rotates anti-clockwise as viewed from the North Pole. The sun rises in the east and sets in the west because of this rotation which also produces the apparent movement of the moon, planets and stars in the night sky.

A year is the time the earth takes to move one complete circuit orbit around the sun.

It takes 365·25 days to do this.

So why is it hotter in summer? The answer is that in summer the heat radiation of the sun is concentrated on a smaller surface area of the earth, as shown in the diagram. The earth is tilted at an angle of 23° to its orbit around the sun. This causes the seasons. The diagram shows that when the earth is at point A in its orbit, the northern hemisphere is towards the sun—it is summer. At point B the northern hemisphere is away from the sun—it is winter. As the earth moves in its orbit the seasons change.

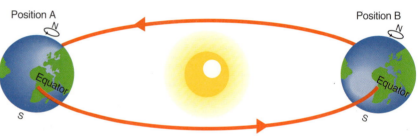

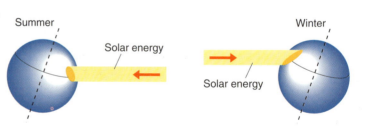

SOLAR ECLIPSE

An eclipse of the sun takes place when the moon is in a straight line between the sun and the earth. The moon then blocks the light of the sun, and part of the earth goes dark until the earth and moon move out of line.

Solar eclipse

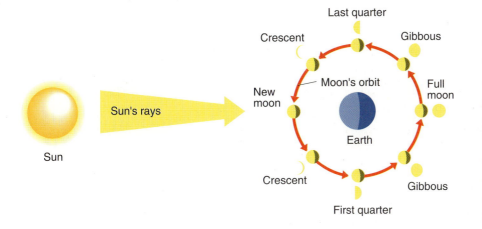

THE MOON

The earth's moon is one of the six largest moons in the solar system. The moon is a natural satellite of the earth. The moon orbits the earth every 28 days. It also rotates about its axis every 28 days, and so the same side of the moon is always visible from the earth. How much of the moon we can see changes depending on the angle from which we see it. We see different shapes or phases of the moon. We can see the moon only because it reflects the light of the sun.

A satellite is any object in orbit around the sun or any other planet. The earth is a satellite of the sun. The moon is a satellite of the earth.

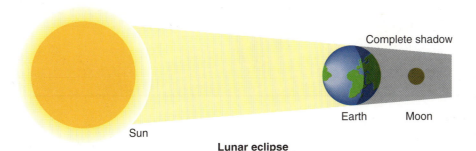

Lunar eclipse

LUNAR ECLIPSE

An eclipse of the moon takes place when the earth is in a straight line between the sun and the moon. The earth then blocks the light of the sun, and the shadow of the earth moves across the moon.

THE MOON CAUSES TIDES

The gravitational pull of the sun keeps the earth in its orbit around the sun. It also pulls the water in the oceans towards it, causing the water to rise. The moon is much smaller than the sun, but it is closer to earth and so has a greater pull on the oceans. This causes the rise and fall of tides twice a day as the earth rotates.

Very high tides—"spring tides"—happen when the sun, moon and earth all lie approximately in a straight line. The gravitational pulls of the sun and moon add to each other, causing the seas to rise higher than normal.

Very low tides—"neap tides"—happen when the sun and moon are at right angles to each other. Their gravitational pulls partly cancel each other, causing a lower than average tide.

Spring tides and neap tides happen twice each month.

BREAKTHROUGH SCIENCE

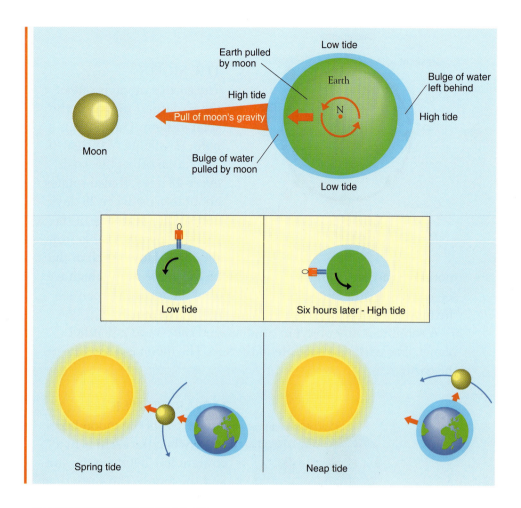

SUMMARY

- A galaxy is a group of stars held together by gravitational pull.
- A light year is the distance travelled by light in one year.
- The solar system is the sun and the nine planets orbiting around it.
- Earth is the only planet in the solar system with water, oxygen and a temperature range capable of supporting life.
- A day is the time the earth takes to rotate about its axis.
- A year is the time the earth takes to move one complete circuit around the sun.
- An eclipse of the sun takes place when the moon is in a straight line between the sun and the earth.
- An eclipse of the moon takes place when the earth is in a straight line between the sun and the moon.
- A satellite is any body in orbit around another body.
- Spring tides happen when the gravitational pulls of the sun and moon add to each other.
- Neap tides happen when the gravitational pulls of the sun and moon partly cancel each other.

QUESTIONS

Section A
1. What is the name of the galaxy to which the earth belongs? _____
2. What is the solar system? _____
3. What is a solar eclipse? _____
4. What is a light-year? _____
5. Spring tides happen when the gravitational pulls of the sun and moon _____ to each other.
6. How long does it take the earth to orbit the sun? _____
7. What is a lunar eclipse? _____
8. Neap tides happen when the gravitational pull of the sun and moon _____ each other.
9. How long does it take the moon to orbit the earth? _____
10. How long does it take the moon to orbit the sun? _____
11. A day is the time it takes _____.
12. The greenhouse effect is caused by the gas _____.

Section B
1. "Day and night, summer and winter, are caused by the movement of the earth relative to the sun." Explain this statement. What causes an eclipse of the sun?
2. (a) Name the two largest planets of the solar system.
   (b) Name the two planets closest to the earth.
   (c) Name a planet other than the earth that has a planetary moon.
   (d) Which planet is furthest from the earth?
3. (a) What is a galaxy?
   (b) Explain how an eclipse of the sun occurs.
   (c) Explain the difference between a solar and a lunar eclipse.
   (d) Explain how the tides on earth are caused by the moon.
4. Explain (i) a day, (ii) a year, in terms of the movement of the earth.
   How does the tilt of the earth's axis account for the seasons of the year? Give a brief outline of the life cycle of any star.
5. (i) The moon is a <u>satellite</u> of the earth and it <u>orbits</u> the earth every 27·3 days. It is the chief influence on our <u>tides</u>. Occasionally there is an <u>eclipse</u> of the moon.
   Explain the <u>underlined</u> terms.
   (ii) Compare the earth and one other planet under the headings: Distance from the sun, Diameter, Range of temperature, Atmosphere.
6. (i) The earth is the only planet capable of supporting life as we know it. Why is this?
   (ii) Explain briefly how the sun's energy is produced.
   (iii) A red giant and a white dwarf are two stages in the life cycle of the sun. Give a brief outline of this life cycle.

WEATHER CHAPTER 62

WATER IN THE ATMOSPHERE

The earth is heated by radiation from the sun. This heat is not absorbed by the atmosphere, but heats up the land and the seas. These in turn heat the air above them. The air near the surface is warmer than higher up. Convectional currents are set up by the hot air, and these lead to winds.

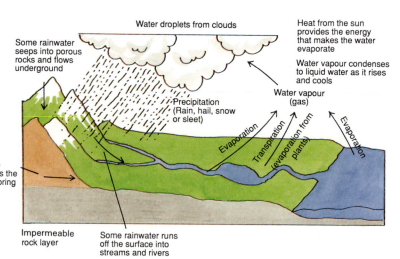

Water in the seas and on land evaporates when heated by the sun. Evaporation increases when there is a wind.

The temperature of the air drops by about 1°C for every 200 m you go up.

HUMIDITY

The air can hold only a certain amount of water vapour. The amount depends on the temperature. Hot air can hold more moisture than cold air. When the air holds as much water as possible, it is saturated. The amount of water in the air compared to the saturation amount is called the humidity of the air. If the air holds half the amount it could hold when saturated, we say the humidity is 50%. Clearly, humidity is related to the temperature of the air. We measure humidity with a hygrometer.

EXPERIMENT 62.1

AIM: TO SHOW THE EFFECTS OF HEAT AND WIND ON EVAPORATION

Method
1. Weigh a damp cloth. Hang the cloth on a line.
2. Weigh the cloth again after 10 minutes. Find out how much water has evaporated.
3. Repeat these steps with the following additions:
   (a) blow cool air over the cloth with a hairdryer
   (b) blow hot air over the cloth with a hairdryer.
   Result: More water evaporates on hot and windy days.

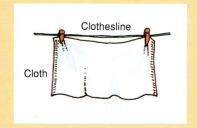

EXPERIMENT 62.2
AIM: TO TEST FOR WATER VAPOUR

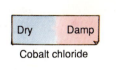
Cobalt chloride

Method
1. Dry some cobalt chloride in a cool oven. It is blue when dry.
2. Leave it in the air.

Result: It turns pink when water vapour is present.

EXPERIMENT 62.3
AIM: TO SHOW THE EFFECT OF HEAT AND WIND ON CONDENSATION

Method
1. Place a mixture of salt and ice in a polished tin. Seal the tin and find its mass.
2. Leave for 10 minutes in still air. Find the mass again. Calculate the increase in mass due to condensation.
3. Repeat steps 1 and 2, but this time place the sealed tin in front of a cold fan.
4. Repeat steps 1 and 2 with a hot-air fan.

Result: The mass of condensed water is greatest for still air, next greatest for cold air and least for hot air.

Conclusion: Heat and wind reduce condensation.

HOW CLOUDS ARE MADE

Air with water vapour in it is colourless. We do not see the water vapour in the air as it evaporates. The moisture-filled air moves along with the winds until mountains or a cold air mass make it rise higher. This cools the air, and some of the water vapour condenses into tiny drops which continue to float along with the wind. These tiny drops are visible as clouds. When the air is cooled further the tiny drops condense to form bigger drops. These fall as rain, sleet or snow.

Air cools as it rises upwards. Water vapour condenses to form clouds.

BREAKTHROUGH SCIENCE

Clouds also form when hot moist air rises. As it rises it expands (because of a drop in air pressure) and cools. Water vapour in the air condenses into tiny drops and makes clouds.

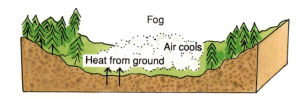

CLOUDS CAN DISAPPEAR ON HOT DAYS

On hot days the sun heats up the air and the small water drops turn back into water vapour. The clouds then disappear.

FOG

Fog usually forms at night or in the early morning when humid air cools. Water vapour in the air above the ground cools and tiny water drops form. These are light enough to float in the air, and are visible as fog. Pollution in the air mixes with fog to turn it into smog.

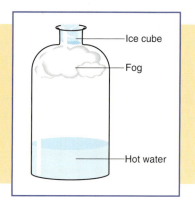

EXPERIMENT 62.4
AIM: TO MAKE FOG

Method
1. Warm up a large bottle. Half-fill it with hot water.
2. Place a large ice cube in the mouth of the bottle.

   Result: The water vapour condenses and forms a fog.

FROST

On cold clear nights the ground can cool to below freezing point. Water vapour in the air condenses onto the ground as a light coating of ice. This is frost. Farmers and gardeners need to know when frosts are likely, to protect their crops from damage.

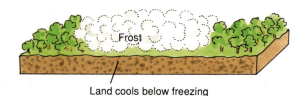

EXPERIMENT 62.5
AIM: TO MAKE FROST

Method
1. Place a tin can in a freezer.
2. Remove it after about 15 minutes.

   Result: After a minute or so the water vapour in the air forms frost on the can.

APPLIED SCIENCE · WEATHER

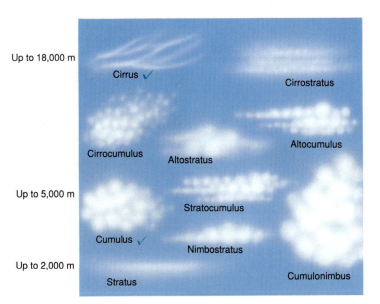

THERE ARE DIFFERENT TYPES OF CLOUDS

Stratocumulus
There are flat low-level layered grey clouds that give drizzle. "Stratus" means a layered cloud.

Cumulus ✓
There are thick woolly clouds with white tops that give heavy showers.
"Cumulus" means a cottony or billowing cloud.

Nimbostratus ✓
There are grey, layered, "black" clouds that give persistent rain. They are part of low-pressure or cyclonic weather.

Cirrus ✓
These are long, narrow patchy white clouds. Cirrus clouds that become thicker as they approach are the start of a warm front.

THE SUN—POWERHOUSE OF THE EARTH
Nearly all energy on earth comes from the sun. Just a little comes from the heat in the core of the earth.

The sun heats the earth by radiation. This heats the land and seas. It does not heat the air it passes through. Some of this heat is radiated out into space by the earth. We notice the fact that heat is radiated out from the earth at night, because this makes the night cooler than the day. The amount lost in this way depends on the atmosphere. The atmosphere acts as a blanket around the earth and keeps in much of the heat. You have probably noticed that cloudy nights are rarely very cold, because the clouds keep in the heat absorbed during the day. Clear nights are often frosty, as a lot of heat is lost.

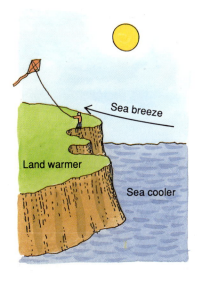

Land and sea breezes
Land and sea breezes are caused by the sun's heat. Land heats up faster than the sea during the day. This causes the air over the land to expand and rise, leading to a drop in pressure which causes the cool air over the sea to move towards the land—a sea breeze.

At night, the land cools faster than the sea. The lower pressure over the warmer sea causes the cool air over the land to move towards the sea—a land breeze.

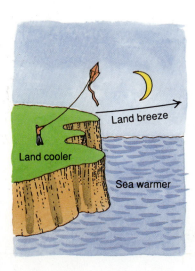

The greenhouse effect

The amount of water vapour and carbon dioxide in the air helps the earth to retain heat. The atmosphere lets in short wavelength infra-red rays which heat up the earth. The earth radiates longer wavelength infra-red rays. Much of this radiation is absorbed by the water and carbon dioxide in the atmosphere. This helps to keep the earth warm, so the atmosphere acts as a kind of blanket. Without this blanket the earth's average temperature of 15°C would drop to −28°C.

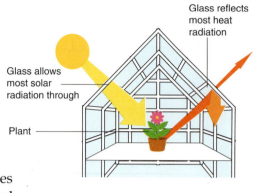
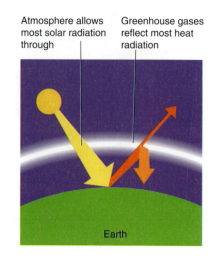

Scientists are concerned that too much carbon dioxide is being released into the air by the burning of fossil fuels. A lot of carbon dioxide would increase the blanket effect of the atmosphere and would cause the earth to get much warmer. This would affect our weather patterns.

Because weather patterns change naturally over long periods of time, it is difficult for scientists to tell whether or not there is a global warming from a greenhouse effect.

HOW GASES BEHAVE

We live at the bottom of a sea—a sea of air. The air has weight and so exerts pressure. We are not aware of this atmospheric pressure because **it presses equally in all directions**. We have already found this out by experiment.

The atmosphere is the layer of air that surrounds the earth.

MEASURING PRESSURE
The mercury barometer

One way of measuring atmospheric pressure is to see how much mercury it will support in a closed tube. Pressure is measured in newtons per square metre (N/m^2). The newton per square metre is the **pascal** (Pa), so $1\,N/m^2 = 1\,Pa$.

Normal atmospheric pressure is sufficient to support a vertical column of mercury 76 cm high, or a column of water about 10 metres high! Another way of saying this is that normal atmospheric pressure is 1000 hectopascals.

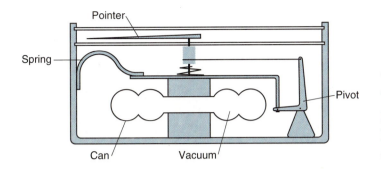

The aneroid barometer

The aneroid barometer does not contain liquid. It is made from a corrugated box that has been partly evacuated. When the atmospheric pressure increases, it presses down the top of the box. This, through a system of levers, causes a needle to move over a scale. The opposite happens when atmospheric pressure gets lower, i.e. the needle moves in the opposite direction.

Atmospheric pressure decreases with increasing altitude. This is because the higher you go, the less air there is pushing down. The density of the air also decreases with increasing altitude.

Atmospheric pressure depends on the height (altitude) at which you measure it. An instrument called an **altimeter** is used to measure altitude. It does this by measuring the pressure.

EXPERIMENTS WITH GASES

A gas trapped in a container has a volume—the volume of the container. It also exerts a pressure on the container and has a temperature. You would find it very difficult to keep track of changes in pressure, volume and temperature, so we experiment with pressure and volume but keep the temperature fixed. Later we shall experiment with volume and temperature and keep pressure fixed.

EXPERIMENT 62.6
AIM: TO INVESTIGATE HOW PRESSURE CHANGES THE VOLUME OF A GAS
(KEEPING THE TEMPERATURE FIXED)

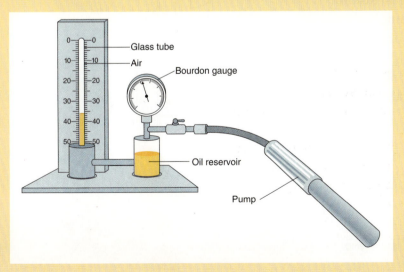

Method

1. Set up the apparatus as shown. Record the pressure of the gas on the pressure gauge and the volume of the gas.
2. Increase the pressure on the gas with the pump. Record the pressure and volume again.
3. Repeat step 2 until you have six or seven sets of readings.
4. Fill in the table at the top of the next page.

BREAKTHROUGH SCIENCE

Pressure	Volume	Pressure x Volume

Result: You will notice that as the pressure increases, the volume decreases. This was first discovered by Robert Boyle, and is called Boyle's law. Scientists call this an inverse relationship. This means that as one goes up, the other goes down.

Conclusion: Boyle's law—For a fixed mass of gas at a constant temperature, the pressure is inversely proportional to the volume.

The chemist Robert Boyle (1627–1691) was a son of the Earl of Cork. He made discoveries in chemistry and is known for his gas law (Boyle's law). He was a founder of the Royal Society. Boyle insisted that experimentation was an essential part of scientific proof and influenced many other scientists.

EXPERIMENT 62.7
AIM: TO INVESTIGATE HOW TEMPERATURE CHANGES THE VOLUME OF A GAS (KEEPING THE PRESSURE FIXED)

Method
1. Set up the apparatus as shown. Record the temperature of the gas (use the thermometer) and the volume of the gas.
2. Increase the temperature of the gas. Record the temperature and volume again.
3. Repeat step 2 until you have six or seven sets of readings.
4. Fill in the table on the next page, and draw a graph of temperature (*x*-axis) against volume (*y*-axis).

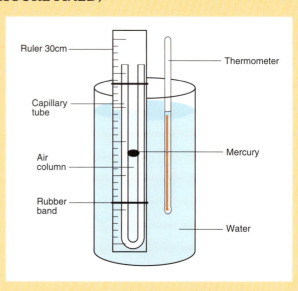

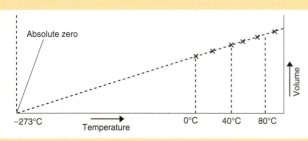

Temperature	Volume

Result: You will notice that as the temperature increases, the volume also increases. This was first discovered by Jacques Charles, and is called Charles' law. Scientists call this a direct relationship. This means that as one goes up, the other also goes up.

The straight-line graph when extended tells us that the gas should have zero volume at a temperature of –273°C. This temperature is called absolute zero, and is the starting point of the Kelvin scale of temperature. You add 273 to a Celsius temperature to get the equivalent Kelvin temperature. **Example:** 27°C = 300 K

Conclusion: Charles' law—For a fixed mass of gas at a constant pressure, the Kelvin temperature is directly proportional to the volume.

ABSOLUTE ZERO

A substance can be heated to any temperature: scientists have produced temperatures of millions of degrees. However, no matter how you try, the lowest temperature you can cool a substance to is absolute zero (0 K).

Example 1

A bicycle pump contains 200 cm³ of air at a pressure of 1,000 hectopascals. If you increase the pressure to 2,000 hectopascals without changing the temperature, what will be the volume of the gas?

$$\text{Use Boyle's law: pressure} \times \text{volume} = \text{constant}$$
$$P_1 V_1 = P_2 V_2$$
$$(1{,}000)(200) = (2{,}000) V_2$$
$$V_2 = \frac{(1{,}000)(200)}{2{,}000}$$
$$= 100 \text{ cm}^3$$

Example 2

A balloon has a volume of 500 cm³ at room temperature (17°C). What will be the volume of the balloon at a temperature of 75°C, when the pressure remains constant?

Use Charles' law:

$$\frac{V_1}{T_1} = \frac{V_2}{T_2} \qquad T_1 = 17 + 273 = 290 \text{ K}$$
$$\qquad\qquad\qquad T_2 = 75 + 273 = 348 \text{ K}$$

$$\frac{500}{290} = \frac{V_2}{348}$$

$$V_2 = \frac{(500)(348)}{290} = 600 \text{cm}^3$$

WEATHER FORECASTING

"Red sky at night, shepherd's delight." This saying was probably used to forecast weather long before scientific forecasting. It uses the appearance of the sky to predict the weather for the following day. Forecasters still use the appearance of the sky, but also take a lot of scientific measurements and record weather patterns over long periods of time. Forecasters have weather recording stations at many locations around Ireland. They also get measurements by radio signals from weather balloons, from aircraft and from ships, as well as satellite photographs that show approaching weather systems.

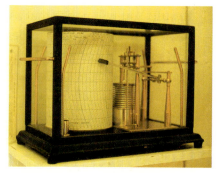

Barograph

Barometer

A barometer measures atmospheric pressure. Meteorologists (weather forecasters) also use a barograph. This is a barometer attached to a device that traces an ink line on a rotating piece of paper. Barographs give a continuous record of pressure changes over a period of time.

Thermometer

A maximum and minimum thermometer records the temperature range over a time period. The thermometer is usually kept in a Stevenson screen to prevent its being affected by rain or direct sunshine.

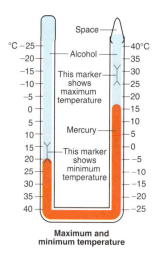

Maximum and minimum temperature

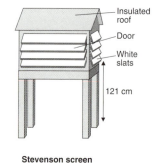

Stevenson screen

Rain gauge

A rain gauge has a funnel that collects rainfall over a period of time. The funnel drains the water into a storage container. The rain is poured into a measuring cylinder to record the rainfall in millimetres.

Measuring the contents of a rain gauge

Applied Science · Weather

Anemometer and wind vane

Campbell–Stokes sunshine recorder

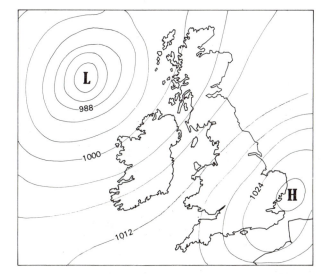

Weather map showing high pressure off England and low pressure in the Atlantic

Anemometer and wind vane

The anemometer measures wind speed. It does so with a set of rotating cups attached to a shaft. The faster the wind, the faster the shaft turns.

Wind direction is measured by a wind vane. The wind vane must be positioned where it is not affected by buildings or the contours of the ground.

Hygrometer

We have already seen that the hygrometer measures the humidity of the air. It compares the temperature of a dry thermometer with that of a thermometer covered with a wet cotton gauze. A set of tables convert the readings of the hygrometer to a humidity reading.

Campbell-Stokes sunshine recorder

The length of sunshine can be measured with a Campbell-Stokes sunshine recorder. The solid glass sphere focuses the sun's rays on a piece of card. When the sun is shining, it blackens the card. The amount of blackening is measured to give the hours of sunshine.

Areas of high pressure bring settled weather

An area of high pressure is an anti-cyclone. It is a large area of stable high-pressure air. The air sinks and warms up, so that any clouds turn to water vapour. The sky is usually clear and sunny with high pressure. The winds in an anti-cyclone blow clockwise.

Areas of low pressure often bring rainy weather

An area of low pressure is a cyclone or depression. It is a large area of rotating unstable low-pressure air. This forms when an area of warm air (from the south) and an area of cold air (from the north) move past each other. The warm air slides over the cold air, and the cold air moves in beneath the warm air. This produces a swirling movement and starts the air rotating. The warm moist air turns to rain when it meets the cold air. The sky is cloudy and rainy in low-pressure areas. The winds in a cyclone blow anticlockwise.

SUMMARY

- Humidity is the amount of water in the air compared to the amount when the air is saturated.
- Clouds form when water vapour condenses into tiny drops which float in the wind.
- The atmosphere is the layer of air that surrounds the earth.
- Boyle's law: For a fixed mass of gas at a constant temperature, the pressure is inversely proportional to the volume.
- Charles' law: For a fixed mass of gas at a constant pressure, the Kelvin temperature is directly proportional to the volume.
- You add 273 to a Celsius temperature to get the equivalent Kelvin temperature.
- A barometer measures atmospheric pressure.
- An anemometer measures wind speed.
- A hygrometer measures the humidity of the air.

QUESTIONS

Section A

1. A _____ measures atmospheric pressure.
2. Humidity is _____.
3. State Boyle's law. _____
4. An anemometer measures _____.
5. State Charles' law. _____
6. Clouds are formed when water vapour _____.
7. You add 273 to a _____ temperature to get the equivalent _____ temperature.
8. A _____ measures the humidity of the air.
9. The greenhouse effect is caused by _____.
10. Cumulus and cirrus are types of _____.

Section B

1. In the water cycle, water evaporates all the time from the surface of the earth. Describe a simple experiment to show the effect of temperature on the rate of evaporation. Name two measurements you would take if you were recording the weather.

2. (*a*) What is meant by the humidity of the air?
   (*b*) Name the instrument you would use to measure humidity.
   (*c*) What does a barometer measure?
   (*d*) If a barometer were brought to the top of a mountain, how would the reading change?
3. "The earth's atmosphere is in a state of constant change. Local changes in the atmosphere determine our weather." What is the atmosphere?
4. Fog and frost commonly occur in winter. What is meant by each of the underlined terms? What is meant by the greenhouse effect? Explain how this effect occurs.
5. (i) Compare three different types of cloud under each of the three headings: altitude, shape, effects on the weather.
   (ii) Name three things you would measure regularly when keeping a record of the weather in your area. Describe with the aid of a diagram an instrument you would use to measure one of them.
6. (i) Describe a simple experiment to show the relationship between the pressure and the volume of a fixed mass of gas where the temperature is kept constant.
   (ii) The volume of a gas is 108 cm^3 at a pressure of 76 cm of mercury. If the pressure drops to 72 cm of mercury but the temperature remains constant, what will be the new volume of the gas?
7. (i) Describe a simple experiment to illustrate how clouds are formed.
   (ii) Name the factors that affect evaporation and describe a simple experiment to show one of them.
8. (i) Describe an experiment to find the relationship between the volume and the temperature of a fixed mass of gas where the pressure is kept constant.
   (ii) At a temperature of 0°C, the volume of a fixed mass of gas is 80 cm^3. What will the volume of the gas be if the temperature increases to 27°C without a change in pressure?

BREAKTHROUGH SCIENCE

FOOD — CHAPTER 63

NUTRITION

You need to eat many different foods in order to provide your body with the substances it requires. Your body uses these substances to make new cells, to repair damaged cells and to give you energy.

A BALANCED DIET

Your diet is simply the food you eat each day. A diet that supplies your body with all the substances it needs is called a balanced diet.

A balanced diet must contain enough of these substances to supply your body's needs: **carbohydrates, proteins, fats, vitamins, minerals, fibre, water.**

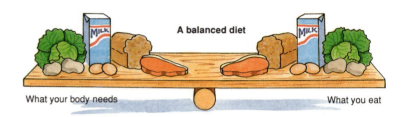
A balanced diet
What your body needs — What you eat

YOUR BODY NEEDS CARBOHYDRATES FOR ENERGY

Foods containing mainly **sugar** or **flour** are **carbohydrates**. They all contain either sugar or starch (flour is wheat starch). Carbohydrates contain the elements carbon, hydrogen and oxygen, which is why they are called carbohydrates.

One of the simplest carbohydrates is **glucose ($C_6H_{12}O_6$)**. Glucose molecules are soluble in water and taste sweet. They are small enough to pass through the walls of your intestines and into your blood. Glucose is a reducing sugar.

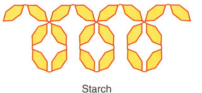
Starch molecule

Sucrose molecules

Glucose molecules

The common sugar you use—**sucrose**—has a molecule made of two simple sugar molecules joined together.

A starch molecule is a long chain of thousands of simple sugar molecules.

Your digestive system changes all carbohydrates into simple sugars, mainly glucose. Respiration releases the energy in glucose.
Carbohydrates are needed for energy.

TESTS FOR CARBOHYDRATES

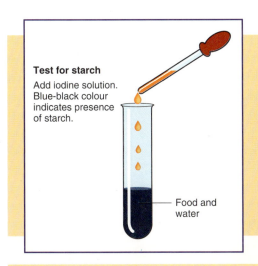

Test for starch
Add iodine solution. Blue-black colour indicates presence of starch.

EXPERIMENT 63.1
AIM: TO TEST FOR STARCH

Method
1. Turn the food into a fine paste by adding water and mashing it.
2. Add a drop of iodine solution to the food.
   Result: The iodine will turn a blue-black colour when there is starch in the food.

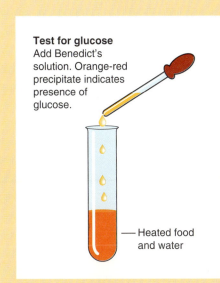

Test for glucose
Add Benedict's solution. Orange-red precipitate indicates presence of glucose.

EXPERIMENT 63.2
AIM: TO TEST FOR SIMPLE SUGARS

Method
1. Make a paste of the food.
2. Add some water to the paste and shake to dissolve any simple sugars.
3. Add some Benedict's solution to the sample in a test tube. Shake to mix the food with the Benedict's solution. Heat the tube gently in a water bath. Do not heat once the colour has changed.
   Result: The solution will turn a yellow-red colour when sugar is present. When there is only a little sugar in the sample, the solution will turn a yellowish green.

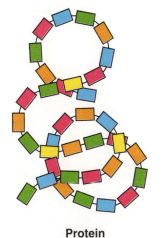

Protein

YOUR BODY NEEDS PROTEINS FOR GROWTH AND REPAIR OF CELLS

Meat, fish, eggs, milk, cheese and most vegetables are good sources of proteins. Proteins always contain the elements carbon, hydrogen, oxygen and nitrogen (some proteins also contain sulphur).

Proteins are long chains of about 500 amino acid molecules. There are 20 different amino acids.

Your digestive system breaks down protein chains into amino acids. Your body then links these amino acids together in a different order to make new proteins for your body. You use proteins to make new cells, to repair cells and to produce enzymes and antibodies. Enzymes and antibodies are proteins.

Proteins are used for growth and repair of cells.

BREAKTHROUGH SCIENCE

EXPERIMENT 63.3
AIM: TO TEST FOR PROTEINS (THE BIURET TEST)

Method

1. Make a paste of the food. Add a small quantity of water.
2. Add an equal volume of 10% sodium hydroxide solution.
3. Add three drops of 1% copper sulphate solution. Cork the test-tube and shake it gently.

   Result: A violet colour appears when protein is present. This colour may not last very long.

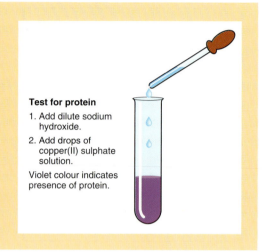

Test for protein
1. Add dilute sodium hydroxide.
2. Add drops of copper(II) sulphate solution.

Violet colour indicates presence of protein.

YOUR BODY NEEDS SOME FATS

Butter, margarine, fat on meat, fish oil, animal oils and vegetable oils are all fats. Fats contain only the elements carbon, hydrogen and oxygen. However, they are not carbohydrates. Fat molecules contain a glycerol molecule attached to three fatty acids. Fats are **insoluble in water**.

Your body breaks down fats into glycerol and fatty acids, and uses fats as a source of energy when all your carbohydrates have been used. Fats also form an insulating layer underneath your skin to keep you warm.

If you eat too much fat, particularly animal fat, it can cause a build-up of cholesterol and other fatty substances in your arteries—this is arteriosclerosis. When this happens in the heart arteries, it blocks the blood supply to the heart and can cause a heart attack.

| *Fats are used as an energy store and for heat insulation.*

Fat molecule

Fatty acid and glycerol molecules

EXPERIMENT 63.4
AIM: TO TEST FOR FATS

Method

1. Squeeze food in a piece of brown paper.
2. Warm the paper gently to dry off any water.
3. Hold up the paper to the light.

   Result: You can see a translucent spot where the food is fat.

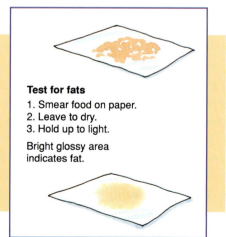

Test for fats
1. Smear food on paper.
2. Leave to dry.
3. Hold up to light.

Bright glossy area indicates fat.

YOUR BODY NEEDS VITAMINS TO STAY HEALTHY

Sailors on long voyages used to suffer from many diseases. Some of these were caused by bacteria and viruses. However, one disease, scurvy, happened only on long voyages. In 1747, Surgeon James Lind found that sailors who ate oranges or lemons did not get scurvy. Later, scientists discovered that certain essential substances are needed to maintain good health—these are vitamins.

Vitamins are chemicals your body needs in small amounts to keep it working properly. We get the vitamins we need from the food we eat. If you do not have enough of a vitamin you develop a **vitamin deficiency disease**. A balanced diet supplies your body with all the vitamins it needs. Your body can make vitamin D, the "sunshine" vitamin, from cholesterol in the skin by using the ultra-violet rays of the sun.

Vitamin	Common source	Function	Deficiency disease
A	Fish, milk, green vegetables	Maintains healthy eyes	Eye problems, night blindness
B group	Liver, eggs, yeast, wholemeal	Involved in respiration and nervous system	Beri-beri: wasting muscles
C	Potatoes, citrus fruit, other fruit, vegetables	Maintains healthy skin, teeth, gums	Scurvy: bleeding gums and skin
D	Liver, fish oils, milk, eggs	Maintains healthy bones and teeth	Rickets: deformed bones

YOUR BODY NEEDS MINERALS TO STAY HEALTHY

Did you know that you have iron in your blood? Your body needs iron and other minerals in order to remain healthy. You don't have pieces of solid iron in your bloodstream; you have iron and other minerals as dissolved salts. Your body needs only small amounts of these mineral salts. A balanced diet supplies your body with all the minerals it needs.

Some mineral salts your body needs are listed in the following table.

Mineral	Common source	Function
Calcium	Milk, cheese, bread	To make bones and teeth
Iron	Liver, egg yolk	To make haemoglobin (red blood cells)
Iodine	Seafood	To make thyroxin

Some other elements your body needs are **sodium, potassium, sulphur, magnesium, copper and zinc.**

FIBRE IS AN IMPORTANT PART OF YOUR DIET
Many foods your body needs also contain fibre (roughage). Fresh fruit, vegetables and brown bread contain a lot of fibre. Fibre is mostly made of plant cellulose. Cellulose is an indigestible carbohydrate which helps the muscles of your digestive system to push the food along quickly. Your muscles do not work well when you lack fibre in your diet—this causes constipation. Doctors have found that people with diseases of the digestive system often lack fibre in their diet.

Fibre-rich food

YOUR HEALTH AND YOUR DIET
Different people need different amounts of food
Even though you are the same age and size as your best friend, you may need to eat more food. You may play games more often, or do more work at home. As long as you eat a lot of different foods including dairy foods, meat, fish and eggs, vegetables, fresh fruit and bread, you are getting all that your body needs. This is a balanced diet.

Balance between the food we eat and the energy we use up

Are burgers and chips bad for you?
Burgers and chips contain protein, carbohydrates and fats. All are part of a balanced diet, but burgers and chips contain a lot of animal fat. Once in a while this is OK, but if you eat only this sort of food every day your diet will have too much fat and will lack some vitamins and minerals. Too much fat causes overweight and also heart disease.

Too much food makes you fat
When you eat enough to give your body all the energy it needs, you stay the same weight. If you eat more than your body uses, you put on weight. Your body stores the extra food as fat in your body.

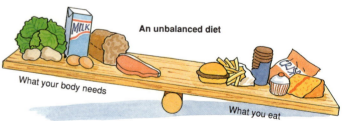

"Dieting" can be dangerous
Many "diets" are unbalanced, and lack one or more of the essential substances your body needs. Some young people get worried about getting fat. Some get anorexia nervosa. They eat very little and lose a lot a weight. They also suffer from mineral and vitamin deficiency. Anorexia nervosa is a serious illness, and needs special medical treatment.

You can keep your weight in balance by moderate exercise and by eating a balanced diet.

FEEDING THE WORLD
World food supply

The world can grow more than enough food to feed all its population. A problem is that it cannot always grow the food in the areas that need it most. Another problem is that transportation networks of air, rail, road and sea are not developed in areas that suffer from famine. Food shortages in Somalia cause famine, while food shortages in India do not. Why? Because India has a well-developed sea and rail system that can transport food to areas of need. Problems of supply and distribution must both be solved in order to free the world from famine.

Causes of famine

Food shortages alone do not cause famine. In fact, during most famines, including the potato famine in Ireland, food was available but was too expensive for poor people.

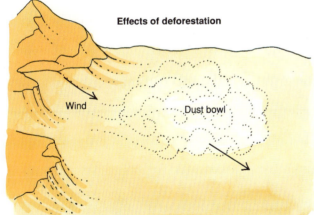

Effects of deforestation

Wind — Dust bowl

1. Famines start when drought, floods, disease or pests destroy the food supply in a region. In areas of drought, the crop failure can go on from year to year.
2. Bad land management makes famines worse. Clearing forests and other vegetation can cause soil erosion and lead to desertification (land turning into deserts).
3. Third-world debt also contributes to famine. Many third-world countries are forced to grow "cash crops"—crops for export to earn cash to pay their debts to rich countries. Much of this money was lent by rich countries to pay for weapons bought from the rich countries.
4. War makes famine worse. Most famines in the past century happened during wars. 1·5 million people died during the civil war in Nigeria in 1967. Recent famines in Ethiopia, Somalia and Sudan have all been made worse by war.

Malnutrition

Millions of people die from the effects of famine each year. Some die of starvation, but a greater number die of diseases such as cholera and typhus. Millions more are left crippled for life by diseases caused by malnutrition.

Malnutrition is due to a seriously unbalanced diet—one that lacks essential protein, minerals or vitamins. Malnutrition permanently stunts the growth and development of millions of children in third-world countries.

SUMMARY

- A diet that supplies your body with all the substances it needs is a balanced diet.

Food type	Needed for	Contains the elements
Carbohydrates	Energy	C, H, O
Proteins	Growth and repair of cells	C, H, O, N, S
Fats	Energy storage and heat insulation	C, H, O

- Your body needs vitamins to stay healthy.
- Your body needs minerals to stay healthy.
- Fibre keeps your digestion working properly.

Test	Method	Positive result
Starch	Add iodine solution	Blue/black colour
Glucose	Add Benedict's solution	Red/orange colour
Protein	Add sodium hydroxide solution and 3–4 drops of copper sulphate solution—heat	Violet colour
Fat	Rub on brown paper	Translucent spot

- Malnutrition is due to a seriously unbalanced diet.

QUESTIONS

Section A

1. Protein is used for _____.
2. The Biuret test is used to test for _____.
3. Malnutrition is caused by _____.
4. Iodine solution is used to test for _____.
5. Fats are used for _____.
6. Benedict's solution is used to test for _____.
7. Carbohydrates are used for _____.
8. Three causes of famine are (i) _____, (ii) _____, (iii) _____.
9. Calcium is needed in your diet for _____.
10. Iron is needed in your diet for _____.
11. Rickets is caused by a deficiency of _____.
12. Potatoes are a good source of vitamin _____.

Section B

1. What chemical test would you use to test a piece of bread for starch? What effect does this test have on the colour of the bread?
2. What are vitamins? Name two vitamins required by humans. Why are minerals important in the diet?
3. State the main functions of protein in the body. Describe, with the aid of a diagram, an experiment to test for the presence of protein in a food sample.
4. Fill in the blanks in the following summary of food tests

Test for	Method	Positive result
Starch		
Glucose		Red/orange colour
	Add sodium hydroxide solution and 3–4 drops of copper sulphate solution. Heat.	
Fat		

5. Fill in the blanks in the following table.

Mineral	What it does	Foods it is found in
Calcium		Milk, cheese
	Is part of haemoglobin in blood	
Iodine	Needed by _____ gland	

6. Choose your answer to (i), (ii) and (iii) from among the following is a list of foods: chocolate, cake, peas, carrots, nuts, cabbage, jam, bread, tomatoes, lemons.
    (i) Which foods contain large amounts of carbohydrate?
    (i) Which foods contain large amounts of protein?
    (iii) Which foods contain large amounts of fats or oils?
7. This table shows the amount of energy (measured in kilojoules) used in a day by a six people.

	Male	Female
Adult (office work)	12,000	9,500
Adult (heavy work)	20,000	12,500
Pregnant mother		10,000
Breast-feeding mother		11,500

On a particular day, each of the men took in 17,000 kJ in the food he ate and the women took in 11,000 kJ.
(i) State what effect you would expect this energy intake to have on the weight of each of the six people.
(ii) Why do you think there a difference between the energy a pregnant mother needs and the energy a breast-feeding mother needs?

8. The nutritional information label shown gives the composition of a particular food from the following list: chocolate biscuit, butter, crisps, potatoes, milk, bread, tomatoes, lemons.
   (i) Name the food that best suits the information on the label.
   (ii) Which of the foods contain large amounts of fats or oils?
   (iii) Name two of the foods that are good sources of vitamins.
   (iv) Name two of the foods that are good sources of minerals.

Nutritional Information per 100 g			
Protein	11·15 g	Carbohydrate	41·9 g
Fat	1·5 g	of which sugar	2·1 g
of which saturates	0·3 g	starch	39·8 g
polyunsaturates	0·5 g	Salt	1·35 g
monosaturates	0·2 g	Dietary fibre	4·09 g
Cholesterol	0·0 mg	Kilocalories (kJ)	215 (913)

9. Meat, fish, butter, eggs, cheese. From this list, name a food that is:
   (*a*) a good source of protein
   (*b*) a good source of calcium
   (*c*) a good source of iron.

FOOD PRESERVATION   CHAPTER 64

Our food is also food for bacteria, yeasts, and moulds: Micro-organisms that live on and in our food. These begin to live and grow on our food if it is warm and moist. Bacteria increase rapidly in warm, moist conditions. These bacteria change the taste and appearance of our food when the food "goes off". Some bacteria also produce poisonous substances in food. Clearly, we need to do something to make our food last longer without spoiling. Food preservation treats food so that it stays fresh longer.

Food preservation usually starts with the washing and preparation of your food. You then cook the food at temperatures high enough to kill all bacteria. Once it is cooked, you must avoid any further contamination by bacteria. The food should be covered and stored in a fridge to prevent bacteria from growing.

Refrigeration means storing food at temperatures below 4°C to prevent bacteria growing on the food.

Bread mould consists of micro-organisims that have grown on moist bread

Refrigeration prevents the growth of bacteria

APPLIED SCIENCE · FOOD PRESERVATION

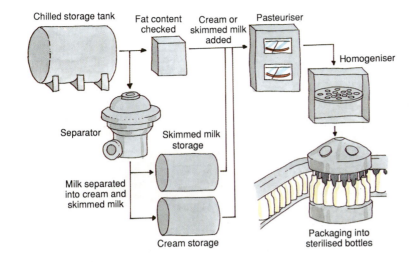

PASTEURISATION
Pasteurisation of food and drink kills micro-organisms but does not affect the taste and appearance of the food. Our milk supply is pasteurised. *Pasteurisation means rapidly heating milk above 72°C for 15 seconds.* After pasteurisation, the milk is cooled immediately to below 10°C. Fruit juice can also be pasteurised.

FREEZING
Storing food at temperatures below −18°C preserves it for a long period of time. Low temperatures do not kill bacteria, but they slow down the growth of micro-organisms. Other chemical reactions that spoil food are also slowed.

Meat, fish, bread, and some fruit and vegetables can be frozen. Home freezers store food in this way.

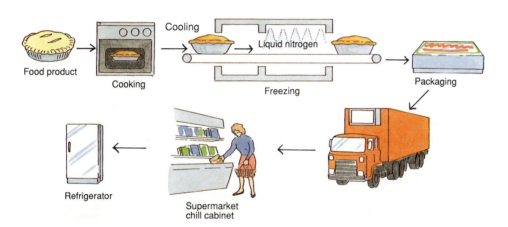

CANNING
Canning preserves processed food in airtight containers. Meat, fish, fruit and vegetables are preserved by canning.

Heat, salt, sugar syrup or some other method is used to kill micro-organisms that would spoil the food during storage. The airtight can then prevents recontamination of the food. Canned food can be stored at room temperature for many months without spoiling. Unlike frozen foods, no special storage place is needed for cans.

DEHYDRATION

Water is essential if bacteria and other micro-organisms are to multiply. All grains and cereals are dried to prevent them from spoiling. Dried milk powder is a major dairy food produced in Ireland. Fish and meats were dried in ancient times to preserve them. Food is also dehydrated by freeze drying. Food science has developed many freeze-dried foods that regain their original form when water is added. Packet soups are examples of freeze-dried foods.

Dried foods

CHEMICAL

Micro-organisms cannot multiply in foods with high levels of salt, sugar or spices. Fresh fish and meat are preserved by salting. Fruit is preserved in sugar syrup. Curries and other Asian foods are preserved in a strong mixture of spices. Smoking meat and fish helps to preserve them as well as adding to their flavour. The smoke treatment is not enough to preserve the food for a long time. It is used in addition to some other form of preservation, such as salting or refrigeration.

Pickling food in vinegar preserves it. Beetroot, onions and gherkins are preserved in this way. The acidity of the vinegar prevents micro-organisms from growing. Many other chemicals such as sodium benzoate and sulphur dioxide destroy micro-organisms in food, but their use is restricted because of their effects on health.

Fruit preserved in syrup

EXPERIMENT 64.1
AIM: TO FIND A SUITABLE METHOD OF PRESERVING BEETROOT

Method

1. Take three slices of freshly cooked beetroot.
2. Place one slice in a jar of vinegar (chemical). Place another slice in a plastic bag and put in a freezer (freezing). Place the third slice on a metal dish and put in a warm oven at 110°C for two hours (drying).
3. Examine each type of food. Would you like to eat it?

    You should now know why beetroot is stored in jars of vinegar. But remember that different foods need different methods of preservation.

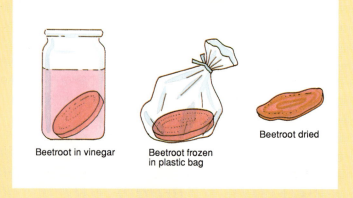

EXPERIMENT 64.2
AIM: TO SHOW THE EFFECTS OF SALT AND SUGAR SOLUTIONS ON PLANT CELLS

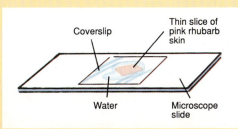

Method
1. Take a young stick of rhubarb and peel a thin slice of the pink skin. Place this on a slide.
2. Make up three slides as follows: (*a*) add a drop of water; (*b*) add a drop of strong salt solution; (*c*) add a drop of strong sugar solution.
3. Cover each with a cover slip. Examine the cells under low power with the microscope.

Result: The sugar and salt solutions draw the water out of the cells. This happens to any bacteria in the food also—bacteria cannot live in these conditions.

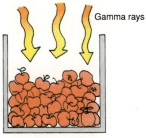

IRRADIATION
Radiation from radioactive isotopes such as cobalt 60 can kill all or some of the micro-organisms, insects and parasites on food. Very high levels of radiation sterilise foods. Fresh fruit treated with radiation will keep for a much longer time than untreated fruit. Some imported fruit is treated in this way. Irradiation can change the appearance and flavour of meat, seafood, and fresh fruit and vegetables, but does not make food radioactive.

EXPERIMENT 64.3
AIM: TO SHOW THE EFFECTS OF MICRO-ORGANISMS ON FOOD

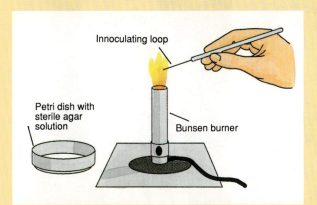

Method
1. Get a number of food samples: fresh fruit, old fruit, milk, bread, etc.
2. Prepare a number of petri dishes with sterile nutrient agar.
3. Sterilise the inoculating loop in a Bunsen flame. Transfer the food sample to the petri dish using the loop. Seal the dish.
4. Repeat this process to transfer each sample to a petri dish.
5. Place the dishes upside down in an incubator at 20°C for two days.

Result: The micro-organism growths produced will be clearly seen on the food samples.

The effects of micro-organisms on food can be seen by sealing small samples of various foods in plastic bags and leaving them in a warm place. After a few days the micro-organism growths can be seen.

FOOD ADDITIVES

A food additive is any substance added to food during processing to preserve it or to improve its colour, texture or flavour. Some people think every food additive with an E number is harmful. All the E number means is that the substance has been tested for safety by the EU. Many substances with E numbers are natural and are found in apples, oranges and other foods.

Food colourings (E1...)
Food colours are used to replace colours lost in food during cooking. They are also used in sweets to make them bright and attractive. Many food processing companies now use natural food colours from plants such as carrots and beetroot to replace artificial colours.

Preservatives (E2...)
These are chemicals that kill or slow down the micro-organisms that spoil food.

Antioxidants (E3...)
Antioxidants are used to keep fats and oils from going off (rancid) and to prevent discolouration of canned meats and fruit. Vitamin C (E300) is an antioxidant used in canned fruit.

Emulsifiers and stabilisers (E4...)
Emulsifiers are added to get oil and water in foods to mix together. Eggs are used in baking as a natural emulsifier. Stabilisers stop the emulsion from separating out. Lecithin is used as a stabiliser in salad dressings and chocolates. Pectin and gelatine, both natural substances, are added to thicken jams and jellies.

Sweeteners (E42... or E6...)
The artificial sweeteners saccharin, sorbitol and aspartame are used as sugar substitutes in diet drinks and diabetic foods.

Flavourings are the largest group of additives. They include salts, spices and other natural and synthetic flavours.

Minerals and vitamins are sometimes added to food to replace losses during processing or to provide additional nutrients. Iodine is often added to salt, vitamin A to margarine and vitamin D to milk.

ADVANTAGES AND DISADVANTAGES OF FOOD ADDITIVES

Advantages	Disadvantages
1. Food poisoning would probably be more common if food additives were not used. Many food additives are preservatives that kill bacteria.	1. Some people are allergic to dairy foods, others to shellfish. These are allergic reactions to natural substances. Some food additives can cause allergies. Artificial food colourings such as E102 (yellow tartrazine) can cause allergic reactions in children, including hyperactivity.
2. Some foods would not keep as long without additives. This would lead to higher prices and shortages of food.	2. Some food additives have been found to be harmful. The artificial sweetener cyclamate was widely used in soft drinks until it was found to be harmful and banned.
3. Foods with additives are tastier and more attractive. Pre-cooked foods are often more convenient.	3. Food additives can destroy vitamins in food. Sulphites, used as preservatives in food, destroy vitamin B.

WHAT SHOULD I EAT?

Many food additives are natural substances that have been used in cooking for centuries. You should have no cause to worry about these additives. If you eat a good balanced diet with a variety of foods, you reduce the amount of any one additive. Unless you have an allergy to a particular additive, you can eat anything in moderation. Remember that a lot of salt or sugar in your diet could harm your health.

SUMMARY

- Micro-organisms spoil food.
- Some bacteria cause food poisoning.
- Food preservation keeps our food from spoiling.
- Refrigeration stops bacteria from growing.
- Pasteurisation means the rapid heating of milk above 72°C for 15 seconds, followed by immediate cooling to below 10°C.
- Canning preserves processed food in airtight containers.
- Dehydration preserves food by removing the water essential to growing micro-organisms.
- Chemicals preserve food by killing micro-organisms in food.
- Irradiation preserves food by killing micro-organisms.
- An E number means that an additive is approved for use by the EU.
- E1 = colourings, E2 = preservatives, E3 = antioxidants, E4 = emulsifiers and stabilisers, E42 or E6 = sweeteners.

QUESTIONS

Section A

1. Food colourings have E numbers starting with a _____.
2. Food poisoning is caused by _____ in food.
3. Pasteurisation is _____.
4. Anti-oxidants have E numbers starting with a _____.
5. Dehydration means _____.
6. Emulsifiers and stabilisers have E numbers starting with a _____.
7. Food exposed to _____ can be stored for long periods without spoiling.
8. Preservatives have E numbers starting with a _____.
9. Refrigeration means _____.
10. Dehydration preserves food because _____.
11. Food is spoiled by _____.
12. Sugar and salt preserve food by _____.

Section B

1. (a) Name three methods of preserving food for human use.
   (b) Explain how one of the methods you have named preserves food.
   (c) E110 and E124 are added to gooseberry jam. Why?
   (d) E300 is also added to the jam. What does this do to the jam?
2. (a) Name two types of food additive. What information is given by the E number of an additive?
   (b) A food contains the additives E471, an antioxidant, and E450, a stabiliser. What do these additives do, and why are they used together?
   (c) Give one advantage and one disadvantage of using food additives.
3. Name a suitable method of preserving each of the following foods: (a) pears, (b) fish, (c) meat.
4. (i) What effect does oxygen have on fats and oils in foods? How can this effect be prevented?
   (ii) A bottle of orange squash has E102 added to it. What does this additive do to the squash?
5. Give one advantage and one disadvantage of preserving food by irradiation.
6. Name two ways of preserving meat. Name four products produced by the dairy industry in Ireland.

APPLIED SCIENCE · PROCESSING OUR FOOD

CHAPTER 65 PROCESSING OUR FOOD

FOOD FROM MILK—THE DAIRY INDUSTRY

Many micro-organisms—bacteria and fungi—are found naturally in food. Given the right temperature and enough time, they increase and multiply. These micro-organisms can cause disease or can spoil the food.

Pasteurisation of food and drink kills micro-organisms, but does not affect the taste and appearance of the food. The process was developed by the French scientist Louis Pasteur in 1860 to preserve wine and beer.

Milk pasteurisers have a number of thin stainless steel pipes. Milk is pumped in one direction through some of the pipes, and hot water is pumped through nearby pipes in the opposite direction. The milk is rapidly heated above 72°C for 15 seconds. After pasteurisation, the milk is cooled immediately to below 10°C.

Even pasteurised milk will go sour after a few days. However, milk can be turned into longer lasting foods such as butter, cheese and yoghurt.

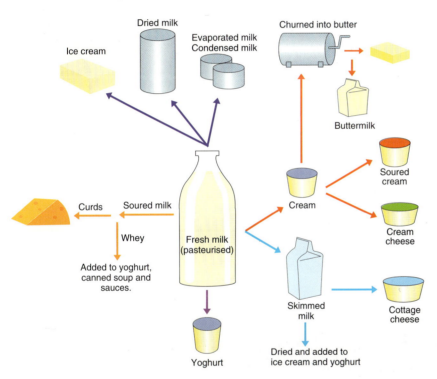

Cheesemaking

People have been making cheese for over 4,000 years. Cheese is made by clotting soured milk with rennet (a digestive enzyme found in the stomach of a calf) and then draining off the liquid whey.

To make your own cheese

In a cheese factory, milk is pasteurised to kill bacteria. After cooling, it is mixed with bacteria that sour the milk to lactic acid, giving the acid conditions the rennet needs. Rennet is then added to the sour milk. About 30 minutes later, the milk clots to form cheese. When the cheese curd is fairly solid it is made into slabs.

409

BREAKTHROUGH SCIENCE

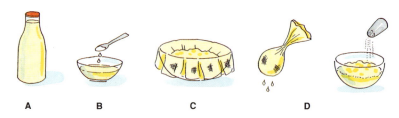

The slabs are salted, drained, and pressed to remove the liquid whey. The blocks of cheese are wrapped and left to mature.

(A) Take one pint of milk and add two teaspoons of lemon juice to sour it.
(B) Add rennet.
(C) Strain through a muslin cloth (the curds stay in the cloth and the whey drains off).
(D) Squeeze and add some salt. Allow to dry and mature.

Butter

Butter is made from milk fat with some salt. Cream (concentrated milk fat) is turned in a special type of churn. This breaks up the fat and causes it to lump together to form butter fat. The liquid buttermilk is drained off. Well-wrapped butter keeps in a fridge for several weeks. Butter may be salted or left unsalted. You can make your own butter by whipping cream with a food mixer until it turns into butter.

Yoghurt

Yoghurt is made from milk soured by a special yoghurt culture. Concentrated milk, or milk with added skimmed milk powder, is heated to about 90°C for a few minutes. It is then cooled to about 40°C. The yoghurt culture (harmless bacteria *Lactobacillus bulgaricus* and *Streptococcus thermophilus*) is added to the milk. Souring and thickening take about 3 hours at 40°C. The bacteria produce the acidity and the yoghurt flavour. Harmful bacteria cannot grow in yoghurt because it is so acidic. Yoghurt lasts for at least a fortnight when kept in a fridge.

EXPERIMENT 65.1
AIM: TO MAKE YOGHURT

Method

1. Add 10 g of dried milk powder to 250 cm^3 of milk in Pyrex bowl. Heat to 90°C for 5 minutes. Cover, and cool by putting the bowl into cold water.
2. Stir five spoonfuls of natural yoghurt (this contains the live yoghurt culture) into 25 cm^3 of treated milk (inoculation).
3. Add this to the rest of the treated milk. Place the mixture in clean plastic cartons. Cover and place in a water bath (or heater) set at 40°C (incubation).
4. Leave for about six hours.

Warning: Do not make yoghurt in a science laboratory or with science glassware, because of possible contamination. Kitchen equipment should be used if you want to eat the yoghurt.

MEAT

The meat industry is a major export industry in Ireland. It uses a variety of methods to preserve the meat, including refrigeration, freezing, canning, and processing with food additives.

Curing pork to make bacon and ham

One method used to preserve pork meat is to salt it to make ham and bacon. Salting pork is called "curing". The pork is put into a strong solution of salt. This kills bacteria, and the high salt content preserves the meat.

Ham and bacon products

Smoking meat

Food is smoked to add to its flavour as well as to preserve it. Meat is hung over the smoke of a woodfire. Chemicals in the smoke are absorbed into the meat, add flavour to it and kill bacteria. Smoking is used in addition to some other form of preservation, such as curing or refrigeration. Sausages, cheeses and fish are also smoked to add flavour to them.

Growth hormones in animals

Implants of growth hormones were used by farmers to increase the growth of animals in their first few years. This produced leaner cattle faster, and also greater milk yields from cows. Many people became concerned about their long-term effects on human health. The EU banned all natural or synthetic growth hormones in animals, because residues were showing up in meat products.

Antibiotics in meat

Animals get sick, just like people. The vet prescribes antibiotics to clear up the infection. Milk, eggs or meat from sick animals must not be used until all traces of the antibiotic have passed from the animal. This should protect us against antibiotic residues. Unfortunately, some farmers add antibiotics to poultry and pig feed to prevent disease. This is bad farming, and fortunately is not very common. The residue of these antibiotics shows up in humans. Also, some bacteria become resistant to the antibiotics. When a doctor prescribes the same antibiotic for a human disease, the antibiotic is less effective against the resistant bacteria.

BREAKTHROUGH SCIENCE

ALCOHOL FROM GRAIN
Brewing

Beer, lager and stout are all brewed from malted barley. Barley is malted by allowing it to germinate to convert starch into sugar. The process is then stopped by heating it. When stout is made, some barley is roasted to give a dark colour.

Brewing involves four steps.
1. Mashing: crushing and heating malt, water and grain to convert starch into sugars.
2. Boiling: adding hops to the mash and boiling it—the liquid is called wort. Hops give beer a bitter taste.
3. Fermentation: adding yeast to the wort. This turns the sugars into alcohol and gives off carbon dioxide gas.
4. Maturing: the maturing process can last from 2 to 24 weeks.

Most beers are then filtered. Beer is usually pasteurised in the container to preserve it.

Fermentation: sugar + yeast ➜ alcohol + carbon dioxide

Wine making

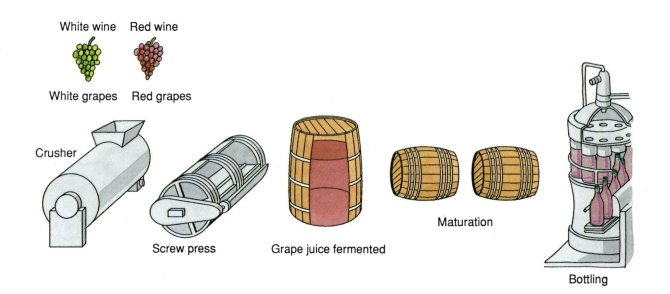

EXPERIMENT 65.2
AIM: TO MAKE ALCOHOL FROM SUGAR USING YEAST

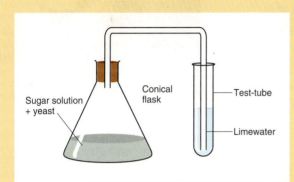

Method
1. Put 200 cm^3 of sugar solution into the conical flask. Add yeast suspension to the solution.
2. Connect the tubing to the limewater solution and leave in a warm place.
3. The limewater turns milky, showing that carbon dioxide is given off during fermentation.

Result: You can smell the alcohol produced.

Fermentation is the conversion, by yeast, of sugar into alcohol and carbon dioxide.

Fermentation is also used in baking bread. Yeast is added to dough and allowed to ferment. The dough is baked and the bubbles of carbon dioxide make the bread rise. The alcohol is burned off in the oven.

Distilling

Alcohol produced from brewing can be concentrated by distillation. The mixture of alcohol and water is boiled in a large copper vessel. The alcohol boils at a lower temperature than water, and turns to vapour. This passes through the condenser or still, and condenses back into a liquid. In this way the concentration of alcohol is increased to 30% or 40% to make whiskey. Whiskey is aged in wooden casks to add flavour.

The concentration of alcohol varies in different drinks:

Beer, lager and stout	4% alcohol
Wine	9%–13%
Sherry and port	16%–20%
Whiskey and brandy	30%–40%

SILAGE

Grass grown to feed cattle in winter is preserved by drying into hay, or can be turned into silage. Hay requires fine weather to dry, and needs a dry barn to store it in. Silage making does not depend as much on good weather.

BREAKTHROUGH SCIENCE

| *Silage is grass preserved by fermentation.*

Grass is cut and chopped by the silage cutter. It is then packed into an airtight silo or silage pit. Molasses and formic acid are added to improve the quality of the silage. New biological additives—special enzymes—are sometimes added. The micro-organisms ferment sugars in the grass into acids, which preserve the grass. The micro-organisms do not work with oxygen present, so the silo must be airtight. The silo must hold any liquid that runs off the silage, as this is acidic and would pollute streams and rivers.

Cattle being fed on grass silage

EXPERIMENT 65.3
AIM: TO MAKE SILAGE

Method
1. Pack finely chopped grass (grass cuttings from a lawn) into the pipe.
2. Leave for 14 days.
3. Let the gas produced bubble through limewater. This shows it is carbon dioxide.
4. Examine the grass after this time. Find out how acid the liquid run-off is (use pH paper).

You could also compare the effects of different additives on the silage by adding molasses to one sample and 10 cm^3 of formic acid to another.

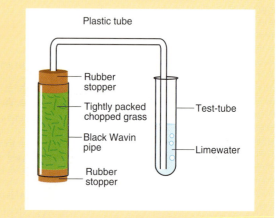

BIOTECHNOLOGY

Biotechnology means using plant and animal substances and micro-organisms to make substances useful to humans. Cheese, yoghurt, beer, wine and silage making are well known examples of biotechnology practised by mankind for thousands of years. Modern biotechnology makes antibiotics to treat disease, enzymes for washing powders and insulin for diabetics.

Recent developments in biotechnology can change the genes of a plant or an animal to produce a new, stronger type. Weather- and disease-resistant crops and lean animals are developed in this way. Scientists have also produced bacteria that eat waste and make gas that can be used as a fuel.

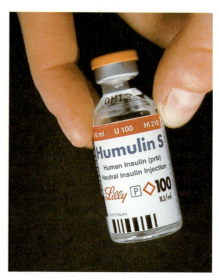

Human insulin

SUMMARY

- Cheese is made by clotting soured milk with rennet
- Yoghurt is made from milk soured by special yoghurt culture.
- Fermentation is the conversion by yeast of sugar into alcohol and carbon dioxide.
- Silage is grass preserved by fermentation.
- Biotechnology means using plant and animal substances and micro-organisms to make substances useful to humans.

QUESTIONS

Section A

1. Name a micro-organism used in the food processing industry. _____
2. Why is rennet added to milk when making cheese? _____
3. Fermentation is _____.
4. Grass preserved by drying is _____.
5. Grass preserved by fermentation is _____.
6. Pork is cured by _____.
7. Yoghurt lasts a long time because _____.
8. When sugar is fermented by yeast, _____ and _____ are produced.
9. Wine is produced by _____ grape juice.
10. Name four substances produced from milk. _____
11. When lactobacillus is added to milk it produces _____.
12. Whiskey is made by _____ a liquid containing alcohol.

Section B

1. What is pasteurisation? Why is it desirable to pasteurise milk and fruit juices? Outline how cheese is produced from pasteurised milk.
2. In an experiment to show how alcohol could be produced by fermentation, four boiling tubes were set up as shown in the diagram. A balloon was placed over the mouth of each tube. The apparatus was placed in a warm place for 30 minutes.

A — Sugar and warm water
B — Yeast and warm water
C — Yeast and warm water and sugar
D — Yeast and ice-cold water and sugar

(i) Which of the balloons do you think will inflate?
(ii) Explain why this will happen.
(iii) What gas causes those balloons to inflate?
(iv) Describe how you would test for this gas.
3. (a) Describe, with the aid of a diagram, a simple experiment to show how you would make silage in the laboratory.
(b) Explain how yeast causes bread to rise.
4. (i) Describe how milk is pasteurised.
(ii) Describe briefly how you would make cheese.
Name three types of cheese.
5. When yeast is added to a warm sugar solution, fermentation takes place.
(i) Name two of the substances produced by fermentation.
(ii) Give one use for each of these substances.
(iii) Yeast is very important in the food industry. Give two examples of the use of yeast.
6. (a) Give two reasons why there is strict control over the use of antibiotics.
(b) State one advantage and one disadvantage of the use of antibiotics in meat production.
(c) Why is the use of growth hormones in meat production banned?
7. Yoghurt, butter and cheese are three dairy products. Describe how you would produce yoghurt in the kitchen for home use. What precautions should you take to ensure that it is safe to eat? Why is yoghurt stored at low temperatures?

ELECTRONICS CHAPTER 66

Before beginning your study of electronics it is essential to revise chapter 13, and in particular to revise the symbols on page 73. Do you own a pocket calculator, a computer, a video recorder, a digital watch? All these things are made by the electronics industry. But what is electronics? What is it all about?

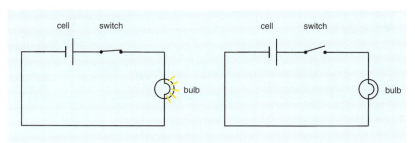

Electronics is about controlling tiny currents. Before we learn how to control tiny currents, let's see how larger currents are controlled.

We can control currents by using a switch, but only in a very limited way. Close the switch; current flows. Open the switch; no current flows.

SWITCHES IN SERIES

Switches are in series when they are connected one after another.

EXPERIMENT 66.1
AIM: TO STUDY SWITCHES IN SERIES

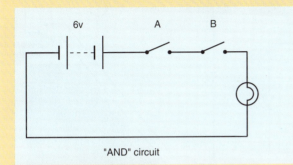

"AND" circuit

A	B	bulb
open	open	off
closed	open	
open	closed	
closed	closed	

Truth table

Method
1. Set up the circuit as shown in the diagram.
2. Close A only. Close B only. Now close both.

Result: The bulb lights only when A and B are closed. Copy the table into your science notebook and fill it in. This type of circuit is called an **and** circuit.

SWITCHES IN PARALLEL

Switches are in parallel when they are connected side-by-side.

EXPERIMENT 66.2
AIM: TO STUDY SWITCHES IN PARALLEL

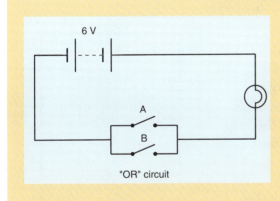

"OR" circuit

Method
1. Set up the circuit as shown in the diagram.
2. Close A only. Close B only. Close A and B.

Result: The bulb lights when either A **or** B is closed. Make out a table in your notebook similar to the one for switches in series, and use this circuit to complete the table.

TWO-WAY SWITCHES

EXPERIMENT 66.3
AIM: TO STUDY TWO-WAY SWITCHES

Method
1. Set up the circuit shown in the diagram.
2. Switch both A and B up: the bulb lights.
3. Switch both A and B down: the bulb lights.
4. Switch one up and the other down: the bulb does not light. This type of circuit is used so that a landing light can be turned on or off using either the upstairs or the downstairs switch.

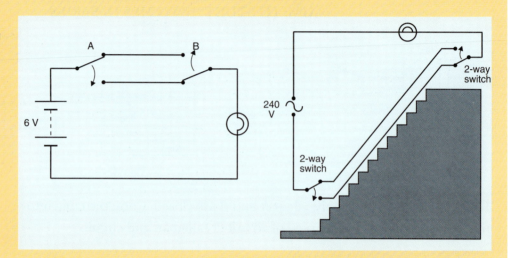

BULBS IN SERIES
Like switches, bulbs are in series when they are connected one after another. Assuming the bulbs are identical, the voltage is divided equally between them so they do not light as brightly as one bulb would on its own.

Another problem is that if one bulb blows, or is removed, the circuit is broken. This often happens with Christmas tree lights.

BULBS IN PARALLEL
In this case, each bulb has the full voltage across it. As a result the bulbs are brighter than when they were connected in series. Also, if one bulb is blown the other stays lighting.

On the other hand, they draw twice as much current from the battery, so the battery runs down faster. The lights in your home are connected in parallel.

RESISTORS
It's time to turn back to chapter 13 once again, to revise resistors. Remember electricity can flow through a conductor such as copper wire quite easily. It can also flow through a resistor, but not so easily.

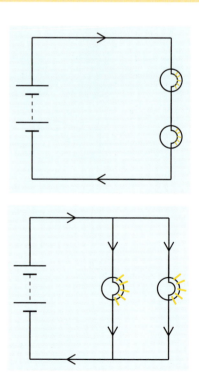

APPLIED SCIENCE · ELECTRONICS

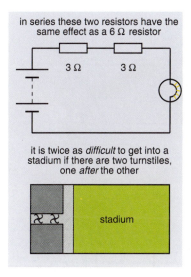

Resistors in series
Two or more resistors in series are simply added together. Current has to flow through both resistors one after the other, so it is twice as hard. Total resistance = 3 + 3 = 6 ohms.

Resistors in parallel
Resistors in parallel have a lower resistance than one of the resistors on its own, because by adding a second resistor you make it easier for electricity to get through.

Remember, according to Ohm's law, if the resistance goes down the current goes up, and vice versa.

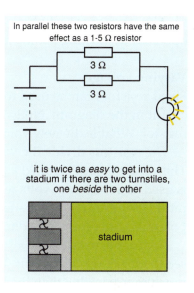

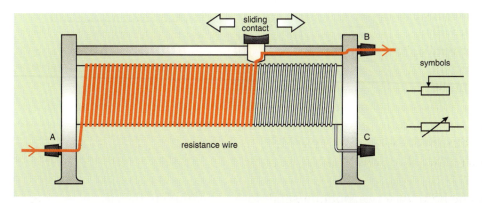

Variable resistor (manual)
The resistors we have used up to now are fixed resistors. The resistor in the diagram is a variable resistor. If the sliding contact is moved to the right the current must flow through more resistance wire, so the resistance is increased. If the contact is moved to the left the current must flow through less resistance wire, so the resistance is decreased. This type of resistor is also called a **rheostat**.

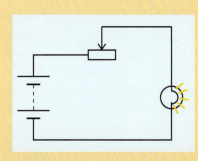

EXPERIMENT 66.4
AIM: TO STUDY THE VARIABLE RESISTOR
Method
1. Set up the apparatus as shown in the diagram.
2. Slide the contact to the right. The bulb dims.
3. Slide the contact to the left. The bulb brightens.

The more compact type of variable resistor in the photograph is operated by turning a knob instead of sliding a contact.

Varying voltage
Fixed resistors can be used to vary voltage. To show this, set up the circuit in the diagram. Note the voltage. Replace B with a 200 ohm resistor. What is the new voltage?

The resistors divide the voltage in proportion to their resistance. This arrangement is called a **potential divider**. A potential divider is used to get a lower voltage from a battery than the battery rating. For example, if a battery is rated 6 volts and you need a potential difference of 3 volts, set up the circuit shown above.

A variable resistor can also be used to vary voltage, as you can see if you set up this circuit. The advantage is that this circuit is much more flexible than the last one. When a variable resistor is used like this it is called a **potentiometer**. It can be used to control the volume of a radio or the brightness of a TV. It can also be used as a dimmer switch, as you will see if you replace the voltameter with a bulb. However, it is not very efficient, so most dimmer switches used in the home are electronic.

Variable resistor

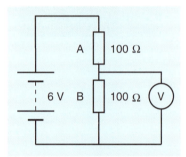

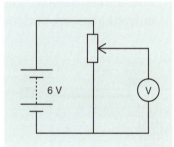

SUMMARY

- When two switches are connected in series, both must be closed to complete the circuit. This is called an **and** circuit.
- When two switches are in parallel, closing either switch will complete the circuit. This is an **or** circuit.
- When two-way switches are used, either switch can turn on or off the circuit from the top or bottom of the stairs.

Bulbs in series	Bulbs in parallel
Dimmer	Brighter
Bulb blows; circuit broken	Bulb blows; circuit unbroken
Battery lasts longer	Battery runs down faster

QUESTIONS

Section A

1. Switches connected one after another are connected in _____.
2. Switches connected side by side are connected in _____.
3. Where in your home would you use a two-way switch? _____
4. The light from two identical bulbs connected in series is _____ than the light from one bulb on its own.
5. The light from two identical bulbs connected in parallel is _____ the light from one bulb on its own.
6. Give one disadvantage of connecting bulbs in parallel. _____
7. Give two disadvantages of connecting bulbs in series. _____
8. Give two advantages of connecting bulbs in parallel. _____
9. Give another name for a variable resistor. _____
10. Draw the symbol for a variable resistor. _____
11. What happens to the resistance of a variable resistor if the sliding contact is moved to the right? _____
12. Two 3 ohm resistances in series have a resistance of _____.
13. Two 3 ohm resistors in parallel have a _____ resistance than one 3 ohm resistor on its own.
14. What is a potential divider used for? _____
15. Give one practical application of a potentiometer. _____

Section B

1. Name the type of simple circuits shown in A and B. What happens in both circuits when the switch is closed? What happens in each circuit if the lamp X blows? How will the brightness be affected if an extra lamp Z is added in each circuit?

2. Compare the two circuits in the diagram under the headings (a) brightness of the bulbs, (b) the rate at which the battery loses energy, (c) what happens if a bulb blows.

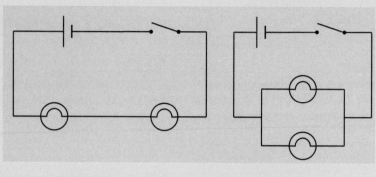

3. Draw a simple circuit diagram containing a two-way switch and a device to show when a current is flowing. Label your diagram. Draw the symbol for a variable resistor.

4. (a) Name the parts labelled X, Y, Z in the diagram on the right. What should be done to the circuit in order that a current may flow? What units are used to measure electric current?
(b) What is A? What is B? State the function of either A or B.

5. Use the information in the circuit diagram below to calculate (a) the potential difference between X and Y, (b) the potential difference between Y and Z. What happens to the lamp shown in the diagram as the sliding contact on D is moved towards Z? Explain your answer.

6. In what units is potential difference measured? What is a potential divider used for? Draw a circuit diagram to show how a potential difference of 1·5 V can be got from a 6 V battery.

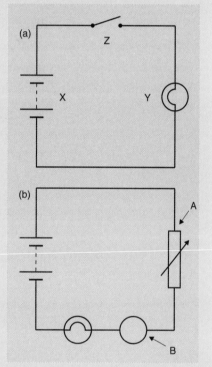

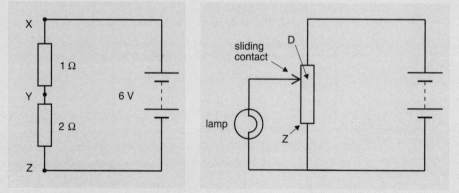

APPLIED SCIENCE · DIODES

CHAPTER 67 DIODES

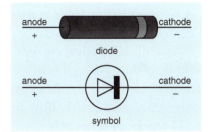

The diode is a very common electronic component. It is usually made of silicon or germanium.

| *A diode allows current to flow through it in one direction only.*

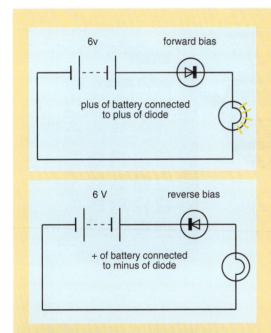

EXPERIMENT 67.1
AIM: TO STUDY THE ACTION OF A DIODE

Method
1. Set up the two circuits shown in the diagram.
2. In the first case the diode is connected in **forward bias**, so it has a very low resistance and the bulb lights. (Note that the arrowhead in the symbol points in the same direction as conventional current.)
3. In the second diagram the diode is connected in **reverse bias**, so it has a very high resistance and the bulb does not light.

CHANGING a.c. TO d.c.

Many electronic devices, such as those in the diagram, can be run off batteries or off the mains (ESB). This is because they have an a.c. adaptor built in to them. The adaptor consists of two parts:
  (a) a transformer, which reduces the voltage.
  (b) A rectifier, which changes a.c. to d.c.
| *A rectifier changes a.c. to d.c.*
A simple rectifier circuit can be set up as shown in the diagram.

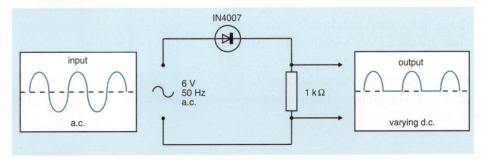

423

LIGHT-EMITTING DIODE (LED)

A light-emitting diode is a diode that gives out light when current flows through it.

As with all diodes, current will flow only when it is in forward bias. The circuit in the diagram shows clearly how the LED works. A resistor is connected in series with it to make sure it is not damaged by too great a current.

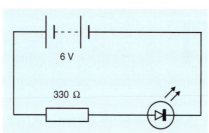
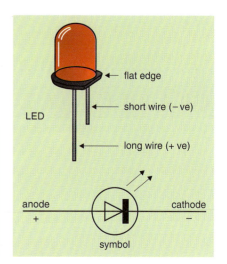

LEDs are widely used to indicate whether electrical devices are on or off. Bar-shaped LEDs are used to form numbers in digital displays.

The simple circuit in the diagram below shows how LEDs can be used to check the polarity of a battery. Can you see how it works?

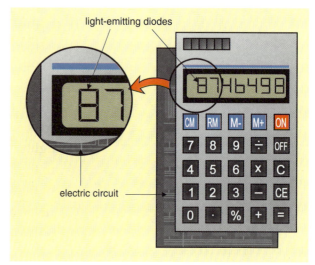

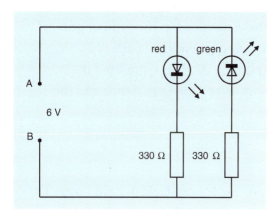

WATER-LEVEL DETECTOR

The circuit in the diagram is used as a water-level detector. When the water rises to the ends of the two wires, the circuit is completed and the LED lights. A buzzer could be used instead of the LED, in which case there would be no need for the protective resistor.

This circuit may not work very well in practice, because water is not a good conductor and even when the wires are very close together the current may be too small to light the LED. What scientists do in this case is to study the circuit and see how it might be improved. We shall see later how this circuit can be improved.

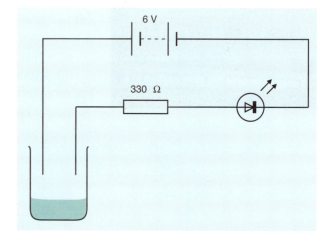

VARIABLE RESISTORS (AUTOMATIC)
Light-dependent resistor (LDR)

An LDR is a resistor whose resistance changes when the amount of light falling on it changes.

When the amount of light goes up the resistance goes down, and vice versa. This can be demonstrated by connecting the LDR to an ohmmeter and varying the amount of light falling on it. (An ohmmeter has its own battery.)

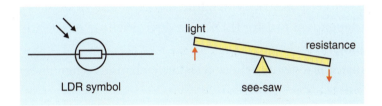

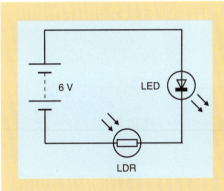

EXPERIMENT 67.2
AIM: TO STUDY AN LDR

Method
1. Set up the circuit shown in the diagram.
2. Note how the brightness of the diode changes as the amount of light falling on the LDR changes.

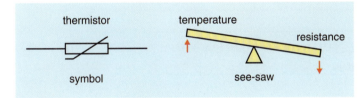

LDRs are used to control street lighting. They are also used in light meters for cameras.

The thermistor

The thermistor is a resistor whose resistance changes when the temperature changes.

As the temperature increases the resistance decreases, and vice versa. The circuit in the diagram can be used to show how a thermistor works. Heat the thermistor and watch how the current changes.

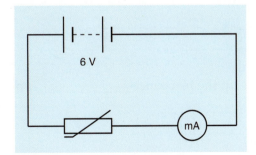

SUMMARY

- A diode is a device that allows current to flow through it in one direction only.
- Diodes can be used in a rectifier to change a.c. to d.c.
- Diodes can also be used to prevent components from being damaged by being connected the wrong way round.
- A light-emitting diode (LED) is a diode that gives out light when current flows through it.
- An LED converts electricity to light.
- A light-dependent resistor (LDR) is a device whose resistance changes when the amount of light falling on it changes.
- A thermistor is a device whose resistance changes when its temperature changes.

QUESTIONS

Section A

1. What is a diode? _____
2. When the anode of a diode is connected to the positive of a battery, the diode is in _____.
3. Draw the symbol for a diode. _____
4. Give one difference between a diode and a resistor. _____
5. A _____ may be used as a rectifier.
6. What is a rectifier used for? _____
7. What is a light-emitting diode (LED)? _____
8. Give one application of an LED. _____
9. The anode lead of an LED is _____ than the cathode lead.
10. Current cannot flow through a diode when it is _____.
11. Draw the symbol for an LED. _____
12. Why must an LED have a resistor in series with it? _____
13. Give two uses of LEDs. _____
14. What is a light-dependent resistor (LDR)?
15. Draw the symbol for an LDR. _____
16. With an LDR, when the amount of light goes up the resistance _____ and vice versa.
17. Give one application of a LDR _____
18. What is a thermistor? _____
19. Draw the symbol for a thermistor. _____
20. With a thermistor, as the _____ increases the _____ decreases and vice versa.

Section B

1. What is device P called? Which lead is the anode? What happens when an electric current flows through P? Draw the symbol for P.

2. Name the device C in the circuit. Will C work if A is closed and B is open? Will it work if B is closed and A is open? What is D? Why is it in the circuit?

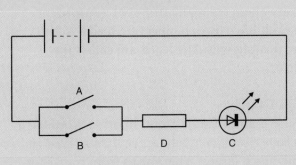

3. Name the parts marked A and B in the circuit on the right. Give one example of the use of B.

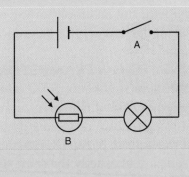

4. Name the device labelled A in the diagram. Through which of the devices A and B will current flow? Explain your answer.

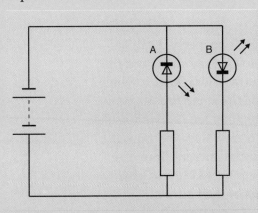

5. What happens when the circuit in the diagram on the right is connected to a 6 V a.c. supply? What is the purpose of the diode? What is this circuit used for?

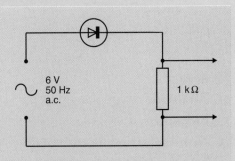

6. What name is given to the type of circuit shown in the diagram below? What is the 330 ohm resistor for? Draw up a truth table for this circuit.

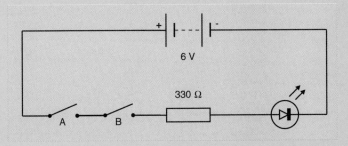

TRANSISTORS CHAPTER 68

Practically every piece of electronic equipment you can name contains transistors. There are many different kinds of transistor, but they all have three connecting leads. In the kind of transistor we are dealing with, the leads are called the **emitter**, **base** and **collector**.

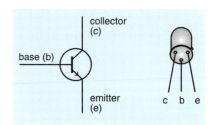

WHAT A TRANSISTOR DOES

What does a transistor do? It **amplifies** current. In other words, if a tiny current runs in the base circuit, a much larger current runs in the collector circuit. How much larger? This varies from transistor to transistor but an amplification of 100 is fairly common (Collector current = 100 × base current).

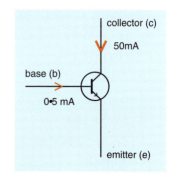

CONNECTING A TRANSISTOR

If a transistor is not connected into a circuit properly it can be damaged or ruined. However, a transistor is very small and the leads are not labelled, so how do you know which lead is which?

At the bottom of the DC108 transistor shown in the diagram, the leads are in exactly the same order as they are in the transistor symbol. Also, the arrow on the emitter terminal in the transistor symbol shows the direction of **conventional current** flow. The **collector** of the DC108 must always be connected **to the plus** of the battery, and the **emitter to the minus** of the battery.

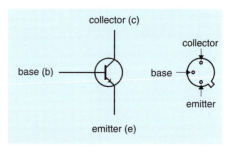

EXPERIMENT 68.1
AIM: TO STUDY HOW A TRANSISTOR WORKS

Method

1. Set up the circuit shown in the diagram.
2. Note that when no current flows in the base the LED does not light, even though the outside circuit **appears to be** complete!
3. Now connect A to the positive terminal of the battery. A current flows into the base. As a result, current flows in the collector circuit and the LED lights.

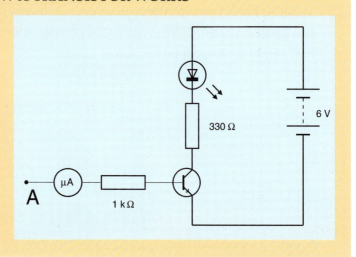

APPLIED SCIENCE · TRANSISTORS

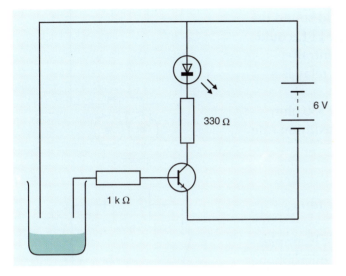

A transistor acts as an automatic switch that is switched on by the base current.

When a resistor acts as a switch it is a high-speed switch. This is one of its great advantages.

WATER-LEVEL DETECTING CIRCUIT

Let us now see how we can make the water-level detecting circuit on page 424 more sensitive by using a transistor. In this case, if even a tiny current flows into the base it will be amplified by the transistor, giving a much larger collector current which causes the LED to light. This circuit can be used to tell if a bath is full, or if your garden is too dry and needs watering.

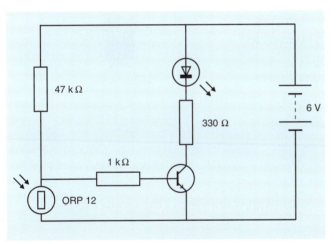

LIGHT-CONTROLLED SWITCH CIRCUIT

Remember that the resistance of a light-dependent resistor (LDR) changes when the amount of light falling on it changes. When the amount of light falling on the LDR is low its resistance is very high, so very little current can flow through it. Instead, most of the current flows into the base of the transistor, switching it on, and the LED lights.

When the amount of light falling on the LDR is high its resistance is low, so the current flows through it rather than into the base of the transistor. The transistor is switched off, and the LED goes out.

This circuit could be used to switch street lighting on at night and off in the morning. It could also be used to switch security lights on at night.

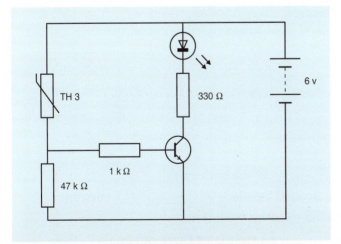

Temperature controlled switch: When the thermistor is cold the l.e.d. is off; when the thermistor is hot the l.e.d. lights up.

TEMPERATURE-CONTROLLED SWITCH CIRCUIT

The resistance of a thermistor changes as the temperature changes. When the temperature is low the resistance of the thermistor is high, so very little current can flow through it. As there is no base current, the transistor is switched off and so the LED does not light.

When the temperature is high the resistance of the thermistor is low. Some of the current flowing through it flows into the base and switches on the transistor. The LED lights.

429

The LED could be replaced by a buzzer which would sound the alarm if the temperature rose above a certain level. The LED could also be replaced by a **relay** which would switch off a central heating boiler when a set temperature was reached. (A relay is a device that switches an adjoining circuit on or off.)

For demonstration purposes in the lab, the LED and 330 ohm resistor could be replaced by a 6 V, 60 mA bulb. The maximum safe current for a BC108 transistor is 100 mA.

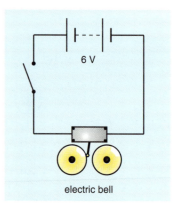

electric bell

TRANSDUCERS

A transducer is a device that converts energy from one form to another.

We have come across a lot of transducers already. For example, a light bulb converts electrical energy to light energy, a microphone converts sound energy to electrical energy, a loudspeaker converts electrical energy to sound energy. There are many other examples to be found in chapter 3. The circuits in the diagram can be used to show how these transducers work. When you look at each transducer, ask yourself two questions. What kind of energy goes into it? What kind of energy comes out of it?

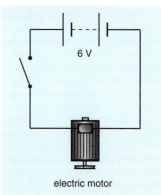

electric motor

SUMMARY

- A transistor amplifies current.
- Connect the collector of the BC108 transistor to the plus of the battery and the emitter to the minus of the battery.
- When a tiny current flows into the base, a much larger current flows in the collector circuit.
- The transistor can be used as an automatic switch which is switched on by the base current.
- A transducer is a device that changes energy from one form to another.

QUESTIONS

Section A
1. A transistor has three leads, called the _____.
2. What does a transistor do? _____
3. The collector must always be connected to the _____ of the battery.
4. In a transistor, the _____ current is controlled by the _____ current.
5. The base must always have a _____ connected in series with it to _____.
6. If there is no current flowing into the _____ of a transistor there can be no _____ current.

7. In a transistor the _____ current is much smaller than the _____ current.
8. In a light-controlled switch circuit, the _____ current to the transistor is controlled by a _____.
9. Give one common use of a light-controlled switch circuit. _____
10. In a temperature-controlled switch circuit, the _____ current to the transistor is controlled by a _____.
11. Give one common use of a temperature-controlled switch circuit. _____
12. What is a transducer? _____
13. Name a transducer that converts (a) electricity to sound _____, (b) electricity to movement. _____
14. In a _____ electricity is converted to light; in a _____ light is converted to electricity.

Section B

1. Identify the terminals x, y, z of the transistor shown in the diagram. Name two components, apart from a transistor, that you would need in order to construct a simple burglar alarm.

2. The diagram shows a temperature controlled switch. Name P, Q, R. Explain how the LED lights up when the temperature rises. What change must be made to the circuit so that the LED lights when the temperature is low and goes out when the temperature is high?

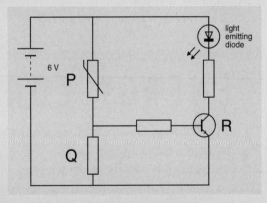

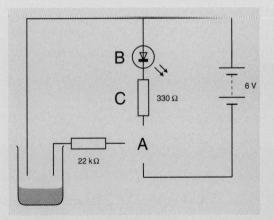

3. Redraw the water-level detection circuit in the diagram, putting in the missing component.
   What is A?
   What is the purpose of B?
   What is the purpose of C?

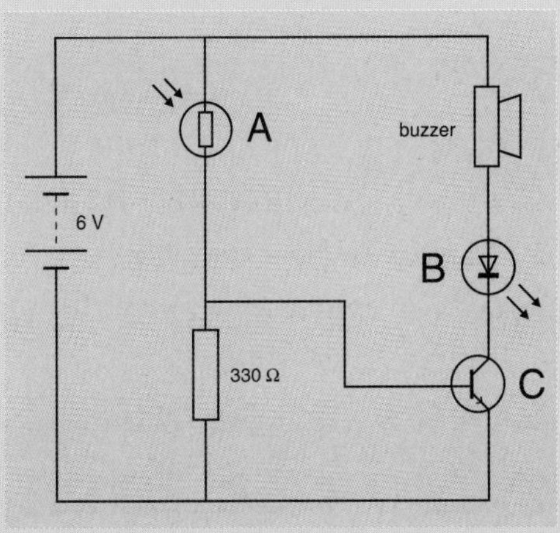

4. In the diagram, what is A? What is B? What is C? Connect up this circuit and put it in a dark press. What happens? Now open the door of the press and shine a flash-lamp on A. What happens now? What do you think this circuit could be used for?

5. Supposing you were in a crowded disco and the lights failed. What do you think would happen? Could the situation be dangerous? How could a circuit like the one in the diagram help?
6. What is A in the diagram? What does it do? If you heated A with a hairdrier or in some other way, what would happen? What could this circuit be used for?
7. Fill in the gaps in this table.

Transducer	Conversion	
	From	To
Microphone		Electricity
	Electricity	Sound
LED		
LDR		
Bulb		

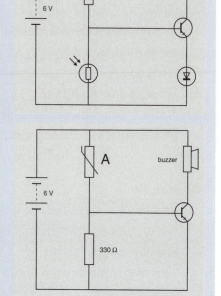

ENERGY CONVERSION CHAPTER 69

Before starting this chapter, revise the chapters on Energy and Energy Sources (pages 7–18).
Energy is the ability to do work.
All our energy comes in the first place from the sun. Plants convert the sun's energy into chemical energy and store it as food. This energy flows along the food chain. Each organism in the food chain releases energy from its food by respiration.

EXPERIMENT 69.1
AIM: TO RELEASE THE CHEMICAL ENERGY STORED IN FOOD

1. Set up the apparatus as shown in the diagram.
2. Light the nut with a Bunsen burner, and hold it under the test-tube.
3. Note the rise in temperature.

Conclusion: The chemical energy stored in the nut has been converted to heat energy.

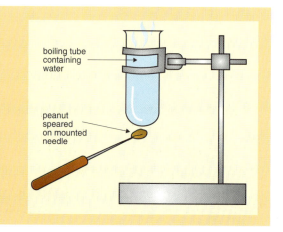

APPLIED SCIENCE · ENERGY CONVERSION

Plants have been converting the sun's energy to chemical energy for millions of years, and animals have been getting energy by eating plants for millions of years. The energy from plants and animals that died millions of years ago is stored in fossil fuels such as coal, oil, turf and natural gas.

STORED ENERGY
Chemical energy is stored in food, fuel and batteries.

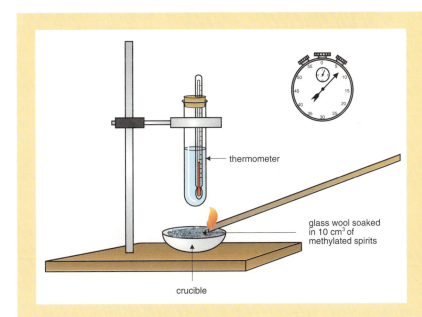

**EXPERIMENT 69.2
AIM: TO EXAMINE THE CHEMICAL ENERGY STORED IN FUEL**

Method
1. Set up the apparatus as shown in the diagram.
2. Light the methylated spirits and allow it to burn for two minutes.
3. Quench the flame and note the rise in temperature of the water.
4. Repeat steps 2 and 3 with other fuels.

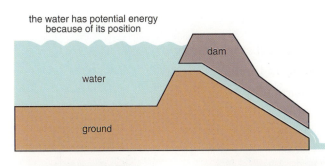

Potential energy is energy stored in an object because of its position or condition.

Nuclear energy is energy stored in the nucleus of an atom.

Nuclear energy is used to make electricity, in medicine and in nuclear weapons. Nuclear energy is dangerous because of the risk of an explosion in a nuclear reactor, and because of the risk of radiation poisoning.

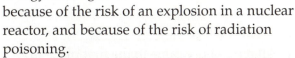

Apart from stored energy, there are other forms of energy such as kinetic energy, heat, light, sound, magnetic energy and electrical energy.

BREAKTHROUGH SCIENCE

Kinetic energy is energy an object has because it is moving.

ENERGY CONVERSION

Apart from photosynthesis, in which light energy is converted to chemical energy, there are many other examples of energy conversion.
- **A swing:** potential energy to kinetic energy to potential energy.
- **Friction:** kinetic energy to heat.
- **A drum:** kinetic energy to sound.
- **Burning:** chemical energy to heat.

There are many more examples to be found in chapter 3.

ENERGY CHAINS

In the following circuits there are a series of energy changes.
- **Circuit 1:** chemical to electrical to heat to light.
- **Circuit 2:** light to electrical to kinetic.

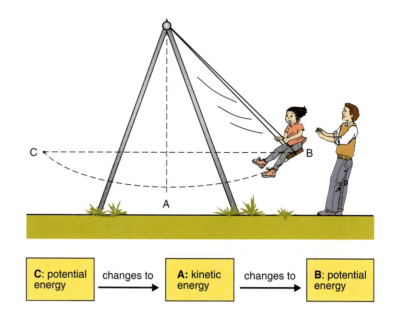

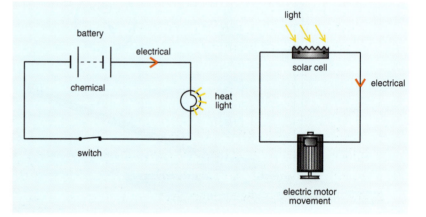

SUMMARY

- Energy is the ability to do work.
- Potential energy, chemical energy and nuclear energy are all stored energy.
- Food, fuel and batteries all have chemical energy stored in them.
- Kinetic energy is the energy an object has because it is moving.
- Burning converts chemical energy to heat energy.

QUESTIONS

Section A
1. Define energy. _____
2. All our energy came in the first place from _____.
3. Plants convert light energy to _____ energy in a process called _____.
4. The energy a body has because of its position or condition is called _____.

5. The energy a body has because it is moving is called _____.
6. When a body falls from a height, its _____ energy is converted into _____ energy.
7. _____ energy is stored in food, fuel and batteries.
8. When we _____ fuel we release the _____ energy stored in it.
9. The energy stored in the _____ of an atom is called _____ energy.
10. _____ energy, _____ energy and _____ energy are all forms of stored energy.
11. A vibrating guitar string converts _____ energy into _____ energy.
12. What kind of energy conversion takes place at A, _____ B? _____
13. What energy changes take place in an electric torch (flash-lamp)? _____

Section B

1. What type of energy has each of the following: a sweet, a stretched elastic band, a moving car, a piece of uranium?
2. What type of energy conversion takes place when you (*a*) rub your hands together, (*b*) dive into a pool, (*c*) light a match?
3. Give examples in which each of the following energy conversions occurs: potential to kinetic, light to electrical, chemical to heat. Describe an experiment to investigate the release of stored energy from food.
4. What kind of energy has the water stored behind a dam? What kind of energy has the water running through the dam? What energy conversion takes place in the generator?
5. Read the following carefully and make a list of all the energy changes that take place. 'After breakfast Mary took out her bicycle and began to cycle to school. It was a cold morning but she was quite warm by the time she reached the bottom of Church Hill. When she reached the top of the hill she was sweating and tired so she let the bicycle freewheel down the other side. When she reached the bottom of the hill she had to brake hard to avoid crashing into the school bus which was parked outside the school.'
6. 'Maura kicked a ball up into the air. She ran across the field and was there to catch it when it came down again.' What energy changes take place here?

ELECTROMAGNETS CHAPTER 70

An electromagnet consists of a coil of wire wrapped round an iron core.

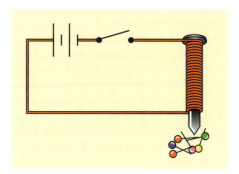

A simple electromagnet can be made as shown in the diagram. When the switch is closed, the nail becomes magnetised and attracts the pins. The core loses its magnetism when the current stops flowing. One application of an electromagnet is the electric bell.

THE ELECTRIC BELL
1. When the button is pressed, current flows and the core becomes magnetised.
2. The armature is attracted towards the electromagnet.
3. This causes the hammer to strike the gong, but it also breaks the circuit.
4. The core loses its magnetism, so the armature springs back into place and the cycle starts all over again.
5. The bell keeps ringing until you take your finger off the button.

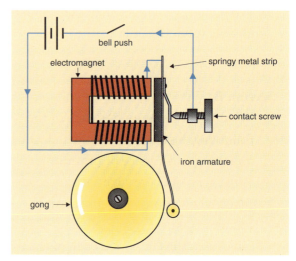

Energy chain in the electric bell
Electrical energy, magnetic energy, kinetic energy, sound energy.

EXPERIMENT 70.1
AIM: TO SHOW THAT A CONDUCTOR WHICH IS CARRYING CURRENT IN A MAGNETIC FIELD EXPERIENCES A FORCE

Method
1. Set up the apparatus as shown in the diagram.
2. Close the switch. The aluminium moves upwards.
3. Reverse the current and close the switch. The aluminium moves downwards.

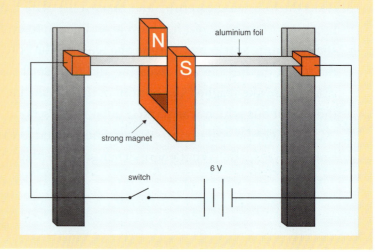

APPLIED SCIENCE · ELECTROMAGNETS

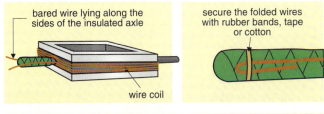

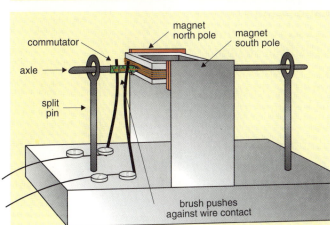

A conductor carrying current in a magnetic field experiences a force.

The electric motor is based on this principle (idea).

THE ELECTRIC MOTOR
A simple electric motor can be built from a kit, as shown in the diagram.

THE GENERATOR (DYNAMO)
In an electric motor, electrical energy is converted to kinetic energy. In a generator, the reverse happens: kinetic energy is converted into electrical energy. Let's have a look at the principle on which the generator is based.

EXPERIMENT 70.2
AIM: TO CHANGE KINETIC ENERGY INTO ELECTRICAL ENERGY

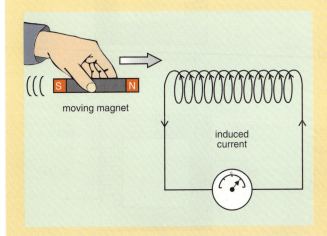

Method
1. Set up the apparatus as shown in the diagram.
2. Push the magnet into the coil. The needle moves to the right, indicating a current.
3. Leave the magnet inside the coil. There is no current.
4. Pull the magnet out of the coil. The needle moves in the opposite direction.
5. Turn the magnet around, and repeat steps 1–4.

Result: A voltage is produced only when the magnet is moving.

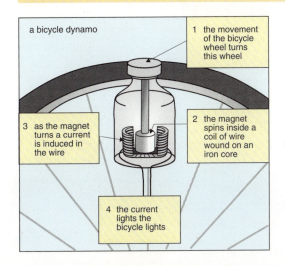

Whenever a conductor moves through a magnetic field a voltage is produced and a current flows in the conductor.

The generator (dynamo) is based on this principle.
A dynamo converts kinetic energy to electrical energy.

Applications
The simplest application of the dynamo is in the bicycle dynamo, but the same idea is used on a larger scale by the ESB to generate electricity.

The English chemist and physicist Michael Faraday (1791–1867) was probably the greatest experimentalist of the nineteenth century. Faraday's work on electrolysis led to his discovery of the laws of electrolysis. He also experimented in electricity, heat, light and magnetism. Faraday believed that since an electric current could cause a magnetic field, a magnetic field should be able to produce an electric current. He demonstrated this principle of electromagnetic induction in 1831. The principle of electromagnetic induction made possible the dynamo, or generator, which produces electricity. The faraday and the farad are named in Faraday's honour.

THE TRANSFORMER

A friend of mine has an old portable black-and-white TV. He uses it in his caravan when he goes on holidays, because it can run off a car battery (12 volts). Years ago he used to use it at home off the ESB supply. But wait a minute! The ESB supplies power to our homes at 240 volts a.c.! Of course it does, but the TV has a transformer, like the one in the diagram, built into it.

A transformer is a device that changes the voltage of an alternating power supply.

The primary voltage and the secondary voltage are related by the following equation:

$$\frac{\text{primary voltage}}{\text{secondary voltage}} = \frac{\text{number of turns in primary coil}}{\text{number of turns in secondary coil}}$$

Since the voltage is reduced, this is a **step-down** transformer. A **step-up** transformer is the opposite, and can be used in a colour TV to bring the voltage up to about 15,000 volts.

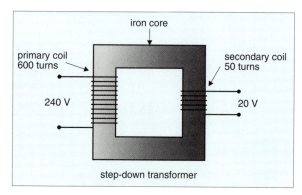

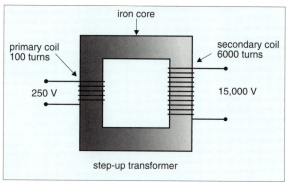

SUMMARY

- An electromagnet converts electrical energy to magnetic energy.
- An electric motor converts electrical energy to kinetic energy.
- The basic principle of an electric motor is that a conductor carrying current in a magnetic field experiences a force.
- A dynamo converts kinetic energy to electrical energy.

- The basic principle of a dynamo is that whenever a conductor moves through a magnetic field a voltage is produced.
- A transformer changes the voltage of alternating current.
- The primary voltage and the secondary voltage are related by the equation:

$$\frac{\text{primary voltage}}{\text{secondary voltage}} = \frac{\text{number of turns in primary coil}}{\text{number of turns in secondary coil}}$$

QUESTIONS

Section A

1. What energy conversion takes place in an electromagnet? _____
2. What energy conversions take place in an electric bell? _____
3. On what principle is the electric motor based? _____
4. Name two common electrical devices that contain electric motors. _____
5. The electric motor converts _____ energy to _____ energy.
6. On what principle is the dynamo based? _____
7. The dynamo converts _____ energy to _____ energy.
8. What exactly is a transformer? _____
9. The primary voltage and the secondary voltage are connected by what equation? _____
10. What is the secondary voltage in the diagram? _____
11. Name two electrical devices that contain a transformer. _____
12. Is the transformer in question 10 a step-up or a step-down transformer? _____

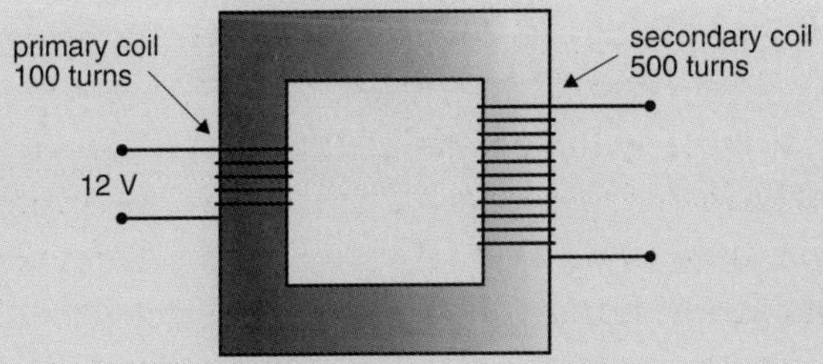

Section B

1. The diagram shows a simple electromagnet. What is the iron inside the coil called? How would you show that a magnet has been formed when the switch is closed? Give one use of an electromagnet. Give one advantage of an electromagnet over a permanent magnet.

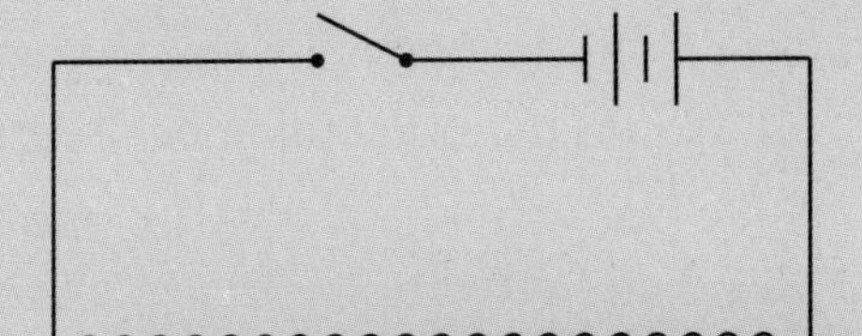

2. A magnet and an electrical circuit are shown in the diagram. What type of magnet is shown? What happens if the strip of aluminium is placed between N and S and the switch is closed? What happens if the current is made to flow in the opposite direction? Name a piece of equipment that makes use of this effect.

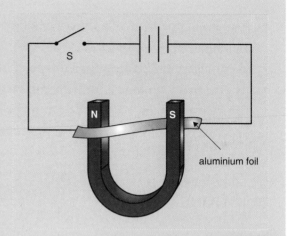

3. The diagram (below right) shows a simple d.c. electric motor. Name the parts labelled A and B. What happens when the switch is closed? Give one practical use of an electric motor.

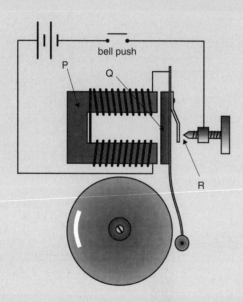

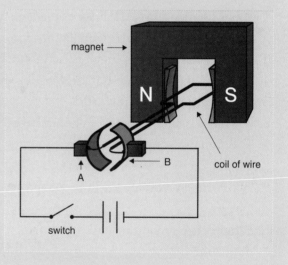

4. The diagram above shows an electric bell. Name P, Q and R. Explain how the electric bell works.

5. The diagram on the right shows a coil of wire connected to a meter. State what would be observed if the magnet were dropped into the coil. Would the same thing be observed if the magnet were held still and the coil moved up towards it? What basic principle is being demonstrated here? Name one device that is based on this principle.

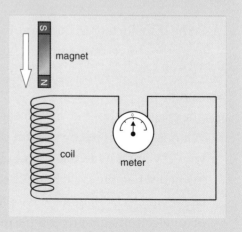

6. 'When school finished it was dark. Mary cycled up Church Hill, freewheeled down the other side and cycled the remaining 500 metres to her home.'
   (a) What does a dynamo do?
   (b) Would Mary find it easier to cycle using the dynamo or without using the dynamo? Why?
   (c) Where on her journey do you think her light would be dimmest? Why?
   (d) Where would the light be brightest? Why?
   (e) Would Mary be tired and hungry when she got home? Why?
   (f) Make a list of all the energy changes that took place on her journey.

7. What effect is being demonstrated in the diagram?
   Copy and complete the table.

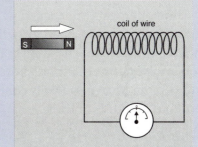

Magnet being pushed in	Needle moves to the right
Magnet inside coil but not moving	
Magnet being pulled out	
Magnet being pushed in much faster	

CHAPTER 71 — HORTICULTURE

Plants provide us with food, timber, clothes (cotton and linen), various medicines and even the paper on which this book is printed. Horticulture is the science of how to grow plants well to produce strong, healthy plants and a good yield of vegetables, fruit or flowers.

Different types of work are done in horticulture.

Nurseries supply young trees and shrubs for fruit growers, gardeners and garden designers; plant-growers produce fruit, flowers and vegetable plants in fields and greenhouses; seed-producers grow plants for their seeds.

SOIL

We usually grow plants in soil. Soil provides raw materials of water and mineral salts (plant nutrients), which are taken in by the roots. Water and mineral salts pass up through the stem to the leaves. In photosynthesis they join with carbon dioxide and with the help of sunlight are made into food. This is stored in the leaves, stem or roots, depending on the plant. The soil also provides anchorage for the plant.

MAJOR PLANT NUTRIENTS

Nutrient	Function	Symptoms of deficiency
(N) Nitrogen	Healthy leaves	Pale green small leaves, stunted growth
(P) Phosphates	Healthy roots	Poor root development, stunted growth, red or purple tinge to leaves
(K) Potash	Good flowering and fruit	Poor flowers and fruit, lowered resistance to disease

Other elements are needed in smaller amounts for healthy plant growth. They include calcium, magnesium, manganese, iron, molybdenum, sulphur, boron, zinc and copper. Most good soils contain the required amounts of these.

Essential plant food

Nitrogen: the leaf maker

Potash: the flower and fruit maker

Phosphates: the root maker

EXPERIMENT 71.1
AIM: TO SHOW THE EFFECT OF DIFFERENT NUTRIENT SOLUTIONS ON PLANT GROWTH

Strawberry runners (March–May) or lettuce plants (April–June) can be used in this experiment.

Method
1. Place a similar plant in each of three pots of vermiculite.
2. Add water (no nutrients) to one pot.
3. Add standard nutrient solution to another pot.
4. Add a 50% concentration of standard nutrient solution to the third pot.

    Check the plants each week. Note the changes in the appearance of the plants.

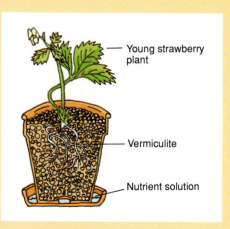

STRUCTURE OF SOIL

Soil particles are formed by weathering of rocks over thousands of years. Rock is broken down into different-sized particles: gravel (largest), sand, silt, clay (smallest).

Soil contains much more than just weathered rock. Mature soil is made up of a number of distinct layers. Each layer is different in its chemical, physical, and biological properties. The topsoil (top layer) contains nearly all the living organisms and has a lot of organic matter. Water passing down through the topsoil carries clay particles with it and deposits them in the subsoil. Subsoil usually has a high clay content. The material under the subsoil is composed of relatively unaltered rocks and stones.

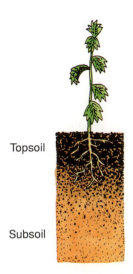

The factors that determine the kind of soil that develops are:
- climate (temperature and rainfall)
- living organisms
- the type of rocks that produced the soil particles
- topography (ground slope and elevation)
- time (hundreds of years).

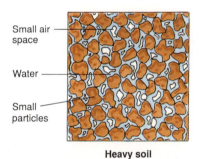
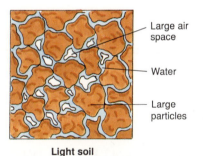

TYPES OF SOIL

Soil is usually described as sandy, clay or loam. Most soils are mixtures of all three in varying proportions. Sandy soil is very loose and will not hold water or nutrients. Clay soil is dense and heavy, sticky when wet, and hard when dry (bricks are made of clay). Loam is a mixture of sand and clay soils, but also contains large quantities of humus that loosens and aerates clay soil and binds sandy soil particles together. Humus also supplies some plant nutrients.

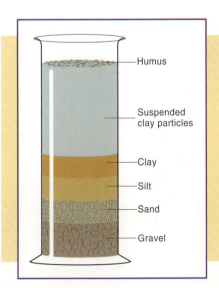

EXPERIMENT 71.2
AIM: TO SHOW THE TYPES OF PARTICLES IN A SOIL SAMPLE

Method
1. Quarter-fill a large measuring cylinder with soil.
2. Add water until it is three-quarters full.
3. Cover with your hand and shake vigorously.
4. Allow to stand overnight.

Result: The result should be as in the diagram.

Repeat this experiment with a number of labelled soil samples from different areas. Note the growth pattern of plants in these areas—does the soil help produce strong growth?

Rub a small sample of soil between your fingers. Does it crumble into dust or hold its structure? Good soil has a stable structure of clumps of soil particles (called "soil crumb") held together by humus.

Soil contains
(1) soil particles
(2) humus (the decayed remains of dead plants and animals)
(3) water
(4) plant nutrients (mineral salts)
(5) air
(6) living organisms.

WATER

Water and nutrients flow from the roots to the leaves of plants. The flow of water from the roots to the leaves is the transpiration stream.

Water is essential for the development of the plant:
(1) to carry water to the leaves for photosynthesis
(2) to carry nutrients up from the roots
(3) to cool the plant.

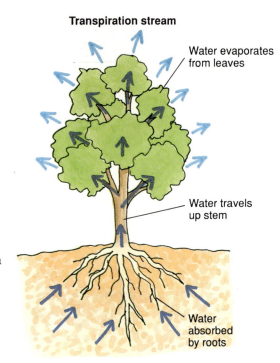

EXPERIMENT 71.3
AIM: TO MEASURE THE PERCENTAGE OF WATER IN SOIL

Method
1. Find the mass of a clean dish.
2. Break some fresh soil into the dish and find its mass again.

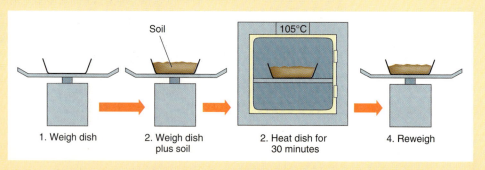

3. Place the dish in an oven at 100°C for half an hour.
4. Remove the dish, cool and get its mass.
5. Repeat steps 3 and 4 until there is no further change in mass, showing that all the water has been evaporated.

Result: Percentage water = $\dfrac{\text{Loss in mass of soil}}{\text{Original mass of soil}} \times 100$

Repeat this experiment with a compost mixture made of 50% peat and 50% sand.

These experiments should be done with each sample at field capacity. Field capacity means that the sample holds as much water as it can and that any more would simply drain out of the sample.

DRAINAGE

Soil with poor drainage becomes waterlogged and lacks the air that roots need to develop properly. If the drainage is too good (sandy soil) the nutrients are washed out of the soil (leaching).

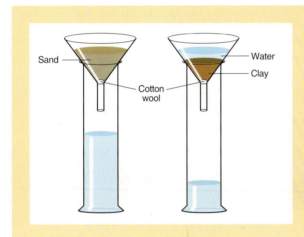

EXPERIMENT 71.4
AIM: TO COMPARE THE DRAINAGE OF SANDY SOIL AND CLAY SOIL

Method
1. Set up the apparatus as shown in the diagram.
2. Pour 50 cm³ of water into each funnel.
3. Wait for five minutes. Which soil has the better drainage?

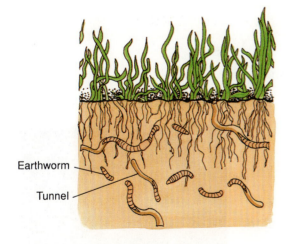

SOIL IS FULL OF LIVING ANIMALS

Many small animals (earthworms, centipedes and millipedes) live in the soil. The soil also contains decomposers (bacteria and fungi) that break down leaves and dead organisms into humus and release nitrogen into the soil. These organisms (and plants growing in the soil) need oxygen for respiration, so air is essential for healthy soil.

The earthworm is the most useful soil organism because it
(1) aerates the soil (burrows into the soil and lets air in)
(2) drags leaves down into the soil
(3) rotates the soil
(4) increases the amount of humus by excreting and by its own body when it dies
(5) improves drainage

BREAKTHROUGH SCIENCE

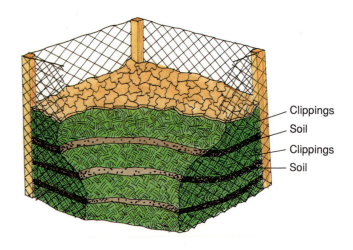

COMPOST HEAP
Gardeners often make heaps of leaves, grass cuttings and hedge clippings. Micro-organisms turn these into humus, which the gardener then digs into the soil. A compost heap should be well aerated and damp but not wet. A mixture of different materials and some added soil speed up the process.

EXPERIMENT 71.5
AIM: TO MEASURE THE PERCENTAGE OF AIR IN THE SOIL

Method
1. Find the volume of a small can by filling it with water and emptying it into a measuring cylinder.
2. Punch a hole in the bottom of the can and fill it with soil as shown in the diagram.
3. Level the surface of the soil in the can.
4. Block the hole and measure the amount of water that must be poured slowly into the soil to fill the air spaces. This gives the volume of air in the soil.

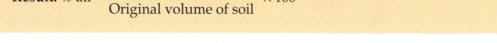

Result: % air = $\dfrac{\text{Volume of air}}{\text{Original volume of soil}} \times 100$

Repeat this experiment with a compost made of 50% peat and 50% sand.

ACID OR ALKALINE SOIL?
The acidity of a soil depends on the rocks from which it was formed. Soils in limestone regions are alkaline. Most garden plants do well in a neutral to slightly acid soil (pH 6 to 7) but not in more acid soils. Heathers, rhododendrons, azaleas and camellias thrive in acid soils. Farmers and market gardeners add lime to acid soil to reduce its acidity. Lime also improves the soil because:
- earthworms and other beneficial organisms flourish in lime soils
- lime discourages some pests, such as slugs and leatherjackets
- it breaks up clay soil.

Compost bin

APPLIED SCIENCE · HORTICULTURE

EXPERIMENT 71.6
AIM: TO MEASURE THE pH OF SOME SOIL SAMPLES

 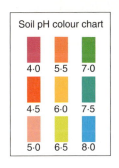

Method
1. Put a small amount of soil into a test-tube.
2. Half-fill the test-tube with distilled water and shake it vigorously.
3. Filter the mixture into a second test-tube and test it with universal indicator paper. Repeat for each sample. Can you spot any connections between the pH of the soil and the types of plant growing in the soil?

Using peat-based compost

PEAT-BASED GROWING COMPOST

Horticulturists usually buy ready-made sterile growing compost. This replaces garden soils that might contain insect pests, weeds, bacteria or viruses that could stunt or kill the plant. Sterile growing compost contains none of these.

Growing compost is a mixture of peat, essential nutrients, some sand and added materials such as vermiculite or perlite. It is sterile, contains all the nutrients necessary for healthy growth, retains moisture and nutrients, and is warm and well aerated.

(Vermiculite is a brown mica material that absorbs water and mineral salts and aerates the compost. Perlite is a similar grey or green material of volcanic origin.)

Most composts are peat-based, although soil-based composts are also made.

Peat-based compost

Advantages	Disadvantages
Sterile, warm, aerated (with sand added), easy to work	Difficult to re-wet if it dries out
Retains moisture and nutrients well	Contains few nutrients (nutrients must be added)

BREAKTHROUGH SCIENCE

HYDROPONICS

Hydroponics is a method of "soil-less" growing of plants. The plants are grown in a solution of essential nutrients dissolved in water.

The plant is kept in a pot of vermiculite, perlite, or small gravel or cinder particles. These give the plant support but do not supply any nutrients. The nutrient solution is fed to the plant continuously, and must contain the same essential nutrient chemicals that are found in fertile soil.

Hydroponics is used to grow plants in areas with poor soil. Hydroponic techniques are also used to grow flowers and vegetables. These hydroponic "farms" are expensive to set up, but give high yields. Every stage of plant growth is monitored and ideal growing conditions are maintained.

Hydroponic "farms" give bigger crop yields, have fewer pest and disease problems, have no need for weeding and give almost total control over plant growth.

EXPERIMENT 71.7
AIM: TO SHOW THE EFFECTS OF A DEFICIENCY OF DIFFERENT NUTRIENTS ON PLANT GROWTH

Seedlings of lettuce, tomato, celery or cereals can be used in this experiment.

Method
1. Place similar seedlings (5–6 cm high) in pots of vermiculite.
2. Add nutrient solution to the pots and grow for a week.
3. Flush out all nutrient solutions with water.
4. Add nutrient solutions to the pots as follows:
   (a) standard nutrient solution
   (b) 50% concentration of standard nutrient solution
   (c) standard nutrient solution without nitrogen
   (d) standard nutrient solution without phosphate
   (e) standard nutrient solution without potassium.

Check the plants each week. Note any change in their appearance.

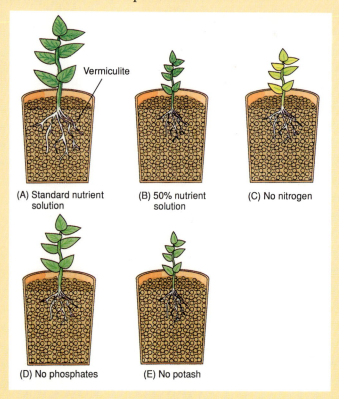
(A) Standard nutrient solution
(B) 50% nutrient solution
(C) No nitrogen
(D) No phosphates
(E) No potash

PROPAGATION OF PLANTS
Sexual reproduction
Many plants flower and produce seeds. A seed is the result of the male gamete from one plant fusing (joining) with a female gamete of another plant of the same kind. The seed germinates to produce a plant with a mixture of the characteristics of the parent plants.

Producing seeds does not always guarantee that new plants will be produced. If too many plants are growing in a small space there will not be enough water, nutrients or exposure to light for all of them to thrive. Many plants have adapted to avoid this by having seeds that are dispersed far from the parent plant. Wind dispersal is probably the most easily observed of these. Dandelion and thistle seeds are very light and attached to a "parachute" which enables them to be blown by the wind over great distances. Sycamore and ash seeds have a wing-like attachment.

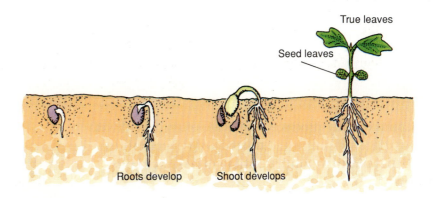

Seed germination
Seeds are usually dispersed in the autumn and lie dormant during the winter. In spring they germinate. In nature plants produce large quantities of seeds, but only a small fraction of them grow to produce fully grown plants. Many seeds are eaten by animals (birds, mice, rats, insects) or are destroyed by bacteria or fungi. Some germinate but die from frost or drought; others are killed by slugs or other pests and diseases.

Germination means the seed begins to grow into a new plant. Water, oxygen and heat are all necessary for germination. Light is necessary only **after** germination. In some plants light inhibits the germination of seeds.

BREAKTHROUGH SCIENCE

EXPERIMENT 71.8
AIM: TO ESTIMATE THE PERCENTAGE GERMINATION OF A SAMPLE OF SEEDS

Any inexpensive seed will do for this experiment: grass, lettuce, carrot, Brussels sprouts, etc.

Method

1. Draw a square 10 cm by 10 cm on a sheet of absorbent paper. Mark a grid of 1 cm × 1 cm squares on this. Place it on wet capillary matting in a shallow plastic tray.
2. Place one seed on each cm square (100 seeds in all). Cover tray with glass and dark paper. Leave in a warm place.
3. Record the number of seeds that have germinated after 7 days and again after 14 days. The percentage germination is calculated from the number of seeds that have germinated

$$\% \text{ Germination} = \frac{\text{Number of seeds that germinated}}{\text{Total number of seeds}} \times 100$$

Commercially produced seeds are often coated with a protective substance and treated with chemicals to keep them from rotting, becoming diseased or being eaten.

Seed dormancy

Many seeds do not germinate even when the conditions necessary for germination (heat, water and air) are present. This is because some seeds develop a tough seed coat which prevents germination. This coat is broken by exposure to low temperatures in winter. Until this happens the seed is dormant and will not germinate. Dormancy delays germination until the spring. Some varieties of seed have a dormancy period of two years. Sycamore and ash seeds are two common seeds that show dormancy.

Tree (ash)

Leaves of ash

Fruit of ash

Leaves of sycamore

Fruit of sycamore

Seed dormancy in sycamore and ash

Advantages	Disadvantages
Better conditions of light and warmth in spring to allow germination and growth	Seed may be destroyed by fungi or eaten by animals
Less competition from annual plants that are still small in spring	Mild winter may not break dormancy and seed may not survive for another year

Scientists have studied dormancy in seeds and have developed techniques to produce rapid germination in dormant seeds. One method is to place the seeds in a bag of damp compost (not wet). Put the bag into a fridge for 21 days. Now germinate the seeds in the normal way. Seed packets usually give directions for breaking dormancy where this is a problem. Scraping the seed coat with sandpaper and soaking in warm water are other methods used.

EXPERIMENT 71.9
AIM: TO INVESTIGATE DORMANCY IN ASH OR SYCAMORE SEEDS

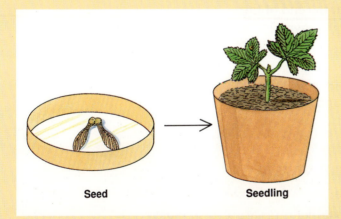

Method
1. Collect green (fresh) ash or sycamore seeds in the autumn (late September–early October).
2. Soak some seeds for 24 hours in warm water. Plant some seeds in seed compost. Leave in a warm place. Note the number that germinate after a few weeks.
3. Allow the remainder of the seeds to dry (until they become brown). Store in a dry jar in a cool place.
4. In spring, plant some of these. Note the number that germinate after a few weeks.
5. Place the brown seeds in a bag of damp seed compost (not wet). Put the bag into a fridge for 21 days. Now plant the seeds in the usual way. Note the number that germinate after a few weeks.

Result: The green seeds germinate, the untreated brown seeds do not germinate, the brown seeds that have been treated to break dormancy germinate.

Gardeners sometimes collect seeds before they have time to develop a tough seed coat and germinate them in the autumn. This is possible because the gardener can protect the plants from frost and other dangers over the winter.

BREAKTHROUGH SCIENCE

ASEXUAL (VEGETATIVE) REPRODUCTION

Some plants produce young plants without sexual reproduction. Strawberry plants grow runners. Where the runner strikes the soil a new strawberry plant grows. The young plant is an exact copy of the parent. This is called asexual or vegetative reproduction. Blackberry bushes, potatoes and daffodil bulbs reproduce asexually. Stem and root tubers, bulbs, corms, runners, and techniques using cuttings, grafting, and tissue culture are forms of vegetative propagation.

Propagating trees from seed often takes a long time. Vegetative reproduction is an economic way of propagating these plants.

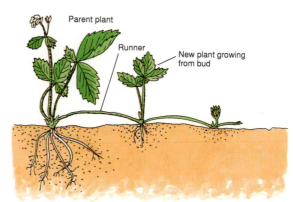
Wild strawberry with runners

Growing plants from cuttings

Plant cuttings (pieces of stem, leaves or roots) can grow into an entire plant because of their ability to regenerate their missing parts. This is used by gardeners to produce many identical copies of a particular plant. There are two main techniques of propagating stem cuttings—softwood and hardwood. Softwood cuttings are stem cuttings taken from newly grown stems before they have time to go woody. Hardwood cuttings are taken at the end of the growing season from young stems that have just gone woody.

Softwood cuttings

These cuttings are usually taken in spring, although softwood cuttings of geraniums (pelargonium), succulents and cacti will root at any time in the growing season.

EXPERIMENT 71.10
AIM: TO GROW A GERANIUM (PELARGONIUM) FROM A SOFTWOOD CUTTING
(Busy lizzie (impatiens), tradescantia and chrysanthemums are also suitable.)

Method

1. Cut a number of pieces of softwood 3–10 cm long from a healthy parent plant. Trim the cuttings back to just below a leaf node. Remove any flowers and side buds.

1. Stem cutting 3-10 cm long

2. Remove lower leaves

3. Dip in hormone rooting powder and plant in potting compost

APPLIED SCIENCE · HORTICULTURE

2. Remove all but 2–3 leaves at the growing tip.
3. Dip the cuttings into hormone rooting powder and shake off the excess. Rooting powder usually contains a fungicide which protects the open cut from attack.
4. Plant them in prepared potting compost. Firm the compost around the cuttings. Water with a fine mist and place in a propagator. A simple propagator can be made with a pot, some sticks and a plastic bag.

Cuttings root best at a temperature of 20°C with good light (not strong sunshine). After about six weeks roots will have grown and the new plants can be repotted individually. Why do you think you were asked to remove most of the leaves? Why do you have to keep the cuttings in a propagator?

Hardwood cuttings
Hardwood cuttings are used to propagate a wide range of trees and shrubs. They are taken in late autumn from the current year's growth. They are usually rooted outdoors and take a long time to root.

EXPERIMENT 71.11
AIM: TO GROW A PLANT FROM A HARDWOOD CUTTING

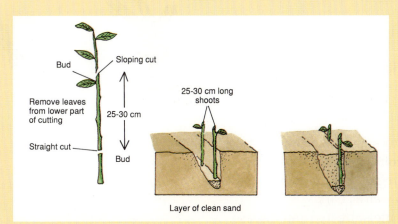

Your choice of blackcurrant, gooseberry, griselinia, privet, escallonia, willow or poplar will do for this experiment.

Method
1. In late autumn cut pieces of young shoots 25–30 cm long from a healthy parent plant. Trim the bottom of the cutting to below a bud.
2. Remove the lower leaves.
3. Select a sheltered part of the garden. Plant the cuttings in a trench 15 cm deep with a layer of sand at the bottom. Space the cuttings 15 cm apart. Rooting powder may be used, but cuttings of these plants will root without it.
4. Fill in the top soil and firm it down well. Keep the area free of weeds.

Leave for a year until the following autumn. The cuttings will then have produced roots and can be transplanted to their final growing position. Why do you think you used sand in the bottom of the trench? Why do you have to keep the area free of weeds?

Micro-propagation
A new propagation technique is cloning (producing identical copies) of plants using tissue culture. A small number of cells cut from part of the parent plant produces thousands of identical plants. Plants that once took years to propagate conventionally are now freely available as cloned copies of a single plant.

Grafting
Grafting is a method of plant propagation that inserts a bud or a stem of one plant (a **scion**) into a root, stem, or branch (the **stock**) of another plant so that they join together and grow as one plant. Grafting is normally done with woody perennial plants only.

Grafting is used
- to increase the numbers of plants of a particular variety where it is difficult to grow from seed or cuttings, e.g. ornamental varieties of ash, spruce, birch and Japanese maple
- to produce dwarf or miniature fruit trees (apple varieties are grafted onto dwarf and semi-dwarf rootstocks)
- to ensure easy pollination (two varieties of apple can be grafted onto the one rootstock; the flowers of each can then cross-pollinate the other)
- to adapt a plant to soil or other growing conditions that would normally stunt growth
- to increase resistance to insects and diseases
- to repair damage to a tree.

How to graft
1. The cambium, or inner bark, of the stock and the scion must make contact to allow the tissues to join. Good clean cuts are needed for this. The cambium is the region of growth cells that produce xylem and phloem tissue as well as bark.
2. Cut surfaces must be protected from moisture loss and disease. (Polythene film is wrapped round the graft and the graft is coated with a layer of special grafting wax.)
3. The stock and the scion must be compatible (closely related)—an apple can be grafted to an apple or crab apple.

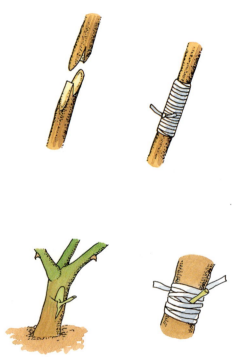

Grafting

APPLIED SCIENCE · HORTICULTURE

EXPERIMENT 71.12
AIM: TO DEMONSTRATE GRAFTING

Your choice of apple, ash or birch will do for this demonstration.

Method:
1. Pot up suitable rootstock in late autumn. Allow to dry off slightly for a week before grafting in early February.
2. Select suitably sized scion stems about 15 cm long.
3. Clean rootstock and scion where cuts are to be made. Use a sharp knife to make a slanted cut in the rootstock 5 cm from the ground. Make a similar cut in the scion.
4. Place the scion on the rootstock and ensure the cambium layers (just under the bark) are in close contact. Bind with rubber banding. Cover all parts above ground level with grafting wax. Cover the soil in the container with a plastic or cardboard disc.

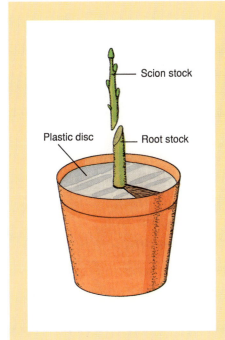

Why do you bind the graft with a rubber band? What do you think is the purpose of the wax covering the graft? Why do you cover the soil with a cardboard disc?

Examine a number of ornamental shrubs and fruit trees and see if you can tell whether they were propagated using grafting.

Other types of grafting

A tongue graft is often used with stocks of 2 cm or less in diameter. A V-shaped cut is made into the rootstock and the scion is cut to fit into this. The stock and scion are then fitted together so that the cambium layers are in contact. They are then tied together with a rubber band and coated with grafting wax. This type of graft is used in propagating ornamental shrubs and apple and pear trees.

Natural grafting happens when branches are tied together so that they will grow naturally together. This helps strengthen the top of the tree.

SUMMARY

- Horticulture is the science of how to grow plants well.
- Nitrogen, phosphates and potash (N, P, K) are three major plant nutrients.
- Humus is the decayed remains of plants and animals.
- Decomposers in the soil break down humus to provide nitrogen for plants.

- Leaching means that minerals are washed out of the soil.
- Good soil should contain humus, water, mineral salts, air, soil particles and living organisms.
- Water is essential for photosynthesis, to carry nutrients and to cool the plant.
- Soil that holds as much water as it can is at field capacity.
- Growing compost is a sterile mixture of peat, essential nutrients and materials such as sand, vermiculite or perlite.
- Hydroponics is "soil-less" growing of plants in a solution of essential plant nutrients.
- Water, oxygen and heat are all necessary for germination.
- Dormancy delays germination of seeds.
- Vegetative reproduction produces young plants without sexual reproduction.
- In asexual or vegetative reproduction the offspring are exact copies of the parent.
- Softwood cuttings are taken in spring from newly grown "green" stems.
- Hardwood cuttings are taken in autumn from young "woody" stems.
- Grafting joins a scion of one plant to the stock of another plant so that they grow as one plant.

QUESTIONS

Section A
1. The decayed remains of plants and animals are called _____.
2. Name the three major plant nutrients. _____
3. What is hydroponics? _____
4. Decomposers are _____
5. The three conditions necessary for germination are _____.
6. What is vegetative reproduction? _____
7. Good soil should contain _____, _____ and _____.
8. The delayed germination of seeds is called _____.
9. What are softwood cuttings? _____
10. Grafting joins a _____ of one plant to the _____ of another plant.
11. Cuttings taken in autumn from young "woody" stems are _____.
12. What is a compost? _____
13. Soil provides plants with _____ and _____.
14. 50 seeds were planted. Only 30 seeds germinated. What was the percentage germination? _____
15. Name the type of soil that has poor drainage _____

Section B

1. (i) Describe a simple experiment to show the percentage germination in seeds.

   (ii) An experiment to show the presence of micro-organisms in the soil is set up as in the diagram.
   - (a) What will be the result of the experiment? Give a reason for your answer.
   - (b) What is sterilised soil?
   - (c) What are micro-organisms?

2. (i) Describe how any two of the following may be propagated: blackcurrant, rhubarb, strawberry, apple tree.

   (ii) You are about to go on holidays. Describe a method of ensuring that potted plants have adequate amounts of water while you are away.

3. Sulphate of ammonia is a salt that is soluble in water and is often used as a garden fertiliser.
   - (a) Name the three major elements needed for healthy plant growth.
   - (b) Why is it important for a fertiliser to be soluble in water?
   - (c) Which of the major elements for plant growth is supplied by sulphate of ammonia?
   - (d) Bonemeal (made from animal bones) is used as a fertiliser by gardeners. What nutrients are found in bonemeal?
   - (e) Name one way that gardeners recycle dead plant material in the garden.

4. In an experiment to estimate the number of earthworms in different habitats with similar soil types, a class obtained the following results.

Habitat	Number of earthworms/m^2
Vegetable garden	100
Grazed pasture	315
Lawn	28
Mixed woodland	460

   Explain, by giving scientific reasons, the differences in the number of earthworms found.

5. (i) Describe, with the aid of a diagram, an experiment to show the percentage of water in a soil sample.

   (ii) (a) What chemical elements are found in a fertiliser such as 10:10:20?
   (b) Name two other elements needed by plants.

6. Examine the graph shown in the diagram, which shows how temperature varies in soil at different depths during a sunny day in August.

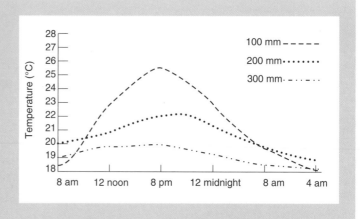

(i) At what time did the soil at 100 mm reach its highest temperature?
(ii) What was the highest temperature reached by the soil at 200 mm?
(iii) Which side of a hill would you expect to heat the faster?
(iv) What is the effect of soil temperature on plant growth?

7. A plant is described as a "hardy annual". What does this mean? How soon after germination could you expect the plant to flower? What kind of conditions does the plant need to grow well?

8. The diagram shows a softwood cutting prepared for rooting.
    (a) What is the purpose of the plastic bag?
    (b) Why were the lower leaves removed?
    (c) Why does the compost need to be well-watered?
    (d) Why was sand added to the compost?
    (e) Why is the cutting kept in a warm place while rooting?

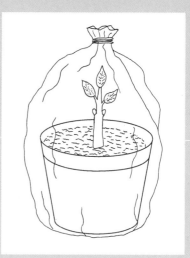

9. Outline the procedure to determine the pH of a sample of soil. Give two advantages and two disadvantages of peat-based potting compost. How could you overcome the disadvantages?

10. Describe how you would take a hardwood or a softwood cutting from a plant and make it grow. Describe an experiment to measure the germination rate of seeds.

APPLIED SCIENCE · GROWING PLANTS

CHAPTER 72 GROWING PLANTS

ASTER *Colour Carpet Mixed*
For beds and borders, also as an edging plant.

QUICK GUIDE		
	FOR INDOOR SOWING	FOR OUTDOOR SOWING
WHEN	• March to April	• May
WHERE	• In boxes of moist compost	• In a prepared seed bed where to flower
HOW	• Sow thinly • Cover with fine layer of compost • Firm gently • Cover box with glass or polythene and shade with paper • Keep at 15-18°C (60-65°F) • Seedlings appear 12-18 days	• Sow thinly in rows 25cm (10in) apart • Cover with ½cm (¼in) fine soil • Firmly gently • Keep moist
CARE	• Remove glass and paper when seedlings appear • When large enough transplant 5cm (2in) apart in boxes of compost • Keep at 10-15°C (50-60°F) • Stand outdoors for a few days end May to June (avoid frosts) • Transplant 25-30cm (10-12in) apart in flowering position in June	• When large enough thin to 25-30cm (10-12in) apart • Seedlings removed may be transplanted 25-30cm (10-12in) apart
FLOWERS	• July to October	• August to October

TIP: Remove dead flowers to prolong flowering
Keep moist and weed free
Performance subject to growing conditions

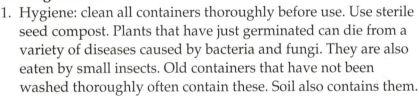
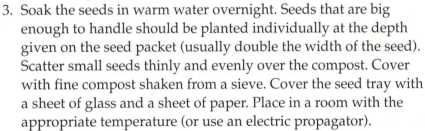

DIFFERENT TYPES OF FLOWERING PLANT

When you buy a packet of seeds, the packet contains a description of the plants the seeds produce. Plants that germinate, grow and flower in a single growing season (usually March to October) are **annuals**. Some of these annuals can tolerate frost and are called **hardy annuals**. Others cannot tolerate frost and must be grown indoors and planted out when danger of frost has passed. These are **half-hardy annuals**. Plants that grow for one season and then flower in the second season are **biennials**. Plants that flower each year and die back over the winter are **perennials**. The seed packet also gives the height of the mature plant, suitable soil and position (sunny, semi-shade or shade) and instructions on germination.

HOW TO GROW PLANTS FROM SEED
Sowing indoors

1. Hygiene: clean all containers thoroughly before use. Use sterile seed compost. Plants that have just germinated can die from a variety of diseases caused by bacteria and fungi. They are also eaten by small insects. Old containers that have not been washed thoroughly often contain these. Soil also contains them.
2. Prepare the seed tray as shown in the diagram. Water with a fine mist spray.
3. Soak the seeds in warm water overnight. Seeds that are big enough to handle should be planted individually at the depth given on the seed packet (usually double the width of the seed). Scatter small seeds thinly and evenly over the compost. Cover with fine compost shaken from a sieve. Cover the seed tray with a sheet of glass and a sheet of paper. Place in a room with the appropriate temperature (or use an electric propagator).
4. Check the seed tray daily. Wipe and turn the glass. Do not allow the compost to dry out but do not overwater. Water with a fine mist spray.
5. When the first seeds germinate remove the paper and place the seed tray in good light. Turn the seed tray regularly to give all seedlings equal light. Raise the glass cover gradually to allow good ventilation. A "mini-propagator" can be made with a pot and a polythene bag. Follow the instructions given above for germinating seeds.

BREAKTHROUGH SCIENCE

EXPERIMENT 72.1
AIM: TO INVESTIGATE THE EFFECTS OF DENSITY OF SOWING AND DEPTH OF SOWING ON GERMINATION OF SEEDS

Any inexpensive seed will do for this experiment: grass, lettuce, carrot, Brussels sprouts, aster, nasturtium, etc.

Method

1. Prepare a number of small seed trays. Sow the selected seeds (in the way advised on the packet) thinly in one tray. Sow densely in the another tray. Note the number that germinate after a few weeks. Note the appearance and growth of the seedlings after germination.
2. Repeat this procedure sowing seeds at different depths. Again note the rate of germination and growth of seedlings.

Result: Densely sown seeds are likely to be tall and spindly and to suffer from "damping off" disease and die. Seeds sown at depths greater than advised will probably germinate but will not have strong growth, as the stem is likely to be weak by the time it reaches light.

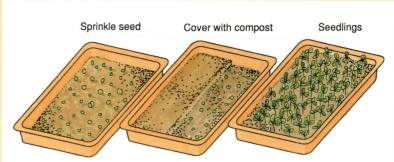

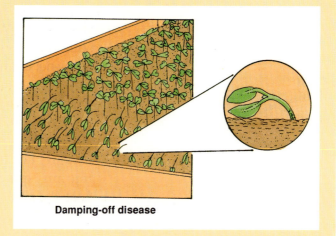

Seedling problems

If seeds are sown too densely: seedlings must compete for light, nutrients and water. They often grow tall and spindly. Because of poor air circulation around them, fungal diseases may grow on stems causing the seedlings to die.

If there is not enough light: seedlings grow tall weak stems.

If seeds are sown too deeply: seedlings grow tall and spindly.

If seedlings are eaten: slugs, woodlice and other pests are present.

APPLIED SCIENCE · GROWING PLANTS

Pricking out seedlings
When your seedlings have grown two true leaves you should transfer them into seed boxes or individual pots. This gives them space, light and nutrients to thrive.

Loosen the compost around the seedling with a small stick. Make a hole with a dibber in the compost of the seed box. Lift the seedling by the leaves and transfer carefully to the seedbox. Water gently.

Hardening-off seedlings
You harden off seedlings to prepare them for life outdoors by gradually moving them to a cooler place indoors. Then move them outdoors during the day, and eventually leave them out permanently.

Sowing outdoors
You must prepare the seedbed (the place the seeds are sown) thoroughly if seeds are to germinate properly outdoors.

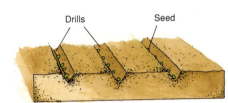

1. Dig the top layer of the soil, break down big lumps and rake the soil until it is fine. Rake in a dressing of general fertiliser to ensure the seedlings will have enough nutrients to thrive. If the soil is dry, water it.
2. Mark out the line to be sown with twine. Make a drill in the soil with a stick.
3. Sow the seed thinly (at the depth advised on the packet). Cover the seed lightly and label the row with a waterproof marker on a plastic label.
4. Thin out the seedlings at the pricking-out stage and leave the strongest-growing ones at the spacing recommended.

Sowing time
The time of year you sow depends on where you live. In general, sow early in the south and west and late in the north and east. Outdoor sowing is affected by the weather, pests and diseases in the soil, and will not produce as many plants as indoor sowing.

CARING FOR YOUR PLANTS
Gardeners ensure strong growth by planting only in sites suitable for each plant and by protecting them from the worst effects of disease, pests and weather.

Plants need water, nutrients, air, space, good light and adequate temperature to grow. They also need protection from diseases, pests and bad weather. Plants that lack one or more of these essentials do not thrive. These essentials are the **limiting factors** on plant growth.

Water

Water is an essential raw material for photosynthesis and other life activities in plants. Plants have different requirements for water. Cacti can live with little water. Lettuces have a constant need for water and will rapidly wilt when they lack water. Plants grow roots towards a water supply (hydrotropism). Gardeners control the water supply by **irrigation** and **drainage**. Irrigation is used to supply water to plants where they do not get enough from the soil. Drainage is used to ensure that plants are not waterlogged. Waterlogging deprives the roots of air and encourages fungal diseases that rot plants.

Overwatering plants can cause fungal diseases such as blackleg

Light

Plants have different requirements for light. Some plants (snowdrops and bluebells) thrive in shady conditions, while others (most bedding plants) prefer a sunny spot. House plants also differ in their requirements for light. Ivy needs some shade, rubber plants need bright light and geraniums need sunny conditions.

In a glasshouse the amount of light is reduced enormously by dirty glass. Clean the glass regularly to ensure proper light. It may be necessary to provide some shade for plants in glasshouses in very sunny weather. Commercial growers use artificial lighting in greenhouses to get plants to grow and flower in winter.

Nutrients

All plants need the essential nutrients. Well rotted manure, garden compost or general fertiliser will add these to the soil. Particular plants often need one nutrient in large amounts. Plants like lettuce and grass grown for their leaves need a good supply of nitrogen—the leaf maker. Flowering and fruiting plants need a good supply of potash, and root vegetables need potassium. Rotation of crops—changing the type of plant grown in a plot from year to year—evens out the demand on the soil for the three main nutrients. It also prevents the build-up of diseases or pests of a particular plant.

Space

Plants need room to grow both above and below the ground. The requirement for space is related to the need for light, water and nutrients. Plants are usually spaced in proportion to their size—large plants need more room because of their root spread as well as their branch spread. Gardeners ensure that plants have adequate space to grow by removing weeds and thinning out other plants.

APPLIED SCIENCE · GROWING PLANTS

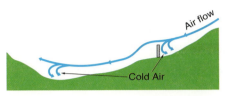

Some areas of the garden are colder than others.

Temperature
Most plants can survive in a wide range of temperatures, but for vigorous growth temperatures above 15–18°C are needed. The higher the temperature (in good light conditions), the better plants grow. Heated greenhouses are used to encourage plant growth in cold weather. Very high temperatures (35–40°C or higher) in unventilated greenhouses can kill plants rapidly. Half-hardy annuals are killed by frost and so are not planted out until any danger of frost has passed.

Air
Air is needed by all plants for both respiration (oxygen) and photosynthesis (carbon dioxide). Commercial growers sometimes use gas heaters to increase the concentration of carbon dioxide (and the temperature) in a greenhouse. Carefully controlled levels increase photosynthesis.

PLANNING A GARDEN
Plants thrive when the soil pH, water, nutrients and light are suitable. Plants that need shade and a lime soil will rarely thrive in acid soil in full sunshine. The right plant in the right place will grow and flower well.

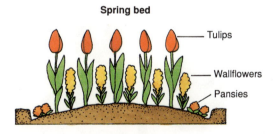

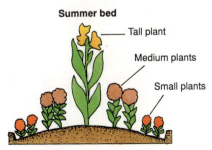

In general, plant tall plants (fully grown size) at the back of a flower bed and smaller plants to the front. Take the colours of the flowers and time of flowering into account to give a colourful garden over the full growing season.

Gardeners use annual bedding plants grown indoors and later planted out. One-year-old biennial plants are often used as spring bedding plants. Sweet William and wallflowers are two examples.

Annual plants are used as summer bedding plants. These include salvia, sweet peas, petunias, marigolds and dahlias.

Perennial flowering plants such as geraniums, fuchsias and primulas are mixed with annuals in flower boxes or hanging baskets. Boxes and baskets require special care to ensure adequate water and nutrients for the plants.

Remove dead flowers from plants (dead-heading) to prevent them from forming seeds. This produces a further bloom of flowers.

You should sow all root crops (carrots, parsnips, turnips, radishes, etc.) where they are to grow and thin them out later. Transplanting these causes the roots to fork and proper edible roots are not formed. Hardy annuals are often grown outdoors as well. Half-hardy plants, house plants and plants grown from expensive seed are all germinated and grown on indoors, and transplanted out when danger of frost has passed.

EXPERIMENT 72.2
AIM: TO GROW A PLANT TO MATURITY FROM SEED

Any inexpensive seed will do for this experiment: grass, lettuce, aster, nasturtium, cabbage, Brussels sprouts, etc.

Method

1. Prepare a tray of seed compost.
2. Water with a fine mist spray.
3. Germinate the seeds as described on the seed packet.
4. Prick out when the first true leaves grow and transplant to a seed box.
5. Harden off and plant out in prepared ground.
6. Remove any weeds. Water thoroughly (in the early morning) when there is drought.
7. Look out for pests and disease and take appropriate action.

GROW YOUR OWN POT PLANT

EXPERIMENT 72.3
AIM: TO GROW A GERANIUM (PELARGONIUM) TO MATURITY

(The plants you propagated in experiment 71.10 can be used.)

Method

1. Put some drainage stones in the bottom of a 15 cm pot. Fill with potting compost with 20% perlite added. The perlite absorbs water and keeps the compost moist. Geraniums grow best in relatively small pots.
2. Transplant the geranium from the propagating pot. Firm the compost around it gently. Water with a fine mist. Keep in shade for a few days to let it establish itself in the new pot.

General care

Place in a sunny position indoors. Geraniums thrive in a sunny position with low humidity. They can survive winter temperatures as low as 5°C if kept fairly dry.

APPLIED SCIENCE · GROWING PLANTS

Stopping

Stopping
Pinch out (remove) the growing tips of the main and side shoots of the plants. This encourages the plant to develop side shoots and form a compact bushy plant with many flower heads.

Feeding
Feed the plant with liquid tomato feed (rich in potash) every 10–14 days. A nutrient solution rich in potash is needed to produce a good show of flowers. Water as required. Keep the compost damp but on the dry side.

Remember: More potted plants are drowned than die of drought!

Dead-heading

Deadheading
Remove the dead flower heads regularly. This stops seeds forming and encourages good flower growth. Remove damaged or diseased leaves to prevent a build-up of infection. Any pests can also be removed when deadheading.

Geraniums can get a fungal disease (blackleg) which blackens and rots the stem. This happens when the plant is kept in wet humid conditions. A dry position, good drainage and careful control of watering will prevent this.

You should care for other house plants in a similar way. Take account of the particular needs of the plant for light, temperature and water. House plants should be fed regularly during their growing season but not during their dormant season. Dust on a house plant can be removed by giving it a gentle shower on a warm day. Allow to dry thoroughly. Many house plants also benefit from being put out in the light during warm summer days.

GRASSES
The most common garden plant is probably grass. Grass is the best ground cover plant for amenity areas that people walk or play on. Grass is used as the playing surface for most outdoor sports - it's a cheap, hard-wearing surface that cushions you from injury. Grass is also used to feed cows and sheep and as silage to feed cattle in winter. Grain (grass seed) is grown on about three-quarters of the world's arable land. Three of these grains—wheat, rice, and maize—are the staple food of most of the world. There are about 8,000 species of grass in the world. Grasses grow on all continents and in all climates. Species of grass are found in the sea, in marshes, in deserts and in Arctic areas.

BREAKTHROUGH SCIENCE

Lawns

Three types of grass are commonly used in growing lawns. These are bents, fescue and ryegrass. Lawns are sown with a mixture of grasses with different growing habits and strengths to provide different wearing surfaces. Ornamental lawns and golf greens contain bents (browntop bent grass, *agrostis tenuis*) and fescues (creeping red fescues, *festuca rubra*) that require a great deal of care and attention. Tougher wearing lawns and sportfields are a mixture of bents, fescues and perennial ryegrass (*lolium perenne*) that are hardwearing and require much less attention. New varieties of grasses have been developed for use in amenity areas. Many of these are strong dwarf grasses that require less cutting. "Troubadour" is a variety of dwarf perennial ryegrass used in lawns and sports grounds.

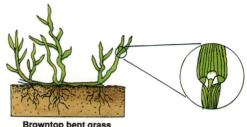
Browntop bent grass
Agrostis tenuis

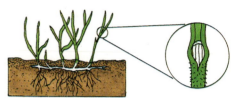
Creeping red fescue
Festuca rubra rubra

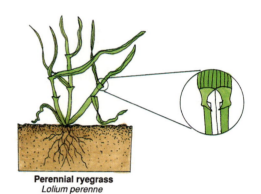
Perennial ryegrass
Lolium perenne

Tufted grass   Stolon   Rhizome

Type	Uses	Advantages	Characteristics
Browntop bent grass	Fine lawns, golf greens and minor component of common lawns	Grows well in most soils, except very light (sandy) soils	Fine-leafed perennial grass which spreads by underground stems (rhizomes) and overground stolons
Creeping red fescue	Fine lawns, golf greens and minor component of common lawns	Grows well in most soils, except very heavy soils	Fine-leafed perennial which spreads by rhizomes
Perennial ryegrass	Sports grounds and common lawns	Grows well in most soils	Broad-leafed grass

A good mixture of grasses for an amenity area should:
- be hardwearing
- regrow well after hard wear
- be drought-resistant
- grow in a wide range of soils
- grow few tough flower stalks
- tolerate shade
- be disease-resistant
- tolerate frequent mowing
- blend with other grasses.

EXPERIMENT 72.4
AIM: TO INVESTIGATE THE GROWING HABITS OF BENTS, FESCUES AND RYEGRASS

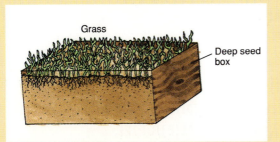

Method
1. Prepare three deep (20 cm) seed trays (tomato boxes) with seed compost.
2. Sow grass seed in them as follows:
   (a) bent at a rate of 30 g per square metre
   (b) fescue at a rate of 40 g per square metre
   (c) ryegrass at a rate of 40 g per square metre.

Water with a fine mist spray and germinate as already described. When the grass has grown, transplant (as a carpet) into an established lawn. Water well until established. Record the growth of each variety and effects of cutting, wear and tear, etc.

Naturalised meadowland
An area of grassland left uncut for a few years develops into a complex habitat. This habitat has many flowering plants and a wide variety of animal life. A meadow can be grown from seed using a special mix of meadow flower and grass seeds. The meadow should not be cut, to allow plants to seed. Dominant weeds must be removed by hand. This type of grassland habitat is the natural habitat of the corncrake. This bird has become extinct in many parts of Europe. The corncrake survives mainly in areas of the west of Ireland where there is little mechanised hay-cutting.

HARVESTING AND CARE OF CUT FLOWERS
We give flowers to others to celebrate birthdays, anniversaries and other happy occasions. Flowers bring the beauty and colour of nature into our homes. It doesn't make sense to spend months growing plants and then neglect the flowers when you cut them. Proper harvesting and care of cut flowers can make them last longer.

Time to harvest
Most flowers are best harvested in the early morning. Flowers with fully formed, coloured buds, ready to open, should be chosen. These will open fully later and will last longer than ones cut when already open. Cut with a sharp secateurs or scissors and take as long a stem as possible. Remove the lower leaves from the stem.

Conditioning the flowers

Plunge each stem up to the bud in tepid water. Cut the stem under water. This provides a clean surface, free of air bubbles, for the stem to take in water. Woody stems should be crushed at the base to expose more stem surface. This helps the stem to remain rigid and not go limp. Leave the stem plunged in water until needed for use.

EXPERIMENT 72.5
AIM: TO INVESTIGATE HOW HARVESTING FLOWERS AT DIFFERENT TIMES AND DIFFERENT STAGES OF MATURITY AFFECTS THEIR LIFE

(Daffodils are probably the cheapest and most easily grown flowers for this purpose.)

Method

1. Cut some daffodils (in the **same** stage of maturity) in the early morning, mid-day and evening of a bright day.
2. Condition the flowers as described. Leave in labelled containers in a cool place. Record the condition of the flowers each day to determine when each sample deteriorates.
3. Cut some daffodils at **different** stages of maturity early in the morning.
4. Condition the flowers as described. Leave in labelled containers in a cool place. Record the condition of the flowers each day to determine when each sample deteriorates.

Care of flower arrangements

Flowers deteriorate from the minute they are cut. Loss of water by transpiration causes them to lose rigidity and wilt. This can be delayed by proper conditioning and by placing flower arrangements in a cool place. Adding special chemicals to the water controls the growth of micro-organisms in the water and supplies nutrients to the flowers. Replace the water regularly— each time cut (or cut and crush) the stem under water.

APPLIED SCIENCE · GROWING PLANTS

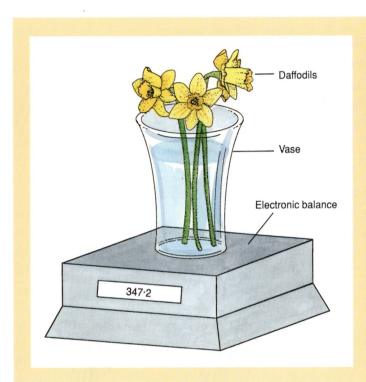

EXPERIMENT 72.6
AIM: TO SHOW THAT CUT FLOWERS LOSE WATER.

(Daffodils can be used again.)

Method
1. Cut some daffodils in early morning.
2. Place the flowers in a vase **without water**. Find the mass of the flowers.
3. Find the mass of the flowers the following day.

Result: The flowers have lost mass.

If you cover the vase with a plastic bag, condensation will form on the inside of the bag.

How can you prove that this is water?

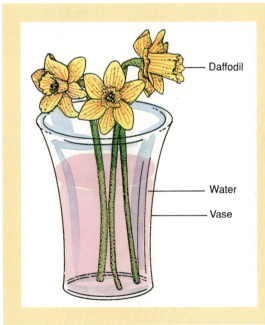

EXPERIMENT 72.7
AIM: TO SHOW THAT CUT FLOWERS ABSORB WATER.

(Daffodils can be used again.)

Method
1. Cut some daffodils in early morning.
2. Condition the flowers as described. Add some food colouring (red or blue will show up well) to the water in a vase. Place the flowers in a vase. The dye will be clearly seen in the petals of the flower the following day.

469

EXPERIMENT 72.8
AIM: TO INVESTIGATE HOW ADDING CHEMICALS TO VASE WATER AFFECTS THE LIFE OF CUT FLOWERS

(Daffodils can be used again.)

Method

1. Cut some daffodils in the same stage of maturity in early morning.
2. Condition the flowers as described. Make up four solutions as follows:
   (a) 1 litre of water
   (b) 12 g sucrose in 1 litre of water
   (c) 20 cm^3 of household bleach in 1 litre of water
   (d) 12 g sucrose and 20 cm^3 of household bleach in 1 litre of water.

Leave three daffodils in each solution in a cool place. Record the condition of the flowers every three days to determine when each sample deteriorates.

Does this experiment give you any clues as to why some people add sugar and aspirin tablets to vases of flowers?

(a) Water   (b) Water + sucrose

(c) Water + bleach   (d) Water + sucrose + bleach

MULCHES HELP CONTROL WEEDS AND MOISTURE LOSS

Weed control is a constant problem for all gardeners. Weeds grow faster and stronger than most other garden plants. Weeds must be removed to allow slower growing plants to thrive.

One method of weed control is to apply a layer of weed-free material (a **mulch**) on top of the soil. Any weeds in the soil that germinate will die for lack of light. Bark, grass cuttings, garden compost, peat, gravel and black polythene are all used. Mulches also help to control moisture loss from the soil.

Mulch	Advantages	Disadvantages
Gravel (4–6 mm)	Prevents weed germination, reduces moisture loss from soil	Gravel must be kept free of leaves and other organic material to prevent weeds from growing, conditions suit many pests
Bark (medium grade)	Prevents weed germination, decomposes to form humus, reduces moisture loss from soil	May cause fungal disease, conditions suit many pests, can be blown away, produces soft growth susceptible to frost
Leaves	Prevent weed germination, decompose to form humus, reduce moisture loss from soil	May cause fungal disease, conditions suit many pests, produce soft growth susceptible to frost

Mulches are an effective alternative to weeding. A mulch does not disturb the soil as does hoeing or hand weeding. Organic mulches (bark, grass cuttings, garden compost and peat) add beneficial humus and control moisture loss from the soil.

Mulches are an effective alternative to weeding and also control moisture loss from the soil.

EXPERIMENT 72.9

AIM: TO INVESTIGATE THE EFFECTS OF DIFFERENT MULCHES ON (A) MOISTURE LOSS FROM SOIL AND (B) WEEDS

Method

1. Fill five containers with soil.
2. Cover the surface of the soil in the containers as follows:
   (a) 2 cm layer of medium-grade bark
   (b) 2 cm layer of peat
   (c) 2 cm layer of gravel
   (d) layer of black polythene
   (e) uncovered (control)
3. Find the mass of each container.
4. Place in a well aired room for a few days.
5. Find the mass of each container and compare with the previous mass.
   Which surface covering was best at controlling moisture loss from the soil?
6. Plant the containers into the soil outdoors. Leave for four weeks during the growing season. Note the numbers of weeds growing in each container.
   Which surface covering was best at controlling weeds?

(a) Bark (b) Peat (c) Gravel
(d) Polythene (e) Control

DISEASES AND PESTS THAT KILL PLANTS

Plants are stunted or killed by a variety of diseases and pests. Fungi, bacteria and viruses affect plants. You already know about some diseases, such as the fungal disease blackleg in geraniums and potato blight. These diseases are prevented by spraying with a fungicide (a chemical that kills fungi) and by growing varieties of plants that are resistant to fungal attack. Plant pests include underground pests that eat roots, and slugs, snails and caterpillars that eat stems, leaves and fruit. In damp weather slugs and snails are the most numerous and destructive of these pests.

APHIDS

Aphids are a common garden pest. This insect has numerous species, from white aphids to green aphids (greenfly) to black aphids. Their life cycles are all similar and infestations of aphids are controlled in the same way. Aphids have a sharp tube-like mouth (stylet) which they use to suck sap from plants.

Life cycle

The aphid's life cycle starts in spring when it hatches from an egg that has survived the winter. Only wingless females are produced. These females produce more wingless female offspring without being fertilised by males (parthenogenesis). These aphids are born alive without any egg stage. Populations of aphids increase rapidly in this way. After a few generations (when the food supply of the infested plant is no longer adequate) winged females are produced and fly off to live on other plants. In this way aphid colonies are formed in suitable conditions of food supply and weather. In the autumn, as the weather cools, both winged males and females are produced. The males mate with and fertilise the females. Females lay eggs that survive the winter. These hatch out in spring to start the cycle again. In mild winters some aphids survive and infestations occur early in the growing season, as the populations increase rapidly once conditions are right.

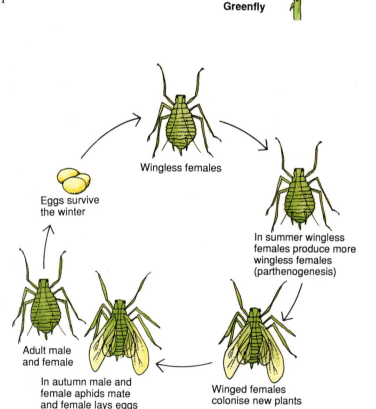

How aphids colonise new plants (locomotion)
Winged aphids fly off to start new colonies on other plants. Aphids also are blown by winds over enormous distances. Aphids secrete a sugary substance (honeydew) that attracts ants and other insects. Some ants keep aphids and "milk" them for their honeydew. The ants protect aphids from other predatory insects. Ants may even take aphids off plants and move them to "better grazing" on other plants.

Season of activity
Large numbers of aphid colonies develop in suitable conditions of food and weather. These are usually in late spring, summer and early autumn. In warm humid conditions (which also produce strong growth of plants) aphid populations increase rapidly. In sheltered and mild areas winter infestation is also possible. Aphid populations are controlled naturally by weather conditions (temperature, rain, wind) and by predators.

Black bean aphid

Host plants
There are over 200 species of aphid in Ireland. Aphid species tend to attack specific plants. Blackfly is the black bean aphid, which attacks only bean plants. Blister aphids produce blister-like swellings on blackcurrant leaves. Woolly aphids live on apple trees. There are lettuce aphids, raspberry aphids, strawberry aphids, gooseberry aphids and aphids on rosebushes.

DAMAGE TO PLANTS
Primary damage
Aphids are a major pest to many crops and ornamental plants. Their large populations seriously weaken plants by sucking sap. This often produces distorted leaves and damages photosynthesis.

Secondary damage
Aphids transmit many plant diseases, particularly plant fungi and viruses. Plant viruses are a major problem as they can be controlled only by burning the infected plants. The honeydew that aphids secrete is often colonised by the black fungus called sooty mould.

Chemical control of pests
Insecticides are used to kill insect pests. Some insecticides (malathion) kill by contact: you have to spray onto the insects to kill them. Others coat the leaves and the insects die when they eat the leaves.

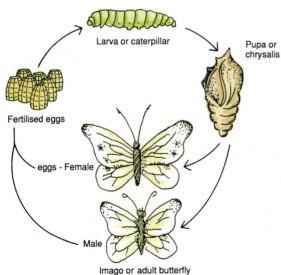

Breakthrough Science

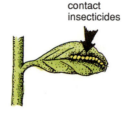

Other insecticides are systemic (dimethoate) —they enter the plant's system and kill any insect that eats part of the plant.

Disadvantages
Some pesticides kill bees and reduce the yield of fruit that depends on bees to pollinate it. Insects also develop a resistance to insecticides. A build-up of insecticides in the food chain may cause problems in the long term.

Biological control of pests
Aphids can be controlled biologically using predators that eat greenfly. Ladybirds (a beetle) and their larvae (young) feed on aphids and are an important natural biological control of their populations. Scientists have developed artificial pheromones (chemical substances released by female insects) to lure male insect pests into traps. These traps are used to control codling moth caterpillars that infest apples and other fruit. Biotechnologists are also working to produce genetically altered plants that are resistant to specific insects or viruses.

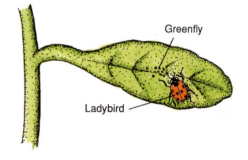

Beer traps for slugs and snails
Slugs and snails can be controlled without using harmful chemicals. A simple trap can be made by filling a jam jar one-quarter-full of stale beer. Place in a hole in the ground and cover with a slate as shown. A number of these will reduce the slug and snail population. Garden birds also help to control these pests.

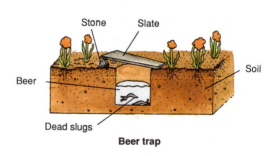

Beer trap

Integrating chemical and biological control of pests
Integrated pest management tries to combine biological controls with limited use of pesticides to control pests. Resistant varieties of plants, crop rotation, planting and harvesting at times that reduce pest problems, natural predators and pheromone traps are combined to reduce pest populations.

Project: A long-term study of aphid infestation on a selected plant
You can do this study easily if you have some plants that suffer from aphid infestation growing in your garden. Record the weather conditions over the spring and summer months. Record the level of infestation on various dates and the changes. Note the strength of growth of the affected plant and changes in this during the infestation. You could also test the effectiveness of biological control by collecting ladybirds and transferring them to the affected plant. Compare this with another plant that you have sprayed with insecticide.

SUMMARY

- Annuals are plants that germinate, grow and flower in a single growing season.
- Biennials are plants that grow in one season and then flower in the second season.
- Perennials are plants that flower each year and die back over the winter.
- The limiting factors on plant growth are water, nutrients, air, space, good light and adequate temperature to grow.
- Plants thrive when the soil pH, water, nutrients and light are suitable.
- One-year-old biennial plants are used as spring bedding plants.
- Annual plants are generally used as summer bedding plants.
- Stopping: pinching off the growing tips of shoots to encourage a compact bushy plant with many flower heads.
- The three types of grass commonly used in lawns are bents, fescues and ryegrasses.
- Mulches are an effective alternative to weeding and also control moisture loss from the soil.
- Plant pests include underground pests that eat roots, and slugs, snails and caterpillars that eat stems, leaves and fruit.
- Large aphid colonies are formed in suitable conditions of food supply and weather.
- Aphid populations are controlled naturally by weather conditions (temperature, rain, wind) and by predators.
- Integrated pest management tries to combine biological controls with limited use of pesticides to control pests.

QUESTIONS

Section A

1. Annuals are plants that _____.
2. Name three limiting factors on plant growth. _____
3. What is a hardy annual? _____
4. Stopping means to _____. This causes the plant to _____
5. What type of plants is used as spring bedding plants? _____
6. What is a mulch? _____
7. Name three plant pests. _____
8. What is a half-hardy annual? _____
9. Plants that flower each year and die back over the winter are _____.
10. Summer bedding plants are usually _____
11. Name three types of grass used in lawns. _____

12. Biennials are plants that _____.
13. Aphid populations are controlled naturally by _____.
14. Integrated pest management means _____.
15. Dead-heading means _____. This is done to ensure _____.

Section B

1. (a) Outline by means of a simple diagram the life cycle of the aphid (greenfly)
   (b) A lawn seed mixture may contain seeds of the following grasses: fescue, bent, ryegrass. Choose two of the above grasses and give the characteristics of each.
   Which of the above grasses would be the most suitable for seeding a hard-wearing surface, e.g. a football pitch?

2. (i) Outline the steps you would take to ensure that summer bedding plants survive when you plant them out.
   (ii) Describe how caterpillars may be controlled in a garden.
   (iii) State three ways of ensuring that cut flowers will last.

3. (i) Outline by means of simple diagrams the life cycle of the cabbage white butterfly. Describe a method of biological control for this pest.
   (ii) What is the best time to cut flowers? Describe how you would make cut flowers last longer.

4. (i) A plant is described as a "hardy annual". What does this mean? How soon after germination would you expect the plant to flower? What kind of conditions does the plant need?
   (ii) Gardeners sometimes use chemical and biological means to control pests. Distinguish between chemical and biological control. Outline two advantages and two disadvantages of each.

5. Your friend has a healthy pelargonium plant (geranium). Describe the steps you would take to grow a plant from a cutting taken from your friend's plant, under the headings:
   (a) taking the cutting
   (b) rooting the cutting
   (c) growing the plant
   (d) caring for the plant.

6. You want to make a garden bed of summer-flowering annual plants. Describe how you would grow the plants for this bed, under the headings:
   (a) germination
   (b) pricking out
   (c) hardening off and transplanting to bed
   (d) caring for the plants to give a long flowering season.

7. Any restrictions of space, temperature, water, nutrients and light limit the growth of plants.
   (i) Describe how limited space may affect the growth of a plant.
   (ii) Describe an experiment to show the effects of a deficiency in nutrients on the growth of a plant.
   (iii) Glasshouses can provide near-ideal conditions for plant growth.
   Explain how growing plants in a glasshouse can produce strong growth.
8. What is a mulch? List two advantages and two disadvantages of using mulches. Name two materials commonly used as mulches. Describe an experiment to show the advantages of using mulches.
9. (i) List three factors that affect the lasting time of cut flowers.
   (ii) Describe an experiment to show the effect of solutions of sucrose and bleach on the lasting time.
   (iii) How could you show that cut flowers transpire?
10. You want to grow some cabbage plants in your vegetable garden. Describe how you would do this, under the headings:
    (a) germination
    (b) transplanting
    (c) controlling pests and weeds.

CHAPTER 73 MATERIALS SCIENCE

People have always used the materials they found around them. A gardener will use a piece of stick lying in the garden to make holes to plant seedlings. The materials found around us include the wood and stone we use to build houses; wool and cotton to make cloth; and rock ores that we melt to make metals. Even man-made substances like plastics and metal alloys are made from raw materials found in nature.

Materials science is the study of the properties of the substances we use to make things. It helps us to choose suitable materials for each object we make. When designing a saucepan, we make the body of the saucepan of metal—a good heat conductor. We make the handle of plastic or wood—a good heat insulator.

Breakthrough Science

HOW DO WE IDENTIFY MATERIALS?

We use a great variety of materials in our daily life. Most of the things we use contain a number of different types of material.

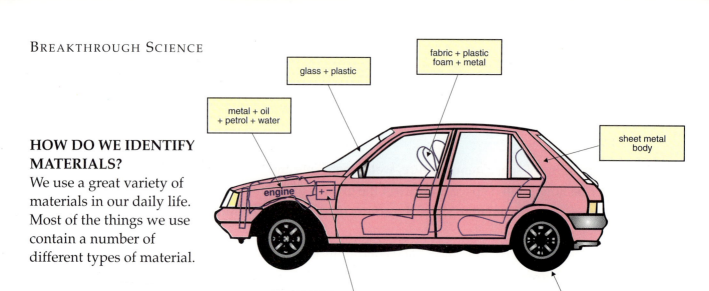

A car contains the following materials:

Body	Sheet steel galvanised and painted
Engine	Iron, aluminium alloy, copper, plastics, rubber, oil, petrol, water
Electrics	Lead, acid (battery), plastic and copper (wiring), bronze bearings in motors and generators
Wheel	Rubber, steel, alloys
Windows	Glass and plastics
Upholstery	Fabrics and foam plastics

Each material is suitable for the purpose it is used for: the sheet-steel body makes the car strong and rigid; the rubber air-filled tyres can change shape and help absorb shocks.

We classify the materials we use to help us identify them.

Hydrocarbons are materials that contain the elements hydrogen and carbon. They are used as fuels and as raw materials for plastics. Hydrocarbons include coal, waxes, tar, oil, petrol and gas.

APPLIED SCIENCE · MATERIALS SCIENCE

Plastic bucket

Clothes made from synthetic textiles

Copper water cylinder

Plastics are man-made (synthetic) substances made from hydrocarbons and other raw materials. Plastics are made by getting a large number of small molecules to join together (polymerise) to form long chains. Polythene, PVC (polyvinyl chloride), nylon, polystyrene, polyester, polyurethane and teflon are common plastics.

Textiles are made from yarn. Yarn is made from fibres twisted together to make long strands. The fibres can be natural (wool, silk from animals; cotton, linen from plants) or synthetic filaments (nylon, polyester, acrylic). Yarn is made into cloth (fabric) by weaving or knitting.

Metals are generally strong, shiny, good conductors of heat and of electricity. Common metals include iron, copper, lead, zinc, tin, aluminium.

Others materials: We use many other types of material: glass, ceramics, wood, man-made wood products such as plywood, etc.

Hydrocarbons	coal, waxes, tar, oil, petrol and gas
Plastic	polythene, PVC (polyvinyl chloride), nylon, polystyrene, polyester, polyurethane and teflon
Textiles	wool, silk, cotton, linen, nylon, polyester, acrylic
Metals	iron, copper, lead, zinc, tin, aluminium
Other materials	glass, concrete, rubber, paper, leather, ceramics, wood and wood products

Project: Try to identify the materials used in the following items, from your knowledge of the classes of material listed above: a cassette tape, a CD, a pair of shoes, a denim jacket, a pair of roller skates, a bicycle. Can you say how each material is particularly suited to the purpose for which it is used?

MIXTURES OF MATERIALS

Alloys are mixtures of metals with different properties from the parent metals. Many alloys have been known since ancient times: bronze is a mixture of copper and tin and is much harder than either of these metals. Bronze was used for cutting tools, swords, spears and ploughs before iron was discovered.

Steel is a mixture of iron and carbon that can be made much harder and stronger than iron. Stainless steel is a rust-resistant alloy made from a mixture of iron, chromium and nickel.

Breakthrough Science

Coins were once made from gold, silver and copper. Gold and silver are no longer used in coinage as they are too expensive (£1 worth of gold or silver would make a very small coin). Most "silver" coins are now made from an alloy of 75% copper and 25% nickel.

Mixtures of fibres are also used in making yarn to give different qualities to cloth. Carpets made from 80% wool and 20% nylon are cheaper and have better wear resistance than pure wool. A blouse or shirt made from a mixture of cotton and polyester has better crease resistance than a pure cotton one. A jumper made from a mixture of acrylic and wool has better shrink resistance than a pure wool jumper.

Project: Try to identify the mixtures of materials used in emulsion paints, epoxy glues, and knives and forks.

Look at the care label on each item of your clothes. Identify which of them are mixtures. Can you tell why each substance in the mixture is used?

WE USE MANY DIFFERENT MATERIALS

Teflon is a plastic developed in 1938. It does not burn, can withstand high temperatures and is resistant to most chemicals. At first it was used in the chemical industry only. The range of uses for this material has grown over the years. Many pots and pans are now coated with Teflon to prevent food from sticking. Teflon is also used to make low-friction wheel bearings.

The first surfers used heavy wooden surfboards. Modern surfboards are much easier to use because they are made from lightweight fibreglass and polyurethane. Windsurfing is a modern development of surfing that uses modern materials for lightweight sails and masts.

New materials often bring about improvements in the design or performance of clothes, tools, machinery or other things we use. They may also lead to a range of new uses that weren't thought of before.

We have already identified some of the materials that are used in making a car. Try to identify the materials used in other objects you see or use every day.

APPLIED SCIENCE · MATERIALS SCIENCE

Item	Materials used
Buildings	Concrete, brick, stone, wood, slate, plastics, steel girders, aluminium
Machinery	Steel, various alloys, rubber, plastics
Clothes	Natural: wool, cotton, silk; synthetic: rayon, polyester, nylon, etc.
Furniture	Wood, plastic, glass, stone, metals
Kitchen utensils	Wood, plastic, glass stone, metals
Toys	Wood, various fabrics, plastic, metals
Jewellery	Metals, precious stones, other stones, glass, plastics, wood
Packaging	Corrugated cardboard (paper), shredded paper, aeroboard (expanded polystyrene)

As you can see from this table, one substance can be used to make a number of different items. Aluminium foil is used as kitchen foil, aluminium is used to make windows and door frames and soft drinks are sold in aluminium cans. The kitchen foil is a very thin sheet of aluminium, drink cans are a bit thicker and windows are made from much thicker pieces of aluminium.

Polythene is used to make dishes, buckets and other kitchen utensils. It is also used to make plastic sheeting, drainage pipes and plastic bottles.

USE MATERIALS SAFELY

Most things we use are designed for a particular purpose and for use in particular conditions. Household electrical sockets and connectors are designed for indoor use only. They are dangerous if used outdoors for electric mowers or hedgecutters. Special weatherproof connectors are made for use outdoors.

Electric-powered garden machines must be used with great care because of the danger of cutting fingers or toes or cutting through the flex. The instructions that come with these machines give advice on the safe use of the machine. Read the instructions **before** you use the machine. Why do you think you are warned not to use these machines in wet weather?

You can prevent accidents by reading the instructions that come with a new machine or piece of equipment.

You might think that household cleaners are harmless. Read the label on any container of household bleach. Do you still think it is harmless? Do you think that under the kitchen sink is a safe place to keep these chemicals? The labels on household cleaners (the most common chemicals in the home) tell you how to use these chemicals safely.

SAFETY SYMBOLS

Flammable	A substance that burns freely when the other two conditions for fire (heat and oxygen) are present. Petrol is flammable.
Toxic	A substance that harms a person in some way. Toxic substances range from materials of low toxicity that cause little damage to toxic substances that kill.
Corrosive	A substance that burns the skin (and other parts of the body). Corrosive substances are particularly harmful to the eyes or if swallowed. Acids are corrosive.
Radioactive	A substance that gives off harmful radiation. Radioactive materials should be stored in special containers far away from normal working places.
Harmful	A substance that damages your health in some way. These substances should be handled with care. Glass wool can cause an irritating itch and is classified as harmful.
No smoking	A safety sign to tell you that smoking is not allowed. It should also remind you that many of the chemicals in tobacco smoke are toxic.

flammable substances

toxic substances

corrosive substances

radioactive substances

harmful substances

no smoking

Project: Examine the labels on as many household chemicals (cleaners, medicines, paints, etc.) as you can find. List the substances contained in them and the safety warnings and symbols given on the labels.

Collect as many instruction leaflets for common household appliances as you can find. Read the instructions. List the appliances and the safety warnings given.

YOU HAVE TO PROTECT MATERIALS

Most things will decay and deteriorate unless you treat them to prevent it. Fabrics and wood are attacked by a number of organisms. Pests such as the larvae (caterpillars) of moths eat some fabrics; woodworm (larvae of various beetles) eat into wood. Various fungi attack both wood and fabric and cause them to deteriorate. Some metals corrode and weaken in moist air. Many plastics become brittle in strong sunshine.

Fabrics

Most synthetic fabrics are resistant to insect and fungal attacks. Natural fabrics are sometimes treated with chemicals to prevent attack. Modern mothproofing uses chemicals without noticeable smells.

APPLIED SCIENCE · MATERIALS SCIENCE

Fixed furnishings such as curtains and carpets are more likely to be attacked than clothes. Furnishings in damp conditions (or near wet windows) are attacked by a fungus that rots the fabric. Clothes that are washed and worn frequently are unlikely to suffer attack.

Wood

Hardwoods used in furniture have a good natural resistance to pests. Furniture may also be treated with preservatives and finishes that include insecticide chemicals. These repel or kill pests. Softwood used for roofing and fencing is treated with special chemicals to prevent insect and fungal damage. A chemical commonly used for this is creosote. Take care when you use creosote, as it is toxic.

Keep wood dry to prevent attack by fungi such as wet and dry rot. Painting and varnishing help keep the wood dry. Damp-proof courses prevent rising damp, and good underfloor ventilation helps to keep wooden floors dry.

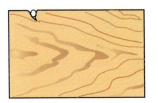

1 beetle lays eggs in wood

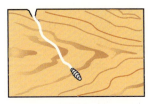

2 woodworm bores holes in wood

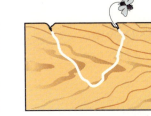

3 woodworm changes to pupa near the surface

4 pupa hatches into beetle

Metals

Iron reacts with damp air and forms a flaky iron oxide. Iron and iron alloys (steel) rust. This weakens anything made of iron. Replacing rusted machinery and protecting buildings against corrosion costs many millions of pounds each year.

EXPERIMENT 73.1

AIM: TO COMPARE HOW EASILY DIFFERENT METALS, IMPURE METALS AND ALLOYS CORRODE

Method

1. Set up the water bath containing a number of different metal strips as shown in the diagram.
2. Leave the metal test strips to stand for a few days. Examine the metals for signs of corrosion.

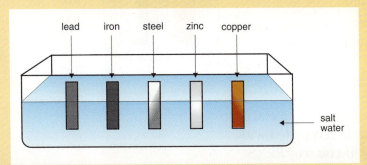

Try this experiment using salt water instead of tap water. Does this help explain why salting icy roads in winter causes rusting in cars?
Repeat the experiment with two of each sample—one painted and one unpainted.

483

How to prevent rust

Some metals rust faster than others. This is often due to impurities in the metal. Mild steel rusts faster than pure iron, but mild steel is more flexible than iron and can be used to make car bodies.

Steel and other metals are protected against corrosion by
(1) painting them: this stops the air and water from coming into contact with the metal surface
(2) covering them with a layer of grease, wax or plastic material
(3) coating them with a layer of another metal.

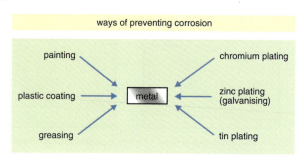

Galvanised steel is steel coated with a layer of **zinc**. This is done by dipping the steel in hot zinc or by electrolysis.

Did you know that 'tin' cans are made mostly of steel? They are made of thin sheets of steel covered, inside and out, with a layer of tin. The tin prevents corrosion of the metal.

Protecting metals by electroplating

Electroplating coats a metal surface with a thin layer of a less reactive metal by electrolysis.

EXPERIMENT 73.2
AIM: TO COPPER-PLATE A PIECE OF STEEL

1. Prepare the steel by rubbing it with emery paper and then cleaning it with methylated spirits.
2. Place the steel strip and the copper strip in a beaker of copper sulphate. Connect the steel strip to the negative pole of the battery and the copper strip to the positive pole.
3. Switch on the current and allow it to flow for about 30 minutes. The steel strip is now covered with a layer of copper. It is copper-plated.
   It is important to keep the current flowing small (about 0·5 A) to make sure the copper plating sticks to the steel.

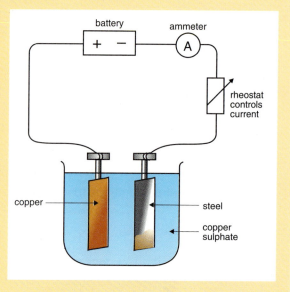

Electroplating is used to improve the appearance of a metal or to protect it against corrosion. The surface of the item is plated with silver, nickel, chromium, zinc, gold or copper. If you see "EPNS" written on knives and forks it means "electroplated nickel silver".

APPLIED SCIENCE · MATERIALS SCIENCE

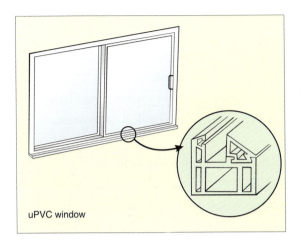
uPVC window

Plastics

Plastics are not attacked by pests and do not corrode or rust. Weathering by exposure to sunshine (ultraviolet light) is the principal cause of deterioration in plastics. Many plastics become brittle and crack easily when exposed to sunshine for a long period of time. Ultraviolet light and weathering break down plastic fertiliser bags into smaller pieces but do not return the materials to the soil—it is **non-biodegradable**. Most plastic packaging is a problem because it is non-biodegradable. Environmentalists encourage people to recycle packaging materials, to use less packaging and to use biodegradable packaging. Most household waste consists of packaging.

EXPERIMENT 73.3

AIM: TO COMPARE THE EFFECTS OF EXPOSURE TO WEATHER ON PLASTIC BAGS AND PAPER BAGS

Method
1. Tie down a number of paper bags and plastic bags outdoors in an exposed spot.
2. Leave them in place over the summer months.
3. Examine the remains of the samples.
4. Leave the bags in place and examine them again in late October.

Result: The paper bags have broken down into the soil. The plastic bags are still there.

Conclusion: Paper is biodegradable; plastic is non-biodegradable.

Material	Attacked by	Effect	To prevent attack
Plastics	Sun	Makes plastic brittle	Special chemicals added to plastic
	Heat	Softens plastic	Use heat-resistant plastic
Textiles	Fire	Black smoke, poisonous fumes	Special treatment
	Fungi	Cloth rots	Chemical treatment or keep dry
	Insects	Cloth eaten	Chemical treatment
	Washing	Cloth shrinks	Follow washing instructions
Metals	Damp air	Metal weakened	Coat with paint, wax or grease, galvanise or electroplate
Wood	Insects	Wood weakened by insects boring holes	Use resistant wood or insecticide chemicals
	Fungi	Wood rots	Keep dry, use preservative chemicals

SUMMARY

- Hydrocarbons are materials that contain the elements hydrogen and carbon.
- Plastics are made by getting small molecules to join together (polymerise) to form long chains.
- Textiles are made from natural fibres (cotton and wool) or synthetic fibres (nylon, polyester, acrylic).
- Mixtures of fibres are also used in making yarn to give different qualities to cloth.
- Metals are generally strong, shiny and good conductors of heat and of electricity.
- We use many other types of material, such as glass, ceramics, wood and man-made wood products like plywood.
- Coins are now made from an alloy of copper and nickel.
- New materials bring improvements in the design and performance of clothes, tools, machinery and other things we use.
- You can prevent accidents by reading the instructions before using a machine.
- The labels on household chemicals tell you how to use these chemicals safely.
- Natural fabrics are treated with chemicals to prevent attack by pests and fungi.
- Softwood is treated with creosote to prevent insect and fungal damage.
- Metals are protected from corrosion by coating with grease, wax, plastic or paint, and by electroplating.
- Electroplating coats a metal surface with a thin layer of a less reactive metal by electrolysis.
- Many plastics become brittle in strong sunshine.
- Paper is biodegradable; plastic is non-biodegradable.

QUESTIONS

Section A

1. A chain of small molecules joined together is a _____.
2. Materials that contain the elements hydrogen and carbon are called _____.
3. Name two synthetic fibres. _____
4. A substance is labelled as flammable. What does this mean? _____
5. Natural fabrics are treated with _____ to prevent attack by pests and fungi.
6. A material is strong, shiny and a good conductor of heat and of electricity. This material is likely to be a _____.
7. Softwood is treated with _____ to prevent _____.
8. Coins are made from cupro-nickel. This is an _____ of _____ and _____.
9. Name two natural fibres. _____
10. You can protect metals from corrosion by _____.

11. What happens when plastics are exposed to bright sunshine? _____
12. Coating a metal surface with a thin layer of a less reactive metal using an electric current is called _____.
13. Name two biodegradable materials. _____
14. The fabric of curtains is found to be decayed. Name two things that may have caused this. _____ .
15. Name three common plastics. _____
16. What is a non-biodegradable material? _____
17. Any substance that burns the skin is called a _____ substance.
18. Name a rust-resistant alloy. _____
19. You find some small holes in furniture. What is the likely cause of these? _____ How would you protect the furniture from further damage? _____
20. What is galvanising? _____

Section B

1. (i) Plastics, textiles, metals and timber are different types of material. Name two other types of material and give a typical use of each.
   (ii) The diagram shows two hazard symbols. State what hazard is described by each symbol.
2. (a) Use words from the following list to answer the questions below: OAK, GLASS, POLYTHENE, COAL, COPPER, WOOD.
    (i) Name a metal.
    (ii) Name a textile.
    (iii) Name a plastic.
    (iv) Name a timber.
   (b) Give one use of plastic in your home.
   (c) Give one use of timber in your home.
3. (a) Use words from the following list to answer the questions below: METAL, PLASTIC, TEXTILE, TIMBER.
    (i) What type of material is chipboard?
    (ii) What type of material is polystyrene?
    (iii) What type of material is cotton?
    (iv) What type of material is copper?
   (b) Name a natural material from this list: POLYTHENE, WOOL, NYLON.
   (c) Choose a synthetic (man-made) material from this list: SILK, LEATHER, POLYESTER. Identify the following fabric care labels.

4. (i) When materials are mixed the resulting mixture may have improved properties. Give an example to support this statement and state the improvement produced.

(ii) Explain what is meant by biodegradable material. Name one biodegradable material and one non-biodegradable material used for packaging.

(iii) Bottles in the laboratory may carry the following warning (hazard) symbol. What does the symbol tell you?

5. | Material | Use |
   | --- | --- |
   | Wood | Roofing |
   | Plastic | Piping |
   | Metal | Windows |
   | Textile | Curtains |

   Select one material from this table.

   (i) Give a suitable example of this material.

   (ii) What is the origin of the example you have given?

   (iii) What property of this material makes it suitable for the stated use?

   (iv) Describe an experiment to examine the effects, if any, of weather on the material you have chosen.

6. (a) Use words from the following list to answer the questions below:
   METAL, PLASTIC, TEXTILE, NYLON.

   (i) What type of material is linen?

   (ii) Which type of material is usually got from oil?

   (iii) What type of material is iron?

   (iv) Name a synthetic (man-made) material.

   (b) Identify the following wash care labels.

APPLIED SCIENCE · TYPES OF MATERIALS

CHAPTER 74 — TYPES OF MATERIALS

PLASTICS, TEXTILES, METALS AND TIMBER
Higher level students study two of these types of materials; ordinary level students study one.

PLASTICS

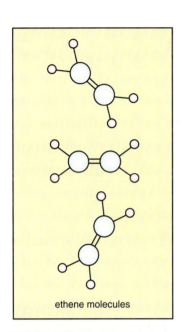
ethene molecules

You probably have many things made from plastics: biros, rulers, set squares, cassette tapes, Walkmans and radios. Electric plugs and sockets all contain some plastics. Other plastics are not so obvious. Nylon, polyester and acrylics are all plastics that are produced as fibres and made into cloth. Many paints and varnishes are also plastics. Foamed plastics have gas blown into them to form a spongy substance used in furniture and insulation materials. Many plastics can be heated and formed into different shapes.

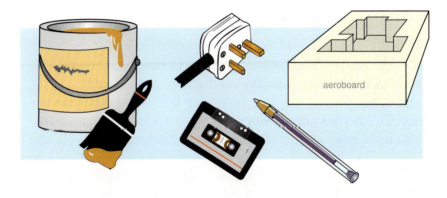

PLASTICS ARE GIANT MOLECULES
Plastics are **polymers**. That means they are made of long chains containing many thousands of similar molecules joined together. This is a bit like joining thousands of paperclips together to make a long chain. The different plastics are made from different basic molecules. Polyethene, also known as polythene, is a chain of many molecules of ethene. Polystyrene is made from styrene molecules. Latex rubber from the rubber tree is a natural polymer.

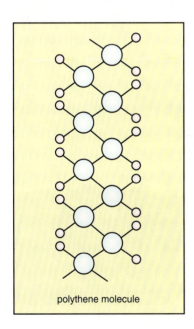
polythene molecule

The simple molecules that are used to make polymers are made from hydrocarbons. Hydrocarbons are fossil fuels. The energy of the sun over thousands of years was stored up by photosynthesis in plants that turned into fossil fuels. Fossil fuels are non-renewable.

Breakthrough Science

Plastics are made by first producing the simple molecules from which the polymers are made. These simple molecules are called **monomers** (mono means one). These monomers are then heated with a catalyst to get them to join together to make polymers (poly means many).

WE USE MANY DIFFERENT PLASTICS

Alexander Parkes invented celluloid, the first synthetic plastic, in 1856. Celluloid was used as a substitute for ivory in billiard balls and piano keys, and later in photographic film.

Polythene is a soft plastic made from the monomer ethene. Polythene is widely used in kitchen ware, drainage pipes, refuse sacks and black sheeting for silage.

Polystyrene is a polymer of styrene. Polystyrene is a clear, hard, brittle plastic. Foamed polystyrene is better known as aeroboard. It is used for heat insulation in buildings, in heat insulating coffee cups and in flotation devices in boats.

Nylon is used to make combs, brushes, machine gear-wheels, electric plugs, ropes and tights.

PVC (polyvinyl chloride) is used to make rainwear, suitcases, floor tiles and insulation on copper wire.

Perspex (acrylic) is used to make the plastic lenses for car tail-lights, safety glasses, illuminated signs and baths and showers.

Teflon is a polymer of tetrafluoroethylene—a molecule containing carbon and fluorine. Teflon can withstand temperatures above 250°C, is resistant to nearly all chemicals and does not burn. It is also used as a low-resistance, low-wear material in wheels and gears.

Some plastics are produced as fibres. These include polypropylene (used to make ropes) and nylon (used in clothes and carpets).

Polythene jug

Polystyrene rawlplugs

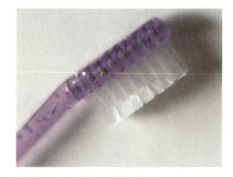

Nylon toothbrush bristles

Nylon fibre tights

Perspex car tail-light

PVC suitcase

APPLIED SCIENCE · TYPES OF MATERIALS

EXPERIMENT 74.1
AIM: TO INVESTIGATE THE FLEXIBILITY AND HARDNESS OF PLASTICS

Method

(A) Flexibility

1. See how easily you can bend a piece of plastic.
2. Place a sample of plastic on two supports as shown.
3. Add masses to the centre of the sample. Find the mass needed to get the centre to bend by one cm. Repeat with a number of different plastics, e.g. polythene, polystyrene, Teflon and perspex.

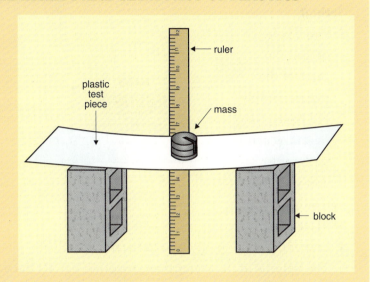

(B) Hardness

1. See how easily you can scratch a sample of plastic with a sharp nail. Compare this with a number of other plastic samples.
2. Now try to cut the sample with a scissors. Repeat with a number of different plastics, e.g. polythene, polystyrene, Teflon and perspex.

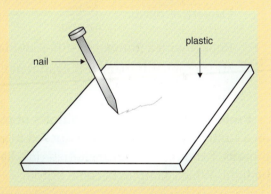

Safety and sports spectacles
Plastic lenses are lighter than glass and don't shatter on impact. Plastic is flexible, but scratches more easily than glass.

Breakthrough Science

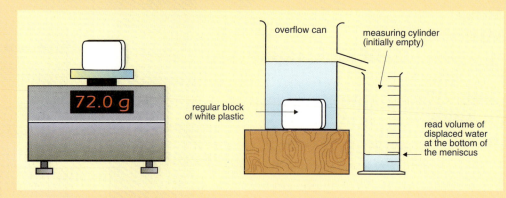

EXPERIMENT 74.2
AIM: TO FIND THE DENSITY OF A PLASTIC

Method
1. Use a balance to find the mass of the plastic sample.
2. Fill the overflow can with water and wait until it stops dripping.
3. Place a graduated cylinder under the spout.
4. Lower the solid gently into the water. If it floats, push it under the water with a set of dividers.
5. The volume of the water that overflows equals the volume of the plastic.

$$\text{Density of plastic} = \frac{\text{Mass}}{\text{Volume}}$$

Repeat with a number of different plastics, e.g. polythene, polystyrene, Teflon, polypropylene and aeroboard (foamed polystyrene).

Project: Place a number of samples of plastic packaging outside on a sunny windowsill. Leave exposed to sunlight for several months. Examine to see which materials have been affected most by light. Which samples become brittle?

PLASTICS ARE INSULATORS OF HEAT AND ELECTRICITY

Plastic insulation is used in electrical equipment. Plugs, sockets, switches, and the insulating bodies of electric kettles, fan heaters, radios, hair-driers and drills are all made from plastics. Some of these plastics break down when heated over the years. Some light-bulb holders give off a "fishy" smell when this happens. Heat-resistant plastics should be used to replace these.

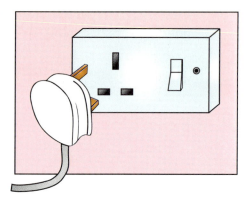

Plastics are also good heat insulators. The insulating properties of plastics are improved by foaming them—blowing gas into the plastic as it cools and making a foam. Aeroboard is foamed polystyrene, and is used in cavity walls to reduce heat loss from homes. Foamed polyurethane is also used as heat insulation.

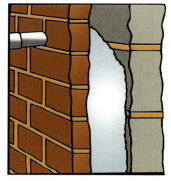

APPLIED SCIENCE · TYPES OF MATERIALS

EXPERIMENT 74.3
AIM: TO COMPARE THE INSULATING PROPERTIES OF PLASTICS

Method

1. Fill a metal can with hot water. Find the temperature of the water with a thermometer.
2. Place the can in a polystyrene container.
3. Record the temperature every minute.
   Draw a graph of the temperature of the water (on the y-axis) against the time (on the x-axis). Repeat the experiment by placing the can in a beaker and surrounding it with (*a*) polyurethane foam and (*b*) cotton wool. Draw a graph (using the same scales) in each case. Compare the graphs.
   Which material shows the slowest drop in temperature?

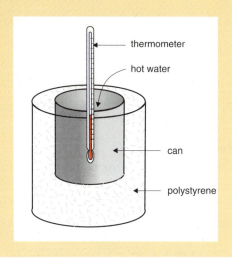

Plastics can be dangerous when they burn

Plastics are made from hydrocarbons. Hydrocarbons are fuels and are highly flammable. Fortunately, the long chains of molecules in polymers are usually much less flammable. Some plastics burn, giving off dense black smoke and toxic (poisonous) fumes. This makes fires involving these materials very dangerous. Some plastics such as aeroboard melt when they catch fire and drip hot burning plastic. Foamed plastics previously used in furniture caught fire rapidly and were a major fire hazard. Many deaths were caused by fires involving foamed plastics in furniture, wall coverings and ceilings. The flammability and fire risks (dense smoke and toxic fumes) of materials in buildings used by the public are evaluated by architects, engineers and planners. A material such as aeroboard is safe when used as cavity wall insulation but is a fire hazard when used as ceiling tiles. Safety standards are laid down that restrict or forbid the use of materials in certain circumstances or locations. Even low-flammability plastics may be restricted if they give off toxic fumes when heated. Plastics can be specially treated to make them less flammable (flame retarding).

The fabrics and foamed plastic cushioning used in household furniture must be specially treated to meet fire resistance standards. Labels on furniture give information on its flammability.

Warning: Do not attempt any practical work on flammability of plastics.

Breakthrough Science

Advantages of plastics	Disadvantages of plastics
Easily moulded and coloured	Soften and melt when heated
Do not rust or rot	Plastic packaging damages environment (non-biodegradable)
Low density	Some plastics become brittle when exposed to sunshine (ultraviolet light)
Good heat and electrical insulation	Some plastics are a fire hazard and may produce toxic fumes and dense smoke if burned

Recycling

Plastics are made from a non-renewable resource. Some plastics are recycled by using them to produce other materials. Transparent polyester bottles used for soft drinks are recycled to make polyester filler for sleeping bags and pillows.

TEXTILES

Textiles are used to make clothes, to upholster furniture, and to make curtains, bedclothes, rugs and carpets. The material or fabric used in the textile industry is made by weaving or knitting yarn. Yarn is made from short strands of fibres twisted together to make long threads.

Natural fibres are used to make yarn.

Wool is spun from the hairs of various types of sheep, goats, camels, alpacas, llamas and even furry angora rabbits. Cashmere is a wool made from Kashmir goathair.

Cotton is the fluffy white seed of the cotton plant. Cotton grows in places with hot climates, like Egypt and the southern states of the USA.

Linen is made from fibres of the flax plant. Irish linen is a high-fashion fabric produced mainly in Ulster. The oldest fabric known is a piece of linen about 7,000 years old found near Robenhausen in Switzerland.

Silk is made from silk threads woven by the caterpillar of the silk moth—the silkworm. Silk has been produced in China since 2600 BC.

Fibres of hemp, jute, and sisal are used to make rough cloth, sacking, matting and ropes.

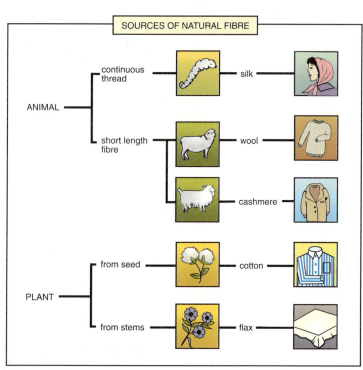

APPLIED SCIENCE · TYPES OF MATERIALS

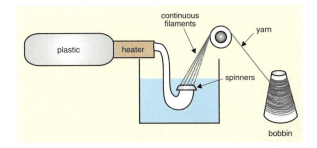

Synthetic fibres are used to make yarn.

All natural fabrics are made from plant or animal fibres. These take time to grow. Man-made or synthetic fibres do not have this disadvantage.

Nylon, polyester, lycra, terylene and acrylic are synthetic fibres. They are made from simple molecules produced from crude oil. These molecules are made into polymers (long chains of simple molecules). This plastic material is forced through tiny holes to produce long fibres. These fibres are woven into yarn.

Synthetic fibres are usually good heat insulators, do not absorb moisture, stretch easily and are easy to dye.

Rayon was the first man-made fibre. It was developed as an artificial silk by Count Hilaire De Chardonnet in 1884 from natural cellulose. Nylon was the first fully synthetic fibre. Different types of synthetic fibres can be manufactured by controlling the conditions. Nylon and polyester fibres can be made extremely fine for women's tights or thick for reinforcing car tyres.

Acrylic fibres are woven into a woolly yarn used for jumpers (Acrilan and Orlon). Kevlar is a synthetic fibre used to make bullet-proof jackets.

We classify fabrics by the fibres used in them.

Fabric	Origin of fibres
Wool	Natural animal
Silk	Natural animal
Cotton	Natural plant
Linen	Natural plant
Nylon	Synthetic
Polyester	Synthetic
Acrylic	Synthetic
Polyester/cotton	Mixture
Wool/nylon	Mixture

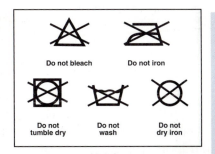

Project: Read the care label on as many different kinds of fabrics as you can find, in shirts, blouses, skirts, jeans, underwear, jumpers, jackets, etc. Sort them into natural and synthetic. Separate the natural fabrics into those of animal and plant origin.

MAKING FABRICS

Spinning yarn
The short fibres (natural or synthetic) are first straightened by combing. The fibres are then drawn into a continuous yarn by spinning. Single threads are often twisted with others to form thicker yarn. This yarn is then made into fabrics or cloth by weaving or knitting.

Weaving cloth
The hand loom is the oldest way of weaving cloth. Hand looms are still used to make Donegal tweed. A simple loom is a made from a wooden frame. Threads are wound lengthways on the frame parallel to each other. This is the warp. The weaver then passes thread over and under alternate warps. This is the weft. The rows of weft are pushed in tightly by a comb.

Dyeing, printing, and finishing
Some fabrics are made from dyed yarn. Other fabrics are coloured by dyeing or have patterns printed on them after weaving. Fabrics are then finished. Finishing includes pre-shrinking of cotton and wool, increasing flame resistance and making cloth water repellent.

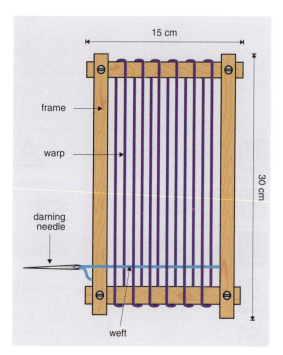

Project: Collect some raw fleece (wool). (Bits of fleece can be found on bushes or wire in sheep pastures.) Use an old comb or brush to untangle this and pull it into straight lengths. Draw out and twist the combed fibres into lengths of yarn. If you have enough yarn you can weave it into fabric using a simple home-made loom as shown in the diagram. (You can also do this with lengths of old wool.)

Wind the warp tightly onto the rectangular loom. Use a large darning needle to pass the weft thread over and under alternate weft threads. Push the weft firmly into place. Pass the weft back through the warp in the opposite way (under and over). Continue this until you have a length of cloth.

DIFFERENT FABRICS HAVE DIFFERENT PROPERTIES
The basic reason we wear clothes is to keep us warm and dry. In our climate clothes need to be good heat insulators. Outerwear (jackets and coats) also needs to be waterproof. You can find some properties of fabrics just by feeling the fabric. Compare the feel of a number of fabrics and record your observations. You can also test the dyes used for light-fastness. Divide a sample of brightly coloured cloth into two parts. Expose one part to bright sunshine for a few weeks. Compare the two parts.

Fabric	Feel	Light-fastness
Wool		
Cotton		
Nylon		
Acrylic		

EXPERIMENT 74.4
AIM: TO COMPARE THE INSULATING PROPERTIES OF FABRICS

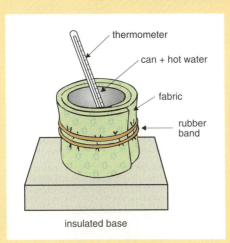

Method
1. Fill a metal can with hot water. Find the temperature of the water with a thermometer.
2. Wrap a fixed length of cotton fabric around the can. Hold the fabric in place with an elastic band.
3. Record the temperature every minute.
   Draw a graph of the temperature of the water (on the y-axis) against the time (on the x-axis). Repeat the experiment using wool, nylon, acrylic, etc. Draw a graph (using the same scales) in each case. Compare the graphs. Which material shows the slowest drop in temperature? Which is the best insulator?

EXPERIMENT 74.5
AIM: TO COMPARE THE WATER REPELLENCY OF FABRICS

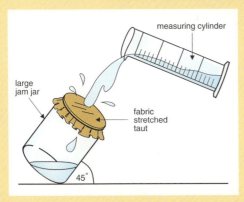

Method
Cut a number of samples of fabric to the same size.
1. Pull each sample tightly over the neck of a jam-jar and hold it in place with an elastic band.
2. Find the mass of the jar and fabric.
3. Put the jar at an angle of 45°. Pour 250 cm^3 of water onto the fabric slowly.
4. Allow the jar to drain for one minute, then reweigh it. Calculate the increase in mass of the jar and fabric. Repeat for each fabric sample. The sample with the least increase in mass has the greatest water repellence.

BREAKTHROUGH SCIENCE

ABSORBENCY

Your sweat glands produce water that evaporates from your body. This happens even on a cold day. This moisture is absorbed by your clothes. Underwear and stockings made from absorbent material are comfortable because of this. The fabric used to make towels must be particularly absorbent in order to dry you.

Fungi and bacteria grow on moist skin. Absorbent materials keep your skin dry and help prevent these infections.

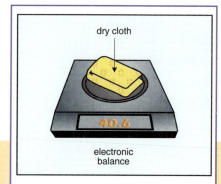

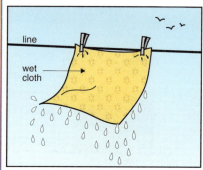

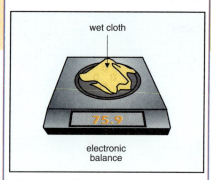

EXPERIMENT 74.6
AIM: TO COMPARE THE ABSORBENCY OF FABRICS

Method

Cut a number of samples of fabric to the same size.
1. Find the mass of a sample.
2. Put the sample into a dish of water and allow to soak.
3. Remove the sample and allow to drain on a line for one minute.
4. Reweigh the sample and calculate the increase in its mass. This is the mass of water absorbed by the sample.

Repeat for each fabric sample.

FLAMMABILITY

Some fabrics catch fire and burn rapidly. Children have been seriously burned by nightdresses catching fire. Materials that catch fire easily are a fire hazard. There are restrictions on the use of these materials in certain types of clothes and in furniture.

Brushed synthetic fabrics are good heat insulators. They are used to make children's toys and night-clothes. Scientists have found ways of reducing the flammability of these fabrics by treating them with special chemicals. Flame-proofing a fabric reduces the risk of the fabric burning. Special flame-proofed fabrics are used in the clothes worn by racing-car drivers and by firemen.

Project: Examine the labels on a variety of clothes. Include as many examples of children's night-wear as you can find. Which ones have any mention of flammability or flame resistance?

 Warning: Do not attempt any practical work on flammability of fabrics.

WILL IT WEAR WELL?

Have you ever worn holes in the elbows of your jumper? If you cycle a bike the seat of your pants wears thin too! This happens because cloth wears when it is constantly rubbed. Some fabrics wear out faster than others.

Denim is a heavy, durable cloth made from cotton. It is used to make jeans, skirts, jackets and children's clothing. Denim is usually pre-shrunk and is sometimes rubbed to soften its tough texture and make it appear worn. Softer denim is made from polyester and cotton. Denim is an example of a hard-wearing fabric. Clothes worn for heavy work need to be strong and to wear well.

EXPERIMENT 74.7

AIM: TO COMPARE THE DURABILITY (RESISTANCE TO WEAR) OF FABRICS

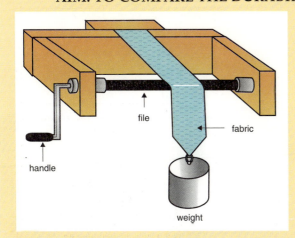

Method

Cut a number of samples of fabric to the same size.

1. Place the fabric sample over the abrasive roller (roller with a very rough surface) and connect the weight to the end of the fabric.
2. Turn the handle on the roller. Count the number of complete turns.
3. Continue until a hole wears in the cloth. Repeat for each fabric sample.

RECYCLING

Fabrics are recycled by the sale of good-quality clothes, blankets and furnishings in second-hand shops. Worn-out clothes are used to make cleaning rags for industry.

METALS

About three-quarters of all elements are metals.

Common metals: iron, aluminium, copper and lead.
Precious metals: gold and silver.
Radioactive metals: uranium, plutonium.

Breakthrough Science

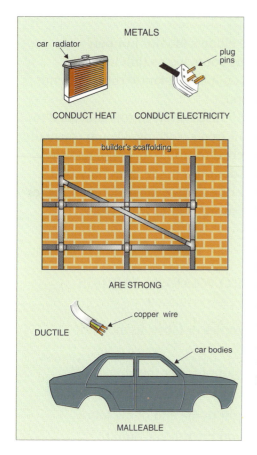

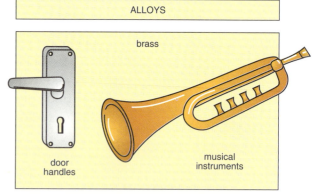

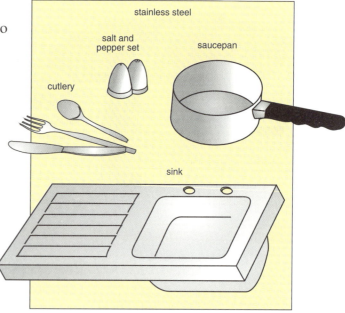

Metals are hard, strong, good conductors of heat and electricity, and resistant to shock. Metals are used in buildings to strengthen them. Metals that can be hammered into sheets are malleable. Those that can be drawn into wire are ductile. Metals can be worked to give a polished surface or metallic lustre.

Alloys

Alloys are mixtures of metals. Different alloys can be made with different properties. Alloys can be very hard, tough and corrosion-resistant. Common alloys include brass (copper and zinc), bronze (copper and tin) and stainless steel (iron, chromium, nickel and carbon).

Aluminium is alloyed with silicon and copper. The copper makes the alloy stronger and the silicon makes the molten metal flow easily into a mould.

Where do we get metals from?

Gold is so unreactive that it is found free in nature as nuggets of gold. Silver and copper are sometimes found free in nature. These metals are very unreactive and do not form compounds easily. Because gold, silver and copper are unreactive, they are often used to make coins and jewellry.

We know from chemistry experiments that the alkali metals potassium and sodium are very reactive—they react violently with water. Calcium and magnesium are also very reactive. You would hardly expect to find lumps of these metals in the ground.

Reactive metals such as these are found only as metal compounds or ores. In fact most metals are reactive and are found in nature as compounds. Rocks and minerals containing metal compounds are called metal ores.

Reactive metals are never found free in nature.
Unreactive metals are often found free in nature.

EXTRACTING A METAL FROM ITS ORE

The commercial process of extracting pure metals from the rocks that contain the ore is long and laborious. Tons of rock are crushed and treated chemically to concentrate the ores that contain the metal. The pure metal is then extracted from the concentrated ore. Some ores contain only 2% of the metal. A tonne of ore produces only 20 kg of metal!

Copper was mined at Avoca (Wicklow), silver, lead and zinc at Silvermines (Tipperary), and lead and zinc at Tynagh (Galway). The lead and zinc mine at Navan (Meath) is the only large-scale mine in operation in Ireland.

Copper was used by humans as long ago as 5,000 BC. We see from the following experiment why copper was easily produced using ordinary fires.

EXPERIMENT 74.8
AIM: TO MAKE COPPER FROM COPPER ORE AND CHARCOAL

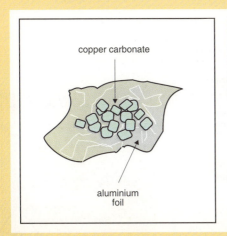
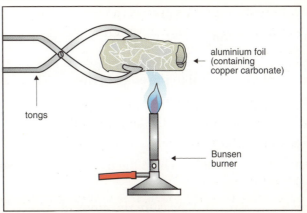

Copper carbonate ($CuCO_3$) is used in place of the natural ore malachite ($CuCO_3$–$Cu(OH)_2$).

1. Mix 2 g of copper carbonate with 2 g of charcoal in a piece of aluminium foil.
2. Wrap the mixture in the foil. Heat it strongly from above and below for 5 minutes.
3. Pour the contents of the foil into cold water.
   Filter the water. The residue contains brown flakes of copper.

Copper carbonate reacts with charcoal when heated to produce copper metal and carbon dioxide.

$$2CuCO_3 + C \rightarrow 2Cu + 3CO_2$$
copper carbonate   charcoal (carbon)   copper   carbon dioxide

A much hotter fire is needed to smelt iron. The Hittites first produced iron about 1,500 BC. Iron tools and weapons are much stronger than bronze and quickly replaced them. The Iron Age gradually replaced the Bronze Age.

DO ALL METALS HAVE THE SAME PROPERTIES?

Plumbers bend copper pipes easily by hand, but they need to use special tools to bend steel pipes. Solder used to make circuit boards for computers melts easily, but the tungsten used in light bulbs can be heated to very high temperatures without melting. You need to test samples of different metals to compare their properties.

How they feel

Most metals bend or flex when a force is applied to them. They usually return to their original position once the load is removed - they are elastic. Flexibility means the extent to which they bend under a load or force.

EXPERIMENT 74.9

AIM: TO COMPARE THE FLEXIBILITY OF METALS

Use equal-sized rectangular strips of copper, iron, aluminium, lead and zinc.

Method

1. Clamp a piece of metal to the edge of the bench. Fix a ruler as shown.
2. Attach a pan to the end of the test piece. Add masses to the pan.
3. Measure the amount by which the metal bends.
4. Increase the force and measure the amount of bending again.
   Repeat with the other metals. Which metal is the most flexible? Are all the metals elastic?

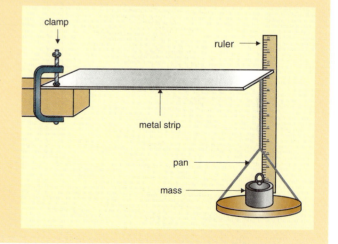

Hardness

You might think that all metals are hard. What about lead? Lead is easily shaped. You will have to test metals for hardness. Hardness tells you whether a metal is hard or soft and how easily you can scratch it.

An easy way to compare the hardness of metals is to see if a metal will scratch a piece of another metal.

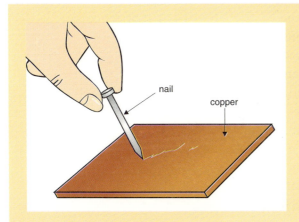

EXPERIMENT 74.10
AIM: TO COMPARE THE HARDNESS OF METALS

Use copper, iron, aluminium, lead and zinc.

Method
1. Try to scratch a piece of copper with an iron nail.
2. Repeat this test on the other samples.
3. Now try to scratch each sample with a piece of copper.
4. Repeat on the other samples.

EXPERIMENT 74.11
AIM: TO FIND THE DENSITY OF A METAL

Method
1. Use a balance to find the mass of the metal.
2. Fill the overflow can with water and wait until it stops dripping.
3. Place a graduated cylinder under the spout.
4. Lower the metal gently into the water.
5. The volume of the water that overflows equals the volume of the metal.

$$\text{Density of metal} = \frac{\text{Mass}}{\text{Volume}}$$

Find the density of a number of different metals, e.g. copper, aluminium, iron.

EXPERIMENTS TO SHOW REACTIVITY OF METALS

The reactivity of metals is found by comparing their reactions with water and dilute acids. Some metals, such as sodium and potassium, react violently with water and acids, while iron, copper and silver show little reaction. We place the metal elements in order on the activity series of the metals:

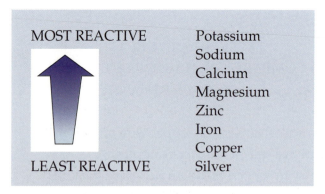

As well as using the reactions of metals with water and acids, you could test a sample of one metal in the salt solution of another.

A more active metal displaces the ions of a less active metal from a solution of its salt. You will see this if you do the following simple experiment.

EXPERIMENT 74.12
AIM: TO SHOW THAT A MORE ACTIVE METAL DISPLACES A LESS ACTIVE METAL FROM A SOLUTION OF ITS SALT

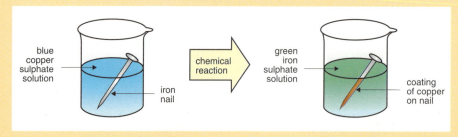

1. Clean the surface of an iron nail with fine sandpaper and wash with methylated spirits and distilled water.
2. Make up a solution of copper (II) sulphate.
3. Leave the iron nail in a test-tube of copper (II) sulphate solution.

   Result: A copper coating forms on the nail.

   Conclusion: Iron (active) displaces the ions of copper (less active) from the solution of copper (II) sulphate. Iron enters the solution as iron ions and copper ions leave the solution as copper metal.

   copper sulphate + iron ➔ iron sulphate + copper

METALS ARE GOOD CONDUCTORS OF HEAT AND ELECTRICITY

EXPERIMENT 74.13
AIM: TO COMPARE THE HEAT CONDUCTION OF COPPER, IRON AND ALUMINIUM

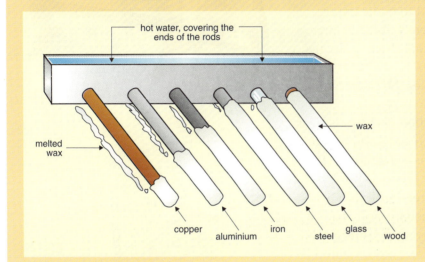

Method
Pour boiling water into the container: the wax will melt on the best conductor first.
Result: The wax will melt on the copper first, then on the aluminium. The wax will melt on the iron last.
Conclusion: Copper is a better conductor than aluminium, which is better than iron.

EXPERIMENT 74.14
AIM: TO COMPARE THE ELECTRICAL CONDUCTION OF COPPER, IRON AND ALUMINIUM

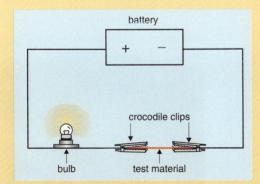

Method
1. Set up the test circuit as shown.
2. Connect the crocodile clips to the material being tested.

Result: The brightness of the bulb gives an indication of how good a conductor each metal is.
Conclusion: Copper and aluminium are better conductors than iron.

RECYCLING

Metals are a non-renewable resource. It takes a lot of water and energy to produce metals from their ores. Recycling aluminium saves 95% of the energy needed to produce a new can. Precious metals are recycled even in small quantities. Iron, copper, lead and aluminium are recycled. 200,000 tons of iron is recycled at the Irish Steel plant in Cork.

TIMBER

Trees grow naturally in a variety of climates. Mankind cultivates many types of tree for particular purposes. Trees are grown to produce wood for building and furniture and as a raw material for manufactured boards such as plywood and blockboard, and to make paper. Wood has the advantage that it is a renewable resource. Replanting programmes can ensure that there is always an adequate supply of wood products.

The type of trees planted determines the characteristics of the wood you get.

Hardwoods are cut from broad-leafed deciduous trees. The wood is generally hard, e.g. beech, oak, ash, teak and mahogany. Hardwoods are used to make furniture, doors, window frames and sports equipment such as hurleys.

Softwoods are cut from conifers (trees with thin or needle-shaped leaves). The wood is usually soft, e.g. lodge-pole pine, Sitka spruce, Norway spruce and Scots pine. Softwoods are used in building, e.g. flooring and roof joists, flooring boards, doors and window frames, and in fencing. Softwoods need protection to prevent fungi and pests from rotting the wood.

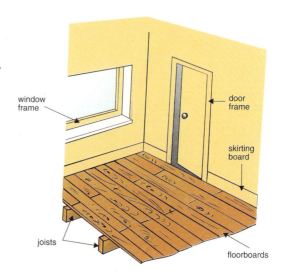

FORESTRY IN IRELAND

Forestry in Ireland has recovered rapidly in the past 50 years. However, the area under forest is only one-fifth of that in most other European countries.

Date	% area under forest
1100	50
1600	12·5
1800	2
1900	0·5
1990	5

Forests are a valuable resource. Home-grown wood replaces costly imported woods. Employment is provided in the forests themselves, in sawmills and furniture making. Forests are a resource for education and research, a habitat for many plants and animals, and an attractive amenity for tourism. Forest parks provide leisure and recreational facilities.

State and EU grants have encouraged the private planting of forests. Almost 98% of all forest trees are conifers. Plantings of broad-leaf trees such as ash, sycamore, birch and alder are encouraged. In our climate these trees grow almost as fast as conifers. The woods produced by broad-leafed trees are valuable for furniture and wood crafts.

HOW STRONG IS WOOD?

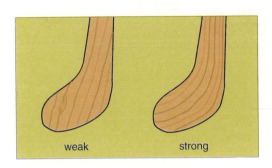

Did you ever try to chop up a block of wood into smaller pieces? You probably know that if you try to cut along the direction of the grain (parallel to the grain) it is much easier than cutting across the grain. This tells you that a tree has different strengths in different directions. Trees can withstand forces at right angles to their length and bend in the wind in this way. However, vertical force such as a bolt of lightning along the direction of the grain can split a tree.

EXPERIMENT 74.15
AIM: TO COMPARE THE STRENGTHS OF THIN STRIPS OF WOOD ALONG DIFFERENT DIRECTIONS.

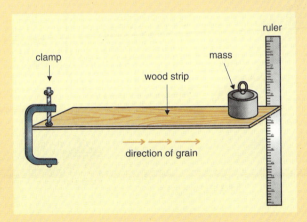
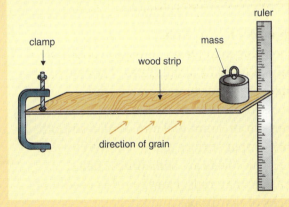

Method
1. Set up the apparatus as shown.
2. Add weights to the end of the strip. Measure the amount by which the strip bends.
   Do this experiment with a piece of wood set (*a*) parallel to the grain and (*b*) perpendicular to the grain.

Breakthrough Science

EXPERIMENT 74.16
AIM: TO COMPARE THE BENDING STRENGTHS OF WOODS

Use equal-sized rectangular strips of different softwoods, ash, oak and other woods as available.

Method

1. Support a thin board by two blocks at either end.
2. Add weights to the centre of the board. Continue to add weights until the board breaks. Record the total force required to break the board.

Repeat for different types of wood using boards of similar dimensions.

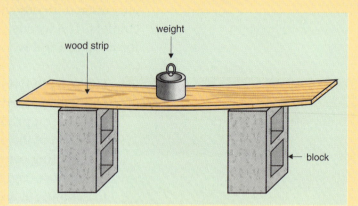

MAN-MADE TIMBER PRODUCTS

Paper

Wood is turned into pulp before being made into paper. Large quantities of softwoods are used to make paper.

Hardboard

Hardboard is made from chips of wood that are pulped to make small fibres. The fibres are mixed with hardening agents and compressed into sheets. Sheets vary in thickness from 3 mm to 9 mm, with a smooth polished upper surface and a textured lower surface. The sheets are sometimes coated with plastic materials. Hardboard is used for the backs of wardrobes and kitchen presses, panelling in caravans, etc.

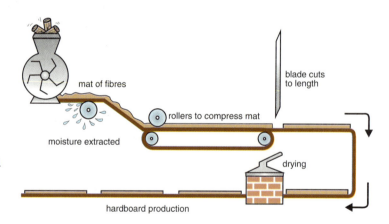

Hardboard is usually made from small branches and waste material from other wood processing.

A modern type of hardboard, MDF (medium-density fibreboard), is manufactured in a similar way with a synthetic resin added to bind the fibres together. MDF is used in furniture as it can be machined and finished to a high standard.

APPLIED SCIENCE · TYPES OF MATERIALS

single-layer chipboard

triple-layer chipboard

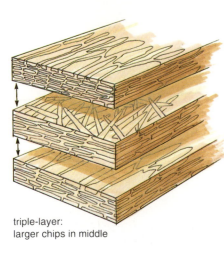
triple-layer: larger chips in middle

Chipboard

Chipboard is made from similar raw materials to hardboard and MDF—wood chips and sawdust. The chips are graded and set in glue. This is compressed into sheets. Cheaper grades have similar-sized chips right through the board. Better quality chipboard has finer chips at the sides with coarse chips at the centre. Plain unfaced chipboard is a low cost board used for temporary partitions. Chipboard faced with wood or plastic veneers is used in furniture manufacture.

Plywood

Plywood is made by gluing a number of thin sheets (laminae) of wood together. A variety of woods are used for the laminae. The layers are glued face to face with the grains of alternate pieces at right angles to each other. Interior grade, exterior grade and marine grade plywoods are made using different glues. Plywood is made in a wide range of thicknesses. Plywoods are veneered with hardwoods and plastics.

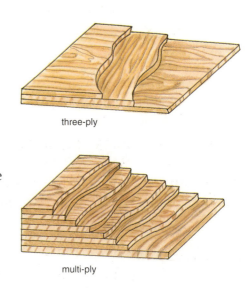
three-ply

multi-ply

Plywood is used to make furniture (veneered plywood), flooring, crates and packaging, and in boat-building.

hardwood veneered panels · boat · chair · head of table tennis bat · bathroom cabinet · Tea chest · veneered TV cabinet

509

> **EXPERIMENT 74.17**
> **AIM: TO INVESTIGATE THE STRUCTURE OF PLYWOOD**
>
>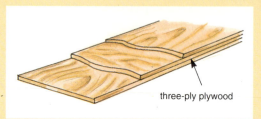
> three-ply plywood
>
> **Method**
> 1. Soak a small sample of internal grade plywood in water for a few weeks.
> 2. Pull the sheets apart.
>   Examine to determine (*a*) the direction of the grain in alternate sheets, (*b*) whether or not the sheets are of the same wood.

Blockboard

Blockboard is made from strips of wood 12–25 mm wide. These strips are glued together and covered on both faces with a thick veneer. Blockboard is used to make furniture.

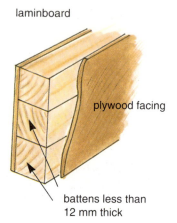

laminboard / plywood facing / battens less than 12 mm thick

THE PROPERTIES OF WOODS

Hardwoods and softwoods have different properties. Even the properties of hardwoods vary a lot. Ebony is a relatively dense dark wood, while ash is a light-coloured wood with a lower density. Each type of wood has uses that depend on its properties. Ebony is used for carving and to make keys for pianos; ash is used to make flexible hurleys and furniture. Woodworkers examine woods for hardness, strength, colour, grain, water absorbency and density.

WOOD MUST BE SEASONED

Wood contains water when it is felled and cut. Much of this water is dried out by allowing the wood to season. Wood is seasoned naturally by allowing it to dry in the open air. This can take a long time. Wood is also artificially seasoned by kiln drying. Most wood products need wood that is seasoned. The boards are then sawn and planed to the shapes and sizes required.

Even seasoned wood absorbs water if it is exposed to rain —wood is hygroscopic. Doors and window frames absorb moisture and expand when wet. This causes them to stick in rainy weather. When the weather is dry the doors and windows dry out and shrink to their previous size. Many finishes for wood doors and windows are waterproofing agents that prevent this from happening.

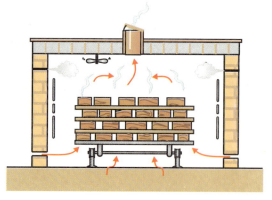

Kiln drying. Ends are painted to prevent cracking.

Advantages of seasoned wood

Seasoned wood:
- is lighter
- is resistant to fungal attack
- is stronger
- will not warp or shrink
- is easier to saw, plane and sand
- is easier to finish (glue, paint, stain or varnish).

EXPERIMENT 74.18
AIM: TO COMPARE THE DENSITIES OF OVEN-DRIED WOOD AND FRESH (GREEN) WOOD

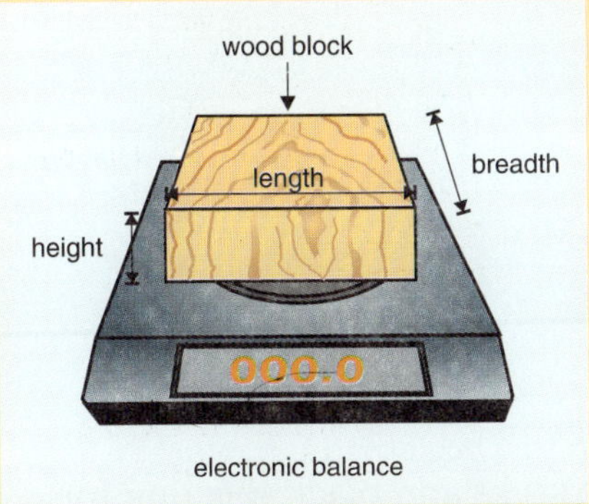

Method

1. Use a balance to find the mass of a block of oven-dried wood.
2. Measure the length, breadth and height of the block with a vernier callipers.
3. Calculate the volume of the block
   Volume = length × breadth × height

Result: Density of wood = $\dfrac{\text{Mass}}{\text{Volume}}$

Repeat with a block of green wood.

Conclusion: The density of wood varies with the moisture content.

TO SHOW THAT WOOD EXPANDS WHEN IT ABSORBS WATER

Soak the oven-dried block in a bucket of water overnight. Measure the length, breadth and height of the block again.

Conclusion: The block expands when it soaks up water.

BREAKTHROUGH SCIENCE

EXPERIMENT 74.19
AIM: TO MEASURE THE MOISTURE CONTENT OF WOOD

Method

1. Use a balance to find the mass of a block of oven-dried wood.
2. Place in an oven at 105°C for 24 hours.
3. Find the mass again. The mass of water = loss in mass.

$$\text{Moisture content} = \frac{\text{Mass of water}}{\text{Dry mass of wood}} \times 100\%$$

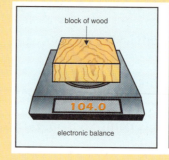

RECYCLING

Good-quality furniture lasts for centuries. Other furniture is repaired and resold. Paper, made from trees, is also recycled. Replanting of forest areas after timber is felled is another form of recycling.

SUMMARY

Plastics
- Plastics are giant molecules.
- Plastics are made of long chains of similar molecules joined together.
- Polythene is used to make kitchenware, drainage pipes, refuse sacks and black sheeting for silage.
- Foamed polystyrene is used as heat insulation in buildings, in heat insulating coffee cups and in flotation devices in boats.
- Plastics become brittle when exposed to sunlight.
- Plastics are good insulators of heat and electricity.
- Some plastics burn, giving off dense black smoke and toxic (poisonous) fumes.
- Plastics can be specially treated to make them less flammable (flame retarding).
- Plastics have low density, are easily moulded and coloured and do not rust or rot.

Textiles
- Yarn is made from short strands of fibres twisted together to make long threads.
- Wool, cotton, silk and linen are natural fibres.
- Nylon, polyester, lycra, terylene and acrylic are synthetic fibres.
- Synthetic fibres are usually good heat insulators, do not absorb moisture, stretch easily and are easy to dye.
- Fabrics are woven on a loom.

- Clothes that keep out moisture are water-repellent.
- Absorbent materials keep your skin dry by absorbing moisture from your body.
- Flammable fabrics catch fire and burn rapidly.
- Flame-proofing a fabric reduces the risk of the fabric burning.
- Durability is the resistance to wear of fabrics.

Metals
- Metals are hard, strong, good conductors of heat and electricity and resistant to shock.
- Metals that can be hammered into sheets are malleable.
- Metals that can be drawn into wire are ductile.
- Metals can be worked to give a polished surface or metallic lustre.
- Alloys are mixtures of metals.
- Rocks and minerals containing metal compounds are called metal ores.
- Copper carbonate reacts with charcoal when heated to produce copper metal and carbon dioxide.
- Most metals bend or flex when a force is applied to them.
- An easy way to compare the hardness of metals is to see if a metal will scratch a piece of another metal.
- The activity series of the metals gives the relative reactivity of the metals.
- A more active metal displaces the ions of a less active metal from a solution of its salt.

Timber
- Trees are grown to produce wood for building and furniture and as a raw material for manufactured boards such as plywood and blockboard and to make paper.
- Hardwoods are cut from broad-leafed deciduous trees and used to make furniture, doors, window frames and sports equipment such as hurleys.
- Softwoods are cut from conifers and used in building e.g. flooring and roof joists, flooring boards, doors and window frames, and in fencing.
- Forests are a valuable resource.
- Home-grown wood replaces costly imported woods.
- Forestry provides employment in forests, sawmills and furniture making.
- Forest parks provide leisure and recreational facilities.
- The strength of wood depends on the direction of the grain.
- The bending strength of wood varies with the type of wood.
- Chipboard is made from wood chips and sawdust set in glue and compressed into sheets.
- Plywood is made by gluing a number of thin sheets (laminae) of wood together.
- Blockboard is made from strips of wood glued together and faced with a thick veneer.
- Wood is seasoned naturally by allowing it to dry in the open air.
- Wood is seasoned artificially by kiln drying.
- The density of wood varies with the moisture content.
- Wood expands when it soaks up water.

QUESTIONS

Section A
Plastics
1. What is a polymer? _____
2. Three advantages of plastics as materials are: _____.
3. Polythene is used to make _____, _____ and _____.
4. Name a plastic used as a heat insulator. _____
5. Plastics are made of long chains of similar molecules joined together. What are these small molecules called? _____
6. Plastics become brittle when exposed to _____.
7. Flame retarding treatment makes a plastic _____.
8. Polystyrene is used to make _____, _____ and _____.
9. Name a plastic used as an electricity insulator. _____
10. When plastics burn they give off _____.

Textiles
1. A loom is used to _____.
2. Short strands of fibres twisted together to make long threads are called _____.
3. Absorbent materials _____.
4. Fabrics that catch fire and burn rapidly are _____.
5. Wool and silk are natural fibres got from _____.
6. Three properties of synthetic fibres are: _____, _____ and _____.
7. Clothes that keep out moisture are _____.
8. Flame-proofing is used to _____.
9. Cotton and linen are natural fibres got from _____.
10. Durability is _____.

Metals
1. Give three properties of metals: _____, _____ and _____.
2. Metal ores are _____.
3. Malleable means a metal can _____.
4. Which is the more reactive, iron or copper? _____
5. A mixture of metals is called an _____.
6. Copper is produced by heating copper ore with _____.
7. A metallic lustre means _____.
8. You can show that one metal is harder than another metal by _____.
9. Metals that can be drawn into wire are _____.
10. Zinc displaces iron from a solution of iron sulphate. This shows that _____ is more reactive than _____.

Timber
1. Hardwoods are used to make _____.
2. Wood is seasoned by _____.
3. Thin sheets of wood are glued together to make _____.
4. Softwoods are cut from _____.
5. Forestry is important because _____.
6. Hurleys are made from ash because _____.
7. Softwoods are used to make _____.
8. The strength of wood depends on the direction of the _____.
9. Chipboard is made by _____.
10. When wood soaks up water it _____.

Section B
1. Describe with the aid of a diagram **one** of the following experiments:
    (a) to compare the malleability (flexibility) of different metals
    (b) to compare the resistance to wear of different fabrics
    (c) to compare the hardness of different plastics
    (d) to compare the bending strengths of different woods.
2. Describe **one** of the following experiments:
    (a) to compare the densities of different plastics
    (b) to show the thermal conductivity of different metals
    (c) to show the effect of moisture on different woods
    (d) to show the flammability of different textiles.
3. Describe with the aid of a diagram **one** of the following experiments:
    (a) to compare the densities of different metals
    (b) to compare the insulating properties of different textiles
    (c) to compare the methods of cutting of different plastics
    (d) to show the effect of grain direction on the bending strengths of different woods.
4. Describe with the aid of a diagram **one** of the following experiments:
    (a) to compare the hardness of different metals
    (b) to compare the flammability of different plastics
    (c) to show the hardness of different woods.
5. Describe the production of a named textile, **or** describe the production of a named man-made timber board, **or** describe **one** of the following experiments:
    (a) to compare the reactivities of different metals
    (b) to compare the flexibility of different plastics.

6. Answer one of A, B, C, D.
   (A) Name two plastics. Explain how you would compare the heat-insulating properties of two different plastics.
   (B) What is meant by the term 'textile'? Explain how you would examine the resistance to wear of different fabrics.
   (C) Name two metals that are mined in Ireland. Describe an experiment to compare the densities of two metals.
   (D) State the difference between hardwoods and softwoods. Name (i) a softwood, (ii) a hardwood. State a common use of each wood you have named. State a property of each wood that makes it suitable for the use you have stated.
7. Describe how you would do **one** of the following:
   (a) protect plastics from the effects of weathering
   (b) protect fabrics from deterioration
   (c) protect metals from corrosion
   (d) protect woods from the effects of weathering.
8. Describe with the aid of a diagram **one** of the following experiments:
   (a) to compare the strength (ductility) of different metals
   (b) to compare the absorbency of different fabrics
   (c) to compare the hardness of different plastics
   (d) to compare the densities of green wood and oven-dried wood.
9. Write a short account of one of the following:
   (a) the production of a plastic from crude oil
   (b) the production of a textile from raw material
   (c) the extraction of a metal from its ore
   (d) the manufacture of hardboard, chipboard, plywood or blockboard.

INDEX

Absolute zero 389
Absorption 284
Acceleration 30
Acid 226
Acid rain 177
Activity series 252, 504
Adaptation 347
AIDS 368
Air 164
Air sac 289
Alcohol fermentation 412
Alcohol thermometer 49
Alkali metal 197
Alloy 245, 500
Alternating current 87
Alveoli 289
Amino acid 284
Ammeter 72
Ampere 72
Ampere, Andre 72
Amplitude 103
Amylase 284
AND gate 417
Anemometer 391
Aneroid barometer 44, 387
Animal cells 276
Annual 459
Anode 262
Antagonistic muscles 308
Anther 338
Antibiotics 369, 411
Antibodies 296
Archimedes 20
Area 2
Arteries 297
Asexual reproduction 341, 452
Assimilation 284
Asteroid 375
Atmosphere 164, 386
Atmospheric pressure 42
Atomic number 192
Atoms 139

Bacteria 368
Balanced diet 280, 394
Balanced equations 222
Barometer 43, 390
Bases 226
Basic oxides 172
Battery 258
Bedding plants 463
Bell, Alexander Graham 104
Bent grass 466
Biceps 308
Bimetal strip 47
Biodegradable 485
Biomass 15
Biotechnology 414
Birth 315
Black hole 377
Bladder 303
Blind spot 311
Block board 510
Blood 295
Bohr, Neils 209
Boiling 6
Boiling and pressure 65
Boiling point 124
Bonds 210
Bone 306
Boyle's law 388
Brain 310
Brass 245
Breathing 289
Brewing 412
Bronchiole 289
Bronchus 289
Brownian motion 127
Bunsen burner 120
Bunsen, Robert 132
Burning 178

Calcium 200
Calorie 281
Cambium 454

Capillaries 297
Capillarity 186
Carbohydrates 280, 394
Carbon dioxide 174
Care of cut flowers 467
Carnivore 345
Carpel 337
Cartilage 307
Catalyst 169
Cathode 262
Cell electric 258
Cell living 275
Celsius scale 49
Centre of gravity 34
Characteristic of living things 270
Charge electric 98
Charles' law 389
Cheese 409
Chemical change 137
Chemistry 119
Chipboard 509
Chlorophyll 276
Chloroplast 276
Chromatography 161
Chromosomes 319
Circuit breakers 86
Circuits electric 70
Circulatory system 295
Clinical thermometer 50
Clouds 383, 385
Coal 175
Cobalt chloride paper 166
Collector, transistor 428
Colours 115
Combustion 13
Compass 91
Competition 347
Complementary colours 115
Compost 446
Compounds 141
Concentrated solution 148

Condensation 6
Condenser 160
Conduction, electric 70, 505
Conduction, heat 53, 505
Conservation 350
Conservation of Matter 137
Consumer 345
Contraception 316
Convection 55
Converging lens 114
Copper oxide 252
Corrosion 247
Covalent bond 213
Crookes' Radiometer 10
Crookes, William 10
Crystallisation 150
Curie, Marie 16
Current 72
Cuttings plant 452

Darwin, Charles 320
Days 378
Decanting 159
Decomposers 346
Deficiency diseases 280
Deforestation 351
Density 19
Desertification 351
Diaphragm 289
Diet, balanced 280, 398
Digestion 283
Dilute solution 148
Diode 423
Direct current 87
Dispersion 114
Distance-time graph 31
Distillation 160
Domestic wiring 85
Dormancy 450
Dry cell 260
Ductile 245
Dynamo 437

Earthing 86
Echo 106
Eclipse 111, 378, 379
Ecology 344
Edison, Thomas 259
Egestion 284
Egg 314
Einstein, Albert 374
Electric bell 436
Electric current 72
Electric motor 437
Electrodes 259
Electrolysis 261
Electrolyte 259
Electromagnet 93, 436
Electromagnetic spectrum 115
Electron 98, 191
Electron shells 192
Electroplating 262
Element 140
Embryo 315
Endocrine system 312
Endothermic reaction 223
Energy 7
Energy conversion 10, 432
Energy sources 13
Environment 344
Enzymes 284
Equilibrium 35
Evaporation 6
Excretion 303
Exothermic reaction 223
Expansion 46
Eye 311

Fabric 496
Fallopian tube 315
Family planning 316
Faraday, Michael 438
Fats 280, 396
Fermentation 413
Fertilisation, animal 315
Fertilisation, plant 339
Fescue grass 466

Fibre 280
Filtration 156
Fire 178
Fire extinguisher 179
Fission 16
Flammability of fabrics 498
Flotation 21
Flower 337
Fluoridation 188
Fog 384
Food 280
Food additives 406
Food chain 345
Food constituents 280
Food preservation 402
Food processing 409
Food tests 395, 396
Force 24
Forestry 506
Fossil fuels 13, 175
Fractional distillation 161
Franklin, Benjamin 100
Freezing food 403
Frequency 103
Friction 27
Frost 384
Fruit 339
Fuel 13, 175
Fulcrum 36
Fungi 369
Fuse 84

Galaxy 111, 374
Galilei, Galileo 376
Galvanising 249, 484
Gametes 314
Gamma rays 115
Genes 319
Genetics 319
Geotropism 336
Germination 340, 449
Glands 312
Glucose 324
Grafting 454

Index

Grasses 465
Gravity 25
Greenhouse effect 353, 386
Growing seeds 459
Growth 271

Habitat 344
Haemoglobin 295
Halogens 202
Hard water 238
Hardboard 508
Hardwood, cuttings 453
Harvey, William 295
Heart 297
Heat 46
Hedgerow 359
Herbivore 345
Heredity 319
Hertz 103
Hinge joint 307
Hormones 312, 411
Horticulture 441
Human body 277
Humidity 382
Humus 362
Hydrocarbons 13, 176
Hydroelectricity 14
Hydrogen 198
Hydroponics 448
Hygrometer 391

Immiscible 159
Implantation 315
Incisors 285
Indicators 226
Indoor plants 464
Infection 369
Infra red light 115
Ingestion 284
Inherited characteristic 320
Insulation, heat 59, 492
Insulators, electrical 70, 492
Insulin 312
Integrated pest control 474

Interdependence 347
Intestines 283
Iodine 202
Ion 210
Ion exchange 242
Ionic bond 210
Iris, eye 311

Jenner, Edward 370
Joints 307
Joule 7
Joule, James 64

Kelvin scale
Kidneys 303
Kilowatt hour 79
Kinetic energy 9
Koch, Robert 369

Large intestine 283
Latent heat 63
Lavoisier, Antoine 171
Leaf 322
Leeuwenhoek, Anton 275
Lenses 114
Lever 36
Life cycle of aphid 472
Life cycle of butterfly 473
Life cycle of star 377
Ligament 307
Light 110
Light-dependent resistor (LDR) 425
Light-emitting diode (LED) 424
Light year 374
Lightning 101
Limewater 175
Limiting factors to growth 461
Line transect 357
Linnaeus, Carl 273
Lister, Joseph 371
Litmus 130, 226
Liver 283
Lubricant 27

Luminous 110
Lungs 289

Magnet 91
Magnetic effect of current 83
Magnetic field 92
Malleable 245
Mariculture 272
Mass 19
Mass number 192
Materials, properties 477
Matter 5, 121
Measurement 1
Melting 6, 124
Melting and pressure 66
Mendel, Gregor 320
Mendeleev, Dmitri 205
Menopause 316
Menstrual cycle 316
Metals 245, 499
Metals, reactivity 251
Micro-organism 365, 405
Microscope 275
Microwaves 115
Milky way 111
Mineral nutrition 327, 397
Minerals 280
Mixtures 147
Molecules 140
Moments 36
Momentum 32
Monomer 490
Moon 379
Motor nerves 311
Mouth 283
Mulches 470
Muscles 308

Natural fibres 494
Nerves 310
Neutralisation 230
Neutron 191
Newton 24
Newton's third law 26

Newton, Isaac 112
Nitrogen 164
Noble gases 209
Non-renewable 14
Nuclear energy 8, 16, 433
Nucleus 139, 191
Nutrition, animal 284, 394
Nutrition, plant 327, 442

Octet rule 209
Oesophagus 283
Ohm's law 76
Ohm, Georg 77
Omnivore 346
OR gate 417
Orbits 375
Ore 251, 501
Organs 277
Ovary, animal 314
Ovary, plant 338
Ovulation 314
Oxidation 265
Oxygen 168

P.V.C. 490
Pancreas 283
Parallel circuits 73, 417
Pascal 43
Pascal, Blaise 44
Pasteur, Louis 371
Pasteurisation 403
Perennials 459
Periodic table 206
Permanent hardness, water 241
Pest control 473
Petal 337
pH scale 228
Phloem 333
Photosynthesis 324
Phototropism 335
Physical change 134
Pitfall trap 355
Placenta 315
Planets 375

Plant cells 275
Plaque 286
Plasma 295
Plastics 489
Platelets 295
Plywood 509
Pollination 338
Pollution 351
Polystyrene 490
Polythene 490
Potential difference 69
Potential energy 8, 433
Power 26, 78
Pregnancy 315
Pressure 40, 386
Priestley, Joseph 122
Primary colours 115
Producer 344
Protein 280, 395
Proton 191
Pulmonary artery 297
Pulmonary vein 297

Quadrat 356

Radiation, heat 57
Radioactive 16
Rain gauge 390
Rectifier 87, 423
Reduction 265
Reflection, light 112
Refraction, light 113
Renewable energy 14
Reproduction, animal 314
Reproduction, plant 337
Resistors 71, 419
Respiration 289
Retina 311
Roots 322
Rubella 371
Rust 247, 483
Rutherford, Ernest 192
Rye grass 466

Safety symbols 482
Salk, Jonas 370
Salts 234
Satellites 379
Saturated solution 149
Seashore 359
Seasons 378
Secondary colours 115
Seed dispersal 339
Semiconductors
Sense organs 310
Sensitivity 270
Sensory nerves 310
Sepal 337
Separation techniques 154
Series circuit 72, 417
Sex cells 314
Silage 413
Silicon diode 423
Skeleton 306
Skin 304
Small intestine 283
Smoking 290
Soft water 238
Softwood cuttings 452
Soil 362, 443
Solar energy 15, 375
Solar system 111
Solubility 149
Solutions 147
Sound 102
Special care labels 495
Spectrum of light 115
Speed 30
Sperm 314
Spring balance
Spring bedding plants 463
Spring tides 380
Stamen 337
Star 376
States of matter 5, 121
Static electricity 98
Stem 322
Stigma, flower 338

520

Index

Stomata 324
Stored energy 8, 433
Style, flower 338
Sublimation 64, 157
Sun 375
Supersaturated solution 151
Surface tension 186
Suspensions 151
Switch 70, 73
Symbols for elements 142
Synovial fluid 307
Synthetic fibre 495
Systems 277

Teeth 285
Temperature 50
Temporary hardness, water 240
Tendons 308
Testes 314
Textiles 484
Thermistor 425
Thermometers 49, 390
Thermostat 47
Thunder 101
Tidal energy 15
Tides 379
Timber 506
Tissue 277
Tog value 60
Tooth decay 286
Torricelli, Evangelista 44
Trachea 289

Transducers 430
Transect 357
Transformers 438
Transistor 428
Transpiration 333
Trophic level 345
Tropisms 335
Trundle wheel 1
Tullgren funnel 355

Ultra violet light 115
Umbilical cord 315
Universal indicator 228
Universe 374
Urea 303
Ureter 303
Urethra 303
Uterus 315

Vaccination 370
Vacuole 276
Vagina 315
Valency 219
Van de Graaff 100
Vein 298
Velocity 30
Velocity-time graph 31
Vernier calliper 2
Virus 368
Vitamins 280, 397
Volt 69
Volta, Alessandro 69

Volume 2
Voluntary muscles 308

Walton, E.T.S. 140
Water 182
Water cycle 187, 382
Water level detector 424, 429
Water treatment 188
Waterproofing 497
Watt, James 78
Watts 26, 78
Waves 102
Weather 391
Weathering 485
Weight 25
White blood cells 296
Wind energy 14
Wood preservatives 483
Woodland 357
Work 25
World food supply 399

X-rays 115
Xylem 333

Yarn 496
Years 378
Yeast 413
Yellow spot 311
Yoghurt 410

Zygote 314

THE WAR IN THE
CHANNEL ISLANDS
THEN AND NOW

THE WAR IN THE CHANNEL ISLANDS

THEN AND NOW

Winston G. Ramsey

AN
AFTER THE BATTLE
PUBLICATION

© After the Battle magazine 1981

ISBN: 0 900913 22 3

Printed in Great Britain.

Designed by
Winston G. Ramsey
Editor *After the Battle* magazine.

PUBLISHERS
Battle of Britain Prints International Limited,
3 New Plaistow Road,
London E15 3JA

PRINTERS
Plaistow Press Limited,
3 New Plaistow Road,
London E15 3JA

PHOTOGRAPHS
Copyright is indicated on all photographs where known. All present-day photographs (including aerial pictures) copyright *After the Battle* magazine unless stated otherwise.

EXTRACTS
Extract from *Commando* by Brigadier John Durnford-Slater by permission of William Kimber & Co. Limited.

FRONT COVER
The wartime role of the Channel Islands is typified in the Marinepeilstaende und Messstellungen — the huge observation towers unique to the Islands. This is the one to be seen at Les Landes, Jersey.

BACK COVER
Fort George Military Cemetery, Guernsey.

END PAPERS
The occupying forces at work . . . and play (courtesy of the Royal Court, Guernsey and the Bundesarchiv).

FRONTISPIECE
On guard at St. Ouen's Bay. La Rocco Tower in the background. *(See also pages 252-253.)*

DEDICATION
To Marguerite

ABOUT AFTER THE BATTLE

After the Battle magazine is edited and published by Winston G. Ramsey, who has had a lifelong interest in all aspects of the Second World War. Where visits to the battlefields were once a hobby, this developed in 1973 into the production of a quarterly magazine devoted to a growing number of enthusiasts — and editions have since been produced in Dutch, French and German.

Each year he and his family seek out and re-photograph the locations and events depicted in wartime photographs; and as memories fade and places change so the search has required increasing determination and perseverance.

The whole production of *After the Battle* is one which has proved an exciting and enlightening experience which has made the family many new friends worldwide.

Acknowledgements

The intention of this book is to provide a new look for both Channel Islanders and visitors at the five years the Islands spent under German occupation, through the presentation of the story with 'then and now' photographs. I have always been fascinated by past events and find that comparison photographs, like H. G. Wells time machine, can bring a new perception to our recent history.

The book was originally conceived as an article for *After the Battle* magazine way back in 1973 and, in spite of much prodding from Michael Ginns of the Channel Islands Occupation Society, many factors caused it to be deferred so that our trip to the Channel Islands to take the comparison photographs did not finally take place until April 1979. Even then, if it had not been for the generous and unselfish assistance of the members of the Society, this book could certainly not have been produced. The secretaries of both branches helped enormously with picture selection long before we actually set foot in the Islands and, once there, hosted us during our visit. They gave up considerable time to guide us around and indicated many salient features which we would otherwise have missed.

Michael Ginns, and his courageous wife Margaret, put up with endless questions and queries . . . and wet clothes during our stay on Jersey during which time the weather alternated from snow to warm sunshine. On Guernsey, Ken Tough, sometimes leading on his ubiquitous bicycle, guided us expertly around the unsignposted roads of Guernsey and also spent a day tramping the hills and dales of Sark. Colin Partridge and Trevor Davenport performed a similar role during the very interesting day we spent on Alderney. To them all and to the many other Society members we met during our tour, my grateful thanks. Many other people were very generous with their time and gave me every assistance. They have not been named lest someone should be left out but I would like everyone to know how much I appreciate their contribution.

Writing the wartime story of the Islands, knowing that their peoples have a wealth of personal experience and the Society a fearsome fund of knowledge, has been a daunting task. I was therefore extremely grateful for the many corrections, additions and amendments suggested to me during the proof stage of the book's production, in which all our guides participated.

In recognising the assistance of the Society, perhaps I should nevertheless reiterate what I am sure will be immediately apparent to its members: that *The War in the Channel Islands — Then and Now* has been written with a wider, more general readership also in mind, for whom specialist knowledge and personal experience cannot be taken into account.

Surprisingly, the most difficult task of all was the taking of the aerial photographs. Because the Channel Islands lie wholly in controlled airspace, with considerable tourist traffic and inter-island air services, only with the closest co-operation of the air traffic controllers was it possible to complete the photography on time and I am especially indebted to Ralph Bowring in Guernsey and Harry Dalton on Jersey for their sympathetic consideration of my requests. Sark itself lies in a prohibited zone, banned to all aircraft under 2,000 feet, and my special thanks go to Seigneur Michael Beaumont for giving me exceptional permission to violate the island's sacred airspace.

Finally to my wife and son — the former for her continuing tolerance and understanding and the latter for his help and enthusiasm — my special appreciation.

Contents

	PROLOGUE	5
Chapter One	GUERNSEY	6
Chapter Two	JERSEY	48
Chapter Three	ALDERNEY	92
Chapter Four	SARK	117
Chapter Five	THE COMMANDO RAIDS	129
	Operation AMBASSADOR	129
	Operation DRYAD	141
	Operation BRANFORD	146
	Operation BASALT	148
	Operation HUCKABACK	158
	Operation HARDTACK 28	160
	Operation HARDTACK 7	165
Chapter Six	THE POST-WAR PERIOD	168
Chapter Seven	THE CHANNEL ISLANDS TODAY	192
Chapter Eight	THE CEMETERIES	206
Chapter Nine	THE WAR GRAVES	224
Chapter Ten	THE WAR MUSEUMS	234
Chapter Eleven	YOUR HOLIDAY HOTEL	240
APPENDIX	Contents of Reviews published by the Channel Islands Occupation Society.	252

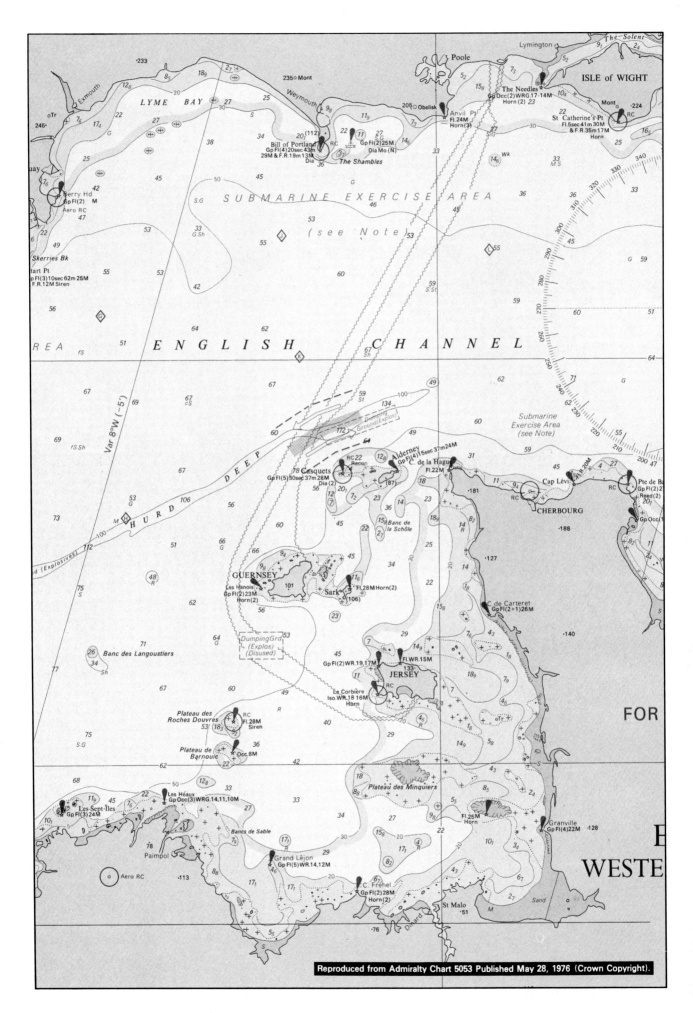

Prologue

The Channel Islands which, geographically, have more of an affinity to France than the United Kingdom, have the unique experience of being the only British territory to have been occupied by German forces during the Second World War. After the invasion of France in May 1940, their proximity to the French mainland precluded any chance of defending them against the invader who could have sat back and shelled them with impunity. Reluctantly, therefore, the British Cabinet took the decision to demilitarise the Islands and thus their peoples, loyal to the British Crown for almost 900 years, although spared the bloody battles of their continental neighbour, were destined to five years of occupation — an event which has left a lasting mark on the Islands, their environment and their peoples.

Grouped more than seventy-five miles to the south of Britain, in the lee of the French Cotentin peninsula, the Islands have always been a favourite holiday retreat for the British. Before the war, the Islands were served by the joint service operated by the Southern Railway via Southampton and the Great Western Railway from Weymouth, the fares (from 45/- to 75/- return depending on class) being identical although the sea journeys to St. Helier on Jersey and St. Peter Port in Guernsey were of unequal length. Travel by air was also possible between Heston and both Islands.

Self-governing and forming two perfectly distinct commonwealths, the Bailiwicks of Jersey and Guernsey have always been free to make their own legislation and to object to any specific Acts of Parliament in the United Kingdom which adversely affect them. In a way, the Islands have achieved the best of both worlds — to accept what suits them best but reject that which might upset the equilibrium. Only in the field of defence have the Bailiwicks been dependent on Great Britain, the nearest to a standing army on the Islands being the formation of various militias in the Middle Ages. Although these militias were supplemented by the stationing of a British Army battalion in each of the Bailiwicks, that on Jersey was withdrawn in the 1920s and Guernsey in 1939. By June 1940, with German troops sweeping across France towards Normandy and Brittany, the British War Cabinet reaffirmed its earlier decision that the Islands had no strategic importance and would therefore not be defended and that they might even prove to be an economic embarrassment to an occupying force.

On June 16, representatives from the two Bailiwicks were invited to London to discuss the implications of demilitarisation and the surrender of the Islands. It was decided that whilst the Lieutenant-Governors of each Bailiwick would be recalled to the UK, the chief civilian administrator, the Bailiff, would remain to ensure that the best treatment possible was obtained for the civilian population.

On June 20 all regular troops who had reached the Islands from France were withdrawn and the two military governors departed the following day. The Imperial General Staff were informed that the military evacuation of the Channel Islands had been completed and that the German government should be so informed. Accordingly, an official press statement was prepared by the Home Office but, for reasons which have never been satisfactorily explained, it was withheld from publication (although the local press in Guernsey had written of demilitarisation the previous day).

King George VI then sent a message of encouragement to the Islanders but the Home Secretary informed the Bailiffs that it should only receive restricted circulation in the interests of 'national security'. The Bailiffs, ever mindful of the vulnerability of the Islands from air attack, complied with the request believing that the Germans had by now been informed of the demilitarisation decision.

Two days later the question of a public declaration of the demilitarisation was again brought before the Chiefs-of-Staff but they only felt that the matter should merely be brought to the attention of the Foreign Office. While the Islanders lay at the mercy of the Germans, perhaps curious to know why the latter had not yet arrived, the British Government was still prevaricating. One MP who raised a question on the evacuation in the House of Commons on June 28 was asked by the Home Secretary to withdraw it as, '. . . we have been at pains to prevent any publicity being given to the fact that a measure of evacuation from the Channel Islands has been carried out and the newspapers in the country have been asked not to publish any statements which would indicate that this evacuation had taken place'.

By now it was too late. For ten days the Germans had been attempting to establish whether the Islands would be defended or if they would be given up without a fight. Their reconnaissance aircraft reported no sign of white flags — the internationally recognised method of surrender — and, in the absence of any clear sign, they now made their move. Late on the afternoon of June 28, Heinkel He 111s were instructed to bomb St. Helier and St. Peter Port and, as a result, some 180 bombs were dropped with hits being scored on the port installations. Altogether forty-four civilians lost their lives in the attack.

Now that it had been forced into making a move, the British Government ordered the BBC to broadcast a hurried announcement to the effect that the Islands were already demilitarised on the nine o'clock news. Then, in an attempt to cover up the real reason for the delay and to blame the Germans for the civilian deaths, the following morning the Home Office released a statement which said that all forces and equipment had already been withdrawn and the Channel Islands had been demilitarised. Although not telling an outright lie, the Islanders and the British public and Press were thereby grossly mislead — the latter into publishing accounts that demilitarisation had not saved the Channel Islands from a particularly barbaric attack!

In spite of what had happened on the evening of June 28, it was not until two further days had elapsed that the Foreign Office asked the United States Ambassador to pass the following message to the German government in Berlin:

'The evacuation of all military personnel and equipment from the Channel Islands was completed some days ago. The Islands are therefore demilitarized and cannot be considered in any way as a legitimate target for bombardment. A public announcement to this effect was made on the evening of June 28'.

CHAPTER ONE
Guernsey

Guernsey's first aerodrome was situated on meadowland to the north of L'Eree Road in the extreme west of the island. In 1939 a new airport *(above)* was laid out about two miles to the south-west — see map. This photo was taken by renowned aviation photographer Charles E. Brown at the official opening ceremony on May 5, 1939 and our comparison picture *below* was taken on February 11, 1981. The aerodrome must have looked much like it did in Charles's photo when the Germans arrived in June 1940.

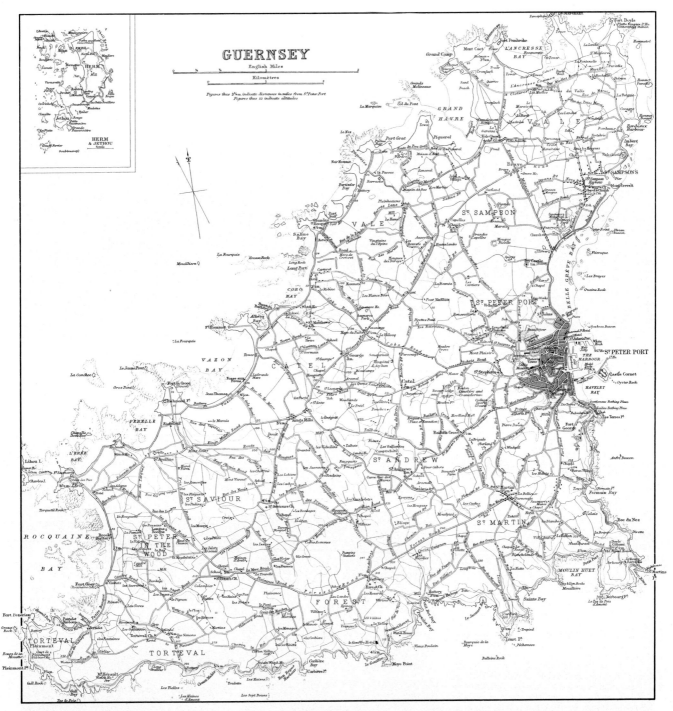

Above: **Bartholomew's pre-war map of Guernsey with the post-war amendment** *below)* **showing the location of the aerodrome (Copyright John Bartholomew & Son, Limited).**

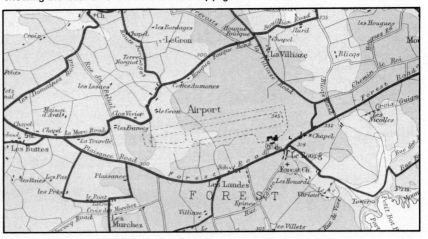

Since 1935, Guernsey had been led by Bailiff Victor Carey in his combined role of leader of the island's parliament — the States — and its Chief Justice. By 1940 he was approaching his seventieth year and, understandably, was not ideally suited to control the panic which ensued from the evacuation announcement on June 19. The Lieutenant-Governor, having been appointed only a few days earlier, was equally unfitted for a period where strong leadership was essential. Although the Attorney-General, Major Ambrose Sherwill, tried to calm the situation in the Royal Court the following day, confusion abounded. There was an immediate run on the banks and it was rumoured that the Bailiff and administrators had already left the island.

It had first been announced that all the population was eligible for evacuation with priority to those of suitable age who volunteered for the services. However, when it became obvious that shipping could not be provided for all the 42,000 islanders, a

vigorous anti-evacuation programme was begun to try to reverse public opinion. A different member of the States (parliament) spoke each day trying to persuade everyone that it would really be better to stay. In the event, some 17,000 people left before the occupation — some forty per cent of the entire population.

Meanwhile, in France, the German Navy, under Admiral Eugen Lindau, was planning the capture of the Channel Islands under the codename GRUENE PFEILE — Green Arrow. Not aware of the Islands' demilitarisation, and in the absence of any visible sign of surrender, units of Luftflotte 3 were detailed to bring matters to a head. St. Peter Port was bombed late in the afternoon of June 28 killing thirty-four. Although light anti-aircraft fire (from the SS *Isle of Sark* anchored in the harbour) was reported by the German crews, there were mixed interpretations as to whether this meant that a subsequent landing would be opposed or not. All the evidence was weighed at a conference in Paris on the afternoon of Sunday, June 30, including a Reuters news agency report on the BBC radio announcement — the German monitoring service having missed the broadcast!

While all this was taking place, although the Kriegsmarine had been given the honour of the capture of the Islands, a Luftwaffe officer seized the opportunity to steal the glory. Detailed to reconnoitre Guernsey on Sunday afternoon in his Dornier Do 17, Hauptmann Liebe-Pieteritz decided to chance his arm and try a landing. Observing that the aerodrome at Guernsey (which had been opened in May 1939) was completely deserted, he came in to land leaving the three other aircraft in his flight to circle overhead. Taxiing up to the airport buildings, he stepped out, pistol at the ready only to find the place unoccupied. Suddenly three Bristol Blenheims appeared overhead and attacked the covering aircraft. Hauptmann Liebe-Pieteritz raced to his aircraft, dropping his pistol in the rush to get airborne, and in the dogfight which followed Leutnant Forster claimed that he shot down two of the RAF aircraft. *(No British aircraft are recorded as being lost in the area on that day — Author.)*

Nevertheless, the point had been established that the Island was undefended and, accepting the Luftwaffe's fait-accompli, Admiral Lindau ordered the immediate occupation of Guernsey the following morning. The naval detachment, which was

Above: **Friday afternoon, June 28, 1940. Smoke rises from burning buildings on St. Julian's Pier, St. Peter Port (Carel Toms Collection).** *Below:* **The same view, thirty-nine years later. The precise piece of wall beside Glategny Esplanade was easily identified.**

Further along on St. Julian's Pier, this plaque records the names of the thirty-four civilians who lost their lives in this raid.

The clocktower at the junction of St. Julian's Pier and the North Esplanade gutted during the raid (Carel Toms Collection).

Today the building has been nicely restored and it now serves as an operations room for radio-controlled taxis.

standing by in Cherbourg ready to launch the invasion, was instead ordered to the nearby airfield to be flown to the island by Luftwaffe transport aircraft. However, when morning came, fog delayed the arrival of the Junkers Ju 52s and the operation was belatedly launched in the afternoon.

The chief of the island's police force, Inspector W. R. Sculpher, had been given the unenviable task of meeting the first Germans to reach Guernsey. He had been rushed to the aerodrome the previous evening when he heard of the landing but had arrived too late. When the news reached him on Monday evening that more aircraft had landed he returned, to be met by the senior German officer Major Albrecht Lanz who asked to be taken to the head man on the island. Inspector Sculpher handed over a prepared letter in German which read: 'This Island has been declared an Open Island by His Majesty's Government of the United Kingdom. There are in it no armed forces of any description. The bearer has been instructed to hand this communication to you. He does not understand the German language.'

Inspector Sculpher then drove with Major Lanz to the Royal Hotel where Bailiff Carey and other senior officials were waiting. After an interpreter had been found, Major Lanz formerly declared that the island was now under German occupation and that he was to be the Kommandant.

Quickly the Germans began to consolidate their position and reinforcements from Infanterieregiment 396 of the 216 Infanteriedivision were shuttled to the island while other units arrived by sea. Firmly, yet without a hint of the brutality which the islanders had expected, the Germans then took over. An immediate step was to call in all firearms which had to be handed in at the Royal Hotel by noon on July 1. However, the Germans were considerate to those who owned souvenirs or trophies and, initially at least, air guns. (One amusing incident occurred in March 1941 when the German authorities discovered a light aircraft in a garage at St. Peter Port. If, they mused, an aeroplane could

Scene of the first part of the drama in the take-over of the island by the Germans was the airport where Hauptmann Liebe-Pieteritz landed on Sunday, June 30 (Carel Toms Collection). *Above:* **This is the entrance to the terminal building as it was and** *below* **during its extension in 1979. Only the tower on the roof remains to be seen as a reference point behind the modern frontage. Author with wife and son begin their tour for the production of this book.**

The invaders arrive. Luftwaffe-style salute on the airfield side of the terminal building (Bundesarchiv). *Below:* Ronnie Cluett of the airport fire service complete with 'jackboots' sportingly stands in for our comparison taken on the extended apron. The new control tower behind his head was built in 1977.

go undiscovered for a year, how many hundreds of weapons could be still in circulation. In the island-wide search for arms and ammunition which followed, one of the directives to the population was that any aeroplanes on their premises also had to be declared!)

On July 2 an ordinance was published which set the scene for control of the island for the next five years. It stated that all island administration was to carry on as before except that where an Act would have previously required the Assent of the British King, instead this would now be submitted to the German Kommandant for approval. Major Lanz used considerable tact in his dealings with the islanders and he decreed that anyone with a grievance could write direct to his office. Church services were permitted as were prayers for Britain and the Royal Family providing these did not include incitements against the occupying power.

Major Lanz soon scored a huge propaganda victory when he asked Major Sherwill to broadcast a message to those Guernseymen who had left for Britain. In his speech he spoke of the great courtesy with which they had all been treated and the exemplary

Left: **An Auto Union/Horch is craned ashore onto the east side of the New Pier. This jetty was built in 1930 at a cost of £148,000 for use by Southern and Great Western Railway ships.** In the background is Castle Cornet (Bundesarchiv). *Right:* **The same spot . . . but a different era. Even the crane has been changed. Now the pier is used by car ferries to the island.**

Above: A well-known and much publicised picture taken by Charles H. Toms of a German military band marching south on Pollet in St. Peter Port. No doubt the wording 'Lloyds Bank' made it an excellent propaganda photograph. *Below:* While there has been a slight change to the bank facade, it is the style of dress of the passers-by that dates our comparison.

Major Albrecht Lanz, the first Kommandant (centre) with Leutnant Mueller (left) and Dr. Maass (R. H. Mayne).

Above: **In Smith Street (to the left of Lloyds Bank in the picture on page 11), German troopers chat up a local girl (Bundesarchiv).** *Below:* **The building on the left, then as now, is the office of the Guernsey Press.**

conduct of the occupying force. Although no doubt true, such remarks coming from the British Attorney-General were seized upon by the German propaganda machine which made great capital out of it to counter British allegations of the cruelty to be expected from the German conqueror.

In August the position of the Channel Islands in the German occupied territories was formalised with their incorporation into the French Department of Manche as a sub-district of Military Government Area A with its HQ at St. Germain. This set-up was a logical extension of Hitler's belief that the Islands were unwilling British colonies. (Even a German lawyer sent from Germany to try to unravel the secrets of the peculiar constitutional background of the Islands failed to interpret this correctly.)

Even more complicated was the command structure concerning the Islands. The civilian administrative side of everyday life was controlled by a Feldkommandant, the Channel Islands coming under the auspices of FK515. The Feldkommandantur consisted of departments which dealt with agriculture, legal matters, transport, etc., and was staffed,

Left: **German officers listen to a band playing outside the vegetable market (Carel Toms Collection).** *Right:* **Market Square in April 1979.**

in the main, by former peacetime German bureaucrats. On the military side, each island had an Inselkommandant (initially Major Lanz on Guernsey and Hauptmann Gussek on Jersey) and usually he was always the senior infantry regimental officer. A department of the Feldkommandantur liaised with the military commander only insofar as to ensure that the left hand knew what the right was doing.

Above the Inselkommandant was the Befehlshaber der britische Kanalinseln (B.d.b.K.) — the Commander of the Channel Islands. Initially, this position was occupied by Oberst Graf Rudolf von Schmettow and he set up his HQ on Jersey on September 27, 1940.

When the 319 Infanteriedivision arrived in mid-July 1941, the commanding general, Majorgeneral Erich Mueller outranked von Schmettow and so became B.d.b.K. but with his HQ on Guernsey. Von Schmettow then became Befehlshaber Jersey. However, as they were both the senior infantry officers, they also held the position of Inselkommandant on their respective islands! (When Mueller was posted to Russia in 1943, von Schmettow was promoted to command the 319 Division and Oberst Heine took his place on Jersey.)

The respective responsibilities of the Feldkommandantur and the occupying Wehrmacht troops proved difficult for even the Germans to understand and several directives had to be issued to state where the dividing line lay. Very broadly, the Wehrmacht was subordinate to the Feldkommandant yet had no right to interfere with the civilian, agricultural or economic life of the Islands. The Wehrmacht could make civilian arrests but, if the offence concerned security, it had to be dealt with by the Feldkommandantur. The Nebenstelle (branch) of FK515 (located at Grange Lodge situated in The Grange) was established on August 9. The first Feldkommandant was Oberst Friedrich Schumacher who was replaced by Oberst Friedrich Knackfuss in October 1941.

The B.d.b.K. HQ in Guernsey was located at La Corbinerie in a large Georgian house called 'The Oberlands' and here in the grounds the Germans constructed two command bunkers and a barrack block — the work being pushed ahead at a frantic pace by the OT twenty-four hours a day.

The organisation problem was compounded as the 319 Division was spread over the three islands and had to maintain a base at Granville on the French coast. In 1942, therefore, an organisation expert, a Colonel Lamey, was detailed to investigate the present set-up. He reported that the whole command structure needed revising and recommended that the 319 Division be re-designated 'Festungsdivision Kanalinseln'. This caused an immediate outcry, not only from the

The Germans commandeered all but a very few private cars on the island. These were taken to the Albert Pier (above) **to await shipment to France (Carel Toms Collection).** Below: **Today, the Pier is still covered with cars every weekday.**

Division but from Army Group D who said it was proud of its number status and that to become a Fortress Division (obviously regarded as an inferior role) would be bad for morale!

As far as the civilian authorities were concerned, the day-to-day duties of the States in Guernsey were undertaken by a Controlling Committee with Major Sherwill as President — a move designed to bypass the Bailiff, who was considered too old for his onerous burden, and to speed up the decision-making process. This Committee met weekly throughout the occupation and a working relationship was established with the Germans — both appreciating that each relied on the other for their survival. The Islanders were, in fact, in quite a strong position for at least three-quarters were involved in working or providing for the military in some shape or form. Both saw that it was in each other's interest to maintain the status quo but this therefore touched on the very fine line between practicality and collaboration with the enemy.

Major Sherwill's dilemma in October 1940 is typical of the difficult decisions which had

Left: **The Morris belonging to the German Press Censor on the island. WH stands for Wehrmacht Heer (Carel Toms Collection).** Right: **After June 1941, the Germans introduced driving on the right — graphically illustrated by our comparison taken on the corner of St. Julian's Avenue and Glatengy Esplanade. The profusion of signs has since given way to a single traffic light.**

Above left: La Corbinerie just off the Oberlands Road, chosen for the HQ of 319 Infanteriedivision and, thus, also of the Befehlshaber der britische Kanalinseln in 1941. Divisional and regimental command bunkers were constructed in the grounds.
Above right: This German barrack block is now used by the nearby Princess Elizabeth Hospital as an emergency building.

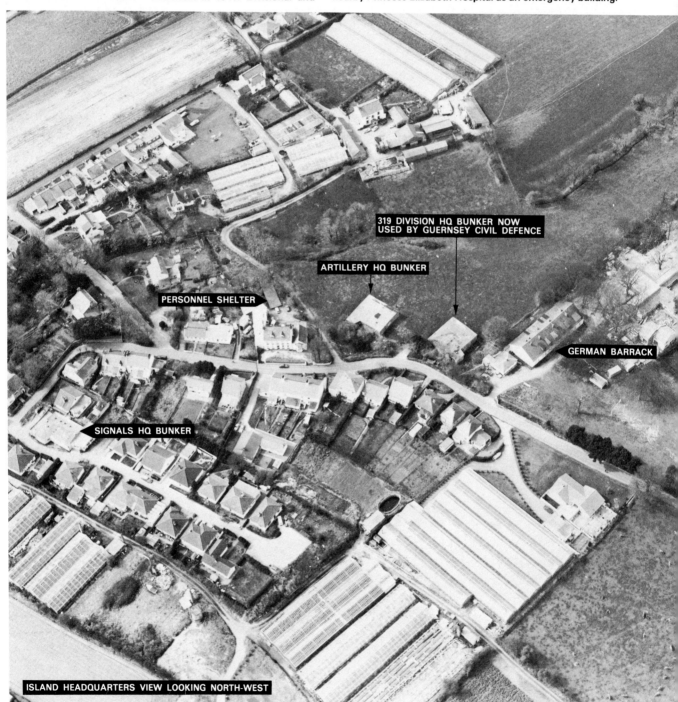

ISLAND HEADQUARTERS VIEW LOOKING NORTH-WEST

to be taken. Two British agents, Second Lieutenant Hubert Nicolle and James Symes had been caught landing on the island. While Major Sherwill was manoeuvring to get them spared a German firing squad, an order was presented to the Committee to endorse various measures against the Jews. Believing all Jews had been evacuated (there were, in fact, still four on the island), and that approval would not mean anything in practice but refusal would materially affect his efforts to help the agents, Major Sherwill signed the order. Later he wrote that, 'nevertheless, I still feel ashamed that I did not do something by way of protest to the Germans.'

Within a few weeks of the German takeover of the Islands, it was obvious to the civilian authorities that supplies of food and provisions would soon be exhausted unless something positive was done. A combined approach from both Bailiwicks to the German Feldkommandant led to the formation of a purchasing commission with Mr. Raymond Falla representing Guernsey and Mr. Jean

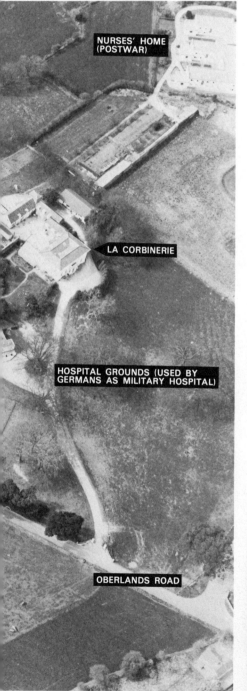

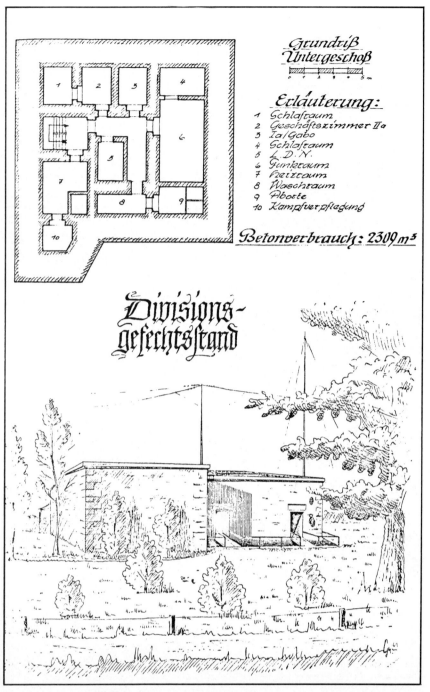

The plan of the Divisional command bunker in the grounds of The Oberlands with our comparison of April 1979. It is now used by Guernsey Civil Defence (Reproduced from the 'Festung Guernsey' books by courtesy of the Royal Court).

Although Oberst Graf von Schmettow was the Channel Islands' first overall commander, his HQ was on Jersey. When Majorgeneral Erich Mueller *(above)* took over as B.d.b.K. in July 1941, The Oberlands became his HQ (R. H. Mayne).

Jouault Jersey. Joined by Mr. Wilfred Hubert, they set up shop in the Hotel de la Duchesse Anne in Granville on the French coast and from there made extensive trips throughout France purchasing essential supplies. Initially, all purchases had to be made in cash and Mr. Falla recalls how he often carried a suitcase full of occupation marks — and sometimes had to spend several hours counting out the notes for goods purchased! The goods so obtained were split between the Islands on a proportional basis according to the population, Jersey receiving two-thirds and Guernsey the remainder.

It was inevitable that the occupation would bring tragedy and one example of the ills which befell the population, which had to cope with German military personnel billeted upon them, was an incident that occurred in Guernsey on November 3, 1941. Unteroffizier Philipp Kuehn was cleaning his pistol on the first floor of Le Carrefour at St. Andrew's and, in doing so, accidentally discharged the round in the breach. The bullet passed through a sofa, through the floorboards and into the kitchen below, where Mr. Ernest Brouard was reading a letter by the window.

Control of all transport was taken over by the Feldkommandantur and all vehicles were assigned to a special organisation. The use of motors by private individuals was limited to essential agricultural duties only. The subsequent shortage of petrol led to all vehicles of more than 12 hp being banned and to the conversion of others to burn producer gas and, by June 1942, there were seventeen such vehicles in use on the narrow roads of Guernsey.

Both command bunkers had moated drawbridges protecting the entrances and roofing tiles for aerial camouflage. This is the regimental command bunker. (The drawing depicts a similar bunker of the same type — in this case the one at Beau Sejour.)

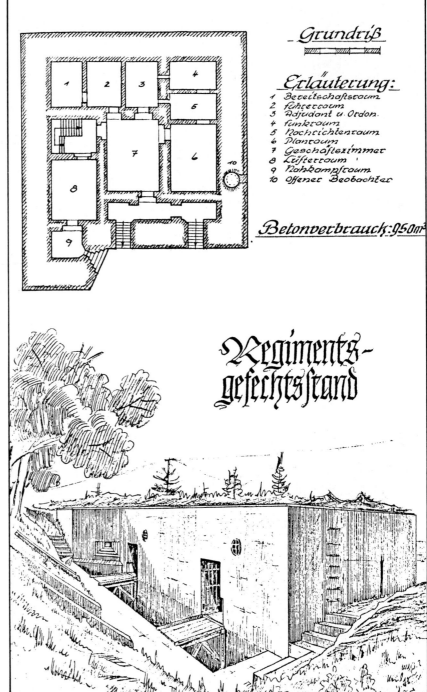

The Nachrichtenstand or Signals Centre for the HQ complex lay on the other side of Oberlands Road down Les Quatre Vents. It now provides a substantial addition to one of the bungalows which were built around it after the war.

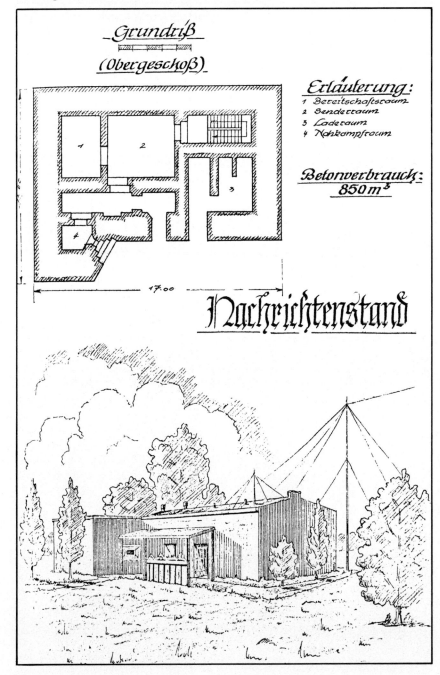

In June 1941, a decree had been issued to make all traffic conform to Continental practice and drive on the right, and yellow lines (which still exist today) were painted across minor roads where they joined major ones to indicate the instruction 'HALT'.

By the end of 1942 a programme of Germanisation of place names for the various defensive positions began. Castle Cornet, situated on the southern arm of the harbour breakwater, became Hafenschloss; Torteval, in the south-west of the island, Spitzkirchen; Mont Crevelt became Krevelberg; Saints Bay, Heiligenbucht; Hougue a la Perre, Gemaeuer; Jerbourg on the south-western corner, Strassburg and Pleinmont on the westernmost promontory was called Westberg.

Not generally known is the fact that all occupied countries had to pay a contribution towards occupation costs. The rate of exchange was fixed at 9.36 Reichmarks to £1 and, for some unknown reason, Guernsey paid a considerably greater share per head of the population than Jersey. As this financial burden was far more than the Islands could sustain from their internal resources, credit was arranged with French banks — the German view being that the costs of their occupation were a British responsibility.

In 1941, Hitler was seized with a sudden fear that his forthcoming attack on the Soviet Union would precipitate the British to take action behind his back in the West. He felt that the Channel Islands might well be the object of such an operation and, accordingly, he issued instructions for the immediate reinforcement of their defences.

Over the next few months, in spite of Hitler's preoccupation with the Russian campaign, a stream of orders arrived from his headquarters on the way the Islands must be defended. In October, he stated that the Heer (Army) unit must be brought up to the strength of a division; all the beaches must be mined; obstacles were to be set up on all open ground to prevent landings by gliders or paratroopers and there should be at least 200 strongpoints on each of the larger two Islands. Anti-tank walls were to be built to restrict the entry of an invader from the beaches and tunnels were to be dug to provide shelter for stores, equipment and personnel during a bombardment.

One significant result of all these orders was the arrival in the Islands of two new organisations. The first was the posting in of 319 Infanteriedivision to replace 216 as the new garrison force. The second was the introduction of the Organisation Todt.

It had been appreciated that local labour resources would not nearly be sufficient enough to carry out the mammoth construction programme envisaged. Also the Hague Convention, which the Germans maintained did not apply to the Islands because the British Empire had not surrendered but to which they were nevertheless acquiescent, prevented the utilisation of civilian workers on military projects. The Germans had got round this to an extent by offering wages and conditions far in excess of others prevailing in the civilian sector, and some people, notably the Irish, had taken up the offers. The Organisation Todt was therefore a natural choice. The OT was basically an organisation formed from German construction companies comprising, in the main, forced or impressed labourers from the occupied countries and, after the victories in the East, Russian prisoners-of-war. The organisation was under semi-military leadership — the brown uniforms of the overseers being increasingly common to the Islands as the war went on. Guernsey had the larger number of workers (some 7,000) followed by Jersey with 5,000-6,000 and Alderney 2,000.

All the construction work was largely dependent on a plentiful supply of aggregate and, in order to transport this to the sites, a

network of 90cm (with a short length of 60cm gauge) railway track was laid out. Those in Guernsey started at the harbour and ran along the eastern coast to St. Sampson's and the nearby quarries; thence north-west to L'Ancresse and the Chouet quarry before turning south-west along the coast to L'Eree.

In addition to Guernsey's big guns, the German defences on the island eventually totalled fourteen coastal batteries and thirty-three anti-aircraft sites and it is suggested that, area for area, Guernsey boasted the most heavily defended territory in Europe. The flak batteries consisted of three sizes: heavy 88mm, light 37mm, and 20mm. The complete German defence system on Guernsey was documented with plans and architects coloured drawings bound in four volumes.

Above: The Fort Saumarez observation tower MP2 can be seen in the background of this typical illustration from the 'Fortress Guernsey' books now preserved in the library of the Royal Court House. The four volumes provide a unique guide to the entire defence system on the island; unfortunately a complete set of similar books (now in the Museum of the Societe Jersiaise) do not exist for Jersey (Royal Court). *Below:* Our comparison of the personnel shelter seen in the foreground, now totally covered over with soil.

Peculiar to the Channel Islands are these huge naval direction-finding and signalling stations, to the Germans Marinepeilstaende und Messstellungen. It was intended that they should be built all around the coastline of Guernsey, Jersey and Alderney to provide a complete monitoring chain. In the event, by the end of the war, only seven had been completed. This is the one on Guernsey at Fort Saumarez on the western coast overlooking Lihou Island. The tower is owned by Mr. J. H. Villiers. Note the tank turret in the foreground (Royal Court).

Chouet Tower, designated MP1 (with the Army Coastal Artillery observation post M2 at the base on the right) stands in the north-west corner of Guernsey. It looks completely safe from this angle . . . but just look at the view from the air!

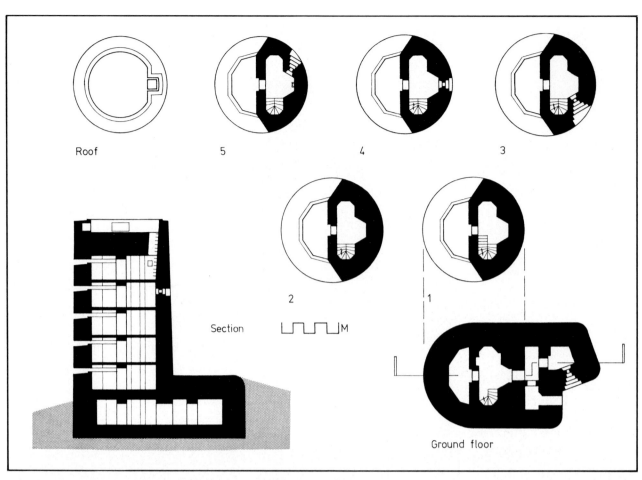

Colin Partridge, Editor of Alderney Journal and an active member of the Channel Islands Occupation Society, has made an extensive study of the fortifications of the Atlantic Wall (published as Hitler's Atlantic Wall) in which his training as an architect has been used to good measure in documenting the Islands' fortifications. This is his plan of the observation and range-finding tower at Pleinmont (MP3) on which the Society's Guernsey branch has obtained a ten-year lease from the owner Mr. F. J. Merrien.

The second love of Colin's life! L'Angle Tower, a naval range-finder post which mounted a Freya radar later in the war. This photograph was taken in 1945 (Public Record Office).

Our rather arty comparison, taken in April 1979, is included by way of a token appreciation from the author for his inestimable help in annotating all our aerial photographs.

In addition to the purpose-built towers, existing buildings were often converted where suitable. This is Vale Mill, originally an old windmill. The cap was taken off and three floors added in concrete, faced with matching granite, before the top was replaced transforming it into MP5 (Public Record Office).

In 1973, the States Public Works department attempted to demolish the German part. Their efforts were unsuccessful and the tower still stands minus its false windows. We leave our readers to make their own judgment about what has been called 'official vandalism'. Did I hear someone say, 'Rubbish!'?

The Islands' vulnerability to invasion was always a prime consideration and every known defence was installed. *Left and above:* Tetrahedra obstruct Bordeaux Harbour (Royal Court).

Saints Bay, always a popular beach for bathing, where for five years during the war a visit could bring death or mutilation. These obstacles were usually mined (Royal Court).

Cobo Bay on the north-western coast renowned for its brightly coloured anemones . . . and its barbed wire. Photo taken near the Rockmount Hotel (Royal Court).

Gun positions were camouflaged in a variety of ways. This one, codenamed Gemauer, is at Tramsheds Corner, La Hougue a la Perne, on the coast just north of St. Peter Port (Royal Court).

Left: Another form of fixed defence were French tank turrets, mounting 37mm guns, set on a 'Ringstand' — a toothed ring embedded in the concrete to serve as a runner. Ringstands can still be seen in odd places on the islands although not at Houmets Lane, Vale, where this picture was taken (Royal Court). *Right:* The present day comparison taken in the front garden of 'Premi' is rather disappointing as a new bungalow completely blocks the view of the Rue de Picquerel.

Above: Fort George dates from 1780 and was maintained until 1926. The Germans utilised the fortress, which overlooks Soldiers' Bay just south of St. Peter Port and re-fortified the old British Clarence Battery position (Imperial War Museum). *Below:* The tangled undergrowth makes a comparison photograph rather difficult. Only the walls of the fort remain standing today, the interior having been gutted to build luxury villas, the construction of which caused much controversy at the time.

Above: Another gun position in Delancey Park was Batterie Sperrber which mounted four 100mm Skodas. The ammunition supply was by hoist from tunnels excavated into the side of the quarry below the guns. This, and all the other photographs in this book credited to the Public Record Office, are from a series taken by Major G. G. Rice during his tour of the Channel Islands in September 1945 (Public Record Office). *Below:* These are two of the four casemates which originally made up this battery — now peacefully overlooking Belle Greve Bay. Bowling greens and St. Sampson's Secondary School lie on the left.

Left: Netting hiding a gun position in the grounds of Les Cotils Home overlooking St. Peter Port harbour (Royal Court).

Right: Comparing the view from the hospital grounds on a misty morning in April 1979.

Above: An alternative form of camouflage was to simulate a civilian building. Here, at Houmet du Nord, a casemate for a 105mm gun is being disguised by the OT as a seaside bungalow (Royal Court and Carel Toms Collection). *Below:* The casemate still exists with a small portion of the woodwork fixed around the embrasure to stop ricochets from an attacker's small arms fire.

The observation and command post for Batterie Dollmann at Pleinmont, *above* in 1945 and *below* in 1979 (Public Record Office).

Backbone of German defence in every theatre of war was the ubiquitous 88mm Flak 37. When sited in open emplacements like those *(above)* on L'Ancresse Common, it could be used in both the anti-aircraft and anti-tank roles. The L'Ancresse Bay defences *(below)* are interesting as they span two centuries: from the 18th century Martello towers to those of the 20th. *Opposite top:* Ken Tough, Guernsey Branch secretary of the CIOS, stands on the same emplacement of Flak Batterie 'Dolmen' 1/292 — now overgrown and just another obstacle to be avoided by golfers. Closs du Valle lies in the background on the right.

Weaponry figured highly in Hitler's thoughts and the largest gun battery constructed on the Channel Islands was situated in Guernsey on an inland site on the western end of the island. The spot chosen was at Le Frie Baton to the west of the aerodrome and the battery, first called 'Nina' (see below) but renamed 'Mirus' after Kapitaen zur See Mirus who was killed on November 3, 1941 by an aircraft on his way to Alderney, was claimed to be the fifth largest on the Atlantic coast. Mirus mounted four 48-ton, 305mm (12-inch) guns having a range of thirty-seven miles. Although the battery faced west, the guns had an all-round traverse and, consequently, were not given the overhead reinforced-concrete protection of the Pas de Calais guns. They fired 800lb armour-piercing shells. Mirus was built by several German firms under OT control with Siemens Schuckert handling the complex wiring of the whole battery. The actual construction of the concrete gun pits and associated shell and ammunition stores was carried out by the Organisation Todt. Mr. Ronald Mauger was employed as a driver by the military transport organisation in Charroterie, St. Peter Port (together with forty other Islanders) and witnessed the entire construction programme which began in the autumn of 1941. He remembers the scene clearly as being like, 'the Israelites toiling in ancient Egypt under the taskmasters of Rameses II with the brown-uniformed OT overseers controlling a working force of some 2,000 excavating the four huge holes to take the underground rooms and associated gun pit. The whole valley swarmed with men and, from a distant viewpoint, they resembled ants hard at work on their kingdom.'

The guns themselves came originally from a Russian battleship *Imperator Aleksandr III* launched on April 15, 1914. After the Russian revolution, the ship was re-named *Volya* (Victory) and early in 1918 she was patrolling in the Black Sea. She ended her life in the French Tunisian dockyard at Bizerta and was broken up in 1935. However, the twelve guns were still usable and were put into store.

In the winter of 1939-40, when Britain and France agreed to give aid to Finland, the guns were loaded on board three ships, the *Karlerich*, *Juliette* and the *Nina*, four guns per vessel, to be shipped to Finland where it was intended that they would, ironically, be turned against their former owners. However, the *Nina* was delayed en route and was captured on the high seas by the German forces invading Norway in May 1940 and

Frank Le Page was a grocer at St. Martin's and we are indebted to his courage and tenacity in taking surreptitious photographs like that *above* of captured Renault FT17 tanks in La Rue Cauchee (Carel Toms Collection). *Below:* We discovered the photographer's vantage point had been the upstairs bedroom of 'Delurd', now owned by Mr. and Mrs. H. Bougourd who kindly allowed us to recreate the scene today.

One of the Russian 305mm guns from the Imperator Aleksandr III negotiates the kink in the road at Le Bourg (specially widened for the occasion) in January 1942 en route to Le Frie Baton, the location chosen for Batterie Mirus just south of Perelle (Bundesarchiv).

taken to Bergen. When heavy artillery was requested for Guernsey, the four guns were transported to Krupp's works in the Ruhr for reconditioning. They were fitted with new mountings and then shipped to Guernsey via St. Malo. The Germans had to tow a floating crane to St. Peter Port specially to lift them ashore. Each barrel and mounting was taken from Cambridge Berth to St. Saviour's on a 24-wheel trailer drawn by four half-tracks. One section of road at Mount Row — Ville-au-Roi had to be widened to enable the convoy to negotiate the corner and all traffic was halted along the route during the operation. After installation, two of the guns were camouflaged by painting to resemble a cottage which revolved the complete 360 degrees with them, and the others with netting. The first shot was fired on April 13, 1942, and each gun fired about seventy shells during the remainder of the war, mainly while training.

On June 8, 1943, No. 1 gun was under repair when enemy ships were sighted on radar at 2.08 a.m. The remaining guns opened fire but at too high an elevation causing one of the barrels to jump completely out of its cradle. Following examination of the damage, it was decided that the modern German propellent was too powerful for these 1914-vintage guns and the charge was consequently reduced. This had the effect of cutting the range from the planned thirty-seven miles to twenty-five although the normal working range was deemed to be only twenty miles.

One humorous incident concerning Batterie Mirus has been related by Herr Willi Hagedorn, formerly Oberleutnant, who was attached to the German Naval Command of the Kanalinseln and who recently returned to the Islands to inspect his former stamping ground. He remembered one day (November 2, 1943) when unidentified objects approaching Guernsey had been picked up on the Fort George radar. All the island's batteries were alerted as the signals were interpreted as being made by a British landing force. The Naval Commander (Seeko-KI) — who had the ultimate responsibility of giving the order to fire at targets at sea — was away in Alderney at the time and the commanding officer of Army Coastal Artillery Regiment 1265, was deputising in his absence. When the blips came to rest in a bay suitable for a landing, the officer panicked and Mirus, together with other artillery, was ordered to open fire. Daylight brought the ultimate embarrassment for there, floating on the sea, were two Allied barrage balloons!

The bend at Mount Row — Ville-au-Roi. This also had to be altered for the passage of the Mirus trailer and remains so today.

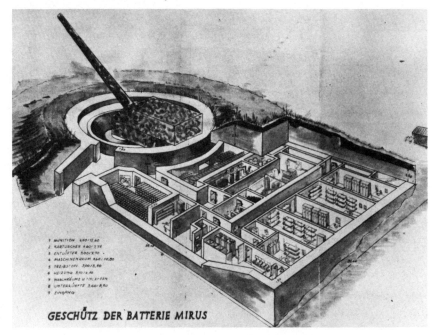

GESCHÜTZ DER BATTERIE MIRUS

Above: Often captioned incorrectly, this well-known photograph shows No. 4 Gun of Batterie Mirus in action sometime after April 12, 1942 when the battery was commissioned (Ullstein).

Below: Today this site off the Rue de L'Arquet is leased by the preservation-minded Richard Heaume (founder of the German Occupation Museum) from its owner Mr. J. H. Lenfestey.

Above: A band plays outside the barrack and canteen block for the Mirus battery in 1943 (W. Schlottmann).

Bottom: Now the building is ivyclad and roofless, just another abandoned wartime building to the casual visitor.

Seeko-KI Kapitaen Steinbach (left) salutes Korvettenkapitaen Max Schreiber, the second battery commander, at the ceremony.

With apologies. Acting the fool thirty-six years later on exactly the same spot.

Left: **No. 1 Gun was one of the two guns disguised as a cottage (Royal Court).** *Above:* **It now lies on land owned by Lieutenant-Colonel Wootton and can be visited by calling at the Tropical Vinery at Les Rouvets.**

Left: **Entry to the gun itself was via this door in the rear of the armoured turret (Royal Court).** *Above:* **The steps in the edge of the barbette pinpoint the same spot for us in 1979. Daffodils grow where once a dragon roared.**

Aided by the ventilator gratings and stains on the concrete, we were able to place Ken Tough in the position of the German gunner for our comparison. The trolley was used to transport bag charges from the underground magazine (Royal Court).

In July 1941, the British Cabinet instigated measures in the Middle East which were to have far-reaching consequences for those on the Channel Islands. During that month, the Cabinet and Chiefs-of-Staff agreed to adopt a firm line concerning the security of the oilfields in Iran — jointly owned with Britain and vital to the Allied cause. The presence of German civilians in the country was a worrying factor and the Iranian Government was therefore asked to expel its German community.

A joint Anglo-Soviet Note presented on August 17 met with an unsatisfactory reply and the invasion of Iran by British and Russian forces was immediately put in hand for the 25th. When it came, Iranian defence was virtually non-existent and a mere twenty-two British troops were killed in the three-day campaign before the Shah announced a cease fire. As a result, the conditions imposed on the Iranian Government were: neutrality in the war; Allied use of the country's communications for the transportation of supplies to Russia and the ejection of all Germans. The Shah abdicated in favour of his son and the new Shah, on the advice of the Allies, restored the Constitutional Monarchy.

All this took the German Foreign Ministry by surprise but when Hitler heard the news he was incensed. He demanded immediate reprisals and the Channel Islanders in their island prisons were seen to be ideal hostages. The Bailiffs were ordered to quickly produce lists of all the 'English' nationals on the Islands. The Germans did not immediately appreciate that 'English' and 'British' did not necessarily mean the same thing and a flood of telegrams went back and forth between the Islands and Berlin — one from the latter even demanding the number of 'Iranian' Channel Islanders! (The sequel to this humorous incident was that one was actually found: a 69-year-old man born in Smyrna.)

Hitler's demand was for ten Britons for every German, but when it was reported that 400-500 Germans had been interned in Iran, the Feldkommandantur doubted if 5,000 British-born residents could be mustered. While the census-takers pored over their figures trying to make up the numbers, notes passed between the British and German

Above: **Charles Brown visited the Channel Islands immediately after their liberation in May 1945. Once there, he strayed from his usual habitat of aeroplane cockpits to record some very revealing and historically important sites and we have included many in this book. This picture shows the other 'cottage' — No. 3 Gun at La Houguette. Note the white cross of surrender laid out on the ground (Charles Brown).** *Below:* **The gun pit is now an unusual playground for children attending the local school built in 1976.**

governments arguing the legality of each others case. Then, although Hitler had signed the order for the deportations from the Islands, the whole matter was inexplicably forgotten.

The story continues next with the action of another government — the Swiss. The following year, the protecting power was engaged in negotiations for an exchange of seriously-wounded prisoners and, perhaps ignorant of the possible consequences, suggested that any Channel Islanders who wanted to go to Britain should be included. The deal proposed came before Hitler for consideration in September and he must have suddenly had a fit when he realised that the people he had ordered to be deported a year before were still enjoying their relative freedom.

An immediate investigation was mounted to find out what had gone wrong and an incredible administrative muddle came to light. Many of the records had been lost in a fire at Hitler's headquarters in East Prussia and there was some speculation that this may have been started deliberately as a cover up. Anyway, it transpired that while the Foreign Ministry had assumed the military authorities were carrying out the deportations, the latter had assumed from an ambiguous telephone call that the matter was to be held in abeyance. Despite the bungling of the reputedly-efficient Teutonic war machine, it lost no time in correcting its error. A priority signal was transmitted to the military governor of France who passed the order on to the Feldkommandant for immediate action. When the civilian administration heard about

The ageless art of wall-painting. These partly obliterated German paintings were photographed at No. 1 Gun site.

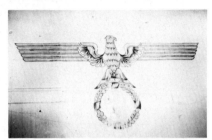

The fourth gun position — that of No. 2 Gun — lies in La Vieille Rue. Purchased by Mr. John Cross, he has incorporated it into the garden of his new bungalow (appropriately named 'Mirus') and, filled with earth, it forms an excellent vegetable plot.

Above: The 'Leitstand' or Battery Command Post for Mirus was located here in this underground control room just off the Rue de la Croix Creve Coeur. A range-finder was mounted on the roof (Royal Court). *Below:* The battery's Giant Wuerzburg radar stood on the opposite side of the road (Royal Court).

Today the command post, owned by Mr. K. Fitzgerald, is used as a combined cold store and party room while only the base of the radar dish remains standing. We deliberately took our comparison photograph from a different angle to avoid foliage which has grown up since the war and now blocks an exact match.

The one feature of the war in the Channel Islands that probably everyone has heard about is the tunnels. Many are the legends about piles of war material sealed up and forgotten, and they have led to at least two tragic deaths (on Jersey). The tunnel beneath St. Saviour's Church in Guernsey, Hohlgang 12 used for munitions and transport, is one of the most interesting as it was never cleared in the post-war scrap drives. It had been opened two or three times prior to our visit — the last being in 1969 by Richard Heaume at his own expense when he cleared it of all the contents. The photo *above* and illustration *opposite* are reproduced from 'Festung Guernsey', the German documentation of the island's defences (Royal Court). Our comparison photograph *below* shows only the very top of the tunnel entrance resealed after Richard's excursion into the unknown.

HOHLGANG 12
MUNITION

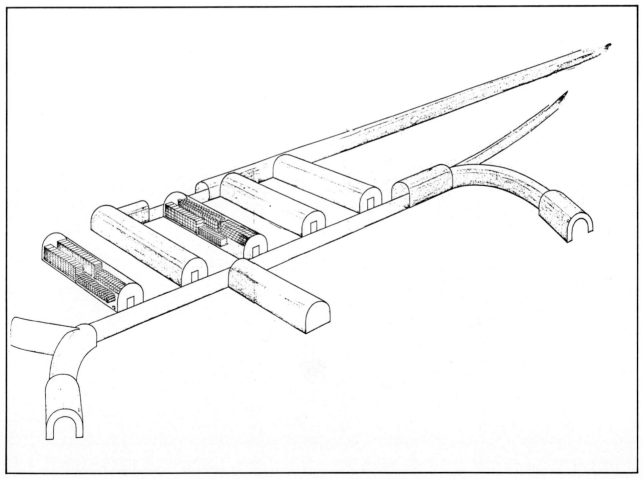

The underground hospital in Guernsey (at La Vassalerie in St. Andrew's) is the largest German construction to be seen today in the Channel Islands. The tunnels run for 1¼ miles and cover an area of 75,000 square feet. Some 60,000 tons of solid rock were excavated during the 3½ years work by the Organisation Todt. At least twenty-three workmen lost their lives in the construction phase although their remains are not set in the concrete of the walls as the more lurid storytellers would have you believe. They were buried at Les Vauxbelets and exhumed and taken to France in November-December 1961. The hospital was only used as such for a short period after D-Day for wounded brought in from Normandy. The hospital complex was linked to an ammunition magazine but the boxes of ammunition had to be kept covered with tarpaulins because of condensation in the summer.

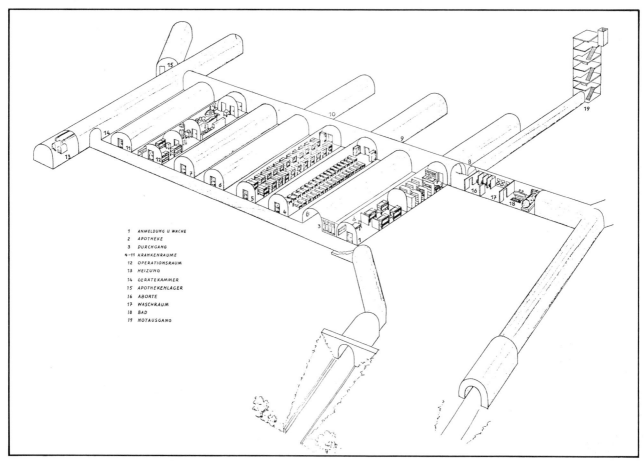

Translation of key: 1, reception and examination; 2, dispensary; 3, connecting passage; 4-11, wards; 12, operating theatre; 13, boiler room; 14, workshop; 15, medicine store; 16, toilets; 17, washroom; 18, baths; 19, emergency exit (Royal Court).

it, it protested vigorously, reminding the Germans of the undertakings given to the Islands in the surrender agreement that the lives and liberty of the population would be guaranteed. However, Dr. Brosch, the Nebenstelle representative on Guernsey and Oberst Knackfuss on Jersey said that nothing could be done as it was a direct order from the Fuehrer.

Newspapers had already announced the categories to be deported and 825 men, women and children left Guernsey on September 26 and 28. (1,200 had already been shipped from Jersey on September 16, and eleven from Sark.)

Life for those that were left continued in its same monotonous round. Following the Nicolle/Symes incident in 1940, all wireless sets had been confiscated as a reprisal but these were returned before Christmas. By June 1942, however, with the BBC broadcasting 'subversive' news to the Continent, the sets were again withdrawn although this only led to the building of crystal sets, the construction methods of which were also given out by the BBC!

There was a scarcity of food although the black market flourished as it did in every occupied country with considerable quantities of provisions being smuggled in from France. By the end of the occupation, a pound of butter was costing £2.50; tea between £20 and £30 a pound and a similar weight in tobacco £112. In any such situation, bartering becomes an accepted way of conducting business but this was outlawed by the small Geheimfeldpolizei unit on Guernsey at the beginning of 1942, possibly to avoid the corruption of the occupying troops.

Another problem for the populace was how to keep occupied. Large tracts of the island were out of bounds, beaches were mined, the curfew restricted travel, and the blackout became total after 8.00 p.m. when the electricity was shut off — all of which contributed towards a dismal existence. The Germans did try on occasions to break the monotony, like the boat excursions allowed to Sark on Easter Sunday and Monday in 1941, although, on the other hand, such concessions were negated by, for example, forbidding a theatrical group to travel to Jersey to perform.

Sabotage or other anti-German acts, even the painting of pro-British slogans on walls, was actively discouraged by the Controlling Committee. With a captive population, reprisals by the Germans would have been all too easy and it was felt that a live-and-let-live attitude would be of more benefit. However, of all the Channel Islanders, Guernseymen were the most active in this sphere which led to a special proclamation being issued specifically to them by the military government HQ in Paris:

'People of your Island! Your destiny and welfare are in your own hands. Your home interests demand that you should refrain from, and to the best of your power prevent, all such actions, which must inevitably be followed by disastrous consequences'.

The first brothel for German forces was opened in Guernsey in February 1942. (There were three altogether: one at St. Martin for officers; one in Saumarez Street, St. Peter Port, for ORs and one in George Street for the OT.) Thirteen Frenchwomen were installed in a house in Saumarez Street and, on the orders of the Feldkommandantur, they had to be provided with ration books by the civilian authorities although the latter argued that the women were part of the military set-up. To add insult to injury, the level of rations awarded was far above that granted to the ordinary civilian as they were classed as 'heavy workers'! Also the doctor in charge of the States Venereal Diseases Clinic was detailed to carry out a twice-weekly inspection (with a once-and-for-all inspection of female OT workers) and, although the doctor refused on the grounds that it was the job of the military,

This is the site of the OT foreign workers camp at Les Vauxbelets, St. Andrew's, now being developed as an 'up-market' housing estate. The workers' cemetery was situated adjacent to the camp.

Above: This is another of Charles Brown's pictures taken in May 1945. Although a pencilled caption indicated that the gentlemen in the photo are Russian PoWs, the location of the supposed camp was somewhat of a mystery. We spent several hours looking for roofs similar to that on the bungalow in the background and in the end had to leave the problem with Ken Tough. After checking every chimney pot on Guernsey he determined that it had been taken on the edge of a camp which once existed in St. Sampson's beside La Rue Sauvage. *Below:* Thanks to Ken's efforts we can let you into the secret — the bungalow is called 'Mon Desir'.

The OT was brought in to add pillboxes and gun emplacements on the ramparts and these have been retained today as part of the continuing history of the castle (Bundesarchiv).

Castle Cornet, founded in the reign of Henry II, formed the cornerstone of Guernsey's defences for over five hundred years and was three times captured by the French. In 1672 lightning struck the magazine and the resulting explosion completely destroyed the medieval buildings. The castle was repaired; heavily fortified in the Napoleonic Wars and again became an important stronghold under the Germans (Bundesarchiv).

The bell in this emplacement, which stands close to the site of the original belfry destroyed in 1672, was presented by the head teacher and pupils of Hautes Capelles School (Bundesarchiv).

Above: This picture was taken on Town Bastion — the stonework enabled us to pinpoint the precise crenel (Bundesarchiv). *Below:* Unfortunately the cannon had been changed to an inferior little beast so we were particularly thrilled to find the original German 20mm cannon still in situ on the top of the Citadel illustrated on the opposite page. Sadly today it is no longer there having since been damaged beyond repair by over-enthusiastic boy scouts.

Above: **The boom is an indispensable part of the defence of every harbour in wartime. Here the boom defence tug tows the suspended steel curtain between White Rock, as the pier in the background is called, and Castle Cornet so closing the entrance in the** *photo below* **(Royal Court).**

Above: **Nothing impinges upon the safe passage of the many yachts that now make the harbour their sanctuary.**

his arguments were overruled. By October 1942, in an attempt to control VD, an order was given to all women who had been so treated to avoid any sexual relations during the next three months' 'quarantine' period. By mistake a copy of the order was given to the Dame of Sark — we can imagine her indignation — although her American-born husband, Robert Hathaway, saw the funny side of it and quipped: 'They must be having a lot of fun and games in Guernsey, which is more than can be said for Sark'!

By 1944, another problem was the increase in crime as both soldiers and OT workers, and some civilians, resorted to theft to obtain enough food to stay alive. After D-Day, when the Islands were cut off from their supply depots in France, the position became progressively worse and it became obvious that the Allies had no intention of trying to capture the Islands but intended to starve them into submission. Such a policy, although meritorious as far as obviating the tremendous loss of life that would have occurred in the attempt to take such heavily fortified islands, had severe consequences for the populace.

Because the supplies of cement and other construction materials had dried up, the Germans began by evacuating all the OT workers which temporarily eased the drain on food supplies. By July, flour for breadmaking on Guernsey was down to forty-four days' supply and stocks of everything else were dangerously low. When the main supply port of St. Malo fell in August, an appraisal of the total Channel Islands' situation in August stated that the troops would run out of food in three months.

By September, food stocks for both troops and civilians were giving cause for concern and, although the position was not yet desperate, in order to prolong what stocks were left, the German Foreign Ministry sent a message through the Swiss to the British Government. This stated that: '. . . on the former British Channel Islands supplies for the civilian population are exhausted'. The Germans proposed two choices: either let the population be evacuated or allow the re-supply of the Islands.

Winston Churchill's answer was short and to the point: the responsibility for feeding the people of an occupied country was entirely that of the Germans. Although his Chiefs-of-Staff and the Home Office wanted to send food, the Prime Minister reiterated that, 'I am entirely opposed to our sending any rations to the Channel Islands ostensibly for the civil population but in fact enabling the German garrison to prolong their resistance. . . . It is not part of our job to feed armed Germans and enable them to prolong their hold on British territory.' His view was that the Germans should be called on to surrender forthwith.

This reply was not one which the Germans

The graffitti of forgotten armies. *Left:* **In the concrete of a Castle Cornet pillbox, six Germans left their mark in 1944.** *Above:* **The following year men came from another army — this inscription can be seen on the Castle Breakwater.**

had been expecting. Hitler declared that there was no question of him surrendering the Islands or of trying to evacuate any civilians. (There were 23,000 on Guernsey and another 39,000 on Jersey.)

While the population starved, German legal experts studied the rules of the Hague Convention concerning the treatment of civilians of an occupied nation. Their view was that there was no obligation under international law for their army to feed the people and that it was not unknown for occupied territories to be provisioned by other nations. Nevertheless, the strong line taken by Churchill led the Germans to believe that the Allies had realised that the position was not as serious as they had tried to make out.

While the Germans argued amongst themselves as to how to reply, pressure was mounting in Britain for the Government to do something to ease the reported plight of the Islanders. Both Parliament and the Press, believing what the Germans had said to be true, pressured Churchill to modify his hard-line approach. By November 7 he had relented and agreed that food parcels and medical supplies be sent through the Red Cross on the understanding that the Germans continued to be responsible for the basic civilian ration.

The British offer was transmitted to the

'Boots-boots-boots-boots-movin' up and down again!' Rudyard Kipling. A German column marches past St. Peter Port Parish Church but their days on the island are numbered (Carel Toms Collection). Today passers-by hurry across the same spot no doubt oblivious to the tramp of former marching feet.

Within their short five-year tenure, the face of the Islands inexorably changed for ever. *Left:* These troopers look out across St. Sampson's harbour to Vale Castle and the stone-crushing plant essential for the construction work (Bundesarchiv). *Right:* A 105mm coastal gun casemate constructed after the wartime photo was taken precludes one from standing on the same spot.

The Red Cross ship, the SS Vega, which performed the 20th century role of Florence Nightingale during the desperate last months of the war (R. H. Mayne).

Germans via the British Legation in Berne and, while the Germans' official reply was awaited, their unofficial action was revealed: they immediately confiscated all the civilian food stocks and reduced the ration to a bare minimum. Churchill was furious when he heard the news but was on the horns of a dilemma as the report had come from a top secret Ultra decoded message. An inviolate principle of Ultra was that no intelligence so received could be used if the resultant action might lead the Germans to believe that their signals were being compromised. Thus Churchill could not reveal that he knew of the confiscation order and had to wait instead, in frustration, for the official German reply.

As early as July, Bailiff Carey on Guernsey had tried to send a letter to the International Red Cross to point out the mounting difficulties the Islanders were facing. By October, when no reply had been received, he assumed (quite correctly) that the Germans had not passed on the letter and the Bailiff then decided the only alternative was to try to send someone to England with a message although he was well aware of the risks. However, before he could act, a group of civilians took matters into their own hands and one of their number, a retired Merchant Navy captain, Fred Noyon, left Guernsey with a friend, ostensibly on a fishing trip, on the night of November 3. The men were picked up by an American ship and they reached London on November 12. By that time, the British Government, through Ultra, was already aware of the situation and, as we have seen, had already taken the decision to send in food parcels but when these arrived it was naturally assumed that Fred Noyon had been responsible for saving the island. (The BBC had broadcast a coded message on a Forces' programme on December 9 to inform the group in Guernsey of Noyon's safe arrival.)

On Wednesday, December 27 a Red Cross Ship, the *Vega*, reached Guernsey. On board were over 100,000 food parcels — more than enough for every civilian on the Islands — together with toilet and medical supplies and cigarettes. The Bailiff of Jersey made his first visit to Guernsey the following day so that he and Bailiff Carey could make plans for its distribution. To give him his due, the German B.d.B.K., Oberst Graf von Schmettow, issued strict instructions to the troops that everything must go to the civilians — a ruling which was strictly enforced even as far as forbidding the acceptance of gifts by his men. On December 29 the *Vega* sailed for Jersey to unload that island's share.

One regrettable result of the delivery was an immediate increase in the crime rate with the theft of food from civilians who were now better off than the Germans. On the pretext of preventing hoarding, the Germans instituted raids on private houses and one direct result of these searches was that much foodstuff was 'confiscated'. At the same time there were some remarkable discoveries and one man was found to be sitting on a cache of 363 tins of provisions and 365 bottles of spirits!

The *Vega* returned in February and thereafter made monthly trips until the surrender. During her six voyages from neutral Portugal she had brought nearly 460,000 food parcels, many sent from Canada and New Zealand, and, by way of a token of their appreciation to the International Red Cross, the Islanders collected £170,000 for the organisation after the liberation.

In February, command of the Islands was transferred to Vizeadmiral Friedrich Huffmeier and it was to him that the Allies

Above: **A frozen moment in time. After years of skimping and saving, the Red Cross parcels arrive. The little ones seem oblivious to the manna which has suddenly arrived (Carel Toms Collection).** *Below:* **Bulging with luxury goods at favourable duty free prices, Le Riche's at the junction of Smith Street and High Street, St. Peter Port presents a new face to the world. The First World War memorial plaque has been replaced by a ciggy advert.**

On May 9, 1945, Vizeadmiral Friedrich Huffmeier, the then-reigning Befehlshaber der britische Kanalinseln, had the ignominious duty of surrendering the Channel Islands. He is pictured here leaving his HQ at the Crown Hotel. For photographs of the hotel then and now see page 250 (R. H. Mayne)

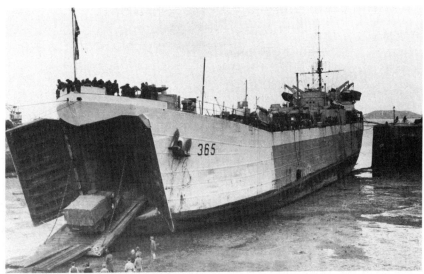

Above: **An LST squeezes between the North and Albert Piers to unload in the inner harbour where a convenient stone ramp** *(below)* **leads up to street level (Carel Toms Collection).**

transmitted a wireless message on May 3 asking him if he was ready to surrender. His reply, sent on the evening of May 6, pointed out that, '. . . the Commander-in-Chief Channel Islands receives orders only from his own Government'.

Meanwhile, German army commands all over Europe were signing individual surrender documents (in Italy as early as April 29) and the overall unconditional surrender of all German forces was signed at Reims at 2.41 a.m. on the morning of May 7. It specified that all hostilities were to cease at midnight on May 8. However, as it was feared by the Allies that Vizeadmiral Huffmeier might not have received this news, a further signal was sent asking him to send envoys to rendezvous at sea four miles south of Guernsey's Les Hanois lighthouse (two miles off the south-eastern corner of the island). The Vizeadmiral replied that as he now had the necessary authority his representative would be at the designated location at noon on May 8.

For months the Allies had prepared contingency plans for the liberation of the Islands, although the operation, codenamed NESTEGG, was dependent on a peaceful takeover. There was no thought of taking the Islands by force of arms. A special task force, No. 135, under Brigadier Alfred Snow (with an attached Civil Affairs Unit) had been detailed for NESTEGG and the Force Commander left Plymouth on HMS *Bulldog* accompanied by HMS *Beagle* to rendezvous with the Germans. The Vizeadmiral's representative was Kapitaenleutnant Armin Zimmermann who was transported to the *Bulldog* by rubber dinghy from his minesweeping trawler. He told Brigadier Snow that he merely had instructions to take the 'armistice' terms back to his chief. The Brigadier pointed out that the meeting had been arranged to accept the immediate and unconditional surrender of the German forces and that if he was not enpowered to sign, then someone who could must return by the midnight deadline. The German agreed but arrogantly warned the Brigadier to move away from the coast otherwise this would be construed as an invitation to open fire. Astounded by Zimmermann's effrontery, Brigadier Snow is reported to have replied: 'Tell Admiral Huffmeier that if he opens fire on us we will hang him tomorrow!'

Relics of the surrender. The rum cask on which Generalmajor Siegfried Heine signed the capitulation of Guernsey at 7.14 a.m. on May 9 on the quarterdeck of HMS Bulldog. Her bell is also preserved with the cask in the German Occupation Museum in Castle Cornet.

In . . . and out. Where the Germans strutted in May 1940 Jolly Jack Tars lead the liberation march past five years later (Carel Toms Collection and Eric Sirett).

It was after midnight when he returned with Generalmajor Heine (since promoted) and, as the overall surrender was then in force, Brigadier Snow decided to steam into St. Peter Port forthwith. At 7.14 a.m. on May 9 the surrender was signed on the quarterdeck of the *Bulldog* and a small advance party under Lieutenant-Colonel Stoneman went ashore to hoist the Union Jack. The Brigadier then transferred to HMS *Beagle* to sail to St. Helier to accept a similar surrender for Jersey.

The Royal Hotel, once used for accommodating senior German officers, now caters for a different clientele.

Superman's performance lasted five years although he staged a brief come back, for one week only, in April 1979! (Carel Toms).

Soon after the outbreak of war in September 1939, Springfield Stadium in St. Helier was commandeered as a First Aid Post. This is the main entrance (Societe Jersiaise).

CHAPTER TWO
Jersey

Unlike Guernsey, throughout the entire period of occupation Jersey benefitted from strong leadership. As soon as the evacuation programme was announced on June 19, 1940, the Bailiff, Mr. Alexander Coutanche, set a powerful example to his 50,000 islanders when he publicly declared his personal decision to stay on. As a consequence, of the 23,000 people who registered for evacuation in the early moments of panic, only 6,600 actually left the island — a mere thirteen per cent.

A Superior Council was set up (the equivalent of Guernsey's Controlling Committee) to deal with the day to day administration of the island pending the arrival of the Germans. One early problem was that of money supply as all those who had chosen to leave for the United Kingdom — both service volunteers and civilians — wanted to draw out their savings from the banks. There was no alternative but to bring in some £300,000 from England to satisfy demand, but this carried the risk that it might easily fall into German hands. At the same time, a move in the opposite direction was taken when it was decided to transfer all securities and negotiable bonds to the United Kingdom for safe keeping. Altogether eighty sacks were shipped to London so denying the invaders a convenient nest-egg worth millions of pounds.

When the Germans came, they adopted the same methods as had been used on Guernsey. The bombing attack on June 28, carried out at the same time as that on the other island, killed ten civilians. Then, early on the morning of July 1 (the day after the successful landing at Guernsey aerodrome), a message was dropped on the airfield calling for the surrender of the island. Dr. Hermann Kindt, a German war correspondent, describes what happened:

'As a result of the information acquired, Admiral Lindau decided to call on Jersey to surrender. It was 2 o'clock in the morning when the summons to surrender, signed by the Admiral, reached the squadron. There were three summonses each for Jersey and Alderney, all in the same wording. As it was a letter of parley, the usual coloured pouch could not be used. New ones were quickly sewed with white pennants attached, cut from the bed linen belonging to the captain of the French squadron whose deserted quarters had now been taken over by the German squadron. It was still dark when the machines started, one for Alderney, and the other, in which was Leutnant Richard Kern, for Jersey.

'The Islands were reached in the early hours. Once more a German plane was droning over Jersey. Only a few islanders were up, but the pouch containing the summons to surrender which had been dropped was seen, found, and taken to the authorities.

'Later it was recounted in the island how surprised the dreamy town of St. Helier was to receive such an early visit. An hotel porter found the letter addressed to the Civil Governor, and took it to the Bailiff. After dropping the notes the two planes returned. The Admiral wished to wait for the signs of surrender. He had stipulated that white flags should be flown as a sign of surrender. It can be imagined how tense the German airmen were during the period of waiting. There could be no peace of mind until such an undertaking had been brought to a successful conclusion.

'Wild rumours were running through the Services. According to one of these, an English cruiser was supposed to have made its appearance in the vicinity of Jersey. In order to clear up this point, Leutnant Kern was sent

Springfield Stadium forty years later. The actual football ground was used as a vehicle depot in 1945 (see page 175).

Above: **The airport at St. Peter's as it would have appeared to Leutnant Richard Kern on his reconnaissance of the island. It was opened on March 10, 1937 — landing facilities on the island before that date being on the 'beach' aerodrome in St. Aubin's Bay between West Park and First Tower — services being conditional on the tide! The new airfield provided four grass runways: N-S, 528 yards; NE-SW and SE-NW 720 yards with the main E-W having a length of 980 yards. An up-to-date terminal building was provided together with control tower which had the latest Adcock direction-finding system and four 1,000 candle-power floodlights could illuminate the entire landing area. It was stated at the time that the airport was the 'most completely equipped of any in the British Isles apart from Croydon'.**

Below: **Jersey Airport in February 1981. The airfield was taken over by the Royal Air Force on May 10, 1945 and handed back to the States of Jersey on October 2 of the same year. St. Peter's Barracks, in the foreground of the pre-war photo (in latter days used for housing homeless people), were demolished after the war and the tarmac runway laid down in 1952 with a length of 1,400 yards. This was subsequently extended in six stages (1956, 1958, 1959, 1960, 1965/6 and 1974/5) to its present 1,865 yards. The terminal building was modernised and extended in 1970. All air traffic in the Channel Islands operates in the controlled airspace of the Channel Island Control Zone up to 19,500 feet with full radar cover as we know to our cost when taking the aerial photographs for this book!**

Left: **Air raid damage to the Star Hotel on the corner of Conway Street, St. Helier from the June 28 attack in which nine civilians were killed outright, the tenth later succumbing from wounds (R. H. Mayne).** *Right:* **Now the Williams & Glyn's Bank with no trace of the spang marks on the wall.**

back to the island. He arrived without being attacked and could see nothing of any active defence. The island lay there just as peacefully as on the previous day. Then, as the machine flew low over the beaches, gardens and town, an idea came to the Leutnant. He saw life going on peacefully below him; he saw the beautiful island, and he had been charged to bring back accurate information. As they flew towards the airport, the beautifully-situated landing ground with its elegant white buildings, he made his decision. He would take the island.

'The plane banked over the flying field and he gave the order to land. What were the sensations of the crew? If the field were mined then it would be all over; if there were means of defence, then it wouldn't be much better for them. But if everything went smoothly — and no one doubted but that it would — then they would be amongst the first Germans to set foot on British soil. Lonely and deserted the machine rolled over the wide, empty field. Then Leutnant Kern gripped his pistol and jumped to the ground. He strode towards the administration building followed by the machine, its machine guns at the ready. Nothing happened. Finally, from the airport building emerged an excited man, who to the astonishment of the newcomers, spoke in German. He took the Leutnant to the telephone and got in touch with the Bailiff.

'Yes, the Bailiff had received the summons to surrender. Why were the white flags not flying? Because the Bailiff had to wait for the decision of the States, and the States had, in the meantime, agreed to unconditional

Leutnant Kern (second from right), looking rather pleased with himself after his solo effort, looks on as Staffel Kapitaen Obernitz meets Attorney-General C. W. Duret-Aubin and the Bailiff Alexander Coutanche. Coutanche was knighted in 1945 for his sterling wartime service, created Baron in 1961 and died on December 18, 1973. Unfortunately the extension of the terminal building makes a comparison photograph meaningless.

Something new for the people of Jersey to get used to — this is Charing Cross, St. Helier, then and now (Bundesarchiv).

Those boots again . . . this time marching down York Street, St. Helier (Bundesarchiv).

surrender. The Bailiff requested that the Admiral be informed of this. Leutnant Kern informed him that the island was under German occupation.

'Shortly afterwards, the planes which had been sent for from the squadron arrived. They were packed with men. While these machines were appearing over the island, the white flags were beginning to be flown. It was a strange sight that met the gaze of the Germans, flying under the blue sky of this summer day, that of a pleasant town from which all forms of white, from sheets to pocket handkerchiefs, were shining. Workmen were engaged in painting white a large section of the airport measuring many square metres.

'When the captain of the squadron landed, the Bailiff, the Government secretary and the chief of police were waiting to receive him in the airport building. The surrender was concluded in a short time, the British officials maintaining a correct attitude. Here also the relations between the troops of occupation and the local authorities have not degenerated. The first orders were given by the captain of the squadron at the airport. The few men who had come with the first machine immediately took over the apparatus used for communications. The rate of exchange was set and a curfew ordered for the island. The Germans then drove in the waiting car to St. Helier, on the way being stared at by the "bobbies" and population. It all happened so naturally that it seemed that the squadron had nothing else to do in the war but occupy the Channel Islands.'

This is the only known photograph of Hauptmann Erich Gussek, the first Inselkommandant of Jersey (he lasted just eleven weeks before being replaced). He is seen greeting Generalfeldmarschall Walter von Reichenau on a visit to Jersey (R. H. Mayne).

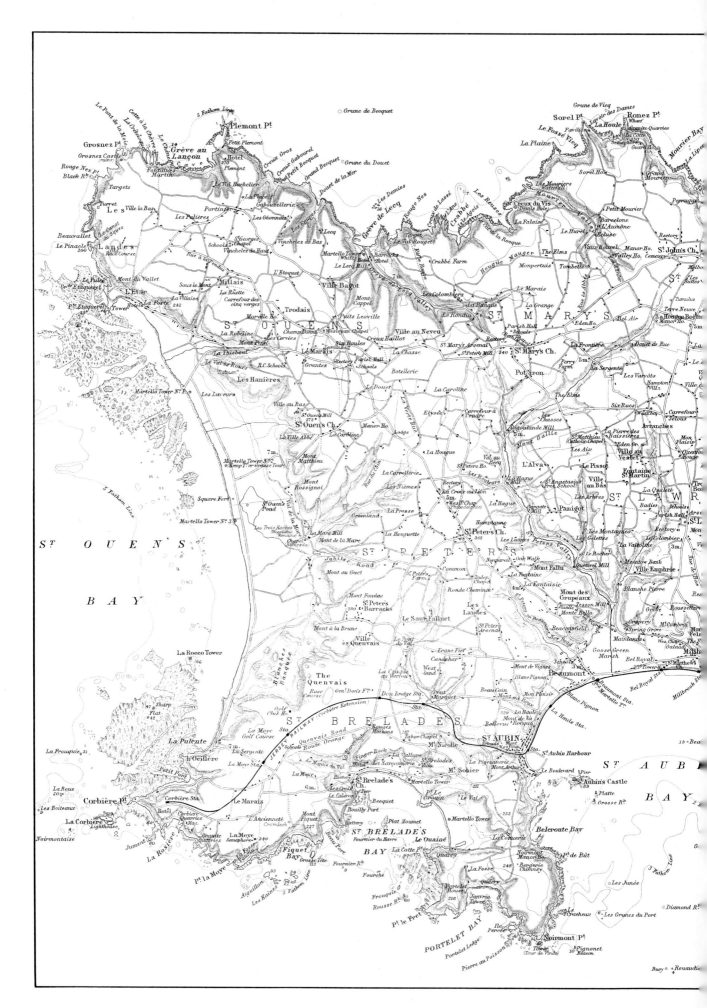

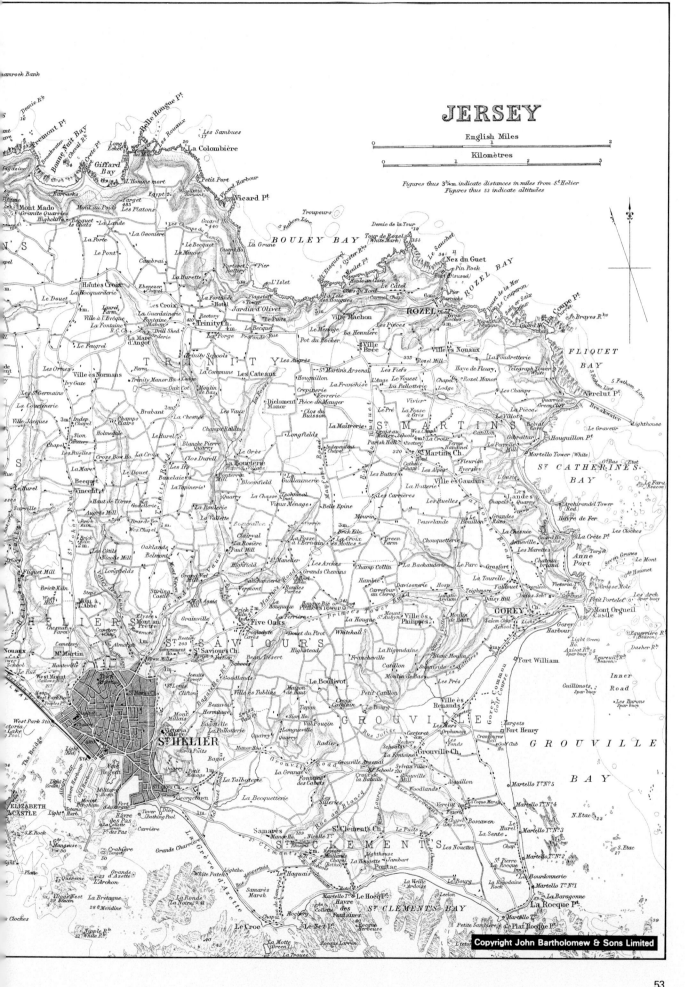

The headquarters for the entire German command of the Channel Islands was located in 'Monaco' in St. Saviour's Road until it went to the Hotel Metropole on April 23, 1941 (Societe Jersiaise).

Monaco stood next to the Jesuit Friends Training College — Maison St. Louis as it was then called. Now it is the site of a multi-storey car park for the Hotel de France.

Left: Although the Town Hall in York Street became the Kommandantur, to the locals it merely became the 'Rat House' in the inevitable corruption of the German Rathaus (Societe Jersiaise). *Right:* Little change in forty years — it's still 'official parking only'.

Feldkommandantur 515 HQ was at Victoria College, Bagatelle Road, St. Saviour's, now Gateway College House (Societe Jersiaise).

Above: A common German defence position was the Tobruk emplacement, seen here giving flanking cover at Resistance Point Third Tower (CIOS). *Below:* A corner of St. Aubin's Bay which has changed little in the intervening period — only the goods on offer have altered at the Gunsite Cafe.

Oberst Graf Rudolf von Schmettow, the first B.d.b.K., had the overall responsibility of fortifying the Channel Islands.

The first ordinance of the occupation of Jersey was published by Hauptmann Erich Gussek, the Inselkommandant, on July 8 along the same lines as that put out in Guernsey six days previously. A working relationship was quickly established by the Bailiff and the German authorities, and Mr. Coutanche later commented that Hauptmann Gussek was perfectly happy to let civil administration continue under his control.

Two days later after the ordinance was published, the Luftwaffe began attacking British shipping in the English Channel in preparation for the invasion of England. Victorious in France, and having taken the Channel Islands in a bloodless coup, the subjugation of Great Britain was seen by the Germans to be only a matter of time. Over the next few weeks, as the air attacks switched their emphasis from coastal shipping . . . to ports . . . to RAF aerodromes and finally to London, few Germans believed that these changes in direction were anything other than the planned moves in Hitler's master plan. However, as summer turned to autumn, the victory, which had earlier seemed to lie almost within their grasp, slowly faded; and as it faded came the realisation that the war would not soon be over.

In the middle of the Battle of Britain, Oberst Schumacher arrived in Jersey as the new Feldkommandant; the overall military commander of the Channel Islands, Oberst Graf von Schmettow, taking up his office in 'Monaco' in St. Saviour's Road on September 27, and it was on him that the responsibility of fortifying the island rested. As Hitler's thoughts turned from England to the Soviet Union, a series of instructions were issued for the installation of a defence system which would eventually absorb one twelfth of the resources allocated to the construction of the

entire Atlantic Wall. Although the Islands only occupied the merest fraction of its 1,500-mile length, by the beginning of 1944 they had absorbed eight per cent of the entire concrete allowance and, more remarkably, an almost equal amount of spoil had been excavated on the Islands as in all the strongpoints on the entire European coastline. It has been said that had the materials and labour used in the Islands been utilised to strengthen the coastal defences on the Continent, these would have been some ten per cent stronger than they were when the Allies invaded in 1944.

Hitler's directive of October 20, 1941 laid down that the Channel Islands were to be converted into an impregnable fortress. By July 1942, there were fifteen 'Festung'-status fortresses on the European coastline but the Islands were not officially designated as such until March 3, 1944.

Experts from the Inspectorate of Fortifications had already been surveying the Islands for the past five months and Dr. Fritz Todt arrived in person to lay down areas of responsibility for the construction programme. These split up the defence work into six separate parts:
1) 319 Infanteriedivision whose role would be that of tactical reconnaissance.
2) Individual troopers would be responsible for the construction of normal field works, i.e. trenches and weapon pits, etc.
3) The Division's engineers were to handle the laying of land mines and distribution of flame-throwers.
4) The Construction Battalion would build those works which were more elaborate than the field type but which were not intended to withstand a prolonged bombardment. The wall of such buildings would have a maximum thickness of one metre.
5) The Organisation Todt would provide the labour for all building in the heavy construction programme covering buildings with walls above one metre thick, quarrying, tunnelling, the construction of roads and railways, organising transport, unloading ships and the supervision of civilian construction firms. To the OT, Jersey was codenamed JAKOB (with GUSTAV and ADOLF for Guernsey and Alderney).
6) The job of the Fortress Engineer Headquarters and Fortress Engineer Construction Battalion was to oversee the heavy construction programme, carry out the mounting of heavy guns, some tunnelling and to supervise the OT.

A prerequisite for the construction programme on the island was an efficient method of transport. In 1936, a disastrous fire at the St. Aubin's terminus of the Jersey Railway and Tramways Company destroyed most of the rolling stock and had led to the closure of the line. Now the line was re-opened and extended under its new OT management although not to a standard gauge, it being a metre to the west of the island and 60cm to the

Construction of a railway to help carry building materials to the sites for the various defence works was essential. The first routing problem which had to be solved with the line to the east of the island was how to cross English Harbour. *Above left:* **This timber-piled bridge, with a section which could be raised in the middle, was the answer** (Evening Post). *Above right:* **Today all trace of the bridge has gone — except for these two solitary fixing bolts on the southern side** *(below).*

The next obstacle to be overcome was Mount Bingham crowned by Fort Regent. This tunnel, which still has the railway lines embedded in the floor, was blasted to make a way through to the Havre des Pas.

Come summer, this promenade (seen here in April 1979) is alive with tourists enjoying a walk along the Havre des Pas beach.

Little do they realise that in July 1943 it carried a German railway to Grouville Common (Evening Post).

east. The OT's intended expansion of the railway was based on an eight-year plan although, in the end, only twenty-one months of this were actually carried into practice.

With due ceremony, the railway was officially re-opened on July 15, 1942 by Oberst Graf von Schmettow. Locomotives had been imported from France and 200 wagons were requisitioned from the French firm of Paul Frot et Cie and repainted with the words 'PAUL GORGASS-POSEN' — the name of the Polish firm which received them. Utilising the original route which ran along the St. Aubin's Bay promenade, branch lines ran down into the harbour; up through Le Perquage to the German power station; across Pont du Val around St. Peter's village to Ronez quarries and to Corbiere — the entire metre-guage system covering some thirty kilometres. In the eastern part of the island, the 60cm-gauge line, after overcoming many difficulties in the harbour area by means of a bridge and tunnel, more or less followed the route of the Jersey Eastern Railway which had been closed in 1929.

Because the Islands had been defended against previous threatened invasions by the French, Martello towers and other early

Another remnant of the Deutsche Eisenbahn on the island is this bridge on the road between St. Peter's and St. Ouen's on the parish boundary.

Left: The old Terminus Building of the former Jersey Railways & Tramways was not used for its original purpose. It was taken over by the Standortkommandant (Garrison Commander) and the Nachschuboffizier (Supply Officer) who was also responsible for all troop travelling arrangements between the Channel Islands (CIOS). *Right:* Still in the travel business in 1980 (M. Ginns).

Above and below: **An Army Construction Battalion at work converting one of the many forts built during the 18th century for its WWII role. This is Fort Henry on Grouville Common. Note the** addition of the searchlight platform, constructed of matching stone. A personnel shelter and ammunition magazine was constructed at its base (Bundesarchiv).

Left: **Behind the fort, overlooking the beach behind it, a 105mm coastal defence gun was installed in a casemate camouflaged to represent a beach bungalow. Note Mont Orgueil Castle in the** background (CIOS). *Right:* Like many of the derelict bunkers which remain on Jersey, this has been bricked up by the Public Works Department.

British fortifications still dotted the coastlines. These were eagerly seized upon by the planners for, not only were they sited in strategic positions, they were also very solidly constructed, usually of massive granite blocks. After modernisation they proved to be excellent locations for fire positions and observation posts.

The largest guns to be installed in the Channel Islands were, as has been described, the 305mms of Batterie Mirus on Guernsey. The biggest emplaced on Jersey were the four 220mm guns of Batterie Roon at La Moye Point which were mounted in open pits giving an all-round field of fire. The batteries of Hindenburg, Ludendorff and Mackensen each consisted of three 210mm guns with a 360-degree traverse while that at Les Landes,

Above: **Mont Orgueil Castle, which dominates Gorey harbour, is Jersey's most cherished ancient monument. Its position on a projecting headland over 200 feet high enables it to command the entire east coast of the island. Its construction was begun in the 10th century by the Dukes of Normandy and, until the prison was built in St. Helier, it served as the island's jail. The Germans used it as a command post and observation point. This huge casemate below the castle served to house a 75mm Pak 40 to cover the harbour (Societe Jersiaise).** *Below:* **The bunker remained until the autumn of 1972 when Demax International from Wolverhampton began a five-week demolition job of blasting and drilling. Total cost of removing a piece of history: £3,125.**

Batterie Moltke, comprised four 155mm weapons. There were three emplacements with 150mm guns — Batterie Schlieffen, Haeseler and Lothringen. The latter is

probably the most well known as it lay at the southern extremity of Noirmont Point. It was a naval battery — Marine Artillerie Abteilung 604 — and, surprisingly, its headquarters was located on Guernsey.

Construction work at Noirmont began in March 1941 on the furthest extremity of the headland between the bays of St. Aubin and St. Brelade. Three of the guns were originally mounted on open, fixed-type emplacements on the edge of the cliff pending the construction of the permanent gun positions. The Germans had intended to replace the Noirmont guns themselves with more modern ones, the same as in the batteries on the other islands, but none were available and the existing guns were remounted (with the addition of a fourth) on the concrete platforms already prepared giving a wide angle of fire over both bays. The battery area was wired and mined for all-round defence with two 75mm guns, one 25mm anti-tank gun and six 20mm anti-aircraft guns together with sixteen fixed flame-throwers and two 50mm mortars.

The bulk of the coastal defence guns were of 105mm calibre and eighteen were installed in reinforced concrete casemates, some of which were converted from earlier constructions, i.e. that at St. Aubin's Fort.

The second most numerous weapon to be installed was a makeshift device improvised by mounting old French tank turrets on concrete emplacements. Sixty-one assorted Hotchkiss, Renault and Somua tank turrets, which mounted 37mm guns, were set on a 'Ringstand' embedded in the concrete which allowed the turret to revolve the full 360 degrees.

Another quite numerous captured piece which was put to good use was the Czechoslovakian 47mm Pak K 36(t) anti-tank gun which was usually sited to provide enfilade fire across vulnerable beaches, a total of twenty-three being installed in Jersey.

Four 75mm Pak 40s were set up in concrete emplacements covering Archirondel, Greve de Lecq, Le Grouin Point and Gorey Harbour while another eight were sited in open positions inland.

Another unusual type of installation was the massive steel turret designed to mount twin MG34 machine-guns. Called a 'Sechsschartenturm', the turret presented an all round field of fire from its six loopholes. The steel itself was over ten inches thick.

There were also another thirty assorted weapons of 37mm or larger scattered throughout the island.

Above: **Naval construction troops together with civilian labourers building a personnel shelter at Noirmont in July 1941 (Bundesarchiv).** *Below:* **Now an integral part of the headland landscape. View taken looking west across Portelet Bay.**

Left: **Heave-ho! An early gun position for a 15cm SK L/45, later abandoned when the guns were remounted on emplacements —** see overleaf (Bundesarchiv). *Right:* **Our solitary 'soldier' contemplates . . . with St. Aubin's Bay in the background.**

Above: Another of the early installations was this rangefinder, temporarily installed in July 1941 (Bundesarchiv). *Below:* It was later replaced by this underground structure protected by reinforced concrete and an armoured cupola. The arms of the rangefinder were removed and thrown over the cliff at the end of war. In the summer of 1981, thanks to the winch on Peter Le Conte's Dodge weapons carrier, the arms were retrieved by the Channel Islands Occupation Society together with other relics.

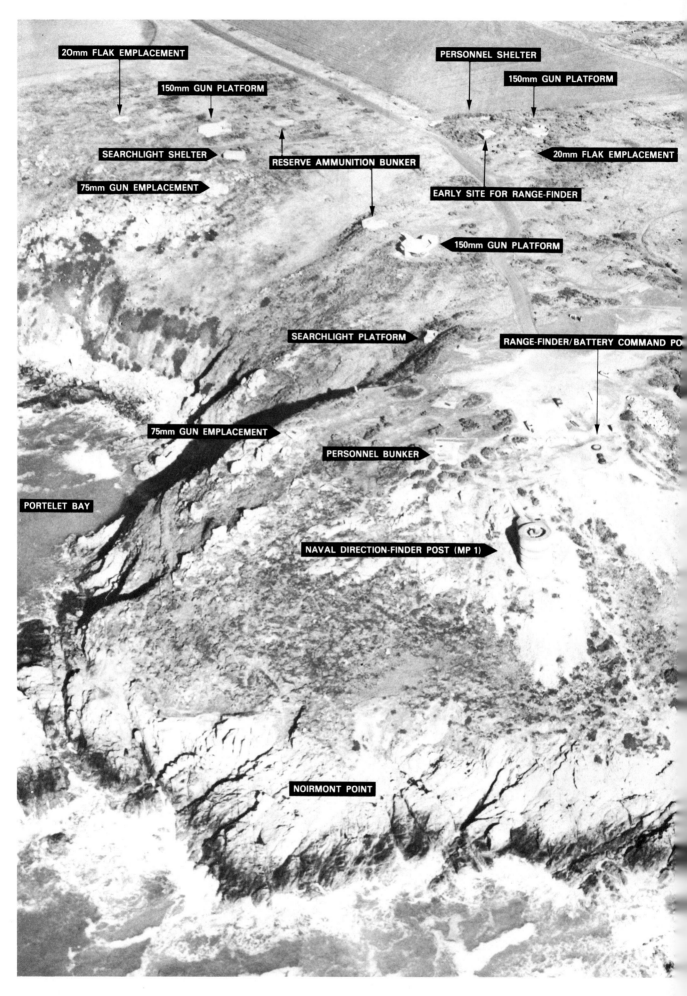

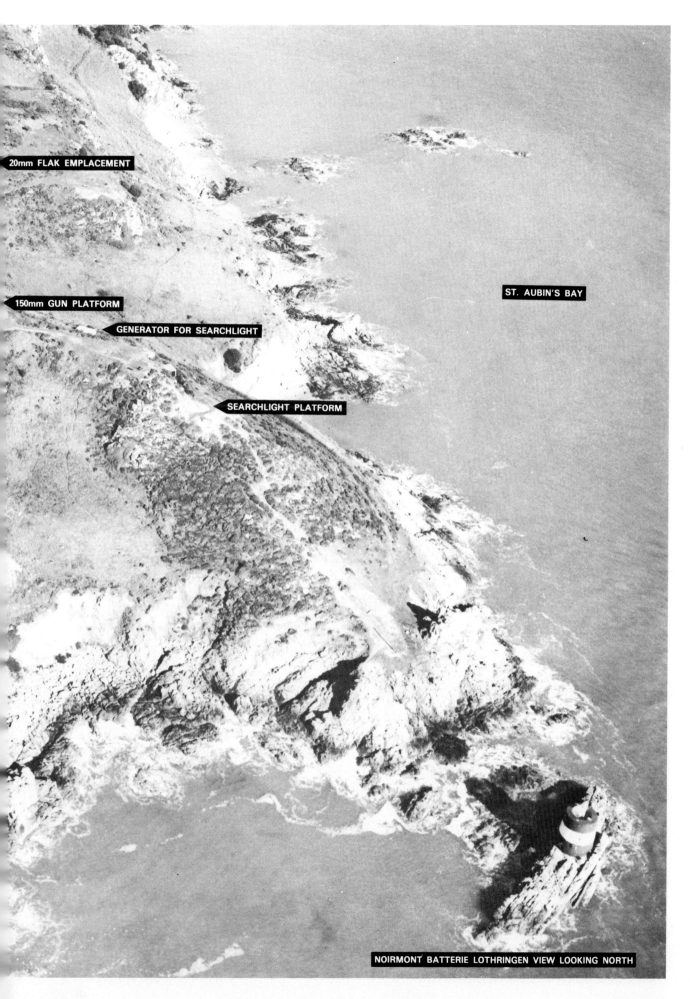

Above: On July 11, 1941 the Oberbefehlshaber West (Commander-in-Chief West) Feldmarschall Erwin von Witzleben paid a one-day visit to Jersey accompanied by the Commanding Admiral in France, Admiral Schultze; his Chief-of-Staff, Konteradmiral Litzmann; the Naval Commander-in-Chief Channel Coast, Vizeadmiral von Fischel and the Naval Artillery Commander Normandy, Hauptmann Fink. The distinguished party visited Batterie Lothringen, the naval coastal artillery at Noirmont, and the Feldmarschall is seen here inspecting a gun crew (J. W. Wallbridge Collection). *Below:* Teddy bears' picnic thirty-nine years later — photo by Michael Ginns in September 1980.

When the Society cleared out the interior *(top)*, Robin Cox found a 'Bulldog' pistol (displayed here by Michael Ginns) in a ventilator shaft. It must have been secreted there by some long forgotten crewman (Evening Post).

Visitors to the islands can now make their own inspection of the Noirmont observation tower and command bunker as the installation was handed over to the care of the Channel Islands Occupation Society in 1977.

These curious stones which can be found scattered over the headland are markers for electricity cables — the arrows indicating the direction of the run and KM possibly standing for Kriegsmarine.

Above: Another Charles Brown special taken of the Noirmont tower in May 1945. The 20mm Oerlikon on the roof still has its covers on. When we showed the photograph to Hugh Dalton, the Deputy Director of Operations for Jersey air traffic control, he wryly commented that, 'they only fly that low today over Jersey . . . once!' *Below:* We settled for this safe shot of the memorial at the Noirmont headland . . . one of the 'musts' on any itinerary for a modern-day visitor to Jersey.

St. Aubin's harbour. The prickly defences on the southern breakwater, but otherwise little changed in forty years (Bundesarchiv).

Gun positions around the island. *Left:* This is the sharp end of a Czech-made 47mm Pak K36 (t) which was emplaced in this naturally camouflaged embrasure in the wall of the Jersey Motor Transport bus garage opposite Castle Street in St. Helier. The date is late 1944 (CIOS). *Above:* Somehow it avoided the scrap man and the the ball mounting still remains today (M. Ginns).

Left: Le Grouin Point situated midway along St. Brelade's Bay with its 75mm Pak 40 (CIOS). *Above:* Note part of the wall of the 1856 firing range on the right.

Left: There was a surprise in store for callers at No. 14 Gorey Pier in 1944 — they had a choice of machine gun fire from ground level or a 37mm Pak 35/36 from the first floor! (Societe Jersiaise).

Right: Well, at least it is probably the most solidly built house in Jersey. Typical of the practical uses to which many former German constructions have been put.

Commercial concerns have also profited by the use of defence works. This 1945 photograph shows the 105mm coastal defence gun of Resistance Point Mole Verclut (Evening Post).

With the gun which formed the main armament removed, the casemate now serves as the entrance to the tunnel and vivier (or crustacean farm) which lie beneath the Gibraltar Rock.

The Germans appreciated that if the Allies invaded Jersey, St. Ouen's Bay (pronounced St. Wans) was one of the most likely landing beaches, with good sand, room to deploy and the direct route to the early objective of any attacker — the aerodrome. Its defence was therefore a priority and, by D-Day, the works would have made any landing a costly business. *Above and opposite:* **Here Army Construction Battalion 158 work on the building of early defences near Les Brayes Slipway (Bundesarchiv).** *Below:* **The anti-tank wall which ran the complete length of the beach was partly built by the firm of Theodor Elsche. Photo taken near La Carriere Point on the section designated Panzermauer 5.**

The low lying beach backing St. Ouen's Bay on the western side of Jersey was one of the most vulnerable stretches of coastline and the Germans made considerable efforts to make it invasion-proof. The Five Mile Road runs parallel with the dunes for the complete length of the bay although its name is a little ambiguous as its coastal distance is only just over three miles. Behind the road, flat, marshy ground stretches for an average of 800 yards before rising in an escarpment, some 300ft high, giving excellent observation along the entire beach. In the middle, a lake, St. Ouen's Pond, formed a natural barrier and two arms were dug from it on a north-south axis to provide a water-filled anti-tank ditch. This protected the central area, and cunningly sited gun positions protected the approaches to the three roads which lead up to the aerodrome. Additional strongpoints, north and south of the anti-tank ditch, provided enfilade fire along the beach. Observation towers, for both visual look out and radar, were provided either end of the bay at La Corbiere and Les Landes.

Les Landes, isolated at the north-western tip of the island, was already a pre-war military training area and the Germans found it ideal for the location of training ranges, camps and for Batterie Moltke. The latter had a primitive beginning in March 1941 when a battery of four 155mm Grande Puissance Filloux — elderly captured French artillery — was ordered to Jersey. Initially the guns were emplaced in makeshift positions consisting of simple, flat, concrete hardstandings allowing an all-round traverse. In 1942, a modernisation programme was begun to convert the site into a full-blown naval battery. The OT set up a labour camp beside St. George's Church for the construction workers and work began on building personnel shelters, underground ammunition stores and shell hoists, etc., linked by concrete-lined corridors. Four new pits were excavated nearer the cliff top and steel cupolas shipped to the island. However, the installation was never finished and the four guns of Batterie Moltke remained on their old positions until the end of the war.

ST. OUEN'S BAY VIEW LOOKING SOUTH-EAST

- AERODROME
- REMAINS OF OT RAILWAY EMBANKMENT
- SITE OF OLD RAILWAY SHED
- RESISTANCE POINT LA MARE MILL
- WATER-FILLED ANTI-TANK DITCH (NOW PARTIALLY FILLED IN)
- RESISTANCE POINT HIGH TOWER (NOW SANDS DISCO)

Left: **Life can be such a drag . . . at least it seems that way for this German overseer as he puffs away near Le Braye. La Rocco Tower in the background was a pre-Martello construction built 1796-1801 near the only safe anchorage on the bay. German gunners used it for ranging shots and it ended the war sadly knocked about (Bundesarchiv).** *Above:* **More recently, the Reverend Peter Manton raised £17,000 for its repair and restoration.**

Strongpoint Etacquerel at the north end of the bay. The casemate conceals a tunnel complex leading back into the hillside (CIOS).

Left: **It's that man again! (Bundesarchiv).** *Above:* **We looked in vain for his dog-ends in 1979.**

Above: 'A' gun of an early Luftwaffe Flak battery overlooking the beach, obviously designed for the dual role of anti-invasion and aerodrome defence (Bundesarchiv). *Below:* **We found the hillside a tangled mass of briars, completely hiding the gun pits.**

Above: If you can detect creases in this photograph of the observation tower at Les Landes, it is because it blew away while taking our comparison photograph at the bottom of the page! (Societe Jersiaise).

Probably the best known defensive construction and one which is unique to the Channel Islands is the multi-storied observation tower. These huge 'Marinepeilstaende und Messstellungen' were initially intended as Army Coastal Artillery OPs but became naval direction-finding and signalling stations of which three were constructed on the Jersey coastline: MP1 at Noirmont (nothing to do with the battery on whose ground it stands); MP2 at La Corbiere and MP3 at Les Landes. It was intended to add a further six towers to give complete radar coverage around the island from the Freya aerials mounted on top. However, because of the lengthy time each one took to construct, the others were never built but the gaps were filled by lesser, improvised positions numbered M1 to M10.

The tunnels were another important facet in the overall plan. Altogether sixteen were planned of which four were virtually completed by 1945 and another three were in an advanced state. They were built primarily for the storage of ammunition, fuel and foodstuffs and as personnel shelters in case of bombing attacks. The island's steep-sided valleys were ideal locations as the tunnels could be driven straight into the sides giving an immediate cover of up to a hundred feet of solid rock. The most widely publicised tunnel or 'Hohlgangsanlage' (literally cave passage installation) is undoubtedly the underground hospital, originally constructed as an artillery barracks and workshops. Only in 1944 was it converted for use as a hospital. Some 14,000 tons of rock were excavated and its tunnels lined with 4,000 tons of concrete.

Scattered all over the island were additional field works and anti-aircraft installations

Above: The Feldmarschall and his entourage (see page 64) arrive at the construction site for Army Coastal Artillery Battery HKB 356, (later renamed Moltke) at Les Landes (J. W. Wallbridge Collection). *Below:* Ken Tough certainly gets around; we last left him in Guernsey with the Russian PoWs (M. Ginns).

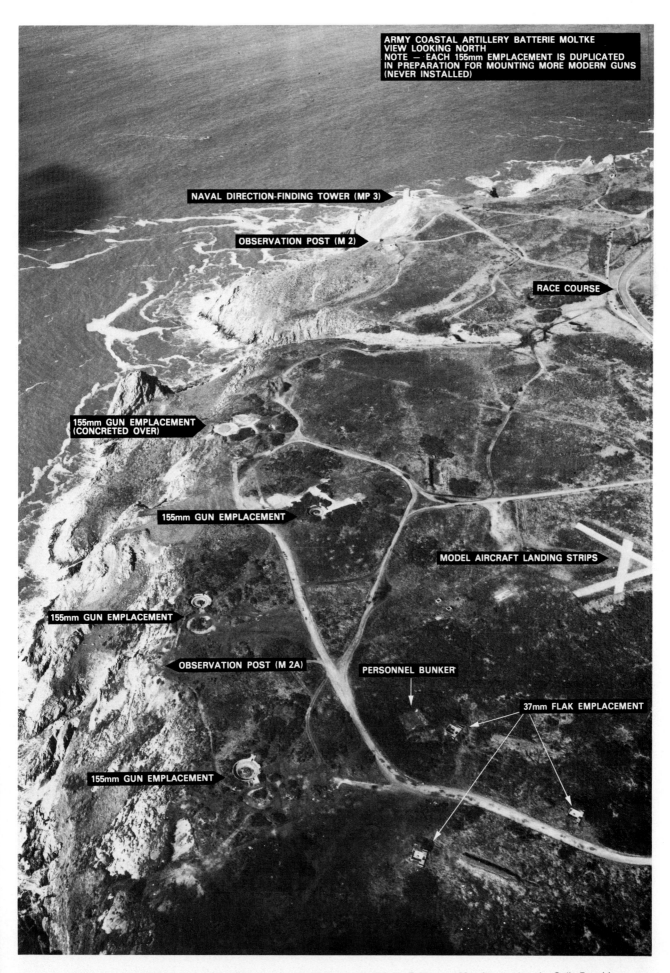

The early and later gun positions are clearly visible in our aerial photograph taken in February 1981. Annotations by Colin Partridge.

What the Valley of the Kings is to Egyptologists, St. Peter's Valley is to students of the Second World War. Lured on by a magnetic attraction for the reputed treasures which lie hidden in its Aryan tombs, the tunnels have been a constant headache to the States authorities endeavouring to keep out intruders and to avoid fatal accidents occurring in their stygian depths.

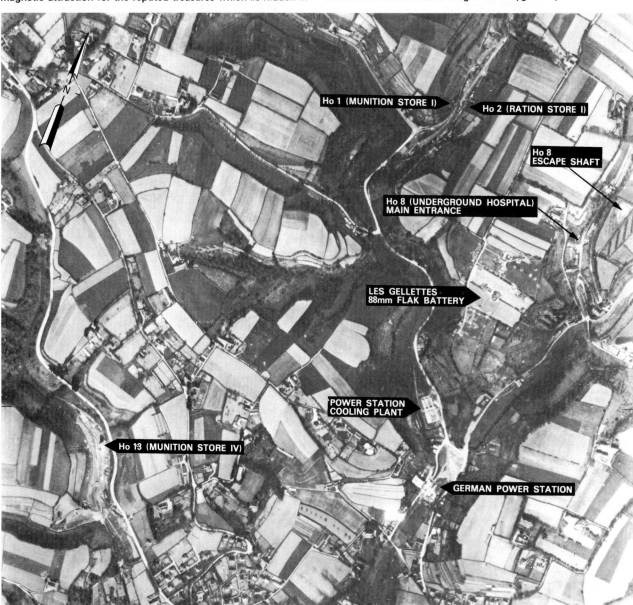

This RAF photograph was taken by No. 541 Squadron on a reconnaissance sortie on February 7, 1945 (Crown Copyright).

Above: This is what all the fuss was about — the guns and equipment sealed up in the tunnels by this man in 1945. His name: Major F. H. M. Sargent, Officer Commanding 135 Field Ordnance Unit (F. Sargent). *Below:* In 1948 this tunnel, Hohlgangsanlage 1, was opened and all the armaments sent for scrap. This same tunnel is now used as a mushroom farm.

The tunnel had three entrances — the RAD badge standing for Reichs Arbeits Dienst (National Labour Service) which helped in the early construction can still be seen over the lower one.

Part of the Ho 1 tunnel complex has been bricked up again but the end of the concrete lining still remains to locate the spot and make an exact comparison possible. Construction was never finished for what was intended as Munition Store I.

which gave the Islands the greatest concentration of firepower on the Atlantic seaboard. This whole defensive system was controlled from a battle headquarters located in the centre of the island near l'Aleval. This 'Kernwerk', as the Germans called it, consisted of three command bunkers, two communication bunkers and another for pumping and storing fresh water. Each of the command bunkers were constructed on two levels, partly underground, with the exterior painted to represent a house. The first time they were used for real was during the Normandy invasion but of course they never really saw service as was intended.

In all it was a mammoth civil engineering programme and, at its peak, the OT had 6,000 workers employed on the island. However, by the end of 1943, many were moved out to work on French coastal defences and bomb damage in Germany, leaving the garrison troops to continue alone. By D-Day most new construction work had ceased.

The population, almost twice as many as in Guernsey, put up with the desecration of their once-beautiful island with fortitude. Outright rebellion against the Germans was obviously impossible so any opposition had to be conducted with subtlety. No better example can be provided than that instigated by the designers of Jersey's occupation postage stamps. Major N. V. L. Rybot received an unpaid commission to produce a stamp to replace the British design. Although his brief had been tightly drawn up by the local Feldkommandantur, he managed to depict the shield of Royal England and to incorporate the letter 'A' in each corner of the penny stamp. There was much speculation as to the meaning of the design especially when 'AABB' appeared on the halfpenny design. Other than perhaps a few cognoscente, the Islanders had to wait until after the liberation before Major Rybot was able to reveal the answer: 'Ad Avernum Adolf Atrox' or 'To hell with you atrocious Adolf'. Likewise the halfpenny design reviled the Italian dictator with the substitution of 'bloody Benito!'

Then, in 1943, Edmund Blampied perpetuated the anti-German trickery when he designed a pictorial set of stamps (printed in Paris) in which he incorporated a 'GR' motif without the Germans realising it.

There were, nevertheless, conscious acts of sabotage like the cutting of telephone cables and over 4,000 Channel Islanders were imprisoned for various offences against the occupying forces. Sixteen Jerseymen died in German concentration camps and nearly a hundred people escaped to France after D-Day, several dying in the attempt (at least one was shot while escaping and another died in a concentration camp).

When Operation NESTEGG was being planned by the Allies for the capture and rehabilitation of the Islands, due consideration was given to the centuries-old traditional rivalry between Jersey and

The real problem was caused by the tunnel designated Ho 2, and built as Ration Store I, on the opposite side of the road. This was the one which had been stuffed full of unwanted German equipment and, in spite of the entrances being sealed up, the tunnel was broken into several times. In 1962, two schoolboys who had managed to get inside were asphyxiated because a fire had been lit inside by other intruders the previous day, consuming all the oxygen. Nevertheless it took ten years before the States authorities finally decided to open it up once and for all, clear it out and re-seal it for good. *Above:* These are the two entrances of Ho 2 today; both now bricked up, covered with earth and grassed over for all time.

Most well known of all is Ho 8 begun as artillery quarters but which ended up as a hospital. Nine Russian workmen do still lie entombed in one of the side galleries which suffered a rock fall which was considered too dangerous to clear, but they are certainly not cemented in the concrete walls as is often rumoured.

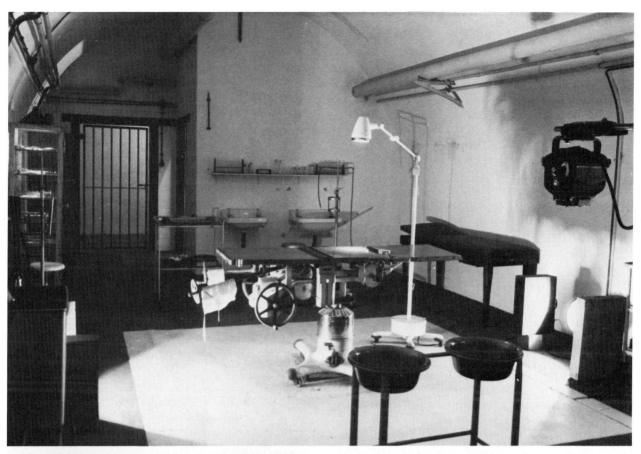

The operating theatre with much of the original equipment.

Above: One of the wards complete with iron-framed beds as if awaiting some ghostly patients. *Below:* A staff rest room.

In a parody of an unfinished symphony, the visitor to this uncompleted galley is assailed by sound effects of picks and shovels — an eerie reminder of the toil which went to build the tunnels.

Above and bottom: **Snow Hill, St. Helier 1941 and 1979 (Bundesarchiv).**

Oberst Friedrich Knackfuss (born 1887) was Feldkommandant of Jersey from October 4, 1941 until February 29, 1944 when he moved to France. He died in a Yugoslavian prisoner-of-war camp on October 7, 1945. It was during his reign that the great postage stamp con was perpetrated (R. H. Mayne).

Admiring the view of St. Aubin's Bay from Fort Regent, St. Helier. The fort was commenced in 1806 on a site purchased by the British Government for £11,280 and subsequently £1 million was spent making the fortress well-nigh impregnable. It was garrisoned until 1926 when the War Office evacuated all the forts in the Channel Islands. *Above:* German super-troopers in 1941 (Bundesarchiv). *Below:* Michael Ginns in 1979. The wall has been altered at this spot but the stonework can still be matched.

Human nature being what it is, it was inevitable that some people would attempt to escape from the virtual prisons which the islands became from 1940 to 1945. Mr. Wilfred Bertram was instrumental in helping many to escape as his farm, East Lynne *(above left)*, was ideally situated facing the sea on Grouville Bay *(above right)*. A small rowing boat could be hidden on the farm and taken across the road at night to be launched although this was not without great risk as Fort Henry with its searchlight overlooked the spot. On January 8, 1945, two American officers escaped from the Allied prisoner-of-war camp on Mount Bingham, St. Helier (situated on the spot *below* where Jersey's learner drivers now try to hit each other). In their case he managed to get them away on the night of January 18/19 in a rowing boat from Gorey harbour. They eventually reached France. For this, and his help in other escapes, Wilf Bertram was awarded the British Empire Medal and the US Medal of Freedom.

Left: This pathetically small dinghy, on display at La Hougue Bie Museum, was used by another escapee Denis Vibert on a three-day voyage to England in 1941. He was picked up by the Royal Navy in a minefield two miles from Portland and eventually became a pilot in Coastal Command. *Above:* A solitary stone gatepost beside the road at Grouville is all that remains of another camp, that of Lager Wick for Organisation Todt workers.

Michael Ginns (whose wife Margaret took this picture for us) commented that when the original picture was taken during the winter of 1944-45, this must have been the coldest and most wretched sentry post in the entire Channel Islands. The photo even looks bleak and dismal. It is an 18th century guardhouse at Plemont adapted to mount a twin MG34 machine gun (CIOS).

Above: Ex-French tank turret mounted at La Haule north-east of St. Aubin. There are two particular points of interest in this picture — the odd piece of isolated anti-tank wall behind the turret and the railway line to La Corbiere on the left (Societe Jersiaise). *Bottom:* Victoria Road now replaces the track and, although the lump of wall is hidden by the hut, see page 201.

Above: **Back to Charles Brown and his magnificent flying machine — now buzzing the direction-finding tower MP2 at La Corbiere.** *Below left:* **The tower was a normal construction in reinforced concrete — the stonework effect has been created by an artist (Societe Jersiaise).** *Below right:* **Only the merest traces of paint remain today, the tower now being occupied as a modern-day counterpart to its former use. It became operational again in March 1977 after a conversion costing £30,000.**

Left: Another direction-finding station was built in 1941 at La Moye, St. Brelade, on the south-western corner of Jersey. A curious structure, built to resemble a house, its code number was designated as M 10 (F. Sargent). *Right:* Today it has lost its roof and the rifle loopholes, once disguised as windows, are now revealed (M. Ginns).

Centre: The war began its final phase with the D-Day landings. This P-47 Thunderbolt bearing invasion stripes belly-landed at the bottom of Jubilee Hill, St. Peter while returning from a raid on Le Mans marshalling yards on June 23, 1944. The pilot, who was taken prisoner, was 2nd Lieutenant W. R. Davis of the 510 Fighter Squadron of the 405th Fighter Group of the US Ninth Air Force (L. Vardon Collection). *Above:* Michael and friends re-enact the scene in August 1979.

Guernsey. Therefore, although theoretically the entire Channel Islands could have been covered in one instrument of surrender, individual documents were prepared which could be signed in both Jersey and Guernsey by the respective Inselkommandanten.

Following the ceremony in St. Peter Port harbour early on the morning of May 9, Brigadier Snow transhipped to HMS *Beagle* for the short sea crossing to Jersey. A telephone call was put in to Bailiff Coutanche warning him that the *Beagle* was on its way and to get Generalmajor Rudolf Wulf (who had taken over from von Schmettow in February 1945) to the harbour.

When the Bailiff arrived at the naval headquarters situated in the Pomme d'Or Hotel, the German general insisted on being accompanied by two of his staff officers. Not to be outdone, the Bailiff said in that case he must also be accompanied by his Attorney-General and Solicitor-General. Meanwhile the *Beagle* had rounded Noirmont Point at 10.00 a.m. and had anchored off St. Helier with Brigadier Snow annoyed to find the Germans not awaiting his pleasure. Unaware of the reason for the delay, impatiently he sent a message ashore that Wulf must be found immediately. In the end it was after midday when the Bailiff and the German Kommandant, accompanied by their prospective staffs, were taken out to the *Beagle* by launch. After the surrender documents (in octuplet) had been signed, Generalmajor Wulf was handed over to Colonel William Robinson.

'I told him I would see him at 4.30 p.m. at the Pomme d'Or Hotel', recorded the Colonel. 'I landed later with a small party and the reception accorded us by the crowd was such a tumultuous one that it was as much as I could do to make the Pomme d'Or in time. However, I did so and then sent for the General.

'I gave him my orders, I told him that the quays and Fort Regent were to be clear by 6 o'clock that evening and that the town, except for the hospital and certain dumps, were to be clear by nightfall, with the exception of a certain number of officers whose services I required. I said that by nightfall today all Germans must be clear of the central southern quarter of the island, leaving their arms and

Above: **The Red Cross comes to the rescue. January 2, 1945 and the first life-saving food parcels are delivered to Martland's Store, Patriotic Street, St. Helier. German Rote Kreuz personnel and members of the St. John Ambulance supervise the unloading (CIOS).** *Below:* **No evidence of a fuel shortage in September 1980 (M. Ginns).**

Left: **Such joy in today's ration-free era of plenty is difficult to appreciate (although the chap with the barrow seems disgruntled with his load). This is Gloucester Street looking from Royal Parade towards the Esplanade (Evening Post).** *Above:* **The contrast on a hot, sunny day in July 1979 (D. Bishop).**

Above: Surrender! White crosses had to be laid out throughout the islands as a sign of submission — an ironic throw-back to 1940 when the Germans requested a similar display. This picture shows the searchlight shelter dousing its light for good at La Corbiere (Evening Post). *Bottom:* La Corbiere lighthouse still flashes at night but only as a maritime warning.

Left: Last German guard duty of the war on Jersey took place on May 9, 1945 at West Park Slipway (Societe Jersiaise). *Above:* Now a useful repository for deck chairs.

ammunition in certain dumps which I indicated to them. I said I would allow the German Commander two per cent of small arms for the use of the guards over the arms dumps as he signified that it was most important that he should be permitted to retain a small proportion of arms in order to maintain discipline among his troops.'

In the afternoon a reconnaissance party, scrambled from the NESTEGG force, arrived under the codename OMELET! HMS *Corby* escorted a landing craft containing about 200 men to St. Helier which beached itself in front of Victoria Avenue. The population enthusiastically welcomed the troops and were thrilled by a flypast of RAF Mosquitos and Mustangs.

Above: **All the pent-up longing for peace spilled forth on May 9, 1945 and the first two British naval officers ashore are carried shoulder high across the Weighbridge (Evening Post).** *Below:* **July 1979. No doubt the tourists waiting for their bus are completely unaware of the joyous scenes enacted where they stand (D. Bishop).**

The following day a symbolic ceremony was held in the Royal Square in the presence of the Bailiff and Colonel Robinson. The National Anthem was sung and people unashamedly wept with emotion. Two days later the main body of Force 135 arrived in the Islands. This time it was the Germans' turn to display white crosses of surrender and, after the landing craft had been unloaded, they began to take on prisoners-of-war.

Brigadier Snow, with the full powers of military commander, went back to Guernsey on May 12 and reached St. Peter Port at 10.45 a.m. Accepting the personal surrender of Vizeadmiral Huffmeier, the Brigadier marched in triumphant procession to the Court House. After lunch he returned to Jersey where a similar ceremony at the Royal Court marked the end of an era. The Royal Proclamation investing him with the power of military commandant was read out and a message from HM King George VI welcomed the Islands' liberation and their return among the free nations of the world.

Above: **The original liberation force consisted of the Somerset Light Infantry but this unit was sent to Arnhem in September 1944. Their place was taken by coastal and anti-aircraft gunners of the Royal Artillery who were otherwise redundant with the removal of the V1 and submarine threat. These men were put through an intense infantry re-training course and then formed Task Force 135. Here the troops pack onto the balcony of the Pomme d'Or Hotel — the former HQ of the Germany Navy in Jersey — as the crowds wait to see the Union Jack unfurled (Evening Post).** *Below:* **The Union Flag still flies from the same flagstaff but the crowds have been replaced by mundane traffic lights and parking meters (D. Bishop).**

Above: Scene on the Esplanade on May 10, 1945. Using every mode of transport available to them, the people of Jersey congregate to watch the landing craft arrive on the beach. Note the Causeway to Elizabeth Castle (Societe Jersiase). *Below:* We discovered the photographer had used the observation post still standing on the corner of West Park.

These same ships beached in St. Aubin's Bay were used to transport the German garrison to the United Kingdom. Some 4,000 were left behind to help clear up the island, the last party finally leaving in July 1946 (Societe Jersiaise).

Above: This is Marais Square in St. Anne where the once renowned Alderney breed of cattle were regularly brought to be watered. At Easter 1942 the occupation forces held their first parade on the island. Spectators are virtually non-existent (Alderney Library).
Below: Alderney is sparsely populated; it being said that when the Queen and the Duke of Edinburgh toured the island in 1957, Prince Philip saluted the cows as they were more numerous than people. This is the deserted square in April 1979.

CHAPTER THREE
Alderney

In June 1940, more than 150,000 Allied troops were evacuated to the United Kingdom from Cherbourg. Inevitably, lying just eight miles from the French coast, Alderney received its share of refugees escaping from the advancing German armies. The first boatloads of escaping civilians began to arrive on June 12, alarming the islanders as to the worsening events outside their sanctuary. Naturally the 1,000 inhabitants felt very vulnerable and their fear was not helped by the lack of clear information as to what was taking place on the neighbouring island of Guernsey.

Like the traditional antipathy of the people of Guernsey for Jersey, that between Alderney and Guernsey, even more deep-seated, had a large bearing on what happened on the island in 1940. The chief administrator on the island was Mr. Frederick G. French and, unlike the other two island leaders, he held the title of 'Judge'. He had no direct telephone link with London and consequently was dependent on the other islands for news. Early on the morning of Sunday, June 16, the remaining troops on Alderney suddenly received orders to quit the island forthwith. When the islanders awoke to find a hasty evacuation in progress, panic, fuelled by rumour, immediately set in. Although Judge French tried

to allay the fears of the population in a meeting held in St. Anne's Parish Church that morning, nothing he said could dispel what the people could see with their own eyes.

The next day, with the arrival of demoralised French troops from the mainland, morale reached rock bottom. The military evacuation escalated and piles of stores and equipment were abandoned with instructions to the Judge to have them destroyed together with all remaining fuel stocks. The crisis was worsened by the news from the crews of the evacuation ships and by the signs of the war approaching the island as columns of smoke drifted above nearby France and the sounds of gunfire could be heard.

Judge French tried desperately to get reliable instructions from Attorney-General Sherwill in Guernsey but not until the 19th (when the opportunity of civilian evacuation was published in Jersey and Guernsey) was the Judge told he was free to do whatever he felt best.

The next day he sent an urgent message to Guernsey by fishing boat demanding that Alderney be afforded the same facilities. Sherwill's reply, received on the 21st, was hardly encouraging stating that, 'we have evacuated masses of our population and expect to complete it by tomorrow night' and concluding with the news that supplies would be sent to Alderney for as long as possible. Fearing a sell-out by the Guernsey administration, when the *Vestal* arrived early on the 22nd to take off Trinity House lighthouse personnel, Judge French gave a message to Captain McCarthy for the Admiralty with a request for ships to take off his people.

This did the trick with surprising speed for, although the *Vestal* had to sail to Guernsey to collect its lighthouse keepers and make a return journey with mail to Alderney, it is believed the Captain sent the message to the Admiralty by radio as six ships arrived in Braye harbour early the following morning (before the *Vestal* had arrived in Southampton). By 11.00 a.m. the exodus was in full swing. Only seven people decided to stay and, after the aerodrome had been obstructed, the wireless transmitter destroyed and some of the military supplies loaded aboard, the last ship sailed at midday. Four days later, the *Alderney* was sent over from Guernsey to collect more of the abandoned stores and to rescue livestock and 248 head of cattle and horses which were brought back and temporarily penned on Guernsey airport, this being the only easily available piece of land large enough to accommodate them.

Following the German occupation of Guernsey and Jersey on July 1 and 2, troop transports were flown to Alderney only to abort their proposed landing when they spotted the obstructions on the aerodrome. Instead two Fieseler Storch spotter aircraft were sent in to land between the obstacles so that the crews could clear the airfield.

Being almost totally abandoned, Alderney did not pose the problems of occupation as did the other islands; in fact the island proved to be a very useful adjunct to the Germans' overall plan for the Channel Islands. Shielded from public view, the island was a ready-made prison camp and thus Alderney became the only part of Britain to achieve the notoriety of having a concentration camp located on its soil.

Gee, — they've got a band! Outnumbering spectators three to one (Alderney Library).

Taking the salute at a spot which one might claim has seen virtually no change during the passage of forty years (Alderney Library).

The island saw the same three construction phases as did Guernsey and Jersey: one, the largely inactive period until Hitler's directives of 1941; secondly, the arrival of the OT construction teams and the frenzied period of fortification, and finally the run down to the end of the war. It was during what might be called the OT-age that Alderney's worst period of history came to be written.

Alderney was largely left to its own devices until February 1941 when the Feldkommandant sent representatives over from Guernsey to report on the possible uses for the island. After inspecting the abandoned houses and piles of British Army equipment still lying around, it was decided that the first move must be a clean-up operation. The few civilians were augmented by men sent from Guernsey and Sark organised with three prime responsibilities. The largest party was to get the agricultural side of the island functioning again as soon as possible. The rabbit population was the greatest cause for concern and a programme of shooting (by the Wehrmacht), trapping and snaring was begun. Cows were brought in from France and every encouragement was given to speed the project so that food could be produced to augment supplies in Guernsey. Unfortunately the farming programme on Alderney never really became as productive as the Germans would have liked, this being largely the result of the earlier months of neglect.

The second party were concerned with the maintenance of property. There had been an unfortunate stage between the evacuation and occupation periods when much looting had taken place. The Feldkommandant, anxious to preserve a good public image for his men and to counter British propaganda, went to considerable lengths to find the culprits and, in the light of evidence supplied by the islanders who had stayed, the blame fell squarely on visitors from Guernsey. All the remaining furniture was removed from the houses which were then boarded up for the duration. Other goods, clothing and personal property, were shipped to Guernsey. Similarly the church plate was removed for safe keeping.

Then there was the breakwater party. Alderney's exposed location means that the island suffers the full force of the Atlantic

Lloyds Bank seems to be a popular backdrop for wartime photos taken in the Islands but in the case of the branch in Victoria Street, St. Anne, it is justifiable as the bank was taken over as the Alderney office of the Feldkommandantur. Here Oberst Schumacher (second from left) inspects the Nebenstelle on May 1, 1941. Sonderfuehrer Hans Herzog is third from left while Dr. Kratzer peeps through from the back row (R. H. Mayne).

Derelict houses backing onto Braye Bay which is the one protected by the mole, originally nearly a mile long. This is one of the pictures taken by Major G. G. Rice during his inspection tour of the Islands in September 1945 (Public Record Office).

Braye harbour in April 1979 with the mole in the foreground. The 'German jetty', partially demolished in 1978, extended from the pier in the centre, parallel to the mole. The row of houses in the pictures below can be seen on the right. Fort Grosnez, in the right foreground, dates from 1853 and later mounted two 8-inch howitzers, five 8-inch mortars, two 68-pdrs and three 32-pdrs as well as two 40-pdr rifled breech-loading guns and a pair of 64-pdr rifled muzzle-loaders.

and, consequently, Braye harbour on the northern coast is only tenable if the breakwater which protects it from the north-western seas is maintained.

It was the British decision in 1844 to make Alderney 'the Gibraltar of the Channel' that first deemed a good harbour essential. The plan was to erect a masonry wall on a rubble mound nearly a mile long on the western side of Braye Bay to provide a huge sheltered anchorage of some 150 acres. After many difficulties and the expenditure of more than £1 million the mole, 4,680 feet long, was completed in 1864. No sooner had work finished than two breaches were torn in it by winter gales. Throughout the remainder of that century, the same alternating pattern of damage and repair continued, the British Treasury becoming increasingly worried by the continuous maintenance costs. Nevertheless, without the help of a breakwater to shield ships at anchor, Alderney, with its chain of thirteen forts, would be useless so there was no alternative but to keep it in good repair.

Thirty-five years later and the houses have been restored to their original styles. The bunker in the foreground for the 47mm anti-tank gun, which fired in an easterly direction from a point near the harbour, still remains.

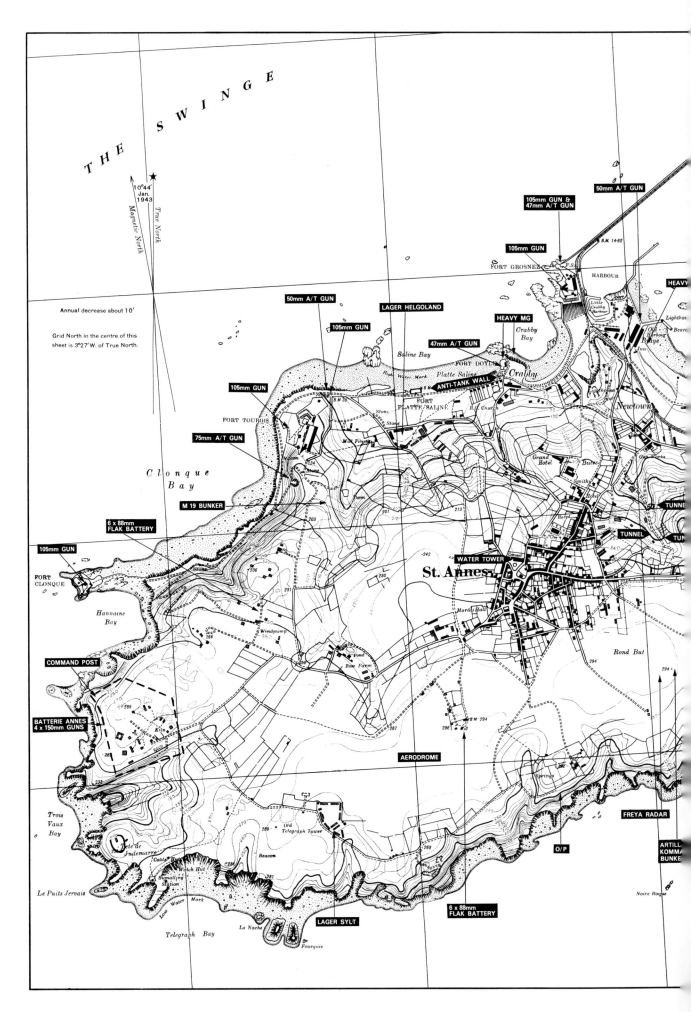

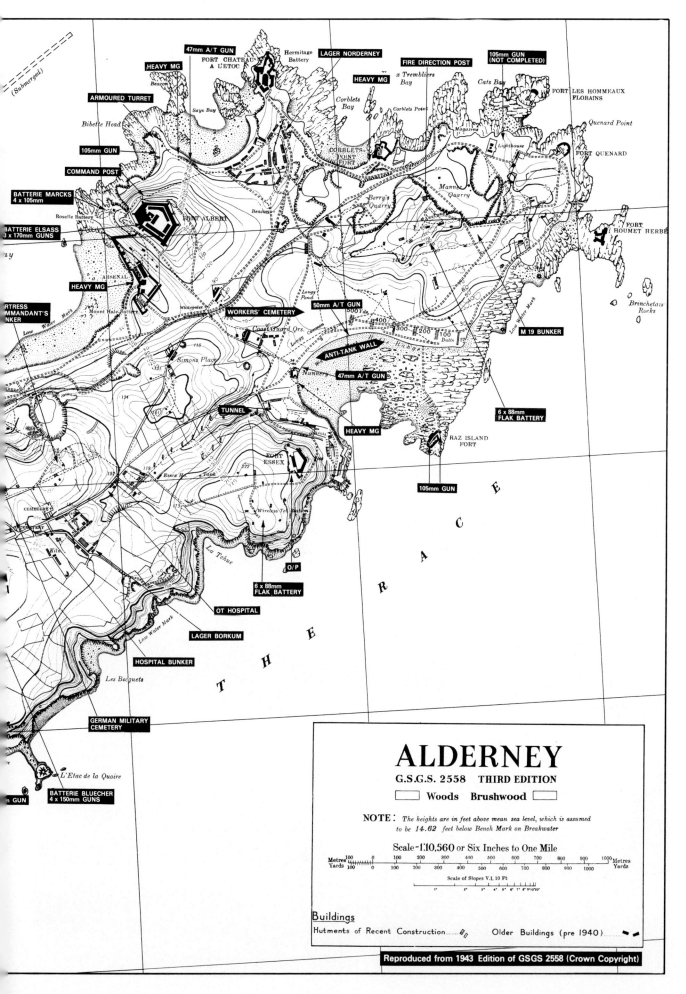

With the advent of steamships, it was felt that its length could safely be reduced without danger and therefore 1,830 feet of the most vulnerable outer end were abandoned. By the time the Germans took over, the majority of this piece lay submerged creating a hazardous 600-yard reef over which ground seas broke very heavily.

Thus it was equally essential to the Germans that a breakwater maintenance party be formed for the continual work necessary on the mole. When the fortification programme began, there was an urgent necessity to increase the dockside facilities for unloading supplies and plans were made to build a new jetty in the harbour. According to one of the Belgians who worked on its construction, the German company involved first tested sections of the prefabricated steel pier in the sea off the Belgian coast before transporting it to Alderney in 1942.

Obviously the Germans had misgivings at having even a small number of civilians roaming free on Alderney and special measures were taken by the Feldpolizei to restrict their movements. Even so, two Guernseymen fishing out of Braye escaped to Weymouth on April 8, 1944.

There were four camps on the island to cater for the workers required for the various construction jobs. Russian OT workers were quartered on the northern coast at Saline Bay in Lager Helgoland while Lager Borkum in the central southern part of the island was used for German technicians and pure voluntary labour from various countries. In addition there were the camps of Norderney and Sylt. The former, near Chateau l'Etoc, was used for European and Russian slave workers while Sylt, the most notorious of them all, reflecting the true meaning of the words 'concentration camp', was organised by the

Above: **Major Rice photographed this well-camouflaged casemate for a 105mm coastal gun and armoured cupola at Bibette Head. The 10.5cm K331 (f) gun had a range of five miles (Public Record Office).** *Bottom:* **Except for the netting, the emplacement remains much as it was with its chunky camouflage of jagged rock.** *Below:* **A Sechsschartenturm was built into the seaward extremity although this has lost much of its rocky cover and is filled with sea water at every high tide.**

Left: This casemate was located in the harbour area for a mobile 50mm anti-tank weapon. The edifice behind it is the stone-crusher — a pre-German construction but nevertheless put to good use by them (Public Record Office). *Right:* The bunker remains today complete with its 'crazy paving' camouflage. The crushing plant was demolished in 1973.

Above: April 1979 photo looking westwards of Batterie Mannez, an 88mm flak battery in the east of the island (see map pages 96-97). Mannez quarry and the mineral railway lie in the right foreground with Berry's quarry behind the battery. Alderney's three coastal batteries (Annes, Bluecher and Elsass) were controlled from the three-storied observation tower (MP3), one battery per floor. Six of these towers were planned for the island but only this one was completed. Corblets Bay lies top right.

Left: Longy Bay, in the south-eastern corner of Alderney, was completely protected by an anti-tank wall. This gate was the one entrance to the beach (Public Record Office). *Right:* The gateway can still be seen today minus its steel-girder gate.

This February 7, 1945 photo by No. 541 Squadron shows the relationship of Batterie Mannez to Lager Norderney (Crown Copyright).

Overgrown foundations on a bleak tract of land backing Saye Bay are all that remain of Lager Norderney.

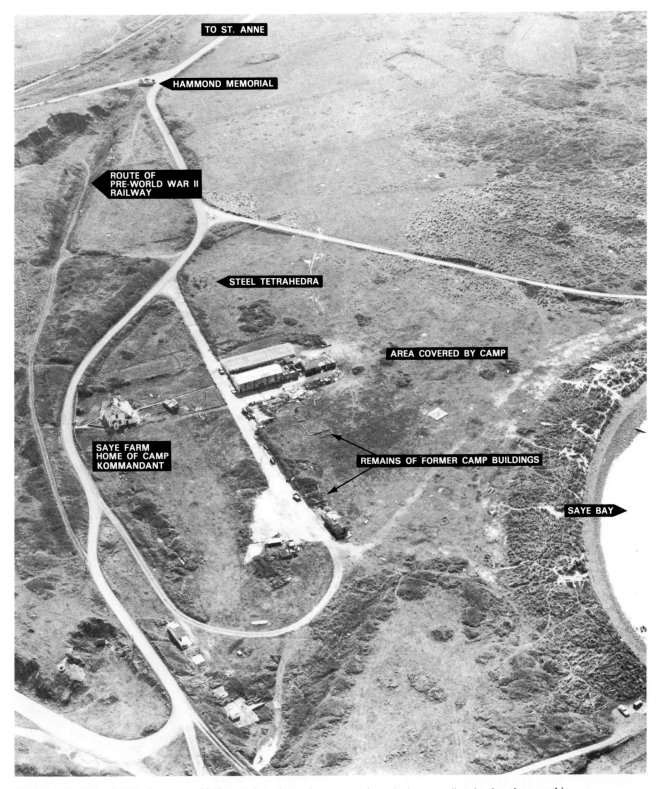

The camp site in April 1979. Compare with the wartime picture from approximately the same direction (north to south).

SS-Baubrigade I from March 1943 to June 1944 as an offshoot of the Neuengamme camp in northern Germany. Sylt lay on the south-western extremity of the aerodrome where the Kommandant's quarters were linked to the camp by an underground tunnel.

Although under the terms of the Hague Convention, it was permissable for civilians to be called upon to perform work for an occupying force, it was specified that this must be of a non-military nature. This dividing line was finely drawn and resulted in many instances where civilians were forced to carry out tasks which directly helped the enemy. One such person was Mr. Gordon Prigent who, following a disagreement with a German on a building site in Jersey, was arrested by the Feldpolizei and transported to Alderney.

When he arrived on the island he discovered to his dismay that he was to be incarcerated, not with the other Jersey conscripts but in Lager Norderney. Mr. Prigent has since described that there were about six other Channel Islanders in the camp, but that the bulk of the inmates consisted of Russians, Poles, Germans and Jews from all over Europe. However, only at the SS camp at Lager Sylt were the inmates forced to wear the blue and white striped concentration camp uniform. Nevertheless, by 1944, supplies of even these became scarce and new arrivals at the camp actually had their white stripes painted on the clothes they stood up in.

Starvation and beatings were part of the daily routine and on June 6, 1944 when the Allies landed in Normandy, the SS made it clear, when they ordered the prisoners to dig their own graves, that they did not wish the concentration camp inmates to be alive in the event of an Allied landing.

The average population in 1943 of all the four camps was between 3,000-4,000 and altogether it has been established that at least 337 foreign workers died or were killed while imprisoned on Alderney.

Above: During Charles Brown's visit to the Channel Islands in May 1945, possibly he was unaware of the true significance of this photograph; and the fact that he had exposed frame 7 of roll 6057 of the only German concentration camp to have been built on British soil. This was the canteen of Lager Sylt. The main entrance is partly hidden behind the chimney with the SS quarters in the background. *Below:* The chimney has gone and the gateposts now mark the track to the VOR airport approach beacon.

Above: This is the base of the SS quarters building — now used by the airport fire service for fire practice. *Below:* Plan of the camp layout, reproduced from the German Occupation of the Channel Islands courtesy of Oxford University Press.

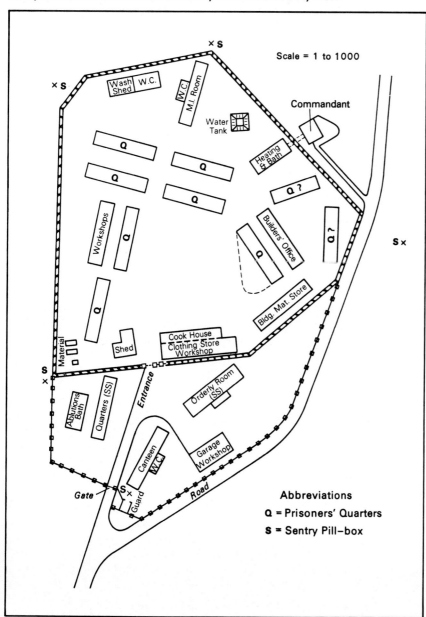

Above: Fire-blackened sentry box on the old western perimeter of the camp and *below,* a twisted barbed wire stake.

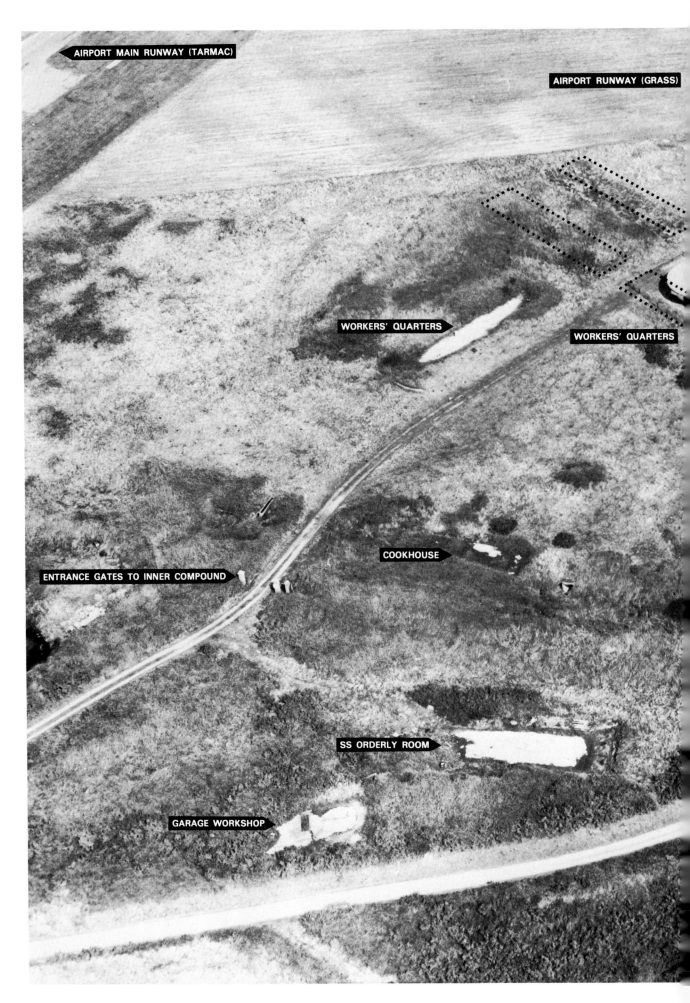

The tunnel entrance *(left)* led from the nearest building inside the wire (the washrooms) underground to emerge behind the Kommandant's quarters *(right)*.

Above: His bungalow once stood on this base with a superb view down the Val de L'Emauve to the sea.

Below: Removed after the war and renamed 'Le Chalet', it now has a new lease of life at Longy Common (Colin Partridge).

This grisly relic was discovered on one of the frames of Charles Brown's Alderney roll of film. It shows a reusable coffin for the burial of slave workers — something oft reported but never proven until now. Denied even the last courtesy of a decent burial, the remains would be carried to the grave and, as the coffin was lowered via the rings on the side, the plate on the end would be operated to drop the body out of the bottom. The coffin could then be used time and time again — perhaps 337 times as this was the number of foreign workers who died or were killed during the war on Alderney. Today, the most prominent relics of the camp are these gateposts which remain as unofficial monuments of man's inhumanity to man.

Above: This is the battery which caused all the fuss in July 1944. Although rather unimpressive to look at, these emplaced 150mm guns (15cm K18s to be exact) outranged the other two Alderney batteries and reached out over fifteen miles. As Cap de la Hague is only eight and a half miles to the east, there was plenty of spare range to harass American positions in France. Being located on prime agricultural land (just east of the town of St. Anne) the positions were completely removed after the war and no trace remains to be seen today (Alderney Library).

The main fortifications these workers constructed on Alderney, just 3½ miles long and 1½ miles wide, were built for the same purpose as the original British constructions — that of blockading the sea approaches to Cherbourg. The three main batteries situated on the island — Annes, Bluecher and Elsass — assisted in closing the gap between Cap de la Hague in France, and Guernsey from where Mirus and the other batteries further south took over.

Batterie Elsass was the first artillery battery to be installed by the Kriegsmarine on Alderney and a site was chosen on the northern ramparts of old Fort Albert which lies to the north-east of Braye Bay. Here a naval construction battalion emplaced three elderly 170mm naval guns in 1941. Nearby, Batterie Marcks had the role of covering the harbour with flanking fire from four 105mm French guns mounted in casemates.

On the south-western headland at the Giffoine, Batterie Annes (formerly Batterie West) was constructed with four 150mm turreted guns mounted in open pits. Batterie Bluecher, just off the Longy Road at La Basse Corvee, was equal in size to Annes and was the battery responsible for shelling American supply routes on the Cotentin Peninsula in 1944. On August 12, the battleship HMS *Rodney* was brought up to deal with Bluecher and one of the Royal Marines on board was Corporal James Mitcheson:

'It was a beautiful summer's day and early in the morning when we dropped anchor on the other side of the Cherbourg penisula. It was about two o'clock when we began firing. We had been told that the German gunners on Alderney were making life awkward for the advancing US troops and that we had got to do something about it.

'All our 16-inch batteries, consisting of nine guns, began firing and we kept up the bombardment for about 2½ hours. Each shell weighed a ton and at the time they were reckoned to cost about £300 each. The rate of fire was one every two or three minutes and we fired about 75 shells.

'We had been unable to come any closer to Alderney because we had been told the Germans had a 9-inch gun with an exceptionally long range which would have outranged *Rodney*. Had we come any closer, they would have got us in range quicker than we could have done them.

'Fortunately, *Rodney* was blessed with having one of the finest gunnery officers in the Navy at the time, a Lieutenant-Commander Peter Larkin, who conducted the bombardment with the aid of a spotter plane. We were delighted when we were told that we had hit our target with our second salvo and had caught the German gun crews napping as they sunbathed around their gun turrets.

Above: Charles Brown accompanied a party of officers on their inspection tour of the island's installations. One port of call was the more formidable-looking Batterie Annes on the Giffoine at the extreme western tip of Alderney. This battery mounted four 150mm guns (SK C/28s) with a range of thirteen miles.

Although the guns and their armoured turrets have long been removed, some fittings do remain, viz. the shell hoists in the niches in the emplacement.

Facial expressions can be so funny. Look at the military policeman (extreme left) looking down at the youthful Kriegsmarine officer (Charles Brown).

'There must have been a general pandemonium as they could have had no idea from where the shelling was coming. They were taken completely by surprise and believed they were the targets of a high-level bombing attack.

'The only unfortunate side to the action, so far as we on *Rodney* were concerned, was that when our captain, Captain Robert Fitzroy, told us on the ship's address system what we were about to do, he mentioned that it was more than likely that we would be destroying the home of the *Rodney's* paymaster, who lived on the island. We never did find out if we did.'

Allied propaganda made much of the fact that the 'island's defences' had been destroyed yet, in actual fact, only one gun had received damage to its carriage. Although the German Army thought it would be impossible to repair, the Kriegsmarine had the gun transported to Guernsey aboard the SS *Robert Muller* which served as the inter-island transport ship. The gun was fitted with a new makeshift carriage in St. Peter Port and tested on the heights above the town. It was then returned to Alderney and subsequently did considerable damage to American fuel dumps near Cherbourg.

When plans were being laid for NESTEGG, it was realised by the Allies that Alderney might pose a different problem than the two main islands. Because of the possibility that the garrison there might well choose to fight it out, Alderney was ignored in the liberation operation and all vessels were warned not to approach closer than fifteen miles.

Information about conditions on the island was scanty and even those Guernseymen who had been on the island could not give an opinion as to whether the Germans would surrender peaceably. The crews and troops of Force 135 must therefore have been rather apprehensive when they sailed across to Alderney in two landing craft from St. Peter

Trevor Davenport, Colin Partridge and the author's wife re-enact the scenario thirty-four years later. Naturally Colin stands in for the Brigadier. Note the attention to detail, like the car parked on the same spot. Nice one Trevor!

The battery command post was on the edge of the cliff. This armoured rangefinder was mounted on its roof (Alderney Journal).

Another No. 541 reconnaissance photo. Note how strips have been dug to obstruct the aerodrome (Crown Copyright).

After the war, it was completely blown out of the ground. The rough water in the channel in the background is the notorious 'Swinge'.

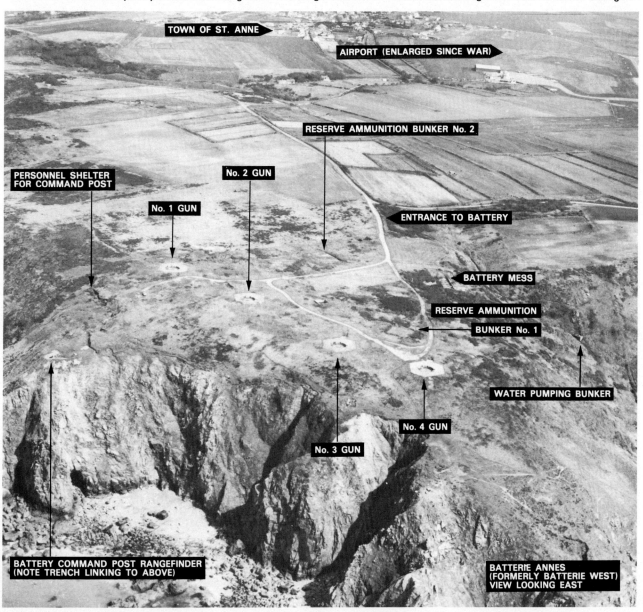

Our low-level oblique 'reconnaissance' photo taken on April 14, 1979 looking eastwards at the former Batterie Annes position.

Parade of German gun crews (and dogs) at Batterie Elsass, the third big gun battery on Alderney, located at Fort Albert. Built in 1858, by 1886 the fort boasted one 7-inch rifled breech-loader, seven 64-pdr rifled muzzle-loaders, four 8-inch mortars and four 68-pdr smooth-bores (Bundesarchiv).

The Germans utilised the old Victorian gun positions to mount their 20th-century firepower. Photo by No. 541 Squadron, this time on February 1, 1945 (Crown Copyright).

The sad remains of the fort after the British Army had finished with it in 1977. Pictured *above* on the ground and *right* our bird's eye view in April 1979.

Port on May 16. As it happened, no problems were encountered and additional landing craft were called in to evacuate the garrison. On May 20, 2,332 prisoners-of-war were taken off, including 170 stretcher cases, in a five-hour operation. Some 500 others were left behind to form a task force to clear up the island and, for six months, these prisoners assisted British troops in the lifting of mines and the removal of war material.

After Jersey and Guernsey were liberated, they were able to begin the changeover to peace with their civilian administrations still intact. Alderney was an entirely different matter and the problem was one which took many years to solve.

Judge French returned to the island accompanied by Brigadier Snow a week after the liberation but decided that, until the minefields had been cleared, it would be

Above: **On their way in. Brigadier Snow smiles for Charles Brown's camera as he walks off the German jetty with his merrie men in May 1945.** *Below:* **Colin also flashes one of his extra special smiles for our comparison in April 1979. Over thirty years of argument and controversy preceded the demolition of the steel jetty in 1978. Both it and its cranes have now all gone — replaced by a single mobile crane.**

Above: On their way out. German troops, now prisoners-of-war, seem cheerful to be getting off the island as they march past the stone crusher to the harbour (Charles Brown). *Below:* Not a very interesting comparison to end this chapter . . . so we have given the last word to Charles *(overleaf).* **We owe so much to Charles and his camera — not only in the field of aviation for which he is best known but for the many other subjects which have been caught by his lens — trains, ships, landscapes and many more before, during and after the war including the previously unknown pictures of his May 1945 trip to the Channel Islands reproduced for the first time in this book. At the time of writing (March 1980) Charles is in poor health but in the happy surroundings of an RAF Nursing Home at Storrington, Sussex. His pictures live on at the Royal Air Force Museum, Hendon.**

Charles E. Brown in 1966 with his battered but trusty Press Palmos. (His Channel Islands photographs were taken on a miniature camera.)

impossible to let any of his people return. He inspected the condition of the houses and, although these were in a reasonable condition, the whereabouts of the contents could no longer be traced. Additionally, little real maintenance work had been done on the breakwater and a large section had been smashed down by the sea exposing the foundations. Judge French did not underestimate the long, hard years which lay ahead as many of the former inhabitants had no desire to return and others were now too old to begin the task of rebuilding a whole community.

By the end of 1945, the mine clearance operation was deemed to be over and the way was open for the rehabilitation programme to begin. Judge French arrived with a small reconnaissance party at the beginning of December and, ten days before Christmas, more than a hundred evacuees returned. Two more batches arrived before the end of the year and six months later the island population had risen to 685 — about fifty per cent of the pre-war total. By this time the German PoWs and British forces had departed leaving the ball fairly and squarely in the court of the islanders.

The Home Office had dreamed up a scheme to base the post-war economy on the Communist-style communal farm. Needless to say it found little favour with the easy-going Alderneymen and, by December 1946, an open feud between the islanders and Judge French led to the Home Secretary making a personal visit to assess the problem. The usual Government remedy was applied as a solution and a Committee was set up to enquire into the state of the island.

As it was vital that the island be set firmly

on the road to self-sufficiency, it was seen that the present heavy Government grants were not the long-term answer. The Committee decided that two things were essential: the provision of a 'healthy balance of economic pursuits' and a reasonable standard of living. The idea was also mooted that Alderney be 'federated' with Guernsey so that the larger island with its greater resources could take over responsibility for major services, always providing

Above: **Charles Brown's tailpiece of Alderney's cattle! Brought in by the Germans after the islanders' stock had been evacuated to Guernsey in June 1940, these Jersey milking cows were pictured being herded along Longy Road in May 1945. Essex Hill with its 88mm flak battery lies in the background.** *Below:* **Today Alderney has a charm all of its own with wild countryside open to the sky . . . and a haven for hikers and those who love the outdoors . . . not forgetting World War II explorers.**

that the inhabitants of Alderney would accept the touchy proposal of being taxed by Guernsey. In the event, this plan was approved in 1948 and the States of Alderney thereafter were allowed to occupy two seats in the Guernsey parliament.

Above: **Parade of the conquerors! The Avenue on Sark most closely resembles the island's 'main street' although there is really no town as such (Carel Toms Collection).** Below: **World-renowned for its ban on motor traffic, tractors were introduced to the island after the war to improve self-sufficiency in agriculture. These now total fifty-six.**

CHAPTER FOUR
Sark

Tiny Sark, five miles east of Guernsey, was more able to keep in touch with the rapid developments of June 1940. The islands had the benefit of the firm, authoritarian rule of Mrs. Sibyl Hathaway, La Dame de Serk, who succeeded her father, William Collings, on June 20, 1927. After the death of her first husband, Dudley Beaumont (grandfather of the present Seigneur Michael Beaumont), in November 1918, she had remarried an American, Mr. Robert Hathaway, in 1929. By virtue of the island's hereditary constitution, supreme authority continued to rest basically with her, and although under Sark law a married woman's property and vote were under the control of her husband, this was something she chose to ignore!

La Dame was in Guernsey when the evacuation panic started and her view of the situation as it affected her island was succinct and to the point: 'We stay and see this island through.' Addressing the island community in

Little did the Germans know what they were up against when they came face to face with La Dame de Serk. Mrs. Sibyl Hathaway with her American-born husband Robert outside La Seigneurie (Bundesarchiv).

St. Peter's Church she said that anyone who wanted to leave was free to do so and a few holidaymakers caught on the island, and some of the British residents, decided to leave. However, all 471 Sarkees decided to stay and calmly await the invader — an example which, if known, must surely have shamed many of those on the other islands.

On July 3, Major Lanz crossed to Sark in the Guernsey lifeboat accompanied by Major Dr. Maass. They landed at Creux harbour on the far side of the island where they were met by the Senechal (the Judge of the Court), William Carre. There and then, the influence of Dame Sibyl reached out to the Germans as they had to walk up the steep hill from the harbour the mile or so to her house, La Seigneurie. There, much to their amazement, she greeted them in fluent German. The visitors, unsure of exactly where the Sark fief stood with regard to the United Kingdom, inquired if the country was neutral. Dame Sibyl was quick to reply to the contrary and to state that the Sarkees were very proud of their British connections.

After lunch, at which we are told the Germans behaved with every politeness and consideration, they departed but the meeting had set the seal on the high regard the German Kommandantur had thereafter for La Dame. The following day a Feldwebel and ten troopers arrived to mark the beginning of the occupation.

The first brush with the new authority came in September 1940 when the Germans banned all fishing as a result of the escape of eight fishermen from Guernsey. As fish was an important part of the diet of the islanders, Dame Sibyl objected. 'We are not Guernsey', she protested stating memorably that: 'Sark is Sark'. Thereafter the island's boats were allowed out, albeit with a guard. Fish caught in excess of the immediate requirements of the Sarkees were smuggled to St. Peter Port where the sympathetic harbourmaster, Kapitaenleutnant Obermeyer was willing to turn a blind eye.

Entries in the Visitor's Book of La Seigneurie for July 3, 1940 (Imperial War Museum).

Although Dame Sibyl retained her health throughout the ordeal of occupation, and the deportation of her husband in February 1943, the strain and loss of weight is evident in these photographs (Bundesarchiv and Carel Toms Collection). La Seigneurie now has a changed entrance. The inscription carved in the stone lintel above the doorway reads: Lat 49° 26' 18" N, Long 2° 21' 48" W.

La Dame turned the tables on the Germans again a few weeks later when they endeavoured to confiscate all wireless sets as a reprisal for a British commando raid on Guernsey *(see Chapter Five)*. When questioned as to why so few sets had been handed in she replied that the islanders had little use for them as few of them understood English!

Dame Sibyl achieved a remarkable measure of independent action for her realm, all of which she was able to obtain without compromising her position or straying over the demarcation line of collaboration. Perhaps because of their inherent ability to cope with annual influxes of tourists yet to retain unspoiled their own simple customs and way of life, the islanders looked on the occupying forces in much the same way. The Germans were visitors, perhaps unwelcome ones, but nevertheless ones which could be tolerated until their holiday was over. As far as the Germans were concerned, their policy could be summed up as: Leave Sark alone.

La Dame only really lost one battle when her firm stand to keep all twentieth-century transport off the island fell by the wayside when the Germans imported two light-tracked vehicles.

At one stage, the Feldpolizei from Guernsey paid a visit to the island when a German Army doctor was found murdered while he slept and his batman was discovered dead at the bottom of a well. It took a considerable effort on behalf of Dame Sibyl to convince the Germans that her people did not commit murder and that there had never been such a crime on Sark for hundreds of years. (Later a German soldier confessed to the double killing.)

The deportations of September 1942 hit Sark equally as they did the other islands although in such a small community the loss was, perhaps, more keenly felt. Fifty-five people were listed for deportation but, on September 25 when the islanders were

gathered at Creux harbour for a final roll call, two people were missing — a retired army officer, Major John Skelton, and his wife. A quick search failed to locate the couple and, somewhat annoyed, the officer in charge despatched the boat without further ado.

Later that evening the full extent of the tragedy was revealed. While the Germans cast their net over the island, the Major's dog was found running loose on the Common near his home. It was the dog which led the way to where Major Skelton lay dead, his wrists slashed and with severe stab wounds in the stomach. Nearby his wife, also with cut wrists, lay dying. Mrs. Skelton was taken to her home and a boat was summoned by the Germans to rush her to hospital on Guernsey. After a long period she recovered, eventually to return to Sark where her husband had been buried in St. Peter's Churchyard.

Following the Commando raid on the night of October 3/4, 1942, measures were taken to step up the defences on the island which until then had been minimal. Some 13,000 mines were sown and the cliffs were closed to the islanders who were subjected to a strict curfew. Additional 'reprisal' deportations were carried out in February 1943 including that of Mr. Robert Hathaway. Although he was in poor health and could no doubt have appealed against the deportation in his official capacity (as Bailiff Carey did on Guernsey) he felt it his duty to go with his people.

On November 22, 1942, nearly 300 miles away at an RAF station in Lincolnshire, No. 49 Squadron received its 'target for tonight' — the area bombing of the city centre of Stuttgart. Nine aircraft at Scampton were detailed for the raid with a routing via Dungeness, Cayeux and Chatillon-sur-Seine. The Lancasters began taking off at 6.16 p.m. One aircraft aborted on take off due to a faulty constant speed unit but the remaining eight aircraft soon disappeared into the 10/10ths cloud over the base. As the round trip would take a little over six hours, the tension among the ground crews relaxed a little until it built up again towards 1.00 a.m.

From 1942, Le Manoir farm became the German headquarters on Sark. Situated in the centre of the island, following the Commando raid in October that year it lay at the heart of a newly-created central defended area. It was an appropriate choice for until 1730 it had been the residence of the Seigneur. It was built by Helier de Carteret in 1565, after Queen Elizabeth I had formed Sark into a feudal holding and appointed him to be its first overlord.

on the 23rd. Eyes strained the eastern horizon . . . one . . . two . . . three . . . four. The phone rang to say another had landed safely at North Luffenham in Rutland. By 2.34 a.m. the sixth aircraft had landed but the crews waited in vain for W4107 piloted by Sergeant E. J. Singleton. When no sign or trace of the missing aircraft had been received, on the

The field where the Lancaster crashed (now locally known as 'Aeroplane Field') still shows the scars of battle where the aircraft ploughed through a dividing hedge. Debriefed after the war, Sergeant McInnes stated that having been hit by flak over the target, they were ordered to bale out. He said he landed about five miles from Stuttgart and spent the rest of the war in Stalag Luft IV at Sagen. The rest of the crew were split up, Sergeant Saunders going to Stalag Luft I at Barth, and Sergeants Corry Hills and Wood to Stalag Luft VI at Heydekrug in East Prussia. Warrant Officer Alexander McInnes was tragically killed after the war the night before his sister's wedding. On July 7, 1946 he had accepted a ride with a friend on a young sailor's motorcycle; they ran into a tree and all three were killed. He was 23 — his grave can be seen in Stonehouse Cemetery, Lanarkshire, Scotland.

All that remains of the Lancaster today.

evening of November 23, the Duty Officer entered a laconic entry in the squadron 540: 'CASUALTIES. The following were reported missing without trace in aircraft Lancaster W4107 on 22/23.11.42. 539435 Sgt. Singleton, Pilot; 576842 Sgt. Wood, F/Eng; 1263894 Sgt. Corry, Nav B; 1048603 Sgt. Pope, Act/AG; 1073261 Sgt. McInnes, WT/AG; 1376451 Sgt. Hills, A/Gnr; 1416688 Sgt. Saunders Act/AG.'

Unknown to the other aircraft, the Lancaster had been hit by flak and set on fire, and was limping home with leaking petrol tanks. Four of the crew had baled out leaving Sergeant L. W. Saunders, the rear gunner, to act as navigator — and to add to the confusion a flare had gone off inside the aircraft. When they reached what they thought was the Isle of Wight they prepared for an emergency landing. Only one field seemed to offer the chance of getting the large four-engined bomber down in one piece and with a rending, tearing noise, the aircraft slithered across the field and through a hedge. One can imagine the amazement of the three remaining crewmen to find that instead of 'England, home and beauty' they had landed on enemy-occupied Sark — the fact that they had achieved the impossible in landing a heavy four-engined bomber on the only suitable piece of ground on the island was cold comfort to the thought of spending the rest of the war in a German prisoner-of-war camp. (A Puss Moth was the first aerial visitor in 1932.)

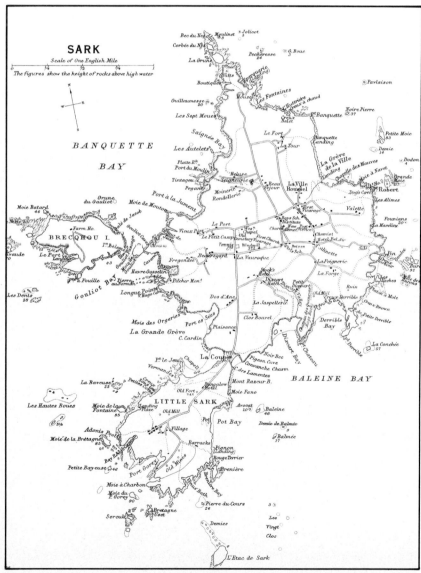

Pre-war map of Sark reproduced by courtesy of John Bartholomew & Son Limited. Aeroplane Field lies to the east of La Seigneurie.

'A pearl set in a silver sea' and 'marooned in the twentieth century' is how Sark has been described by two writers.

This author approached the island on a glorious winter's day in 1981 with special permission to photograph it from the air.

Creux harbour lies on the far side of Sark from Guernsey and, if the weather is fine, visitors are thus afforded a bonus sightseeing trip as their vessel circumnavigates the island. (This is the harbour in use during the war.) The tunnel in the background, bored through solid rock, leads to the Vallee du Creux with its steep climb to the top of the plateau (R. H. Mayne).

In one respect — the food supply — Sark suffered equally with the other islands and even more so when it came to cooking it. There never had been a gas main on the island and there were no supplies of bottled gas. Coal was non existent and the only electricity generators had been requisitioned by the Germans to supply their own installations. As a result, many trees had to be cut down for fuel. What food there was dwindled dramatically during 1943, the garrison suffering even more than the island folk. So desperate was their plight that the soldiers collected shell-fish and shot birds, rats and domestic pets to augment their meagre rations. In September 1944, Dame Sibyl masterminded a raid on the German food store in the village hall which provided just enough to last until the Red Cross parcels began coming early the following year.

Throughout the war, La Dame was able to follow the course of the war from BBC bulletins picked up on her hidden wireless receiver. Thereby she learned that Hitler had committed suicide on April 30 and she realised that the end could not be far away. When the surrender of the German forces was announced on May 8 she raised the Union Jack on the tower of La Seigneurie and flew the American flag in honour of her husband still in captivity. Bailiff Carey telephoned Dame Sibyl from Guernsey advising her of the surrender there and asking her to be ready to receive a representative who would be bringing the Proclamation of Liberation to her the following day. That afternoon she gathered the whole population to listen to the Prime Minister's 3.00 p.m. broadcast. The German garrison were nowhere to be seen.

At 9.00 a.m. on the 9th, Mr. Harold Brache, the Assistant Secretary to the States Supervisor, arrived to be welcomed by Dame Sibyl, Senechal Carre and Mr. William Baker, the island treasurer. Having told the Germans to keep out of the way, that evening the Proclamation was read to the islanders in the school building. Afterwards everyone went

Above: Ken Tough, local historian and secretary of the Guernsey branch of the Channel Islands Occupation Society, acted as our guide during our short but interesting visit to Sark in April 1979. He showed us many places and features which we might otherwise have missed. *Below:* For instance these are the remains of the German defences still to be seen embedded in the slipway beside the tunnel.

The invaders left their mark on Sark. Literally in the photo *above* by painting the crooked cross of Nazism on the tunnel entrance and figuratively *below left* by burning down the Bel Air Hotel! The blaze was caused by the careless use of paraffin stoves — now an empty plot is all that remains. Likewise the Swastikas have been obliterated . . . but see page 197 (Carel Toms Collection).

La Coupee is undoubtedly the most awe-inspiring natural formation in the Channel Islands. It is a natural viaduct, some 260 feet above the sea, linking Great and Little Sark and, as every schoolboy knows, it is an excellent example of the process of converting an isthmus into a strait. Originally there was only a dangerous footpath three feet wide but this was widened sufficiently in the 17th century to take a horse and cart. At least one person has been blown into the sea while crossing in a high wind.

In 1945, German prisoners-of-war were usefully employed widening La Coupee to ten feet and resurfacing the road under the supervision of the Royal Engineers (R. H. Mayne). These plaques *(below)* record the event.

to La Conellerie on the cliffs facing Guernsey where a victory bonfire had been prepared. The honour of putting the torch to the huge pile of brushwood was given to little David Baker aged four — reports were later received that the blaze was visible twenty miles away.

At 2.00 p.m. on May 10, the first British officers arrived by naval launch from St. Peter Port. Colonel Allen was taken to the German Kommandant and, after Dame Sibyl had translated the instrument of surrender, it was signed at Rosebud Cottage to officially end the occupation. Colonel Allen apologised that he could not spare any troops immediately and he asked La Dame if she could cope for another few days on her own. Her reply was characteristic: 'As I have been left for nearly five years I can stand a few more days.'

The following day she gave marching orders to her 275 prisoners-of-war: they were to remove all the mines they had sown in the harbour area, take down the anti-invasion obstacles in the fields and return all the confiscated wireless sets. For the next five days the Germans, in the words of Dame Sybil, were told to 'get down to clearing up the mess you've made these past five years'. Then, on the 17th, all the PoWs were gathered in a field above the harbour ready to be shipped off to captivity.

Thereafter mine clearance was taken over by the Royal Engineers and the Royal Army Service Corps began a shuttle service to re-provision the island. Going in the opposite direction was the transport that the Germans had imported to the island. Although there were protests from some islanders that the new motorised era could not now be swept away, La Dame soon proved otherwise and Sark, under her firm guidance, returned to its unique place in the history of the Islands.

Right: **The official Instrument of Surrender was signed on May 10, 1945 here in Rosebud Cottage which stands next door to the German headquarters building at Le Manoir.**

Left: **The final act. The end for these Germans was by the same way that they came in — down the stone stairway at Creux harbour (R. H. Mayne).** *Right:* **These days the tourist normally departs from La Maseline — not like this to a watery grave!**

Brigadier John Durnford-Slater, who commanded the newly-formed No. 3 Commando on Operation AMBASSADOR. After a very active war, in which he participated in operations from North Africa to the Arctic Circle, he died tragically under the Brighton Belle Express near Haywards Heath on February 5, 1972.

CHAPTER FIVE
The Commando Raids

According to Hilary St. George Saunders, unofficial biographer to the Commandos, credit for their conception goes to a Royal Artillery General Staff Officer, Lieutenant-Colonel Dudley Clarke. In June 1940 he was Military Assistant to Field-Marshal Sir John Dill, the Chief of the Imperial General Staff, and, in the aftermath of the evacuation from Dunkirk, recalled, 'what other nations had done in the past when their main armies had been driven from the field and their arsenals captured by a superior enemy'. Lieutenant-Colonel Clarke later wrote on the evening of June 4 that he tried to marshal his ideas into the outline of a plan: a plan which revolved around the idea of carrying on guerilla warfare against the enemy . . . much as the Boers had done in South Africa and the Jews more recently in Palestine.

While it now seems certain that the idea of a 'Commando' force was born on Tuesday, June 4, 1940, the fact that the Prime Minister had written a memorandum on the same subject to the Chiefs-of-Staff that same day throws some doubt on the origination of the idea. It was immediately after Winston Churchill had delivered his memorable oration in Parliament on the recent traumatic events in France and the dangers which lay ahead for Britain, that he wrote as follows: 'It is of the highest consequence to keep the largest numbers of German forces all along the coasts of the countries they have conquered, and we should immediately set to work to organise raiding forces on these coasts where the populations are friendly'.

Initially referred to as 'Striking Companies', the new organisation came into being with unusual speed. While circulars were being issued to all military commands calling for names to be put forward of volunteers willing to embark on special service of an undefined but hazardous nature, the first operation was launched on June 23. That night a small force in several boats, including Lieutenant-Colonel Clarke as observer, landed in various places on the French coast in the Pas de Calais area. The ease with which the raiders had made unopposed landings was encouraging as was the subsequent morale-boosting communique issued to the general public.

By now, Sir John Dill had approved the title of 'Commando' for the new force which was to consist of ten separate Commandos of 500-odd men each. Leaving the designations of Nos. 1 and 2 for a tentative idea for special airborne commandos, No. 3 Commando was the first to be formed on July 5. Captain John Durnford-Slater, serving as adjutant for the 23rd Medium and Heavy Training Regiment at Plymouth, was selected as its CO, being instantly uplifted two ranks to Lieutenant-Colonel.

The composition of a Commando was specified as 250 ordinary ranks commanded by 247 NCOs (sub-divided into 122 lance-corporals, eighty-one corporals, forty-two sergeants and two WO2s). Officers consisted of twenty-four subalterns, ten captains (each in charge of one Troop) and one major.

Volunteers for the Commandos had to be already fully trained soldiers and inevitably there was opposition from some commanders to a unit which could take their best men and had the smell of that aversion to all regular soldiers — a private army. At the same time, some COs were only too quick to approve the release of their more troublesome types but Lieutenant-Colonel Durnford-Slater commented that, 'we never enlisted anybody who looked like the tough guy criminal type as I considered that this sort of man would be a coward in battle'. However, he was not against accepting minor offenders as the threat of being returned to their units (or RTU as it was termed) was a stabilising force and a unique form of punishment.

Commando troops received exactly the same training as normal infantry except that it was tougher and greater emphasis was put on being able to be independent and highly mobile. The only difference was that the men were not barracked but had to find their own accommodation for which they were given a daily allowance of 6s. 8d. — a small yet important facet of their training to be self-sufficient.

The HQ for No. 3 Commando was located at Plymouth and it was from the nearby harbour at Dartmouth that they sailed for their first operation on the evening of July 14.

GUERNSEY
Operation AMBASSADOR

On July 2 (two days after the occupation of Guernsey had begun), the Prime Minister sent a minute to General Lord Ismay, the head of the Military Wing of the War Cabinet Secretariat:

'If it be true that a few hundred German troops have been landed on Jersey or Guernsey by troop-carriers, plans should be studied to land secretly by night on the islands and kill or capture the invaders. This is exactly one of the exploits for which the Commandos would be suited. There ought to be no difficulty in getting all the necessary information from the inhabitants and from those evacuated.'

Accordingly, four days later, a Guernseyman was taken to the island by submarine to carry out a reconnaissance codenamed ANGER. Formerly a member of the Royal Guernsey Militia but now a Second Lieutenant in the British Army, Hubert Nicolle was taken ashore by collapsible boat by Sub-Lieutenant J. L. E. Leitch and landed on the beach at Le Jaonnet Bay on the south coast. The plan was that Second Lieutenant Nicolle would reconnoitre the proposed landing area on the north side of the island and return to the beach to be picked up two days later. His place on the island would then be taken by two other officers who were also familiar with the locality and who could then guide in the Commandos.

The switch took place on the night of July 9/10 — Second Lieutenants Philip Martel of the Hampshire Regiment and Desmond Mulholland of the Duke of Cornwall's Light Infantry being landed in a similar fashion via submarine and boat.

Second Lieutenant Nicolle reported that there were 469 Germans on the island and that, although machine gun posts had been set up around the coast, the main body of troops were concentrated in St. Peter Port.

On the basis of this information plans for Operation AMBASSADOR were laid. After landing, forty men from H Troop, under Captain V. T. G. de Crespigny, were to create

PETIT PORT.
When the tide serves, this is one of the best bathing places in the Channel Islands, though many steps must be descended.

Illustrations such as this from pre-war guide books or picture postcards aided intelligence on suitable landing places and helped to plan operations against the enemy-held coastline of Europe right up to D-Day. At low tide, the sandy beach at Petit Port would have provided an easy landfall. However, because Operation AMBASSADOR was postponed for forty-eight hours, the forty men of No. 3 Commando which comprised Landing Party No. 1 had to land on the rocks at the neck of the bay. Neither of the other two landing parties made it ashore (Ward, Lock & Company Limited).

a diversion for No. 11 Independent Company who would attack the aerodrome in the parish of Forest. The latter unit was split into two parties. The first comprising twenty men under Captain Goodwin was to land a mile or so further to the west, nearer the airfield, while the second party of sixty-eight men, commanded by Major Todd, was to come ashore at Moye Point directly south of the target.

Meanwhile at Devonport, the destroyers HMS *Scimitar* and HMS *Saladin* were waiting to transport the men and act as escort for the seven RAF rescue launches which would take the troops from the ships to their respective landing beaches. At the last minute, bad weather delayed the operation for twenty-four hours but there was no means of informing Lieutenants Martel and Mulholland on Guernsey. No sooner had this first flaw in the plan become evident than another problem befell the mission.

In his autobiography *Commando*, Lieutenant-Colonel Durnford-Slater explained what happened:

'Since we were to sail from Dartmouth, I had breakfast at the Royal Castle Hotel on the morning of the 14th. I was excited, naturally, at being on the verge of our first operation, a very secret affair of course, and it came as rather a shock when I saw my sister Helen sitting at the next table with her husband, Admiral Franklin. Helen saw me and smiled happily.

' "Hallo, John! What on earth are you doing here?"

' "We've got some troops training in the area," I said. "What are you doing?"

'She said they were down to visit their son at the Royal Naval College. I felt uneasy but tried not to appear so. Fortunately, for the strain was growing, I was called out to the foyer of the hotel a few minutes later. An officer from Combined Operations Staff had just come off the night train from London. 'He said: "Colonel, the whole plan has been changed. Jerry is too strong. He's been reinforced at some of the places where we had intended to land."

'We moved into a bedroom of the hotel and worked out a new plan on the spot. Now we were to land at Petit Port on the south side of the island, just west of the Jerbourg Peninsula and not on the north coast as originally decided. We were to sail at six o'clock that evening. Our role was still to create a diversion for the Independent Company which was to attack the airfield.

'We completed our preparations in the gymnasium of the Royal Naval College, Dartmouth. Many of the weapons had been specially brought from London, as tommy guns and Brens were in very short supply and could only be issued for actual operations. We obtained the help of some cadets from the college who thoroughly enjoyed the work of loading the magazines and helping us in general. We planned the approach with the naval commanders and started to brief the men. Before we realised it, it was a quarter to six, and we had to embark hurriedly in the destroyer. It was a lovely summer evening and as we steamed out of the harbour most of the town was out walking on the quay. I wondered what they thought of our strange-looking convoy.

'I went over the final details with my officers in the Captain's cabin of the *Scimitar* on the way across. We had been so busy all day, dealing with naval officers and obtaining and issuing our special weapons, that this was the first chance our officers had had to discuss it all together.

'Lieutenant Joe Smale's party was to establish a road block on the road leading from the Jerbourg Peninsula to the rest of the island, so that we should not be interrupted by German reinforcements. My own party were to attack a machine gun post and put the telegraph cable hut out of action. Captain de Crespigny was to attack the barracks situated on the Peninsula, and Second Lieutenant Peter Young was to guard the beach. Peter did not relish this job as he wanted more action.

' "All right," I told him, "if it's quiet, come forward and see what's going on."

' "You chaps satisfied with the arrangements?" I asked finally. They nodded. We synchronised our watches. The password for the operation was "Desmond".'

The official report compiled after the raid then explains yet another set back: 'Permission to proceed was received by telephone at 1800. Crash boats Nos. 300 and 301 were found not in a fit condition to undertake the voyage and they had to be left behind. In the revised plan, No. 1 landing (Durnford-Slater) was left undisturbed; No. 2 landing (Captain Goodwin) was halved in size and took only No. 302 crash boat, and No. 3 landing (Major Todd) was allotted crash boats Nos. 303 and 313, the remaining men being put into the whaler of destroyer H54 *(Saladin)*.

'Owing to the trouble over crash boats 300 and 301, certain stores had to be transferred to other crash boats and two of the serviceable crash boats had to make extra trips to embark the men in the destroyers. The convoy started from Dartmouth harbour at 1845 hours and proceeded to sea at the speed ordered. For some reason (possibly an extra trip to collect life belts), crash boat 313 was delayed in starting, and at about 1950 hours she was 5 or 6 miles astern of the convoy. For a time she kept at this distance, so speed of convoy was reduced to 15 knots to allow her to catch up. 313 came up well for a time and then dropped astern again, so the speed of the convoy was still further reduced to 10 knots; 313 then came into her correct position just before nightfall. Speed was then increased to 18 knots.

'Landfall was made at about the time expected and the convoy then went to within about 5 miles of the shore in order to check its positions. Visibility was poor, there was some mist and the moon was obscured. The high cliffs at the western end of the south shore of the island were easily distinguishable but it was extremely difficult to distinguish any points on the coast to the westward. Beyond the ending of the cliffs there appeared to be a number of rocky points protruding with misty

Petit Port in April 1979 with its 'stairway to heaven' still looking exactly as it was forty years ago.

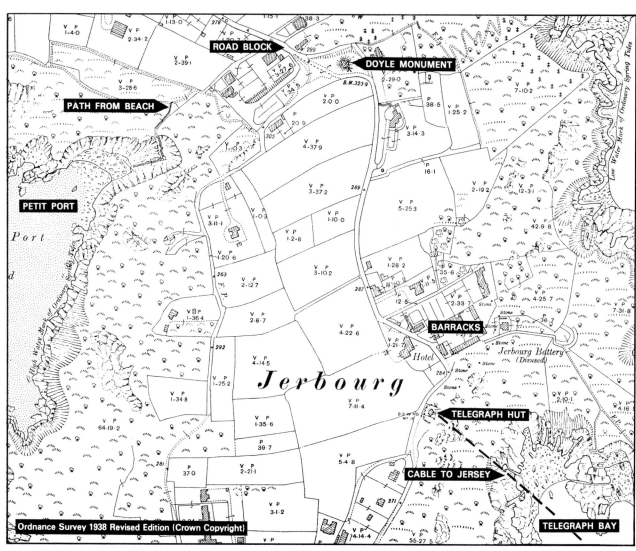

Recreating history. *Left:* **The rocky beach with the stone ramp leading to the staircase.**
Right: **Ken Tough lives up to his name and sets the pace.**

spaces and low lying land in between. Destroyer H21 *(Scimitar — Author)* and crash boats 323 and 324 broke away and passed destroyer H54 on the starboard side. (They were not seen again by H54 until Dartmouth was reached on the return journey.) Just as H21 was disappearing into the mist and slight drizzle ahead she appeared to be turning in towards the coast. When the Captain of H54 was satisfied that he had reached position Y, he stopped, the crash boats came alongside and the whaler was put into the water. The troops were then embarked in the boats, crash boat commanders were shown the points on the shore by the Captain of H54 and the boats had all left the ship by 0045 hours.'

Lieutenant-Colonel Durnford-Slater then describes what happened when they arrived off the island:

'The launches purred away from the mother ship. The naval officers in charge of the launches started off on the agreed course, watching their compasses carefully. My own eyes were on the cliffs and I was astonished to note that we were heading out to sea in the direction of Brittany. "This is no bloody good," I said to the skipper of our launch, "we're going right away from Guernsey."

'He looked up from his compass for the first time. Then he looked back and saw the cliff.

' "You're right! We are indeed. It must be this damn degaussing arrangement that's knocked the compass out of true. I ought to have had it checked."

' "Don't worry about the compass: let's head straight for the beach."

' "Right!"

'About a hundred yards from the beach a black silhouette seemed to approach from our port side. In undertones some of the men murmured, "U-boat!"

'Momentarily my heart sank. What a mug's game this was! Why hadn't I stayed at home in warmth and comfort? Then I realised that the U-boat was only a rock which bore the exact shape of a submarine superstructure.

'At that moment the launches, simultaneously and side by side, hit bottom. As they had not been designed as landing craft, they drew several feet of water. Besides, as the plan had been postponed for forty-eight hours, the tide was not half-way out. It was high. The bottom, instead of being smooth sand as had been calculated, was studded with boulders. I jumped in, armpit-deep. A wave hit me on the back of the neck and caused me to trip over a rock. All around me officers and men were scrambling for balance, falling, coming up and coughing salt water.

'I doubt if there was a dry weapon amongst us. Once on shore, we loosened the straps of our battledress to let the sea pour out. Then, with a sergeant named Knight close behind me, I set off running up the long flight of concrete steps which led to the cliff top, 250 feet up. In my eagerness I went up too fast. By the time I reached the top I was absolutely done, but Knight was even worse, gasping for breath like un untrained miler at the tape. I was exhausted myself and my sodden battle-dress seemed to weigh a ton. My legs were leaden, my lungs bursting. I could hear the squeak and squelch of wet boots as the rest of the troop followed us up from the beach. Fortunately the night was warm.

'I had an idea we were already behind schedule and I led on between a few small houses. We had to be clear of Guernsey by 3.00 a.m. As we passed each house, a dog inside began to bark. Presently there was a chorus of barking dogs behind us.

' "For God's sake, come on," I panted to Knight, who seemed to be slowing down. "We haven't got all night."

'By then I had my second wind and didn't

Where's Ken? Half way up and flagging — this must be Sergeant Knight!

Left: **Second wind.** *Above:* **On the flat at last.**

feel tired again during the operation. My headquarters party was close on my heels: Lieutenant Johnny Giles, CSM Beesley, Knight, two lance-bombardiers and a sapper. Another dog began to bark.

' "Shut up!" Johnny Giles yelled at it and the barking became louder.

' "This is going to alert the whole damn island," somebody remarked ruefully.

'One of the staff officers in London had suggested sending an aeroplane to circle over our operational area with a view to deadening any noise we might make and I had accepted this idea. At this moment I saw the aircraft, an Anson, circling above us at about three hundred feet. He was plainly visible and his exhaust pipes were glowing red.

'The machine gun post, which was the first objective of my little group, was at the tip of the Jerbourg Peninsula, eight hundred yards from the landing place. I went as far as the barracks with de Crespigny. Just before going into the barracks, de Crespigny broke into a house to get information from the householder. I went in with him through the back door. However, the man we found was so terrified that he had entirely lost the power of speech; all he could do was to let out a series of shrieks. We left de Crespigny and began climbing down the cliff. I sent Beesley, Knight and the others to the cable hut. Johnny Giles and I crawled up on either side of the little mound in which the machine gun nest was dug. I carried grenades and a .45 Webley; Giles, a giant of well over six feet, had a tommy gun.

'We jumped to our feet and into the nest, a sandbagged circle. We were both ready to shoot, but I found myself face to face with Johnny's tommy gun; and he with my Webley.

' "Hell!" Johnny said bitterly, "there's no one here!"

'We went down to where the others were cutting the cables leading from the hut. Knight asked me rather plaintively:

' "Please can I blow the place up, sir?" He had a pack of demolition stores on his back and was aching to use it.

' "No. Apparently the Germans don't know yet that we've come. There's no point in announcing it. Just cut the cables."

'We went back to see if we could help de Crespigny's party. It was pitch dark and, as I approached, Corporal "Curly" Gimbert burst through a hedge at me. The next thing I felt was a bayonet pushing insistently through my tunic.

' "Password!" Gimbert hissed.

'He was a big, powerful man. It seemed a long time before I could say anything. There have been worse occasions since, when I've been less scared. At last I remembered the word and let it out with a sigh.

Looking back along the path where it passes the houses (which existed in 1940) and meets the Jerbourg road. Wot no dogs?

The derelict telegraph hut, spared the attention of Sergeant Knight and his explosives in 1940 for us to photograph nearly forty years later.

Above: **This is the road block set up by the Commandos beside the Doyle Monument. It was the job of Privates Fred and Pat Drain to cut the telephone wires on the right-hand side of the road — a seemingly unusual case of two brothers participating in the same raid (Carel Toms Collection).** *Below:* **The Germans carried out their own brand of demolition work on the monument — see page 140. Although it was rebuilt after the war, its replacement no longer contained an internal stairway and observation platform.**

135

The rocky foreshore which made the embarkation such a fiasco. The dinghy with the weapons (and Fred Drain) sank on the right-hand side and Fred and three other men were left behind. When we showed him this picture, it immediately brought back the memory of their later interrogation on this same beach by a German officer who stood in the centre with his back to the sea. Fred felt sure he was leaning against the large rock in the foreground during their grilling.

' "Desmond!" I said.
'Gimbert, recognising my voice, removed the bayonet quickly.
' "All right, Colonel."
'I thought he sounded disappointed.
'When we rejoined de Crespigny, his men had finished searching the barracks. There, as in the case of our machine gun nest, no one was at home. It was past time for the fireworks at the airfield between the Germans and our Independent Company. I listened. The night was still. Ignored, the dogs had stopped barking some time before. I looked at my watch and saw with surprise and some dismay that it was a quarter to three: time to go.
'We formed up on the road between the barracks and the Doyle Column, a monument we had used as a landmark. It was easy to guess from the muttered curses that the others shared my disgust at our negative performance and at the fact that we had met no Germans. George Herbert was particularly upset and begged me to give them a few minutes more to visit some houses nearby which he thought might contain Germans. In this atmosphere of complete anti-climax it was clear that none of us wanted to leave but I called the officers together.
' "We've got to be back on the beach in ten minutes," I said urgently.
'They got their men going on the run. In short order, I herded them like a sheepdog down the concrete steps. Still the enemy showed no sign that he knew of our visit.
'I was last down from the cliff top with Peter Young clattering just ahead of me. Near the bottom I accelerated and suddenly realised that my feet had lost the rhythm of the steps. I tripped and tumbled the rest of the way, head over heels. I had been carrying my cocked revolver at the ready. During the fall it went off, seeming tremendously loud and echoing against the cliffs. This, at last, brought the Germans to life. Almost at once there was a line of tracer machine gun fire from the top of the cliff on the other side of our cove. The tracers were going out to sea towards the spot where I thought our launches must be awaiting us.
' "You all right, Colonel?" It was Johnny Giles' anxious voice.
' "Yes."

'I told him to get on with forming the men up on the beach. I had landed hard on the rocks and was shaken and bruised but there was nothing seriously the matter. I never carried my pistol cocked again.
'Within five minutes my men were all formed up on the beach. I knew now that we were late for our rendezvous — it was ten past three — and that if the destroyers had obeyed instructions they were already steaming towards Britain. Then I saw the dim shapes of our launches about a hundred yards out.
' "Come in and pick us up!" I shouted.
' "Too rough! We've already been bumping on the rocks. We'll stove our bottoms in if we come any nearer."
' "Well, send your dinghy in for the weapons."
'They did. It was a tiny craft, no more than nine feet in length. With each load of weapons went two or three men. As it came in for the fifth run, a high sea picked it up and smashed it against a rock. The dinghy was a total loss and one trooper was reported drowned.
' "The rest of you will have to swim for it," I ordered.
'Fortunately we were equipped with Mae Wests and we all started to blow them up.
'Some of the men began peeling off their uniforms and wading into the sea. Three men came up to me in the darkness. I recognised Corporal Dumper of Lieutenant Smales's road-block party.
' "Could we have a word with you, please, sir?" Dumper said. He seemed a little nervous.
' "What is it?"
' "I know we should have reported this in Plymouth, sir," he said apologetically, "but the three of us are non-swimmers."
'I was ready to explode. The original letter calling for Commando volunteers had specifically mentioned that they must be able to swim. Then I calmed down.
' "I'm afraid there's nothing we can do for you except try to send a submarine to pick you up tomorrow night," I said. *(Conflicts with Private Drain's recollection — see below.)*
' "Thank you, sir," Dumper said. "Sorry to be such a nuisance."
'I removed my tunic and struck out in the water. Some of the men, with more wisdom than modesty, preferred to swim naked. I had the added handicap of sentiment. In my right hand I carried a silver cigarette case which my wife had given me; in my left a spirit flask which had been my father's. A rough sea had come up since our original landing. In these circumstances the hundred yards to the launches seemed endless. For the first fifty, breakers thundered and broke over my head. It took, I suppose, seven or eight minutes to swim out but it seemed hours and I was exhausted. As a sailor bent down from the launch to drag me aboard, the final effort of helping him, to my great annoyance, made me let go of the flask and case. When I was interested again in such matters, I noted that my wristwatch had stopped. I asked the time.
' "Half-past three," the Captain said.
' "My God! We'll have missed the destroyer completely."
'The discipline and bearing of the men during the difficult swim out to the boats was admirable. There was no shouting or panic; each man swam along quietly. The crews of the launches were continually diving in to help the most exhausted men over the last stages of their journey. Altogether, this most difficult re-embarkation was carried out quietly and efficiently.
'With dawn half an hour off, it looked as if we should have to head for home in the launches. This was not a prospect to bring delight. The crews of these boats were brave men, mostly yachtsmen with no service experience. At this point, they seemed unable to reach a decision for further action. The second launch had just broken down: ours threw it a line and had it in tow. There was a

general discussion of the situation by all hands. Even the engine attendant left his recess to chip in.

' "What the hell are we going to do now?" he demanded.

'This was too much for me.

' "For heaven's sake," I snapped, "let's stop the talking and pull out to sea."

'They did as I suggested.

'I was sure that by now the *Scimitar* had gone; a certainty shared by all aboard the launch. It now seemed doubtful that, towing the other launch, we could make it to England, even if we were lucky enough to escape German fighters which could easily nip out from airfields on the French coast. I felt that only a piece of extraordinary luck could save us.

' "May I borrow your torch?" I asked our Captain.

'He handed it to me and I flashed it out to sea, knowing that this was a despairing hope.

'To my delight, a series of answering flashes came back from just beyond the point. The *Scimitar's* Captain, I later learned, had decided to take one last sweep around for us on his way home! He was exposing himself to a tremendous risk of air attack, as daylight was only a few minutes off and the Luftwaffe had many airfields within a few minutes' flying time. Our own air cover of Hurricanes could not be expected at this time to operate so near to the coast of France.

'After blowing up the ailing launch, we transferred to the destroyer.

'Captain de Crespigny, noticing that I was shivering with cold, kindly lent me his tunic. I wore no shirt and put the tunic directly over my bare shoulders and arms. Just before getting to Dartmouth, de Crespigny said:

' "Oh, by the way, Colonel, I do hope everything will be all right."

' "What do you mean?"

' "I forgot to tell you that I've been suffering from scabies," he said.

'I rushed off for a hot bath in the Captain's bathroom. Like the operation itself, nothing came of it. My own tunic, which had my name sewn into the collar, was picked up next morning by the Germans on the beach. Durnford is a well-known name in the Channel Islands and some of the Durnfords there were harried a good deal by the Gestapo who thought that I might still be lying up in the island, harboured by namesakes.

'We arrived back in Dartmouth, safe but distinctly down in the mouth, at eight o'clock in the morning.'

While all this was taking place, the other two landings were proving even more abortive. Captain Goodwin led the second landing party away from the *Saladin* in boat No. 302 at forty minutes past midnight. Although supposedly on course for the landing beach, it seems that this boat's compass was also defective, possibly caused by weapons being stacked too close, and instead the men arrived on another island believed to be Sark. Although they retraced their course, by the time they reached the *Saladin* it was 2.25 a.m. and too late to continue with the operation.

Because of the shortage of boats to transport the main party ashore, the *Saladin's* whaler was used to augment boats 303 and 313 and was taken in tow by the latter. However, soon after leaving the destroyer, it began to leak badly and ship water. Although attempts were made to pump and bail out, the amount of water in the whaler increased and the men had to be taken off and shared between the two air-sea rescue boats, with the whaler tow being taken over by 303. With the extra load, and trying to tow a waterlogged boat, speed was minimal and, at 1.45 a.m., 303 turned back to the destroyer which was reached just over an hour and a quarter later.

Major Todd in 313 with thirty-six men on board had struggled on to try to reach the beach but by 2.00 a.m. he estimated they were still a good twenty minutes from the shore. He had also lost contact with 303 and therefore decided to abort. However, they could not even find the *Saladin* and had no alternative but to set course for England direct. Plagued by engine trouble, at first light they were met by RAF fighters which provided an escort until they reached Devonport at 10.00 a.m.

Winston Churchill, who had expected great things from his latest brainchild, was not amused when he received the report on AMBASSADOR. He sent a scathing directive to the headquarters of the embryo Combined Operations: 'Let there be no more silly fiascos like those perpetrated at Guernsey' and, as a result, Admiral of the Fleet Sir Roger Keyes was brought in to become the new Director of Combined Operations.

According to the official report, four men were lost in the operation (not including Lieutenants Martel and Mulholland). Gunner John McGoldrick of the Royal Artillery was simply reported as 'missing believed drowned' and Corporal Dennis Dumper, 1st East Surrey Regimental Police, and Privates Fred Drain (2nd Bedfordshire & Hertfordshire Regiment) and Andy Ross of the Black Watch as 'missing, probably prisoners-of-war'. Not until we traced Fred Drain in February 1981 were we able to piece together exactly what had happened after the launches departed. Fred, now a successful building contractor in south London, clearly remembered the night . . . and the five years of captivity which followed. It was he, not John McGoldrick, who had been in the boat with the weapons when it capsized in six feet of water and he was only saved from drowning by his brother Pat who was also a Commando on this raid.

Both the Drain brothers had joined the 2nd Battalion of the Bedfs. & Herts. before the war and had volunteered for Commando service shortly after being evacuated from Dunkirk. They were both assigned to H Troop, Special Service Troops, and received very little 'Commando' training other than a few days practice with a small boat.

It had been the brothers' task to cut the telephone wires near the Doyle Column and this had been achieved by Fred standing on Pat's shoulders to reach the step brackets on the post to climb up to the wires.

When the dinghy sank beneath Fred, Pat, a strong swimmer, said he would get him out to the boats but Fred freely admits to a fear of deep water and told his brother to go it alone.

He told us that all the Commandos had been advised that a submarine would surface off La Creux Mahie beach from midnight onwards on the following Wednesday (July 17) and that any men left behind should make their way along Guernsey's south coast to the rendezvous. Having been given additional French money on the beach by those Commandos returning to England, they were in quite good spirits and had every expectation that they too would soon be home. They had a silk handkerchief map and all had compass collar studs and magnetic needles but no arms, except for a .38 Webley & Scott which Corporal Dumper wore as part of his regular attire as a military policeman.

After the promised submarine failed to turn up to take the four men off the island, they made their way to Torteval where they called at the general store (right) owned by Mr. & Mrs. Walter Bourgaize. Here they were hidden for a couple of days, despite the fact that German troops were billeted at the house on the left at the end of the road, before they decided there was no hope of escape and that they were only putting lives in danger.

In April 1979, Walter Bourgaize was kind enough to return to his old shop (now the Torteval Shopper) to be photographed there specially for this book.

The four men climbed the steps back to the road and began walking parallel to it but off it in case a German patrol was sighted. By doing so they came across a small wooden cabin, completely overgrown, about mid-way to La Creux Mahie. It was a sort of potting shed and had recently been used as there were tins of tangerines and some tasty cheese still inside. As this seemed an ideal place to hide up until the submarine was due, they stayed here all through Monday and Tuesday and the best part of Wednesday. To help out with food, Corporal Dumper and Gunner McGoldrick went out across the road and foraged some tomatoes.

Towards evening on the Wednesday they set out for the rendezvous, taking care to keep off the road and move through the yellow-flowering gorse. By 10.00 p.m. they were in position on the rocks on the westernmost promontory of La Creux Mahie about twenty feet from the sea. Gunner McGoldrick had dried out the torch after its ducking on the Sunday and was ready to signal out to sea with three short flashes. Midnight came and went but there was no sign of the submarine. They stayed until well after 2.00 a.m. vainlessly signalling and as their batteries faded, so did their spirits.

From that day until we spoke to Fred at his home in 1981, he had no idea why the submarine had not showed up. As the rescue arrangements for any Commandos left behind had been announced before the operation, naturally the men felt let down. We told him that the fly in the ointment had been the Naval C-in-C at Plymouth who, when he received Lieutenant-Colonel Durnford-Slater's report that men had been left behind on Guernsey, refused to back him up; the record simply stating that: 'for naval reasons the attempt to take the men off later was not possible'.

However, unaware that their fates had been sealed by a decision taken 250-miles away, dispirited, the four made their way back to the shed to work out what to do next. They decided to try to make contact with the locals who might help them get a boat. Arriving at a small village store in Pleinmont Road, they knocked on the door. This was answered by Mrs. Ada Bourgaize who was naturally surprised and not a little alarmed as a squad of Germans were billeted in the house at the end of the road. She recalls that when they said they were Tommies she immediately put them in her garden shed (although Fred Drain recalls they went into a 'back room'). Mrs. Bourgaize then telephoned her husband at work and when he returned to his house, he says the first thing the soldiers asked for was 'a fag'. They were soaking wet and while they took off their wet clothes for his wife to dry, Walter Bourgaize fetched a ladder from the engineering works next door to hide them in his attic.

The men asked if it was possible to get a boat to escape and Mr. Bourgaize said he would do what he could to find out. After lunch he went to the Imperial Hotel at Pleinmont where he knew a man who owned a boat but when questioned he said that he had no petrol for it. Mr. Bourgaize arrived home around 4.00 p.m. and gave the Commandos the disappointing news.

Corporal Dumper was of the opinion that if they could steal an aeroplane, provided he could get it off the ground, he reckoned he would be able to fly it to England. However, they more or less gave up this idea when Mr. Bourgaize said the aerodrome was heavily guarded.

There is some doubt as to the exact period they spent with the Bourgaizes, — it was either one or two days — but they decided in the end that they could not risk endangering the lives of the storekeeper and his wife any longer. The Germans were conducting house to house searches and they regularly came in the shop to make purchases (although oddly enough the store was never searched, even after they had departed).

Fred Drain remembers that they left in broad daylight one morning calmly walking up the road towards the aerodrome. He says he feels sure that although they had not spoken their thoughts to each other, they all realised the game was up and that, secretly, they wanted it all to be over. Forty years later he told us that this must have been at the back of their minds — to walk out boldly in broad daylight — rather than creep away at night when they would stand a good chance of being shot if spotted in the darkness. (This also seems borne out by the fact that, unbeknown to Fred Drain, Corporal Dumper had left all his personal possessions — a crucifix, two penknives, his watch, cigarette lighter and his stud compass together with his revolver — with the Bourgaizes.)

Before they reached the aerodrome, a German patrol picked them up and took them initially to the airfield where they were put in a room until an escort arrived to take them to the civil prison in St. Peter Port. There they were split up — Fred and Andy Ross occupying one cell with Dumper and McGoldrick (who were quite good friends) sharing another.

Next morning, under heavy guard, they were taken from the prison to Jerbourg and back to the beach where they had landed. The tide was out and a German infantry captain,

Corporal Dumper's .38 Webley & Scott, given to Mr. Bourgaize with his personal possessions, is now on display in Richard Heaume's German Occupation Museum located near the aerodrome and thus quite close to the spot where they were captured.

standing with his back to the sea, then interrogated the four men in broken English as to what had actually happened and what their individual roles had been. None of the four intended giving away any information and Fred remembers leaning against a rock idly tossing a small pebble into the air and catching it. This obviously annoyed the German for it brought a swift rebuke to: 'Put avay ze schtone!' Although his manner was harsh, they were not illtreated and the German told them they had recovered all the weapons.

They were then taken back to the prison where they spent one more night before returning to the aerodrome where a Ju 52 was waiting to fly them to Cherbourg. From there, they were taken by lorry to a huge prison at St. Lo mainly occupied by thousands of French Colonial troops although there were about fifty men there from the King's Own Scottish Borderers. The camp was located in a former French barracks on the edge of town and there they stayed in overcrowded conditions until a few days before Christmas when the entire British contingent was cleared out. The subsequent journey by goods train lasted four or five days and ended at Stalag VIII B at Lamsdorf in Upper Silesia near the Polish border. This camp nominally held over 50,000 British troops but the majority were billeted outside in individual working parties repairing roads and working on farms, in factories or down coal mines.

It was at Lamsdorf that Fred Drain was split from the others and thereafter lost contact with them, but over the next four years he tried to escape three times. The last time he got as far as Vienna Central Station having jumped a coal train bound for Italy. With a friend they had burrowed themselves into the corners of an open coal truck one night and lay covered up to their shoulders. Unfortunately they had not realised that the coal in each truck had been sprinkled with chalk so that signalmen, looking down as the train passed them, could easily spot if coal had been stolen or tampered with en route. The black corners were a dead give away and both were yanked out at Vienna. They were held in Vienna Central Gestapo prison for a week before being transported back to Germany.

Fred was finally liberated by American forces at Regensburg after the PoWs had been marched 1,500 kilometres in six weeks away from the advancing Russians. He stayed in the Army; went to Korea and became a Drill Sergeant with the Bedfords until he retired from the Forces in 1958.

After the four Commandos had left the village store at Torteval, Mr. Bourgaize buried Corporal Dumper's pistol in a tin lined with hay underneath his coal heap. Both he and his wife were very frightened as every home was being searched. Three months later, he recovered the tin to have a good look at the pistol but, to his dismay, the hay had 'sweated' in the closed tin and the weapon was pitted with rust. He buried it again beside the chimney stack where it stayed until the liberation.

After the war Mrs. Bourgaize wrote to the address one of the men had given her but she never received a reply. By 1978, just before their Golden Wedding anniversary, they decided to pass on the personal items to their nephew and the pistol was given to Richard Heaume for his museum.

Second Lieutenants Martel and Mulholland (who were to signal to the boats from Le Jaonnet beach by flashing the letter D in Morse if it was safe to come in) were also left on the island. As they were Guernseymen, they managed to find refuge with their families for some time while they tried to make good their escape but, in the end, realised it would be better to give themselves up. Tragically, Lieutenant Mulholland was accidentally killed on September 3, 1945 and was buried in St. Martin's Cemetery.

After he was liberated at the end of the war, Fred Drain carried on with his regiment serving time in Germany, Korea and at the unit's home base at Bedford. He became a Drill Sergeant and very smart he looks too being inspected by the Colonel of the Bedfordshire and Hertfordshire Regiment, Lieutenant-General Sir Reginald Denning. Fred told us he used to announce himself to new recruits as follows: 'My name's Drain. Now let's all have a good laugh and get it over with!'

The Doyle Monument, a hundred feet high, was erected to commemorate the works of Lieutenant-General Sir John Doyle (1750-1834), one of Guernsey's most industrious Governors. Before the war, one could collect the key and a torch from the caretaker at 'Doyle Bungalow' opposite the tower to climb the 133 steps to the observation platform.

Above: The Germans soon put paid to that. Although it served them as an excellent lookout post, and was provided with a canopy against inclement weather, it lay directly in line with the guns of Batterie Strassburg should they wish to fire towards France (Bundesarchiv). The answer was this spectacular explosion (Imperial War Museum and Royal Court). *Below:* This is the gun position with the rebuilt tower today.

We followed up Operation AMBASSADOR during our visit to Guernsey in April 1979. Today Petit Port is one of the few places where it is possible to retrace a Commando raid knowing that one is walking in the exact footsteps of the attackers. The steep stairway at Petit Port, leading from the beach to the Jerbourg Road, remains exactly as it was in 1940 except for repairs to worn steps. It is a stiff climb and one can imagine the exhaustion of the Commandos after climbing the steep stairway in sodden clothing weighed down with equipment. At the top the Doyle Column still stands although this is not the one seen by the Commandos. That monument was later blown up by the Germans and has since been rebuilt but on a more modest scale. At Torteval the general store is little changed except in name (no longer owned by the Bourgaizes) and not far away at Richard Heaume's German Occupation Museum one can see the pistol carried by Corporal Dumper on the first full-blooded raid undertaken by British Commandos in the Second World War.

In a nutshell, what the Special Air Service was to the Middle East and the Desert War, so the Small-Scale Raiding Force was to Europe and the French coast. The SSRF was conceived by Major Gustavus March-Phillipps (below right) — Gus to his friends — who, in the company of his outstanding 2 i/c Captain Geoffrey Appleyard (nicknamed Apple), drove down to Dorset one day in March 1942 to select a secluded headquarters and training base for the forthcoming series of raids to be carried out across the Channel. On the 20th they selected the lovely Elizabethan Anderson Manor (above) near Blandford Forum and it was here that the next three operations to the Channel Islands were masterminded. It was ideally situated for their jumping off ports of Portsmouth, Poole and Portland. Then it accommodated some thirty officers and men; now it is owned by the Isaac family.

CASQUETS

Operation DRYAD

The failure of Operation AMBASSADOR to achieve its stated goals was first-rate ammunition for those to whom the idea of the new Commandos was an anathema. Throughout the summer of 1940 the very existence of the new force was in balance. As Winston Churchill later wrote: 'The resistances of the War Office were obstinate, and increased as the professional ladder was descended. The idea that large bands of favoured "irregulars" with their unconventional attire and free-and-easy bearing should throw an implied slur on the efficiency and courage of the Regular battalions was odious to men who had given all their lives to the organised discipline of permanent units. The colonels of many of our finest regiments were aggrieved. "What is there they can do that my battalion cannot? This plan robs the whole Army of its prestige and of its finest men. We never had it in 1918. Why now?" It was easy to understand these feelings without sharing them. The War Office responded to their complaints. But I pressed hard.'

While the Commando battle was being fought in the higher echelons of the Army, the men on the ground got down to a programme of self examination and hard training to learn from the mistakes of Guernsey. One of the most pressing problems was that of reliable boats; another was the provision of adequate firepower. The smaller the force the harder-hitting it had to be but supplies of Brens and the American-made Thompson sub-machine gun were scarce. One source quotes that there were only forty tommy guns available at the time and all weapons had to be returned to a central store after each operation ready for re-issue.

Another difficulty was keeping the men occupied with a sense of purpose — training was all very well but there had to be operations at the end of it all. Throughout the winter of 1940-41 there was undoubtedly frustration for both officers and men. Then, in March 1941, the period of inactivity was broken when Nos. 3 and 4 Commandos raided the Lofoten Islands, an archipelago off the northern coast of Norway. Other operations in Europe and the Middle East followed and with them a change of leadership with the appointment of Lord Louis Mountbatten as Combined Operations supremo in October 1941.

With the new management came the turning point and thereafter the Commando force really came into its own. Within six months the 'greatest raid of all' — the attack on St. Nazaire in March 1942 — had been planned and executed; described aptly by Lord Louis as 'not an ordinary raid but an operation of war'. In August came Dieppe, comprising several independent Commando operations, and with it the whole original idea of the Commandos as a small force carrying out small raids was changed. The Commandos had outgrown the original reason for which they had been formed and there became a need to go back to basics. The result was that, in addition to the Special Boat Section, the Small-Scale Raiding Force was created under Major Gustavus March-Phillipps, Major J. Geoffrey Appleyard and Captain Graham Hayes.

When the new force visited the Channel Islands in September 1942 it was with a simple yet brazen idea in mind — the capture of a complete lighthouse crew.

Eight miles north-west of Alderney, a helmet-shaped chain of rocks called the Casquets straddles the main Guernsey to Southampton sea route. The rocks are notoriously dangerous to shipping and in 1722, the operators of ships passing the rocks (then called the Casketts) applied to the owner, Thomas Le Cocq, to build a lighthouse and offered him ½d per ton whenever their vessels passed the light. Le Cocq approached Trinity House (the body responsible for maritime safety around the shores of Great Britain) and a patent was obtained in June 1723.

Trinity House had decided that a light of a particular character was necessary to distinguish it from those on the opposite shores of England and France and three separate lights in the form of a horizontal triangle were proposed. Initially these were coal fires burning in glazed lanterns on a building erected on the main islet and were first lit in October 1724.

The lease to Thomas Le Cocq lasted for sixty-one years until Trinity House took over in 1785. Five years later the lights were converted to Argand lamps with metal reflectors; a revolving apparatus was fitted to each tower in 1818 and they were raised 30ft in height in 1854.

In spite of the fact that the rocks were lit, the Casquets have been the scene of many shipping disasters, among them the British man o'war *Victory* in 1744 with a complement of 1,100 and the SS *Stella* in 1899 with a loss of 112 lives.

A diaphone (fog horn) was installed in one of the eastern towers in 1921 and wireless communication in the other tower to the east in 1926. The light was then exhibited solely from the 120ft western tower, the signal in 1940 being three group flashes every thirty seconds. The 184,000 candlepower light was visible for seventeen miles.

As we have seen in Chapter Three, the Trinity House Vessel *Vestal* arrived in Alderney on June 22 to evacuate all its lighthouse personnel. The Casquets light had been turned off at 1445 BST the previous day; the two Venner time clocks had been removed together with the Order Book, Journal and various papers and the doors locked. The keys were then handed to Mr. N. J. Allen on Alderney. In view of the hasty evacuation, he was asked to return to the lighthouse to run the oil from the tanks to waste although was prevented from doing so by adverse weather. The keys were eventually forwarded to London.

Lighthouses were just as important for navigation in war as they were in peacetime; perhaps more so because of the absence of lights on shore. The German policy regarding lighthouses was similar to that adopted by

These rocks eight miles from Alderney have seen the death of many ships. They have been lit since 1723 and were taken over by the Germans in 1940. This was the venue for the debut of the Small-Scale Raiding Force and the plan was brazen and ambitious — the capture of the entire garrison.

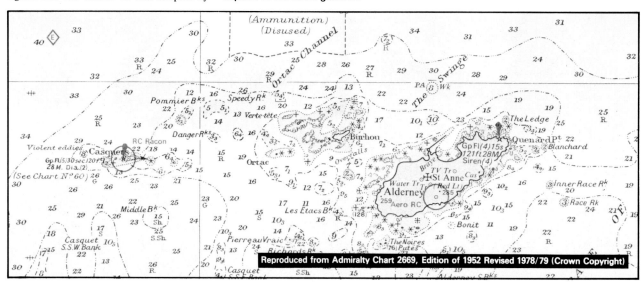

Trinity House for the Royal Navy: a light would only be exposed (at reduced power) when requested by friendly ships.

When the Germans took over the Channel Islands they soon installed their own men in Les Hanois (on the rocks of the same name south-west of Guernsey) and the Casquets. Barbed wire was installed against possible attack. The Casquets was an unenviable posting as the small garrison had to be provisioned by boat from Alderney with all supplies including fresh water and there were some periods, especially in winter, when the seas made any approach impossible for weeks on end.

This then was the target for Operation Dryad. Mounted on the night of September 2, 1942, the story is told from Major March-Phillipps' official report:

OPERATION REPORT

'After innumerable fruitless attempts, this operation took place in spite of what has come to be called DRYAD weather, wind force 3 rising to 4 and sometimes 5.

'MTB 344 and ten officers and two ORs of SSRF (Small-Scale Raiding Force) personnel took part. The MTB sailed from Portland Bill at 2100 hours. In spite of a very careful overhaul, the port engine again gave trouble and the passage had to be made at a reduced speed of some 25 knots for the first 25 miles. It was then possible to increase the engine revolutions, and the normal cruising speed of 33 knots was maintained until within five miles of the objective.

'At 2210 hours a white light, flashing every five seconds was seen, bearing 20 degrees on the port bow. This light only showed for five minutes and then went out. It was probably Alderney at its maximum visibility. Two very bright light beams then appeared on the starboard bow, which might possibly have been a searchlight on Guernsey, and one on the starboard quarter which must have been on a ship. A vertical red beacon was also just visible well on the port bow. This was thought to be on the mainland at Cape de la Hague.

'At 2230 hours speed was reduced to 15 knots, and a red light flashing once every fifteen seconds was seen very fine on the port bow. This was at first believed to be Casquets

Three pictures taken by Sonderfuehrer Hans Herzog from the FK515 office in Alderney of a party of Germans en route for the 'Huetchen' lighthouse, as they called it, in the summer of 1941 (Alderney Library).

but at 2245 hours a rock was reported on the port beam at about one mile rise, and course was altered to close it. This turned out to be the Casquets and a course was laid to approach it from the northward against the tide. The MTB's main engines were then cut off and the silent auxiliary used. The red light was identified as Sark.

'The MTB was manoeuvered to within 800 yards of the rock, where she was anchored with a 45lb Admiralty-pattern anchor and 50 fathoms of 2½-inch rope, and the landing party went ashore in a Goatley-pattern assault craft, paddling four aside, leaving the MTB at 0005 hours.

'Many and conflicting eddies of tide were experienced on the approach, which took considerably longer than was anticipated, probably because the approach was later than had been calculated and the NE-going flood tide was by then running hard. In fact the landing was not made until 0025 hours.

'Unlike plans made for previous occasions the landing was made, not at any recognised landing point, but on the face of the rock immediately under the engine house tower. This was done partly because of the difficulty of finding holding ground for the kedge anchor on the recognised north landing, and partly because it was feared that the landing points might be guarded or be set with booby traps. There was a fairly heavy run on the rocks from the south-westerly swell but the kedge anchor held well and the landing was made without mishap or any harm being done to the boat. A way was then found up the 80ft cliff and any noise made by the party was drowned by the rumble of the surf and the heavy booming of the sea in the chasms and gulleys.

'Meanwhile the boat had been pulled off the rock by the kedge anchor and was held riding the swell between the bow and stern lines about 20 feet off the rock. One officer, Captain Graham Hayes, MC, whose seamanship on landing and the much more difficult operation of re-embarking was admirable, was left in the boat in charge, and another officer minded the bowline and kept watch through the infra-red receiving set on the MTB.

'Coiled dannert wire was met and climbed through on the way up the cliff and the gateway was found to be blocked by a heavy knife-rest barbed wire entanglement, but a way was found over the western wall and the whole party made the courtyard unchallenged. At this point, the order was given for independent action and the party was split up and rushed the buildings and towers according to a pre-arranged plan. Complete surprise was obtained and all resistance was overcome without a shot being fired.

'Seven prisoners, all of them Germans, including two leading telegraphists, were taken in the bedrooms and living rooms. The light tower, wireless tower and engine room were all found to be empty, although the generating plant in the engine house was running, and the watch, consisting of two men, was in the living room. The rest were in bed, with the exception of two telegraphists who were just turning in. A characteristic of those in bed was the wearing of hair-nets which caused the Commander of the party to mistake one of them for a woman.

'The prisoners were re-embarked immediately and taken down over the rocks by the way the raiding party had come up, some of them still in their pyjamas, as time was getting short and it was expected that the operation of embarkation would take some time.

'Re-embarkation commenced at 0100 hours. The wireless was then broken with axes and the buildings and offices searched for papers, documents and code books. The light and the engine room were left intact.

'The following papers were removed:
Codebook for Harbour Defence Vessels FO i/c France, signal books, records, W/T diary, procedure signals, personal letters and photographs, identity books and passes, ration cards, Station Log, Ration Log, Light Log and a gas mask and gas cape.
(These papers were handed over to the military authorities at Portland on return.)

'A thorough search of the buildings revealed the presence of a quantity of arms and ammunition. Each man was equipped with a rifle of the old Steyr pattern and there were two large cases of stick grenades, one of them open. There was also an Oerlikon cannon-shell (small-calibre) gun, loaded and placed against the wall in the living room. If a good watch had been kept, or if any loud noises had been made on the approach or on landing, the rock could have been rendered pretty well impregnable by seven determined men.

'Particular attention is called to the presence of stick grenades in such outposts for they are formidable weapons. The Oerlikon gun and the rifles were removed by the raiding party but the stick grenades and ammunition store were left untouched. It was not possible to remove them in time and no attempt was made to blow them up as it was considered most important to make no noise that might reveal the presence of a raiding party to the mainland.

'Meanwhile the embarkation of the prisoners was proceeding under the direction of Captain Burton and Captain Hayes. This was a particularly difficult and hazardous operation as the slope of the rock at this point was at least 45 degrees and the prisoners had to slide down and be hauled into the boat by Mr. Warren, the bowman, as she rose on the swell. Great credit is due to all concerned that this operation was successful for one mistake might have meant the swamping of the boat which might have brought disaster on the party. When the search party arrived with the papers and arms, the prisoners had all been embarked and it was then decided by the Commander and Captain Hayes not to send for the small emergency dory on the MTB but to load all personnel into the Goatley which was standing up well to the weight. It was decided, however, to jettison the arms and they were accordingly thrown into the sea. (The emergency dory was not sent for because of the distance it would have had to come and the inevitable delay which would have been caused.)

'When the search party was finally embarked at 0110 hours there were nineteen men in the Goatley, which rode the swell admirably, though dangerously low in the water. A tribute must be paid to the Goatley which comes from all members of SSRF. This boat, which is entirely without lines or shape and designed on the principle of the flat iron, has behaved splendidly under all conditions. It has weathered moderate seas and stood up to pounding at rock landings in a way that entirely belies its looks and the natural reactions of any seamen when confronted with such a hull.

'During the operation, the MTB had dragged her anchor to the northward, and Lieutenant Bourne, RNVR, wisely decided to

Captain Graham Hayes had considerable experience at handling a boat as before the war he had been round the world on a Finnish sailing ship, the 'Pommern'. He was a boyhood friend of 'Apple' and had joined him and 'Gus' early in 1941 as the nucleus of a Special Service Unit to operate off the coast of West Africa. Just ten days after the return from the successful Casquets operation, on which his seamanship had been especially commended, he went on another SSRF raid to Normandy (see page 148). The landing party was surprised and Captain Hayes was left behind and taken prisoner. Exactly ten months later, on July 13, 1943, Graham was shot by the Germans — this is his grave in Viroflay New Communal Cemetery near Versailles (J. P. Pallud).

Visitors to the lighthouse today set their helicopters down on the landing pad on the right. If you have the chance to visit the Casquets don't forget your hairnet!

weigh and close the Casquets before the signal was received.

'The Goatley was intercepted about 500 yards off the rock, at 0135 hours and the prisoners, who were very docile, were battened down in the forecastle with Captain Dudgeon and two others, where they gave no trouble. The voyage home in a rising sea, though desperately wet, was without mishap and the MTB docked at Portland at 0400 hours.

'There were two small casualties, Captain Kemp was injured in the leg while embarking and Captain Appleyard, who was acting bowman and the last to leave the rock, sprained his ankle in the descent which had to be made without the assistance of the rope and with the boat well away from the rock. In most cases of difficult landings, bowman must swim out to the boat.

'Great credit is due to Lieutenant Bourne for his handling of his ship in this and the previous operation, both of which were hazardous and difficult undertakings in close proximity to reefs and sunken rocks, and to Captain Appleyard, whose navigation made them possible. Also to Private Orr, a German speaker, who marshalled the prisoners and did much to make the search successful.'

It was discovered that the lighthouse crew had orders to regularly report in to Alderney but, because a routine message had been transmitted just five minutes before the Commandos landed, it was some time before the Germans realised what had happened. With the continued silence, a boat was dispatched to the rocks and it returned to state that the entire crew had disappeared. As the transmitter had been put out of action, it was assumed the men had been captured by force.

An argument then developed between the German Navy and the Army in the Channel Islands as to who was responsible for the defence of the lighthouse. The officer in charge, Obermaat Mundt, was from the 3 Batterie of the Marine Flak Regiment 20 and had been in command for exactly a month. His three radio operators, Funkgefreiters Dembowy, Kraemer and Reineck, were Navy men whereas the three guards, Gefreiter Abel, Kepp and Klatwitter were from the Army.

When Hitler was informed of the raid his first reaction was to order the abandonment of the lighthouse, especially in view of the problems of its defence and provisioning. His order to this effect was issued three days after the operation, ironically on the same day that a new eight-man garrison was taken to the Casquets. However, the Kriegsmarine stated that the lighthouse was necessary for observation and navigation and that it would be adequately protected in future with mines and more barbed wire.

Within two weeks of the raid, Seventh Army HQ reported to Ob. West that the strength of the garrison had been increased to thirty-three men all told: Wehrmacht — one officer, three NCOs and twenty-one troops; Kriegsmarine — one NCO and seven ORs. There was one 2.5cm Pak, five machine guns, one anti-tank rifle and 350 hand grenades and additional wire entanglements and trip wires were in course of installation to restrict any future landings.

The Germans continued to man the Casquets lighthouse for the remainder of the war until May 17, 1945 when two officers and twenty men were taken off as prisoners-of-war. Trinity House left six prisoners on the rock to continue with maintenance of the lighthouse until they were able to replace them with their own men.

The station was electrified in 1952, the light output being increased to 2,830,000 candelas making it visible for twenty-eight miles in clear weather. The light signal has been changed to five (instead of three) white group flashings every thirty seconds with the fog signal comprising two blasts, each of two seconds duration, every sixty seconds. A helicopter landing pad was marked out on a flattish portion of rock, the three keepers being relieved every twenty eight days by one of the Trinity House Bolkow Bo 105D helicopters. Ironically, therefore, although the isolation of the historic lighthouse remains, contact with civilisation has been made simple by the use of machines made by the former enemy.

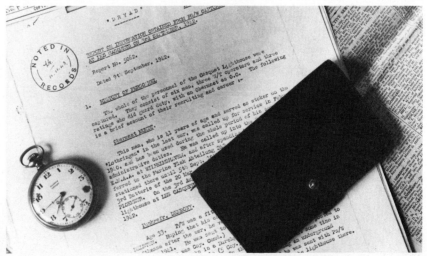

This pocket watch and wallet were taken from Obermaat Mundt by Captain Bruce Ogden Smith (who was in the SSRF with his brother Colin) and he still retains them as unique souvenirs. He also had Mundt's helmet at one time but gave it away some time ago.

BURHOU
Operation BRANFORD

The successful operation by the Small-Scale Raiding Force on the Casquets lighthouse was quickly followed by a visit to another outpost of the Channel Islands five days later. Captain Colin Ogden Smith was in charge when MTB 344 set out from Portland on September 7 with eleven men on board. At 9.05 p.m. course was set for Ortac Rock, three miles north-west of Alderney, and virtually mid-way between the island and the Casquets. The purpose of the mission was to make a landing on and reconnoitre the nearby island of Burhou, half-a-mile long and barely 300 yards wide, to establish its military potential.

The weather was fine with a gentle swell from the south-west and the MTB was making good progress when, at 10.00 p.m., the port engine cut out. There had been problems before over the maintenance of fuel pressure to this engine and, as there was a likelihood of encountering E-Boats where every ounce of speed would be vital, Captain Ogden Smith reluctantly decided to abort the operation. The MTB was turned around and course set for base but they had not been travelling for many minutes before the mechanic, working frantically on the dead engine, managed to get it started. As the fuel pressure now registered normal, Captain Ogden Smith decided to turn about but to return at once if the engine showed further signs of trouble.

Within an hour-and-a-half of leaving England a big searchlight, believed to be located on Guernsey, was sighted on the starboard bow and a few minutes later Alderney light, flashing four every five seconds, was picked up fine on the port bow. Shortly afterwards, a red flash every fifteen seconds identified the bearing of Sark and at 11.00 p.m. red lights ahead indicated the position of Alderney harbour. Speed was decreased to fifteen knots and the sea was scanned for sight of the sentinel-shaped Ortac Rock. Here the full force of the Atlantic meets the rocky outcrops north and west of Alderney and even on a calm day presents an awesome sight. At night the scene was even more confusing to the Commandos with breaking seas on all sides.

A few minutes later a rock believed to be Ortac was spotted about three miles to port with the Casquets on the starboard bow. Course was altered to avoid what appeared to be dangerous reefs until Ortac stood out undisputedly. At this point, the confused and breaking sea was identified as the Pommier Bank and the Danger and Dasher Rocks. Having pinpointed its position, the MTB manoeuvered inside the Burhou reef on a line of 45 degrees magnetic. The anchor was dropped just after the Casquet rocks came into transit with Ortac at fifteen minutes past midnight.

Five minutes later the landing party of six Commandos, under Captain Ogden Smith and Second Lieutenant Anders Lassen (who had also participated in the Casquets raid), embarked in a Goatley to paddle the 600 yards to the shore.

Captain Ogden Smith later described what they found:

'Shore was reached at 0028 hours and the landing made on the reef at a place about 60 yards west of the southernmost point of the island. The rock here is steep and in steps. The boat was held off by kedge anchor. There was no noticeable tidal set and the sea was absolutely calm. The party, less cox and bow who remained with the Goatley, made its way for 60 yards over broken rock which was wet and slippery with seaweed to the rockline above high water. The only building on the island was a house about 400 yards NE of the landing place. This house had been partly

On the night of September 7, 1942, Captain Colin Ogden Smith led the Small-Scale Raiding Force mission to the barren islet of Burhou, three miles north-west of Alderney (see map page 142). Although this particular operation was uneventful, Colin had many more adventures as a Commando until July 29, 1944 when he was cornered in a field at Kerbizec whilst with the Maquis behind enemy lines in Brittany. Colin and two Frenchmen, Gerard de Carville and Maurice Miodon died fighting together — this is their combined grave in Guiscriff Communal Cemetery.

demolished by artillery fire, the roof and the first floor having collapsed inwards, entirely filling the ground floor. There were several small shell craters round. The party divided here, Corporal Edgar taking two men to examine the westward end and myself going with two men to cover the central and eastern part. No sign of recent habitation or any defence works were encountered.'

In Captain Ogden Smith's report he stated that, 'the island is roughly 700 yards long and ranging from 300 yards at the east and west extremities to 150 yards in the centre. There is a broken central ridge of bare granite rock which at the highest point rises 8-15 feet above the ridge line, below is soft grassy soil which is split up by irregular rainwater channels 4-6 inches wide and anything up to 12 inches deep which makes walking difficult and hazardous. The average slope is 20 to 30 degrees. The highwater line is bounded by a more or less continuous band about 15 feet wide of smooth flat granite boulders. There are frequent lengths of smooth granite rock outcrop throughout the island. The foreshore is made up of broken granite rock with pools and long narrow gulleys which can be identified from the air photographs.'

He concluded that, 'landings in similar weather would seem possible anywhere west of the southern reef and east as far as the gulley immediately below the house. Similar conditions exist on the north shore. The higher the state of the tide the better.

'Pack artillery or mortars or loads requiring two or three men are practicable. Wheeled or track guns would present great difficulties as there are no sand beaches and all landings would have to be made over rock. There are a number of places where high-angle guns could be placed, though the ground is very soft where the grass grows. There is sufficient crest clearance except immediately behind the rough ridge rocks.'

During the landing, the new crew on the Casquets lighthouse were seen to communicate with Alderney by lamp (the call signs being four 'A's in Morse), possibly because the wireless transmitter had not been repaired. While the party was still ashore, the lighthouse light came on at 1.00 a.m. flashing three times every twelve seconds and a searchlight appeared for a few minutes at Cap de la Hague on the coast of France. Another light was seen on a ship to the north although Alderney itself was blacked out.

At 1.38 a.m. the Goatley returned to the MTB after exactly an hour ashore. Everything had gone to plan and they moved off for Ortac. Just before they reached the rock, the Casquets light was extinguished but, with a bearing set for Portland, the MTB docked without incident at 4.30 a.m.

Burhou was used by the Germans for a convenient target for the artillery battery on the Giffoine — hence the demolished house — and as far as is known it served no other purpose during the war.

Today the island is owned by the States of Alderney and is retained as a bird sanctuary. A simple wooden hut, furnished with two bunks, has been built in the ruin of the masonry house and is available for rent at a modest 50p per day. Would-be visitors can only land on the island in calm weather and must take their own drinking water. No dogs may be taken to the island and applications to book the hut are made to the Clerk of the States.

Burhou in April 1979. The location of the ruined house visited by the Commandos can be seen in the centre.

SARK

Operation BASALT

No sooner had Captain Ogden Smith returned from BRANFORD than Major March-Phillipps was preparing for Operation AGUATINT to Ste. Honorine in Normandy the following weekend. He set out on Saturday evening, September 12, in the company of Major Geoffrey Appleyard, Captain Graham Hayes and nine Commandos but unfortunately Ste. Honorine was more heavily guarded than anticipated and the Germans more alert. After they had dropped anchor, Major Appleyard, who had injured his leg, remained on the MTB while the others paddled ashore in the Goatley boat. A fierce fight developed ashore and, as the Commandos tried to get back to sea, the Goatley was hit and holed and Major March-Phillipps killed. As a result, Major Appleyard was forced to withdraw to save the MTB leaving behind Captain Hayes and the others.

On September 12 the greatest tragedy of all befell the infant Small-Scale Raiding Force when its mentor, Major 'Gus' March-Phillipps was killed during Operation AGUATINT. Long before the name Omaha entered the history books, the Normandy beach at Ste. Honorine was blooded in battle. Few visitors to the huge American D-Day cemetery on the cliff top at nearby St. Laurent stop at the little village churchyard where the remains of the 34-year-old Commanding Officer, awarded the DSO, MBE and Mentioned in Despatches, now repose (J. P. Pallud).

Undaunted, Major Appleyard immediately began work on the next SSRF raid — Operation BASALT. This was aimed at Sark (which he knew quite well from pre-war visits) and it was intended that a landing should be made to capture prisoners and obtain information on the German defences on the island. Included in the party of twelve men was Captain Ogden Smith and also Second Lieutenant Anders Lassen who had been specially commended for his judgement and seamanship on the raid to Burhou.

As the Commandos left Portland in MTB 344 at a few minutes past seven o'clock on the evening of October 3, little did they realise that the results of this raid were to have worldwide repercussions.

The German strength on Sark at that date comprised one heavy machine gun section; one light mortar group and one anti-tank platoon, all from 6 Kompanie of Infanterieregiment 583 of the 319 Division. The fixed defences consisted of three anti-tank guns, three flame-throwers, one light machine gun and 939 S-mines laid in twenty-two separate minefields. A five-man engineer unit was also stationed on the island (billeted in the Dixcart Hotel) carrying out work on the harbour installations at Creux. The Inselkommandant at the time was Oberleutnant Herdt.

As the MTB approached Baleine Bay its identity was requested by lamp flashed from a lookout post on Little Sark. Major Appleyard replied by Aldis that they were German and seeking shelter in Dixcart Bay for the night which apparently satisfied the post as no further signal was observed. At 11.30 p.m. the MTB dropped its anchor close by Point Chateau where a steep promontory known as the Hog's Back juts out into the bay.

Bombardier Redborn was one of those in the landing party: 'Everything went according to plan; the navigation was excellent. We landed exactly at the right spot. We rowed in and the landing boat was made fast and left with a guard while the rest of us clambered up the steep path which led to the top of the cliff. The job of guarding the boat was not to be envied. Under no circumstances had they to

leave their post before a definite time whereupon they had orders to row out to the MTB whether the landing party had come back or not.

'When we got to the top of the cliffs, we found barbed wire entanglements. The stillness of the night was only broken by the cry of a seagull or when the wire was snapped with cutters. We fumbled around the whole time in the dark...

'When we had gone forward a little — I was in the centre of the file with Corporal Flint and Anders Lassen, with Captain Pinckney bringing up the rear — we heard a German patrol coming. We all dived off the path and the Germans went past without noticing anything. After this, although we did not run into any other patrols, there were many false alarms when we heard sticks snapping and suchlike.

'Major Appleyard thought it best to go into a house and find out the local situation. The houses lay almost a mile inland. The first was empty so we scrambled down and up a little valley to the next one — a big lonely house on its own. We kept watch for a few minutes till all was silent as the grave. We tried every door and window but all were locked so we smashed a window in the French doors, undid the latch and tumbled into the room.

'Downstairs was all empty but Major Appleyard and Corporal Flint who went upstairs were luckier. There they found an elderly lady. I did not see her myself because I and some of the others had to stay on watch downstairs. We had made a lot of noise breaking the window and, as there was always the possibility of an enemy patrol, we had to be prepared to shoot if surprised.

'The elderly lady was wonderful. Although alone in the house and awakened by two men with blackened faces, she remained completely calm and was immediately aware of what it was all about. When the Major came down he said she had given him important information about the gun emplacements and defended positions and so on. She also said the Dixcart Hotel up the road was occupied

We retraced the Commandos' route. Looking from the Hog's Back to La Jaspellerie on the opposite side of the valley.

Today the path along the top of the Hog's Back is well-beaten although flanked by an impenetrable blanket of thorns.

Our route led past Petit Dixcart *(above right)*, the house which the Commandos found to be empty in 1942, before we clambered

through the woods, across a stream and up the hillside in the background to La Jaspellerie *(below)*.

At the time of the raid, La Jaspellerie (Jerseyblick to the Germans) was occupied by Mrs Pittard, youngest daughter of Commander Mardon. During the Germans' investigations into the operation, which they believed had been aided by inside information from someone on Sark, Mrs Pittard confessed that the Commandos had called at her house. The Germans, who were already preparing to evacuate the whole of Sark as a reprisal, rashly promised she would not be taken to Germany if she cooperated. As a result of what she told them, Generalfeldmarschall Gerd von Rundstedt, Oberbefehlshaber West and uncle of von Schmettow, accepted that there had been no prior communication between Sark and England. However, Mrs Pittard had helped the enemy and he had to request a decision from Oberkommando der Heeres (OKH) as to whether or not the word of a German officer should be kept and also if it was necessary to carry out the deportations. The matter was passed on to the highest level in the Oberkommando des Wehrmacht (OKW), it being suggested that if the entire population was evacuated on military grounds, this would solve the dilemma. In the end, this was avoided but both she, the Misses Duckett and Page from the Dixcart Hotel, Mr. Hathaway and others were all deported; the ladies initially to an internment camp at Compiegne near Paris.

Operation BASALT was commanded by Major John Geoffrey Appleyard who must still have been very much affected by the loss of both his commanding officer and his best friend exactly three weeks earlier. As well as the Military Cross and Bar, Geoffrey was awarded the DSO on December 15, 1942 for his part in the cross-Channel raids by the SSRF. He went on to join the SAS Regiment — and was posted missing on July 12/13, 1943 when the Albemarle detailed to drop parachutists to hold a bridge in Sicily (in which he was an observer) failed to return. His memory is commemorated on the Cassino War Memorial in Italy *(right)*. Tributes to 'Apple' abounded; his father's book, simply called 'Geoffrey' is a moving story of a great man lost. Lord Louis Mountbatten wrote: 'He was a grand leader and I was proud to have him in my command.'

A tough-looking Major Appleyard knocks at the French windows. The house was deserted and devoid of furniture. Not wanting to end up in Sark's mini-jail, we decided that breaking in would be taking it too far so we contented ourselves with this shot through the window of the stairs ascended by Major Appleyard and Co.

and used as a headquarters. She understood that there were some Germans at the hotel with sentries in the annexe next door.

'Captain Pinckney asked the lady if she would like to go back to England with us but she said she would not as she did not want to abandon her property. She begged us not to say she had given us information — that we obviously would not do — although the Germans found out about it nevertheless.

'We had already been on the island an hour and we would have to hurry if we were not to go home with our job unfinished. The MTB had orders to leave the island if we were not back within four hours. A corporal was therefore sent back to the boat with instructions to make them wait an extra half-hour.

'We set off towards the town and when we neared what we believed to be the German quarters, Anders and I were chosen to deal with the sentry. We went ahead to see the lie of the land. A little later we came back to tell what we had found out. And we made some

Anders Lassen, the fearless Dane, was a member of 'Gus' March-Phillipps' team from the earliest days. He was an active participant in all the SSRF cross-Channel operations and went on to become a Major with the SAS. He was a formidable adversary and could move so swiftly and silently that it is said he could kill a stag with a knife. In April 1945, the Special Boat Service of the SAS mounted Operation FRY to capture the four islands in Lake Commachio in northern Italy. Anders Lassen landed near Porto Garibaldi and eliminated two German positions with grenades himself as his unit advanced along the causeway. By then his patrol had suffered many casualties and, after regrouping to attack a third position, he threw a grenade which was answered by a cry of 'Kamarad!' As he went forward to take the surrender, he was cut down by a burst of automatic fire yet his dying gesture was to lob another grenade to enable his men to capture the position.

For a man such as Lassen to come through five years of war unscathed and be killed just three weeks before the end was a tragedy, and his reward was the ultimate accolade — the Victoria Cross. They took him to Argenta Gap War Cemetery — chosen by the 78th Division for battlefield burials — and there laid him to rest among 1,249 of his fellow servicemen.

The advance to the Dixcart Hotel. The hotel annexe *(photo right)* is the first building reached along the track from La Jaspellerie and it must have been close by the spot that Anders Lassen attacked the German sentry.

The annexe, a prefabricated style of building not constructed of stone like the main hotel, photographed during our visit in April 1979. The covered passageway connecting it to the hotel on the left was still in existence at the time of the raid.

Major Appleyard's official report on the fight reads as follows: 'The whole party then entered the annexe and a thorough search revealed the presence of five Germans, all sleeping in separate rooms. Their clothes were searched for weapons, pay books, papers etc. likely to prove of value, and the prisoners were then taken out of the house to be assembled under cover of the trees nearby. In the darkness outside the house one of the prisoners, seeing an opportunity, suddenly attacked his guard and then shouting loudly for help and trying to raise an alarm, he ran off in the direction in which it was known there were buildings containing a number of Germans. He was caught almost immediately by his guard, but after a scuffle again escaped, still shouting, and was shot. Meanwhile, three of the other prisoners seizing the opportunity of the noise and confusion, also started shouting and attacking their guards. Two broke away and both were shot immediately. The third, although still held, was accidentally shot in an attempt to silence him by striking him with the butt of a revolver. The fifth prisoner remained quiet and did not struggle. There were answering shouts from the direction in which the prisoners had attempted to escape and sounds of a verbal alarm being given.'

jokes about it: Anders said it would have been better if he had had his bow and arrow!

'We returned to the spot where the sentry was on patrol. As there was only one man, Anders said he could manage him on his own. We lay down and watched him and calculated how long it took him to go back and forth. We could hear his footsteps when he came near, otherwise everything was still. By now the others had crept up so that all caught a glimpse of the German before Anders crept forward alone.

'The silence was broken by a muffled scream. We looked at each other and guessed what had happened. Then Anders came back and we could see that everything was alright. The Major believed that the way was now clear for us to approach the annexe. We had expected to find another sentry outside the hotel and because of this we went in formation but none was seen.

'It was difficult to open the door and we made a great deal of noise before we rushed inside. We were very surprised not to find anyone in the first room — a kind of hallway — and we carefully approached the door on the far side. It was Anders who opened the door and we found it led to a passageway with about six doors on each side. The Major gave orders that each man should take a room and all go in at the same time.

'I rushed into the room allotted to me and heard snoring. I switched on the light and saw a bed with a German asleep. The first thing I did was to draw the curtains and tear the bedclothes off him. Half-asleep he pulled them back again. I got the blankets off a second time and when he saw my blackened face he got a shock . . . I hit him under the chin with a knuckleduster and tied him up. Then I looked round the room for papers or cameras.

'I got him to his feet still half-senseless and out into the corridor where Captain Pinckney, Andy and the others already stood; there were five prisoners all told. I covered them while the others searched the rooms once more and when this was done we took the prisoners outside.

'When we were all outside, it happened. Until then, everything had gone fine but as soon as we were out in the moonlight they began to scream and shout, probably because they saw how few we were. All five of them had their hands tied behind their backs but they were not gagged. As soon as they started hollering we set about them with cuts and blows. Major Appleyard shouted, "Shut the prisoners up!" and this began a regular fight.

'I was not exactly clear over what happened next as I had so much trouble with my prisoner — he had got his hands free and we were fighting. He was just on the point of getting away so I gave him a rugger tackle and we both fell to the ground. He got free again as he was much bigger than me but I grabbed at him and we rolled about in a cabbage patch. One of the officers shouted above the noise: "If they try to get away, shoot them."

'Captain Pinckney's prisoner got free and started towards the hotel shouting at the top of his voice. The Captain went after him and a shot rang out. I had just about had enough of my German; I couldn't manage him so I had to shoot him and found that the others were doing the same with their prisoners. All, that is, except Anders who still stood and held two Germans tightly.

'More shots rang out with shouting and screaming. It was a hell of a rumpus and lights were coming on in the hotel. Anders who had now freed himself of his prisoners wanted to throw some grenades through the hotel windows but Major Appleyard said no, keep them we may need them later. By now Germans were pouring out of the hotel and when we saw how many of them there were, we decided to get away. We still had one prisoner who had seen what we had done to the others and he was stiff with fright and did everything we told him.

'The most important thing now was to get back to the boat as quickly as possible. The island was waking up and the German headquarters was like a wasps' nest. How we ran. We ran until every step hurt but not in a panic, still in open patrol formation.'

When the Commandos reached the Hog's Back, the German was all in and had to be dragged down the cliff. Lieutenant Lassen was the rearguard helping Private Smith who had been wounded. Those on the MTB were about to leave when the Major's party rowed out from the shore. Clambering aboard they left Sark around 3.45 a.m.

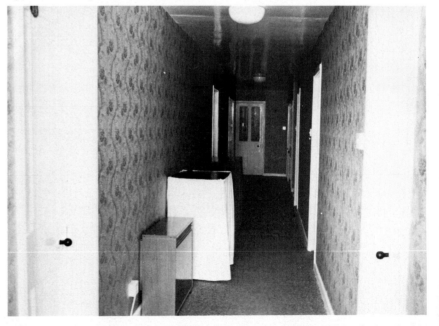

When we took this photo it was difficult to imagine the drama which took place in this passageway thirty-seven years previously.

The fight had taken place at around 2.30-2.50 a.m. Oberleutnant Herdt, the Inselkommandant, was informed by Obergefreiter Schubert, the Orderly Corporal on duty, at 3.00 a.m. (4.00 a.m. German time) who gave the official alert five minutes later when the situation was clarified. Troops and strongpoints were alerted and the Kommandant personally took charge of the search operation. Gefreiter Klotz was found completely naked and he explained what had taken place and how he had managed to escape.

A unit under Leutnant Balga was ordered to the Hog's Back to try to apprehend the Commandos before they could leave the island, and at 3.25 a.m. a customs boat, with a crew of five armed with a machine gun, put to sea to search Derrible and Dixcart Bay. At 3.37 a.m. Oberleutnant Herdt got through to the Regimental Command post on Guernsey, told them of the British attack, and then left to meet up with Leutnant Balga searching the headland. He ordered Dixcart Bay to be sealed off while a second detachment advanced towards Derrible Bay. Following the tracks left by the Commandos, their route was traced and several items of equipment recovered: two Commando knives, a magazine for a sub-machine gun and pistol, a pair of wire cutters, torches, a woollen cap, scarf and several toggle ropes.

By this time news of the raid was speeding up the German chain of command: Division HQ was notified at 3.50 a.m.; LXXXIV Corps HQ at 4.30 a.m. and Seventh Army HQ at 6.10 a.m. At 6.45 a.m. the news had reached Army Group D in Paris and the Chief of the General Staff. Within three hours an eleven-point questionnaire was telephoned from Ob. West:

1) Who is the responsible Inselkommandant or Commander for Sark?
2) What was the state of the weather at the time of the attack?
3) How was the landing or climbing site protected?
4) Was the billet inside or outside a strongpoint?
5) Was the engineers' billet protected?
6) Was the engineer unit and its accommodation known to the Inselkommandant to whom it was designated?
7) Had an alert been raised?
8) At the time of the attack had combat noises been heard and by whom?
9) What communications does the island possess and what is the chain of communication?
10) What is the reason for the lapse in time between the attack at 4.10 and the report to Seventh Army HQ at 7.45? *(German Central European Summer Time — Author.)*
11) What is the current strength of the Sark garrison (unit, troops and whether reinforced)?

The answers were requested to be telephoned through by 11.00 a.m. that day.

Meanwhile the Divisional Kommandant and the Geheimfeldpolizei had arrived on the island and immediately began their investigation at the Dixcart Hotel. The joint owners of the hotel, the Misses Duckett and Page, who amazingly slept throughout the whole operation, were interrogated and a Mr. George Hamon was arrested while inspecting his rabbit traps in the area. The Germans had a strong suspicion that someone had told the Commandos there were troops at the hotel and every effort was made to locate the informant.

Meanwhile, the information requested by Army Group HQ had been telephoned through at 11.45 a.m. — three-quarters of an hour after the deadline and a record of this telephone call has been preserved (reproduced with original German times):

1) Oblt Herd (sic) Kp. Chef. 6/IR 583
2) Misty, starlight at higher level, heavy ground fog, consequently poor visibility.
3) Landing place not precisely established yet, probably Dixcart Bay defended by single S-minefield.
4) To be clarified in due course by personal investigation by Division's Kommandeur.
5) Engineers' billet lay about 500 metres from the coast and 400 metres from company reserve strongpoint.

The fight . . . the pandemonium . . . the shots, and the deaths — it all seemed so unreal this sunny day.

6) The Inselkommandant knew of the presence of the engineers unit. The question of command and billeting is being clarified at the moment.
7) The alarm was given by the Company Commander at 4.10 after rifle fire and calls for help.
8) No battle noise was heard during the attack on the billet.
9) Telephone cable from Inselkommandant on Sark to Regimental Command Post IR 583 on Guernsey. All strongpoints on Sark are linked to one another.
10) 4.47 Report from Company to Regiment.
4.50 Report from Regiment to Division.
5.30 Report from Division to Corps HQ which only referred to cries for help, shots from a house in the centre of the island, wounded, confused, in the course of further investigation, a possible attack considered. Chief of General Staff ordered fuller investigation to establish whether it was an enemy attack, sabotage or a drunken disturbance.
6.40 Second message from Division.
6.45 Corps HQ advised (with consultation).
6.55 Message relayed to Commanding General.
7.00 Call from Commanding General to Operations Section re details of report and presentation.
7.10 Message to Seventh Army HQ.
11) Garrison on Sark: 6/IR 583 reinforced by one platoon 14/583; one heavy MG platoon; one heavy mortar section; three anti-tank guns, three flame-throwers (as fixed weapons) and one light machine gun.

It is interesting to note that a direct answer to question 10 concerning the lapse of time was avoided by simply listing the times of messages. In a follow up teleprinter signal at 1.05 p.m. it was claimed the delay was due to the fact that 'the Division as well as Corps had made its first report only after further clarification of the facts' — whatever that means!

While the reports were flashing back and forth, the situation on the island was being assessed. Two engineers were dead — Unteroffizier Bleyer and Gefreiter Esslinger — and another, Gefreiter Just, was slightly wounded and a fourth man (Obergefreiter Weinrich) was missing believed captured. Oddly there is no mention in any of the official German reports of the sentry attacked by Lassen who later recorded in his diary that this had been the first time he had used his knife.

When Generalleutnant Erich Mueller (CG of 319 Division) discovered that his specific instructions, given barely three weeks earlier to Oberleutnant Herdt, not to billet the engineer detachment away from the Company reserve had been ignored, he was furious and the Inselkommandant was relieved of his command. He was sent for court-martial as was the Orderly Corporal, Obergefreiter Schubert, as it was found that he had failed to pass on a report that the sound of engines had been heard by a lookout at the harbour.

As far as the British were concerned, the implications of the operation were embarrassing and there was no intention of releasing any account of the affair to the Press. However, for the Germans it was a marvellous opportunity for propaganda. Following close on the heels of the Dieppe raid, where Canadian orders had been captured stating that prisoners' hands were to be secured to prevent them from destroying secret papers, this was additional evidence of the alleged barbarity inflicted by Commando troops. Three days after the raid, a German communique was issued declaring that the men had been illegitimately bound and that it was while resisting that two had been shot.

In an attempt to refute the German claim, Combined Operations Headquarters took the unprecedented step of issuing a statement on the raid, carefully omitting any reference to the binding of the prisoners:

'A small-scale raid was made last Saturday night on the island of Sark. It was one of many such operations which are successfully and frequently carried out and about which nothing is normally said.

'But since the enemy have, from ulterior motives, announced the raid, with the addition of inaccurate details, the facts are now issued.

'The main purpose of this raid was to obtain first-hand information about suspected ill-treatment of British residents in the island. As a result of this, these suspicions have now been confirmed by the seizure of a proclamation signed "Knackfuss Oberst Feldkommandant."

'This states that all male civilians (a) not born in the Channel Islands or (b) not permanently resident there, between the ages of 16 and 70, have been deported to Germany, together with their families.

'This deportation took place last week at the shortest notice. Nine hundred men were conscripted from Guernsey, 400 are still to go, and it is expected that there will be more from Jersey than there will be from Guernsey. Eleven men of Sark were warned to go last week, but two committed suicide and only nine left.

'The total British raiding force consisted of ten officers and men and there were no casualties. Five prisoners were taken, of whom four escaped after repeated struggles and were shot while doing so. One was brought back to this country. He has confirmed these deportations and has stated that they were for forced labour.'

When Hitler learned that the men who had been killed had been tied up (the word in the German report is 'Fesselung' which translates as 'fettered' or 'bound') he was incensed. Berlin threatened retaliation and stated that British prisoners-of-war captured at Dieppe would be fettered from mid-day on October 8 and kept chained, 'until the British War Ministry proves that it will in future make true statements regarding the binding of German prisoners or that it has succeeded in getting its orders carried out by its troops'.

Under sweeping headlines in the Press referring to 'NAZI BLACKMAIL', Britain replied that, 'if Germany carries out this threat, the British Government will have to consider its future action'. Almost by way of an aside, on the same day the Government revealed the imbalance of PoWs held by each side: 285,000 Germans and Italians to 115,000 British prisoners.

On October 9, Berlin announced that 1,376 British officers and men had been shackled — Britain retaliating the following day by manacling an equivalent number of German PoWs in Canada. With a danger of the situation getting out of hand, at this point the Swiss Government stepped in as mediator. Their proposal was that both sides should simultaneously unfetter their prisoners. This the British authorities did on October 12 but the German Government refused to release

Both Colin Ogden Smith and his brother Bruce participated in Operation BASALT. (Colin had virtually press-ganged his brother into joining the unit from his cushy number as a CO's driver!) When we spoke to Bruce in his London flat surrounded by relics and mementoes in February 1981, he still had a clear recollection of what took place: 'We had taken a grey-coloured cord with us specifically to tie the Germans up as the purpose of the raid was to bring back prisoners. We were all armed with .45 Colts. In the fight my prisoner got away and when the Germans started pouring from the hotel we ran like hell to the boat — it was every man for himself. The prisoner held us up; he was still in his pyjamas. Afterwards, we never thought any more about the significance of what we had done until the Press took it up. I believe 'Apple' reported direct to the Prime Minister himself who was not in the least worried'.

Bruce added this ashtray taken from the annexe to his collection. Shortly afterwards he was separated from Colin who, as we have seen, was killed in 1944. (His elder brother Tony was captured at Tobruk with the Green Howards.) Later he was decorated for his work in surreptitiously surveying the invasion beaches prior to D-Day. After the war he entered the family fishing tackle business (indeed he had invented a special fishing pack with his father which was included with every RAF dinghy) and London readers may remember his shops in St James's and Cheapside.

Fort George, near St. Peter Port Guernsey, was used by the Germans as a military barracks and was the location chosen for one of their military cemeteries (Royal Court). Badly bombed in 1944, it was sold in 1961 for demolition and redevelopment into high-class accommodation to encourage wealthy immigrants to settle in Guernsey.

their prisoners until a further demand had been satisfied: that the British Government would give an assurance forbidding the binding or shackling of prisoners in any circumstances whatsoever.

Although the situation as far as the innocent prisoners-of-war were concerned had been defused, Hitler was not satisfied. Six days later he issued his 'Kommandobefehl' — an order giving his commanders carte blanche authority to execute any Commandos captured in future raids.

Thus a new phase in the Commando story began. Perhaps the furious German reaction was part and parcel of a desire to curb an increasingly annoying form of warfare — one which kept the defenders guessing as to where the next blow would fall.

Although small in stature, Operation BASALT, as we have seen, was large in its repercussions. It was, therefore, with an extra-special interest that we took the short boat trip to Sark on April 11, 1979 in the company of Ken Tough of the Guernsey Branch of the Channel Islands Occupation Society.

The visitor to the island now lands at the new harbour of La Maseline, dedicated by HRH Princess Elizabeth and the Duke of Edinburgh on June 23, 1949. Construction had originally been started prior to the outbreak of war and took eleven years in all to complete. Nearby a tunnel through the rock face leads to the old Creux harbour, this being the one in use during WWII.

Refusing offers of lifts from the drivers of tractors (fifty-six of which are the only form of motor transport allowed on Sark), a pleasant walk of about a mile brought us to Pointe Chateau on the south-eastern corner of Great Sark. Here the Hog's Back splits Baleine Bay into the two smaller bays of Derrible and Dixcart. The Hog's Back is a perfect description for the place the Commandos landed. Bare of trees, its steep sides dropped almost sheer to the water's edge, where there was no beach — only rocks. The climb up from the sea would be a dangerous one in daylight let alone at night, loaded with equipment. The Hog's Back itself, although long cleared of its wartime defences, remains as it was, the narrow path flanked by prickly briars and blackthorn conjuring up mental pictures of it when once covered in barbed wire and mines.

To follow the route taken by the Commandos, one leaves the trail and crosses a field to reach the first house they visited — Petit Dixcart. It was empty then and Mrs. Patricia Falle, the present owner, was also absent when we called. Across the Dixcart Valley, La Jaspellerie stands out like a signpost, standing all alone overlooking Dixcart Bay and one can easily understand why the Commandos made it their next objective. Pressing on across the forested valley we made a noisy approach to the back of the house.

Although we determined that it was now owned by Miss Anne Pickthall, La Jaspellerie was completely shut up and empty although the French windows looked inviting . . . !

We continued along the track which the Commandos must have taken to approach the Dixcart Hotel. After a couple of hundred yards, a right turning leads down to the hotel. Possibly somewhere near here the sentry was attacked yet, for us, the day trippers milling outside the hotel spoiled the illusion. The hotel is now a popular island watering hole and, somewhat exhausted from our march, we halted to explore the interior of the bar.

Having walked in the footsteps of the Commandos, now was the time to reflect on the wide-ranging implications of what had taken place here thirty-seven years before. The hotel and annexe (which still serves as overflow accommodation) were now bustling with visitors and it was difficult to appreciate on this fine spring day that we were at the place where the binding and subsequent fight with the prisoners had occurred. Two Germans had died on this spot and many Commandos and Special Air Service personnel had been exterminated later as a direct result of what had happened here.

To find out what became of the Germans who had been killed we had to return to Guernsey. We knew that two bodies had been taken to Creux harbour by horse-van borrowed from the Dame of Sark. They were then taken across to Guernsey where they were laid to rest in the military cemetery at Fort George. There, in the steeply-stepped burial ground within the perimeter of the old eighteenth-century fort, we found their graves: Gefreiter Heinrich Esslinger, 30 years old, and Unteroffizier August Bleyer aged 28. Here also we found a possible answer to the mystery of the sentry attacked by Anders Lassen. There was one other death on October 4, 1942 — that of Obergefreiter Peter Oswald. Could this be one of the missing men, we wondered?

Young Captain Pat Porteous, awarded the VC on October 2, 1942, commanded Operation HUCKABACK. He survived the war and now lives in Sussex.

HERM

Operation HUCKABACK

By the end of 1942 the Commando organisation had been expanded out of all proportion to its early beginnings. Prior to the formation of the Army Commandos, it had always been the prerogative of the Royal Marines to provide detachments for amphibious operations such as raids on the enemy coastline and bases. The fact that the Marines had not been used on early operations was largely due to the fact that they were considered an indispensible force for the protection of the United Kingdom should a German invasion take place. They were one of the few trained formations left intact after Dunkirk and they viewed the formation of the new Army Commandos with suspicion. However, Royal Marine Commandos were in operation at Dieppe and thereafter they worked alongside their Army counterparts.

No. 3 Commando, which had participated in the Dieppe operation, suffered grievously in the assault and lost nearly a third of their strength. Consequently they had to go through a period of recovery and reorganisation before being posted to the Middle East. Thus it fell to their traditional rivals — No. 4 Commando under Lord Lovat — to carry out the next visit to the Channel Islands, which occurred on the night of February 27-28, 1943 under the codename of HUCKABACK.

The original concept of the operation was a combined raid on three of the islets: Brecqhou, off the western coast of Great Sark and the two Guernsey 'proteges', Herm and Jethou. However, when this had to be cancelled due to bad weather, the mission was revived with Herm as the sole target.

Herm lies about three miles off the east coast of Guernsey and is about 2,500 yards long and 800 yards wide. At the outbreak of war, the island was owned by Sir Percival Perry who had done everything possible to preserve its natural beauty. Great pains were taken to preserve its environment of which Shell Beach is unique in the British Isles. This beach, over half-a-mile long, is a left-over from a long lost age and many shells found there are of creatures no longer to be found anywhere else in the northern hemisphere. The island boasted a golf course, a small combined shop and Post Office and, in the 1931 census (the last before the war), it had a resident population of forty-eight. Belvoir House was the main residence on the island. Except for Mr. M. Le Page who was left as a caretaker, all the islanders were evacuated in 1940 and, after the Germans took over, they used it for target practice for the light batteries near St. Peter Port. When not so used, officers used to go there rabbit shooting and the only official unit to be stationed on Herm is believed to have been a flak battery.

Operation HUCKABACK was basically a reconnaissance operation designed to establish whether it was feasible to land artillery on Herm in the event of an invasion of Guernsey being necessary. The small force of ten men from No. 4 Commando under Captain P. A. Porteous (OC of No. 62 Commando who had only recently been gazetted for the award of the Victoria Cross for his conduct on Orange 2 Beach at Dieppe the previous August) left Portland in MTB 344. The party was landed by a dory on a rocky beach at the south of the island at 10.55 p.m. — five minutes before the scheduled time. During the three-hour inspection visit, the reconnaissance party made no contact with enemy troops or civilians but did not have time to reconnoitre the cottages and quay on the western side. Sheep were found grazing in the fields and, apart from one or two German notices in one house, the Commandos reported that their was no visible sign of occupation and no evidence of enemy defences. Their report stated that artillery could be landed on Shell Beach.

The force left Herm at 2.00 a.m. and arrived back at base at 5.44 a.m.

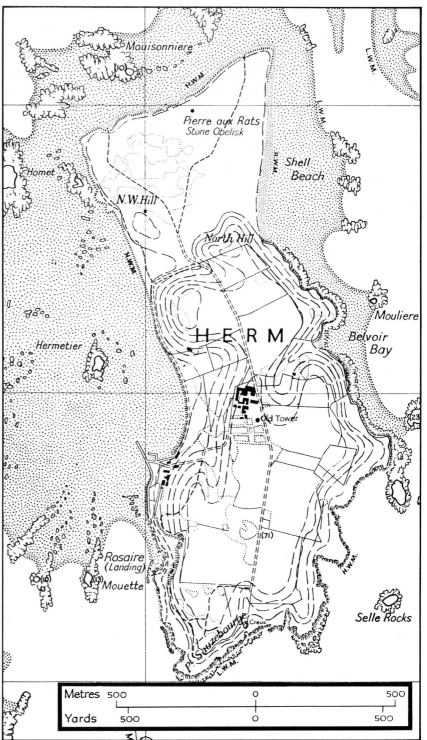

Reproduced from the 1942 edition of GSGS 4377 (Crown Copyright).

Left: **The pre-war method of visiting Herm — peeling advertisement at St. Peter Port harbour.** *Right:* **It's the attention to detail that matters — bird life is one of the attractions publicised today!**

Apart from this one inconspicuous Commando raid, Herm has little Second World War history. The Germans, unaware that the island had had nocturnal visitors, did nothing to improve its defences. Mr. Le Page recalls that he heard footsteps that night so he put the lights out, locked up and prayed that nobody would knock on his door! The island's main claim to fame was that the German flak battery succeeded in shooting down at least one aircraft which crashed in the sea off the east coast on August 24, 1943 — the only trouble being that it was one of their own!

Herm ended the war with its natural beauty unspoiled by concrete additions; and day trips from St. Peter Port, which had always been popular with tourists before the war, were reinstated although the return fare in 1979 was £1.80 compared with the pre-war 3s 6d. The island was purchased by the States of Guernsey from the Crown on December 17, 1946 for £15,000 and is now tenanted by Major A. G. Wood.

The magnificent Shell Beach on the north-east corner of Herm photographed on Saturday, April 14, 1979.

JERSEY

Operation HARDTACK 28

It was unusual that Jersey, the largest of all the islands, had not been the object of a Commando raid until Operation HARDTACK 28 was launched in December 1943. Two raids had been planned previously — Operation TOMATO in September 1940 and CONDOR in July 1943 but both had been cancelled.

Prior to D-Day, a series of small-scale raids were carried out all along the coast of France to gather intelligence and take prisoners who might be able to give information on the local defences. These raids, planned by Combined Operations at its headquarters at 1A Richmond Terrace, Whitehall, were prefixed with the codename HARDTACK and No. 28 was allocated to a reconnaissance to Jersey on December 25-26, 1943.

In the leading role of Father Christmas was Captain Philip Ayton who led his party of nine (including Lieutenant Hurlot, a French officer, and three French Commandos from Inter-Allied unit No. 10) aboard a motor gun-boat en route for the northern coast of Jersey. Anchoring in Petit Port, the party made a safe landing in the small, rocky cove at 10.45 p.m. They scrambled over a wire fence and investigated first two buildings a short distance up the steep valley. One was a stone hut and the other was constructed of corrugated iron and, although they were obviously used by fishermen, both were empty and there was no sign of recent occupation.

The Commandos continued up the valley and about 200 yards from the cove came up against a three-foot-high wire fence on which there were notices in red on white stating: 'MINEN' and 'STOP — MINES' in English. Alarmingly, the notices were affixed facing inland which indicated in no uncertain fashion that the Commandos had just walked through the minefield!

Pressing onward, climbing a steep path, they then arrived at the small hamlet marked on their maps as Egypt. Again there was no sign of recent occupation and all the windows of the houses were broken and many had what looked like shell-fire damage. Taking a footpath which seemed to run around the village they came upon another notice, this one reading in English and German: MILITARY ZONE — CIVILIANS STRICTLY FORBIDDEN.

Moving across country the raiders then made tracks for what they understood was the position of a German observation post. They moved quietly, keeping a close watch for mines although none were encountered. The OP, which also turned out to be unoccupied, consisted of a well-camouflaged concrete strongpoint with two entrances on the landward side and a flight of seven or eight steps leading down to a steel door which was locked. A pillbox faced the sea and the whole position was camouflaged with netting. The slit trenches were in a bad state of repair which seemed to indicate that they were no longer in use.

Having drawn a blank. Captain Ayton decided to try to get more information by calling at a nearby farm and, choosing to approach it directly along the road, the party boldly marched up as if a German patrol. Knocking loudly on the door they received no answer although voices could be heard coming from inside. They knocked several more times and after about twenty minutes a woman opened a first floor window. She was obviously very nervous and when they asked the whereabouts of any Germans, she said she could not help. Instead she directed them to another farm where she said they might get the information they required.

Having little option, the group followed the woman's instructions and approached the second farm down a steep track. Knocking again loudly on the door, it was opened within a few minutes by a man who seemed extremely frightened — at least he stood there, mouth open wide, completely speechless. Another man then came to the door, equally wary of the visitors, but, nevertheless, he invited the Commandos inside. Three did so while the others remained on watch outside.

The two men turned out to be brothers — John and Hedley Le Breton and they proved to be extremely helpful. This is the official summary of the information they were able to give to Captain Ayton:

(a) No Germans are billeted in any farms or houses in this part of the island.
(b) The two men did not know if the observation post at 773819 was occupied by Germans — they had never seen any. Neither

Philip Atterbury Ayton, a Captain in the Argyll and Sutherland Highlanders, was just twenty-two years old when he commanded the small unit of the 2nd Special Boat Section to Petit Port, Jersey on Christmas Day 1943.

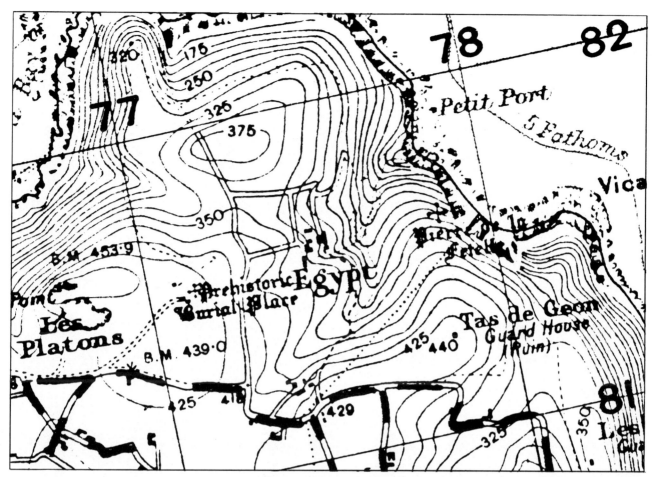

An enlarged extract from the second edition of GSGS 3967 dated 1942. Compare with the reconnaissance photo of approximately the same area taken by our friends of No. 541 Squadron on their sortie of February 1, 1945 (Both Crown Copyright).

did they know anything about the semaphore station at 779813.

They knew that there were a few Germans at Porteret (783804)

(d) The Les Platons strongpoint (770815) was occupied by about 15 Germans and about 10 Russians. They told us that the Russians were prisoners compelled to work for the Germans; they thought they were armed and did patrolling duties the same as the Germans. The sentry in this strongpoint usually stays very near to the centre and does not patrol or go near the perimeter of the strongpoint. They told us that the strongpoint was quite well armed but did not know the actual strength.

(e) They estimated that the total number of Germans on the island were about 1,000.

(f) Most of the German army transport was horsed owing to a severe shortage of petrol. The farmers knew that there were some MT vehicles at St. Helier.

(g) They confirmed that the beach at Petit Port was mined but could not tell us the exact whereabouts of the mines.

(h) The farmers confirmed the presence of mines around the strongpoint at 770815, but did not know the exact whereabouts.

Walking through the minefield on the track from the cove up the valley.

The ghost village of Egypt — still in ruins from when it was once a German battle training ground.

La Geonniere where the English-speaking 'Germans' knocked up Miss Le Feuvre.

The Le Breton farm where the Commandos gleaned their information. After the war, the brothers were awarded a plaque in commemoration of the visit and the help they gave. Its inscription includes the names of those on Operation HARDTACK 28: Captain Ayton, Lieutenant Hulot, Sergeant Roberts and Corporals Houcourigary, Le Stang and Roux.

(i) The Germans in this area did not appear very active. Neither of the farmers had seen any patrols except for one patrol of Russians who had arrested them one night for being out after curfew. The patrol was armed. After explaining they had permission to be out after curfew, they were released.

(j) Curfew in winter is from 2100 to 0700 hours.

(k) There is no resistance movement in the island and the population generally is not hostile to the Germans. The farmers explained that the island was so small that it would be difficult to get up a movement of resistance without German reprisals. The population is very frightened of the Russians who are ill fed by the Germans and they had to forage round the farms for food; but they are also sorry for them as they are suppressed by the Germans.

(l) There was a good deal of resentment against the Germans for having removed half of the available medical supplies on the island (sent from the Red Cross); they also took half of such food as was produced and imported for the island. Feeding presented no difficulty on the farms, but the farmers stated that conditions were quite bad in the town and large villages where the Germans had more control over the population.

(m) The village of Egypt was sacked because the proprietor was a Jew. (According to German explanations.)

(n) The farmers gave us a copy of the local newspaper and also a copy of the German paper.

The brothers then agreed to guide the Commandos to the strongpoint at Les Platons. They said they were pleased to have met them and they expressed a hope that they might return soon. After giving a glass of milk to each man, the farmers then led the way across country to the west. When they reached the eastern perimeter of the strongpoint, the brothers bade them goodbye and returned home leaving the Commandos to reconnoitre the position. When they got within sight of the outer barbed wire, Captain Ayton and Lieutenant Hulot went ahead alone. Shortly, they came to what looked suspiciously like a minefield with small sticks about ten inches high sticking out of the ground at regular, yet staggered intervals. Gingerly, one of the rods was wiggled but without any apparent effect. There was no sign of a sentry and equally no sign of the entrance to the strongpoint. With only forty-five minutes left to reach the MGB before it would automatically depart, Captain Ayton decided to abandon the search for a prisoner and to return to the beach.

'At 0445 hours we reached Petit Port', reported Lieutenant Hulot later, 'but the dory was not there and we moved northward along the cliff to look for it, flashing a torch for fifteen minutes to attract its attention. At about 778820 we came to a three-strand cattle fence which extended down the cliff. Captain Ayton crawled under the fence and had proceeded about five yards on the other side by the time I was at the fence. Suddenly, there was a vivid red flash which lit up the whole area and a loud report which sounded like the explosion of two No. 82 grenades. I first thought that a German patrol had found us and proceeded cautiously forward but saw no one and presumed that Captain Ayton had trodden on a mine. At first we could not find him and searched the cliff side. Then we heard a faint cry for help and I found him lying badly wounded on the cliff side with his foot entangled in some brambles which probably prevented him from falling down the cliff. This could not have been a normally exploded S-mine as only Captain Ayton was wounded by the explosion.

'We had no time to examine the minefield as we thought an enemy patrol would come along to investigate the explosion and we were also anxious to locate the dory. One very small

The German strongpoint at Les Platons is now occupied by the BBC transmitting station.

fragment of metal from the mine was recovered, however, and brought back.

'After this happened, we heard the MGB which was just visible to us (about 400 yards from the shore) and so started to flash a torch again until the MGB answered our signal. There was still no enemy reaction. By this time we had given up all hope of being picked up that night.

'The dory had returned to the MGB but was instructed to return to pick us up and shortly afterwards we saw it making for the shore and so we flashed to show our position. After much difficulty we managed to get Captain Ayton down the cliff and finally re-embarked in the dory at about 0520 hours. Although we had waited for the dory for about half-an-hour after the explosion, there was no enemy reaction whatever.'

So ended the first and only Commando raid on Jersey. Although Captain Ayton was still alive when the unit reached Portland, he died of his wounds in hospital the following day and was buried in Dartmouth Longcross Cemetery. We retraced the complete route taken by the Commandos during our visit to Jersey in April 1979. Today, the track down to the beach provides a pleasant walk with magnificent views. The abandonment of Egypt led to the area being transformed by the Germans into a battle training area — hence the notices spotted by the Commandos. It also became the target area for Batterie Mackensen, located to the north of St. Martin's Church, three miles away, when it fired its 210mm guns in practice across the north-eastern corner of Jersey. A few houses still remain standing today, overgrown and silent, with a profusion of wild life, making this a unique spot well off the normal tourist routes and seldom visited. (Hence the interesting 1977 sequel on page 203.)

Just across the other side of the valley stand new luxury villas in direct contrast to the simple farm visited by the Commandos. Its location is not generally known and it took us some time to find the correct one — Le Champ du Chemin. Like the Commandos, we spoke to Hedley Le Breton first and we could get little sense out of him until his brother told us he was somewhat backward — hence his speechlessness when confronted by the Commandos. John told us that his brother had first thrown a bottle at the visitors with their blackened faces (not mentioned in the official report) and that the Commandos had spoken in French. He also said that his parents slept through the entire episode in an upstairs bedroom!

They told us one very amusing snippet that concerned Miss Le Feuvre — the lady who had first spoken to the Commandos from the window of her farm, Le Geonniere. The next day, when she saw the brothers, she said that she had had a visit in the night from a German patrol but she said she was surprised that they all spoke good English!

Before we left the Le Bretons, we had an opportunity to see the plaque given to the brothers by the Free French after the war. It was accepted on their behalf by the Constable of Trinity on June 23, 1963 when the Free French granite plinth was unveiled in the Royal Parade. The latter is inscribed (in French): 'Erected by the Free French in appreciation of the People of Jersey for assisting France to join the United Kingdom in response to the appeal of 18 June 1940 launched by General de Gaulle.' (De Gaulle had stopped at Jersey airport en route to England in a Whitley bomber.)

We continued our investigation by following the Commando's route to Les Platons to the site of the strongpoint, but this is now occupied by the BBC radio and television transmitting masts and the pillbox is no longer visible.

Captain Ayton, severely wounded, was taken immediately on arrival at Dartmouth to the Royal Naval Auxiliary Hospital and, although everything was done to try to save his life, he died on Boxing Day. He was buried in Grave 136 in the military plot (Section G), at Longcross situated on the Totnes road beyond Townstall Churchyard, one of the twenty-eight Second War burials in the cemetery.

SARK

Operation HARDTACK 7

Planned to take place simultaneously with the raid on Jersey, HARDTACK 7 (numbered out of sequence), aimed at Sark, had an inauspicious beginning.

The story is told here by the commander of the raid Lieutenant A. J. McGonigal:

'At 2345 hours, the MGB arrived at the lowering position off Derrible Point. The dory was lowered and proceeded inshore where the force disembarked on the extreme south point of Derrible Point at 2355 hours. The initial climb to the ridge on this point was short but difficult, and the force took fifteen minutes to negotiate it and assemble on top. From the briefing photographs this had appeared to be the only difficult climb on the route and the rope was accordingly left at this point in case of a hurried return.

'From here the point consisted of a number of pimples connected by narrow ridges. The passage entailed a series of climbs up and down and it was not until 0100 hours that the last ridge connecting the point to the mainland was reached; this consisted of a knife ledge. On one side there was a sheer drop to Petite Derrible Bay, and on the other side a similar drop to Derrible Bay. An attempt to cross this ridge proved unsuccessful. The ridge itself was about thirty feet in length with no footholds, and the edge itself was too sharp to provide a hand grip.

'The patrol therefore attempted to descend and bypass this ridge. As the rope had been left at the initial climb, toggle ropes were used and the crossing was accomplished in a series of stages governed by the length of the toggle ropes. At the other end of the knife ledge, a sheer climb of about thirty feet to the mainland plateau was encountered and was found impossible to climb. Below this there was a sheer drop to the sea.

'As the time by now was 0200 hours, the force returned to the dory looking as it did so for a climb down to the beach. In the dark, however, it was not possible to find one.

'During the whole of this time, although the patrol made a certain amount of noise through loose stones falling over the edge, no signs of any enemy sentries or patrols were seen.

'The force re-embarked in the dory at 0300 hours and proceeded westward to Derrible Bay where Sergeant Boccador and I carried out a reconnaissance of the beach area, in the

The knife ridge, which aborted the first landing, is clearly visible in our shot of Derrible Point taken in February 1981.

north-east corner of the bay. I worked along the cliff edge to the beach moving slowly and carefully since Sergeant Boccador had reported to me that he had seen a sentry patrolling the cliff head over the bay. We eventually reached the end of the rocks.

'The beach was of shingle and was covered with flat slate-like stone. The sea came up to the cliff edge at this point and from what I could see of the bay, appeared to do so all the way round. There were no signs of mines or wire. I found a small box on the shingle beach which I brought back with me. In order to get at this I had to move for a distance of from five to six feet on to the beach and encountered no mines.

'The cliffs at this point appeared quite unscaleable; the only path from the point appearing to run along the ridge and to end at the knife edge. At the latter point, the path appeared to have been cut, either through natural causes, or as the result of demolition.

'Sergeant Boccador and I returned to the dory at 0410 hours, which then embarked with the complete force to the MGB, which was reached at 0425 hours.'

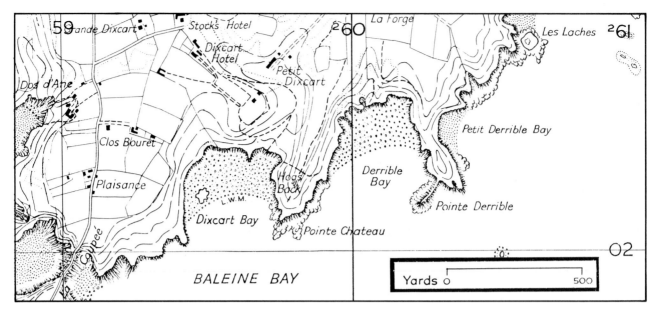

Two nights later the same force returned to Sark to make a second attempt to capture a prisoner. Lieutenant McGonigal continues his report:

'The force landed at point 599021 and, after climbing a 200-foot sheer rock face, met a further very steep slope about 100 feet in height with a shingle, slate, and stone surface. The force followed the eastern edge of this slope and encountered a wire fence consisting of three strands of very thick copper wire and two thinner strands of ordinary wire. This wire was cut and the force proceeded along the top of the Hogs Back, continually searching for mines as it progressed. Plentiful cover was afforded by rock and gorse.

'At point 599024, a path approximately six feet wide was encountered, on either side of which the ground, which was thickly covered with gorse, fell away very steeply. We found that it was impossible to walk through this gorse without making considerable noise and we therefore continued along the path.

'I was leading the patrol and had gone forward some fifteen yards, feeling for mines as I did so, when two mines went off behind the patrol, wounding Corporal Bellamy and Private Dignac. Corporal Bellamy died about two minutes later and Private Dignac received very severe wounds in the body.

'The first mine had exploded about two feet behind Corporal Bellamy, the last member of the patrol, and the second mine about five feet to the left of it. (The empty container was taken from the first hole and brought back with the force.)

'The force then started to carry Corporal Bellamy and Private Dignac out of the minefield. I took the lead, still feeling for fresh mines, and had taken only a few steps when two more mines went up in quick succession in front and to the side of me. (Lieutenant McGonigal himself was injured as a result.) After these explosions, Sergeant Boccador was the only member of the force who remained unwounded. Private Dignac was wounded still further by these explosions and Sergeant Boccador told me that he was dead.

'In view of the fact that my force had sustained such casualties, I decided to leave the two bodies, retrace my steps and return to the boat. No sooner had we started to move, however, than more mines went up all around us. I cannot say how many there were but at the time we had the impression of being under fire from a heavy calibre machine gun. We continued our withdrawal to the dory.

'On our way up, we had hidden a wireless set No. 536 under a rock but we were unable to find it on our return journey and so were obliged to abandon it. It was also impossible

One cannot help feeling that HARDTACK 7 lived up to its name and that everything was done the hard way. For their second attempt, the Commandos switched the landing point to the Hog's Back but, instead of coming ashore at the tip, the map reference and Lieutenant McGonigal's description would seem to indicate that they scaled the almost-vertical cliff face on the eastern side. It was also inevitable, following Operation BASALT, that the area would be mined. This is the view looking north from the topmost point of the Hog's Back across map reference 599024 where the disaster occurred. The map is an extract from the 1942 edition of GSGS 4377 (Crown Copyright).

for us to get down the last sheer twenty feet of rock and to bring the rope with us. Repeated attempts were made to pull it down after we had got to the bottom but it had stuck firmly and so, cutting it as high as we could, we left it and returned to the MGB.'

Lieutenant McGonigal then listed his observations in the same matter-of-fact way:

'(a) The first two mines that exploded were behind the patrol and, although we moved about continuously in advance of the two craters, no further mines were exploded. It would therefore appear that we had reached the edge of the minefield and had been unfortunate enough to explode perhaps the last two mines in the field. It is interesting to note that although Sergeant Boccador and myself were feeling our way very carefully, we felt no contact points nor saw any other signs of mines.

'(b) All the injuries caused by the exploding mines were sustained by those members of the force who were either standing or kneeling. A person lying flat seemed to be immune from them.

'(c) Despite these explosions, no signs of Germans were seen or heard.

'(d) There is a mobile searchlight on the island. We saw it as we were coming in on the MGB. From 2100 hours until 2140 hours it was turned on every ten minutes and appeared to come from the area of Le Creux Harbour. It was flashed from the cliff top on to the water — the length of beam was approximately 500 yards.

'(e) The S-phones we found to be a complete failure.

'(f) The felt soled boots were extremely good.'

Next morning the bodies of the two dead French Commandos, Corporal R. Bellamy and Private Andre Dignac, were recovered by the Germans and buried in the small military section in the churchyard opposite St. Peter's Church. Thus, with an end result worse than on any of the other six raids carried out on the Channel Islands during the Second World War, the last Commando expedition ended in disaster. Altogether, in three year's operations, three Commandos had been killed, two wounded and six men made prisoner; two or possibly three Germans had been killed, one wounded and another eight captured. On balance, therefore, the Commandos possibly won by a hair's breadth but against that one must set the considerable reprisals carried out against the populations — many of whom abhorred the very idea of the raids.

The two French Commandos were buried by the Germans in St. Peter's Churchyard but Andre Dignac was exhumed after the war.

CHAPTER SIX
The Post-war Period

Above: The party's over. The players begin the huge task of clearing up the Channel Islands. Under the supervision of the British Army, former Kriegsmarine personnel cut down the wire barricades on Albert Pier, St. Helier. Note the camouflaged flak position in the background (Charles E. Brown).
Below: Thirty-four years later and the level of the promenade has been raised and steps added to enable one to walk over the roof of the bunker.

After the joyous celebration of the liberation, the Channel Islanders began the task of getting back to a semblance of normal life. For some, however, memories of the previous years, of events and happenings whilst occupied, were difficult to forget and the recriminations started.

In May 1942 a number of Jerseymen had been rounded up and sent to Alderney to work on the German installations there. Those who went were deemed in certain quarters to have volunteered for the work and, as a consequence, were to be considered collaborators. However, letters published in the *Jersey Evening Post* in December 1945 from Mr. M. A. Nicolle and C. Le Vesconte, two of those sent to Alderney, protested that most had been forced to go and that out of one gang of twenty-two, nine were volunteers while thirteen had been conscripted.

A more contentious issue was raised by the arrest of Pearl Joyce Vardon, a 30-year-old ex-Jersey schoolteacher, by the Special Branch on her arrival at Northolt Airport from Germany in February 1946. Miss Vardon taught in a local school until taking up a post in the office of a German building contractor operating on the island. Later she left for Germany and was next heard broadcasting dance music and other items on German wartime radio. She was arraigned before Sir Bertram Watson at Bow Street Police Court charged with 'doing an act with intent to assist the enemy, it being alleged that she joined the German propaganda service in February 1944'. She pleaded guilty at her trial at the Old Bailey the same month (claiming that she had fallen in love with one of the German officers in the occupying force) and was sentenced to nine months' imprisonment.

Another Channel Islander who joined the German broadcasting service and who was brought to book after the war was John Lingshaw who had been one of those deported to Germany after the occupation. He voluntarily made recordings of British news bulletins and was charged at the Old Bailey in 1946 with conspiracy to assist the enemy. On conviction, Lord Goddard, the newly-appointed Lord Chief Justice, sentenced him to five years penal servitude.

An interesting story of collaboration with the enemy (which was to have a fascinating sequel exactly three decades later on the 30th anniversary of the liberation) came to light when one of the first Jersey PoWs returned to the island in July 1945. Gunner P. Silvester joined up in February 1940 but was made prisoner when his ship was sunk by a German raider off the Australian coast. After a five-month voyage in the hold of the *Tannenfels* with a hundred or so other prisoners, they docked at Bordeaux and were transferred to the merchant seamen's PoW camp at Westertimke in northern Germany.

When Gunner Silvester was freed in May 1945 he stated that, 'one day I was amazed to hear two fellows who had not long arrived in the camp running Jersey down for all they were worth. They proved to be two "conchies", whom I had known well by sight in the island. They had apparently been deported from there and sent to us. They were too friendly with the Jerries for our liking and one day they disappeared. I was told they had joined the Britische Frei Korps, a body which the Jerries were always trying to get us to join. They used to pass leaflets round the camp. This free corps was to fight Bolshevism and some actually joined it to my knowledge. These two "conchies" were stated to have joined but I did not believe it until one day some months later one of them showed up at the camp again; he was wearing a German uniform with a white band and a Union Jack

The original plan was to collect the wire and dump it over the cliffs. However, in days when 'environmental pollution' was not even thought of, some far-sighted individual decided it would be better to stuff it into one of the German tunnels and seal it up — the one chosen was Ho 4 at Grands Vaux. *Above:* Ex-German Bedford loads wire at the top of the pier. The vehicle was one of a number captured in 1940 and having served the Germans for five years, they rejoined the British Army (Evening Post).

on the cuff. He told someone that his mate was in Danzig and that he was on leave. I never saw either of them again but I am ready to swear they joined the corps.'

Although it is easy to sit back now and criticise collaboration, it was impossible to live on the Islands during the war and avoid working for the enemy in some shape or form. In a small place like Guernsey, for example, tomatoes were grown for home consumption and for export to the Continent and shopkeepers had to sell to both islanders and German troops.

The fact that some element of collaboration had existed on the Islands rumbled on in the post-war years. When in 1970, Dr. Charles Cruickshank was specially commissioned by the States of Guernsey and Jersey to write the official history of the Channel Islands during the Second World War, it was hoped that his research into documents never before released would banish the skeleton in the cupboard once and for all.

Publication of the work (*The German Occupation of the Channel Islands,* published by the Oxford University Press) was timed for the 30th anniversary of the liberation and the launching ceremony was laid on at London's Imperial War Museum to coincide with the Occupation Exhibition to be opened by HM the Queen Mother at 11.00 a.m. on May 7, 1975.

Unfortunately, the launch of Dr. Cruikshank's scholarly work was seized upon by a Mr. Peter Tombs as a 'whitewash' of the alleged collaboration by the Islanders with the Germans. Mr. Tombs threatened to continue with his campaign and prepared to distribute a leaflet at the Imperial War Museum opening ceremony, co-signed by Eric Pleasants and John Leister, claiming that they had all been handed over to the Germans by the civil authorities for being in possession of stolen German petrol soon after the occupation began.

Although the opening of the exhibition and launch of the book were very successful, the persistent Mr. Tombs declared at the end of the morning that there was still room for debate before the subject could be officially closed and he declared his willingness to face Dr. Cruickshank on a public platform to discuss the way the Islands behaved under the Germans. Mr. Pleasants and Mr. Leister both said they would be prepared to join Mr. Tombs who claimed he was nearly ready to publish his book *The Traitor Isles.*

Dr. Cruickshank answered his critics in a press interview by stating that if there was a 'villain of the piece' it was the British Government. 'Their decision to leave the Channel Islands undefended without immediately declaring them an open city laid the inhabitants open to the possibility of air attack — which did, in fact, come with the loss of forty-four lives — and amounted to criminal negligence'. Dr. Cruickshank later commented: 'If the British Government had immediately evacuated the Islands, the Germans would have been left with no civilian population to grow food or provide services.'

Dr. Cruickshank, who studied at Aberdeen, Oxford and Edinburgh universities, admitted that the emphasis in his research was firmly put on historical documents, rather than on the personal reminiscences of Islanders. 'Thirty years on', he said, 'personal recollections can be unreliable and impossible to substantiate with documentary evidence', although he said that everybody had been given a chance to put their points to him during the project.

'If the 60,000 people in the Channel Islands in the occupation were alive today, they would all find fault with it in 60,000 different ways. They may have been involved in events that were not recorded. All that the serious historian can do is refer to documents in the case of a dispute.'

Asked to comment on the allegations of Mr. Tombs and friends, he dismissed these for the simple reason that Mr. Tombs had not produced any of his evidence. 'The great majority of people did the sensible thing and sat tight, on the advice of their States. Some collaborated, of course, and some wanted actively to resist, but were talked out of it'.

'If you have the equivalent of a 25-million-strong army occupying Britain — and there was one armed German to every unarmed man, woman and child — there was no point in sabotage because you couldn't escape more than five miles. Both Bailiwicks, did exactly the right thing and, by standing up for themselves, got concessions that other governments might have been scared to ask for. At the end of the day the Germans were worse off than the Channel Islanders. It was not a benign occupation for the Germans.'

However, Peter Tombs and his two colleagues were not satisfied and were determined not to let the matter rest. The following month Mr. Tombs announced the formation of CHAT: Cruickshank's Historical

Left: Artillery carriers, their guns silenced for good, await disposal in St. Helier harbour (Charles E. Brown). *Right:* As we have said before, it's the attention to detail that counts. Michael Ginns moving the hook to convince you that time stands still on Jersey.

Amendments Trinity, and claimed it had members in the Channel Islands, England and France. *Jersey Evening Post* reporter Chris Bright delved a little deeper into the background of 'the Trinity' and came up with a remarkable story linking two of the men with Gunner Silvester's allegations of July 1945:

'The rest of the Trinity comprises John Lester, or Leister, now known as Beckwith, and Eric Pleasants, president of CHAT', wrote Chris Bright in the June 23 edition of the paper, 'both of whom were named at the end of the war in connection with their alleged involvement with the German Army's British Free Corps.

'The committee, they say in a joint Press statement, is ''dedicated to the publication of historical truths of the German occupation of the Channel Islands.''

'All three men have already claimed to have books in the pipe-line — Mr. Tombs's *The Traitor Isles* and a joint venture by Mr. Pleasants and Mr. Lester called *We did it our way.*

'Mr. Pleasants and Mr. Leister came to Jersey as conscientious objectors just before the war and claim that it was because the civil authorities handed them over to the Germans for stealing petrol for a getaway that they landed in prison in Europe.

'Post-war reports indicate that it was as prisoners-of-war that they became involved with the British Free Corps, a unit formed by John Amery to fight with the Germans against the spread of Bolshevism. Amery, hanged just after Christmas 1945, was said to have had deep misgivings about the British-Soviet alliance.

'A Jersey gunner on demob leave in the island at the time told an *Evening Post* reporter that he had seen two of the men named in the charge, Leister and Pleasants, in Free Corps uniform at his camp at Bremen,

The 'gang of three' as they would no doubt be called in the idiomatic phrase of today. L-R: Peter Tombs, Eric Pleasants and Mr. Leister/Lester/Beckwith are faced by the 'Sherlock Holmes' of the Jersey Evening Post — Chris Bright at the Imperial War Museum on the occasion of the launch of Dr. Cruickshank's book (Evening Post).

and the allegation was passed on to Army Intelligence.

'On January 4, 1946, Derek John Leister (23), of Camden Town, was one of five men committed for trial at the Old Bailey charged with conspiring with Amery, Eric Pleasant (sic) and a Thomas Haller Cooper, to assist the enemy by joining the British Free Corps. *(The others were Ronald David Barker, Kenneth Edward Berry, Alfred Vivian Minchin and Herbert George Rowlands.)*

'Leister appeared at the Old Bailey on three charges of trying to recruit into the Corps, formerly known as the League of St. George, and on February 20 was sentenced to three years' penal servitude.

'An account of the trial, using pseudonyms', went on Chris Bright in his article, 'appears in a history of the Corps, *Yeomen of Valhalla*, by the Marquis de Slade, the pen-name of J. Eduard Slade. Of the five accused, the only one to receive a three-year

More demilitarisation work photographed by Charles Brown in May 1945. *Left:* This is the signal station at the end of St. Helier's Albert Pier which the Germans had fortified. The wording proclaims: 'Attention! No mooring. Defence Boom'. The position had been prepared for demolition by the burial of artillery shells which were removed by a PoW working party. *Below:* Margaret Ginns looks for shells of a different sort in 1979.

sentence is referred to by the author as Beckwith, which happens to be the name now adopted by Mr. Lester/Leister, who claimed dual Anglo-German parentage.

'The Judge, Mr. Justice Croom-Johnson, had said at the trial that Beckwith, in a way an unfortunate captive after the Germans occupied the Channel Islands, might, by remaining in England, never have become involved in the Free Corps.'

Chris Bright stated that: 'Eric Pleasants confirmed last week that both he and Mr. Lester/Leister/Beckwith were the selfsame people as those named at the Old Bailey trial in 1946. "We've been waiting for someone to bring all this up again."

'However, although Mr. Leister went to jail', the *Post* revelation continued, 'Mr. Pleasants was anxious to point out that though he was named in the charge, he personally was never charged, never appeared in any court in connection with the case and, consequently, was never sentenced.

'He said: "I was spending seven years in Siberia at the time. I had been arrested by the Russians for insurrectionist activities and was sentenced to death. This was later commuted to life imprisonment, but I was released on an exchange agreement after seven years." '

The following year, Mr. Peter Tombs was still at it and still without a publication date for his book. However, on March 10 he proudly announced a breakthrough in his research as he said he had discovered Martin Bormann living quite contentedly working as a farm labourer in Norfolk. . . .

Top: **PoWs clearing St. Catherine's Bay in May 1945 (Evening Post).** *Above:* **At the risk of being a bore . . . children — then and now.**

Above: German prisoners were forced to clear minefields — this one lay just outside the eastern perimeter of Guernsey aerodrome. Charles Brown was again on hand with camera to record the scene. *Below:* Author forces his son to his knees.

Three weeks after the liberation the tremendous job of clearing the Islands of their wartime legacy began. The first priority was the clearance of the minefields which contained 114,000 mines, those on Jersey numbering at least 60,000. The work there was entrusted to the 259th Field Company under Captain R. A. Foster and, utilising German PoW labour, the island was declared free of all mines later that year.

The second priority was the collection of all wartime stores, weapons and ammunition when some 415 guns of all types including 10,000 rifles, 750 machine guns, 150 mortars and flame-throwers, 1,500 pistols and 26,500 tons of ammunition were gathered in by Major F. H. M. Sargent, OC of the 135th Field Ordnance Unit. The majority of the ammunition (6,000 tons) was removed from the tunnels in St. Peter's Valley as well as huge quantities from La Rosiere in Grands Vaux Valley and St. Aubin's railway tunnel. Each gun position had its own store and small arms ammunition was also found scattered in private houses used by German troops.

The third stage in the clearance programme was the actual disposal of the material. The majority of the ammunition was loaded aboard LCTs by German PoWs and taken to the eastern end of Hurd Deep, some 400-odd feet in depth, near Alderney where it was dumped overboard. Some 6,000 tons was considered too dangerous to withstand the seventy-mile journey and this was taken instead to the Five Mile Road on the westernmost extremity of the island, where the escarpment provided a measure of blast protection, and blown up. There was only one serious accident when a fire broke out in a 100-ton dump of 88mm shells but luckily nobody was seriously injured and the one man in the vicinity was able to shelter behind the sea wall.

Disposal of the guns presented a more difficult problem. Where these were firmly casemated, they were left in situ after having their breach blocks removed. Finding the correct wheels and limbers for the other field guns which had been emplaced in fixed positions was another difficulty. In the end, after much labour and improvisation, twenty-five heavy guns were transported to Les Landes and run over the cliff. All the easily movable guns were concentrated at Fort Regent and 155 were taken out to sea and dumped. This itself was no mean feat for the guns could be easily loaded onto the LCTs by dockside crane whereas out at sea, with the flat-bottomed vessel bucking wildly, it was difficult for the mobile crane to neatly drop the material over the side. The release of two tons of metal tended to cause a nasty whip of the jib which was only solved by placing the load on the bulwark and letting it topple over from there into the water.

All Charles' photographs were without captions, or even locations, when we discovered them in his old shop in Longfellow Road, Worcester Park. This mine-clearing series was a gem, but the big problem was: Where was it taken? For us, as with all After the Battle publications, we cannot use a picture unless we have confirmed its location by a personal visit and taken a comparison photograph. Most of the other shots Charles had taken in 1945 were identified by members of the Occupation Society but this group foxed everyone. Because of their sequence on the roll, we were pretty sure they had been taken on Guernsey and the only clue was the stilts of what appeared to be a water tower in the top left of this photo. Such a tower still stands near the aerodrome and, after much searching, we were able to pinpoint the exact spot *(below)*.

This minefield contained Holzminen — one of the most difficult types to detect being made of wood.

All these 88mm flak guns were concentrated at Fort Regent, prior to loading aboard ship for dumping at sea (Imperial War Museum).

War material being taken aboard LSTs at St. Helier *left* and *above* at Gorey (Evening Post and Imperial War Museum).

The 400-foot-deep trench called the Hurd Deep, north of Alderney and the Casquets, became its graveyard (Imperial War Museum).

Another 181 guns were placed in one of the tunnels in St. Peter's Valley and sealed up. It was this event, fully reported with photographs in *The Evening Post* on February 23, 1946, which was to lead to so much heartache over the next three decades. Other weapons were cut up on the island and shipped to the UK as scrap.

The fourth part of the Army's programme was the rehabilitation of the amenities of the island. The clearance of all available land was considered first priority followed by public amenities such as beaches and golf courses. All anti-aircraft landing obstacles, including the forest of upright posts which were erected all over the island in 1944, were removed and the job of clearing hundreds of tons of barbed wire began. Using PoW labour guarded by men from the Royal Artillery under Major H. Harland, the wire was collected with the intention of throwing it over the cliffs. However, it was seen that this was not a good long-term proposal and it was therefore decided to pack it into a tunnel at Grands Vaux which would then be sealed up.

The smaller bunkers and field works were dealt with by blowing them up and then clearing the sites using civilian labour supplied by the States labour department. Where the structure was too massive to remove safely by explosives, it was cut down sufficiently to enable it to be covered by a farmable layer of soil. At Grouville golf course, a military road had to be removed and much landscaping was required to try to bring the course back to its former glory.

As far as the railway was concerned, with little chance of making the line in Jersey a paying proposition, a decision was taken to lift the track. By 1946 work was proceeding removing the rails along the Havre-des-Pas and the replacement of kerbstones and tarmacing of the surface. Then, as soon as the railway trucks and engines had been gathered together at Beaumont prior to their shipment back to France, work began pulling up the lines to St. Aubin and those running over the western part of the island. The German stonecrusher in St. Peter's Valley was brought into operation again on February 5, 1946 to produce the necessary material to make good the roads. All the iron work was dumped at the top of the Old Harbour and later shipped to England.

Above: **Ex-German transport, much of it captured from the British Expeditionary Force in France in 1940 or commandeered locally during the war, was concentrated at Springfield Football Stadium to await WD auction.** *Below:* **Good game! Good Game! What could not be sold was crushed to scrap** (F. Sargent).

Above: Transport is the life-blood of any community — all the more so for the Channel Islands in the early years after the war. The airports had been mined by the Germans by burying bombs beneath the surface of the 'drome and the filled-in craters where these were dug out are still evident in this early post-war picture of Jersey (Charles E. Brown). *Below:* The renewal of the Island's link with tourism, and the subsequent growth of air transport, has changed its face by 1981.

How many visitors would recognise the terminal building which the Luftwaffe occupied as a crew room? At least there is still a slogan on the wall!

Today, the route of the railway can still be followed — Railway Walk and the bridge and embankment at Val de la Mare being obvious reminders to the modern-day tourist. The tunnel constructed at St. Aubin's, to give the trains of the old railway company a straighter run to La Corbiere, still has the old German railway lines embedded in the floor. The interior was enlarged and extended by the Germans who cut galleries at right angles deep into the hillside to provide ammunition storage served by a narrow-gauge track. The tunnels are now used as stores by the Public Works Department.

Whilst any equipment of value to the Army such as radar sets, range finders or searchlights was sent to England, there were still some 185,000 items of furniture and barrack room equipment and 1,200 tons of miscellaneous stores to be disposed of ... from salt cellars to barrack huts.

A series of public auctions were therefore held during 1946 in order to get rid of the equipment. A typical sale was the one held on Tuesday, February 12 at the dump at La Haule. There were not a great number of bidders and most of the articles sold in quantity lots went to trade buyers. There were two prospective buyers from England and the one from Derby made a killing by purchasing several thousand scrubbing brushes at ½d each!

The Government auctioneer was Mr. Harold G. Benest and prices of the various lots make interesting reading (remember 6d equals 2½p): 162 large cans dairy type, 7d each; 1,500 shovels with handles, 5½d each; 1,000 picks without handles, 6½d each; a small selection of smith's tools, Fuller's sets, etc., 22/6; eighty enamel jugs, 4d each — this was one of the best bargains as this type of jug was then selling at 5/11 in the shops. An electric welding machine went for £12; a Lister motor, £40; a lathe, engine and dynamo for £36; six drums of carbide at 17/6 per drum; a damaged Oliver typewriter was knocked down for 7/6; three pairs of rubber boots went for 8/- per pair; a weighing machine in kilos, 32/6; a box containing sixty-five plates, 16/-; 722 small size paraffin cart lamps, 1½d each; a box containing twenty-four rolls of cine films went for £4 5s; a box of several thousand spoons a bargain at 10/-; 4,312 shallow aluminium eating bowls, 3d each and some thousands of hand scrubbing brushes sold at ½d each.

Forty-nine wooden chairs went for 1/3 each; bar-type chairs with metal frames, 2/- each; several hundred fire extinguishers sold en bloc for £3 2s. 6d.; a box of scissors containing about 500 pairs went for £2 15s.; an aluminium copper and boiling machine, £3; galvanised water tub, £4; two boilers at £2 each; one double range in good condition, 10/-; and another in better condition for £7; a small upright stove, 26/-; 280 small white glass lamp globes and twenty-four paraffin lanterns went for 7/-; seven paraffin type

Coming out, though, was the railway line, lifted because there was no chance of viability when competing with the combustion engine. *Above:* Messrs J. Halliwell and H. Assinder, two long standing members of the Royal Jersey Golf Club, prepare to resume their pastime in spite of the 60cm OT railway on Grouville Common (Evening Post). *Below:* German PoWs raise the metre-gauge track in front of Commercial Buildings, St. Helier in February 1946 (Evening Post).

stable lanterns for 11/-; forty-one brass table reading oil lamps at 6d each; and five carbide hurricane lamps for 6/-; some thousands of stable brushes, bought en bloc for 2d each; two damaged stencilling machines with spare parts, 6/-; several hundreds of tin openers, cut-throat razors, knives, etc. for £4 the lot; an assorted lot including fire extinguishing powder and foam, comprising more than 100 tins, 25/-; thirty-six lengths of fire-fighting hose with connecting attachments, £5 10s; two lengths of hose for £1 and nearly 300 assorted plates, cups and saucers for 16/-.

Mines and mishaps have given way to discos and dancing. *Above:* **Minefields once protected High Tower Hotel which was the living quarters for the crew for the nearby resistance point of the same name (Evening Post).** *Bottom:* **Now the decibels of the Sands Disco keep away all but the brave.**

Restitution of missing property figured largely in the newspaper columns when war ended. Typical was an advertisement which appeared in March 1946 asking if anyone knew the whereabouts of an oil painting by Widgery of a Dartmoor scene and a print of pintail ducks by Peter Scott, both requisitioned by the Germans in 1944.

The Services did an admirable job in restoring the island's amenities and, within a year, the major part of the work had been completed. However, disposal of material by sealing it up in the underground tunnels, whilst it seemed an ideal solution at the time, was to leave a continuing problem for the States authorities.

Left: **This American Chevrolet, imported to Jersey before the war, was requisitioned by the Germans; commandeered by the British and crashed by liberation troops on July 28, 1945 on South** Hill. Note the Austin wrecker coming to the rescue and the private car which still bears German insignia of a reconnaissance unit (Evening Post). *Right:* **Now it's GBJ-OK?**

By May 1946 the worst of the clearing up had been done and the first anniversary of VE-Day in the Channel Islands — May 9 — was declared a public holiday. Parades were held to celebrate the occasion. This is the large crowd which gathered at West Park to listen to the Lieutenant-Governor of Jersey reading a message from the King prior to a grand march past.

The tunnels are probably the one aspect of wartime Jersey that everyone will be familiar with. Stories of the buried equipment therein have abounded during the post-war years and have led many individuals to try to break into the entrances. At the same time, the Public Works Department, ever mindful of the dangers, has waged a continual battle to keep these sealed up.

In 1962, two schoolboys were killed by carbon monoxide poisoning after getting inside one of the St. Peter's Valley tunnels and, on another occasion, someone even bored a way through thirty-six inches of reinforced concrete! Finally, in 1973, a decision was taken by the Jersey Public Works Committee that the only way to stop these activities would be to open the tunnels, clear them completely of all war relics, publicising what was found, and then sealing them for good.

On January 31, 1973, the Committee's far-seeing new president, Senator John Le Marquand and Deputy David de la Haye squeezed through a two-foot hole in the first of the tunnels to be opened in St. Peter's Valley to inspect the interior first hand. It was the first time for twenty-five years that anyone had seen inside officially and they were rewarded with the sight of a unique museum. The horseshoe-shaped tunnel (having consequently two entrances) was piled high with rusting helmets, oxygen cylinders, weapon parts, magazines, field kitchens and all the paraphernalia of war.

Senator Le Marquand later admitted that for many years the Public Works Committee had thought the tunnels contained little or nothing of value. Now, he said, the States would select the most interesting items for local exhibition and, thereafter, the rest of the material recovered would become available to collectors who had so often offered to help with a recovery operation. Then the Senator issued a warning to parents to keep their children away while the tunnels remained open and a round-the-clock guard was mounted by the police. Nevertheless two men described as 'dealers', were charged the same night with illegally entering the tunnels!

Two years later, Tony Titterington was still hard at work restoring items recovered from the tunnels ready for display in Elizabeth Castle. Tony's method of working was to photograph each piece, be it ammunition limber, field kitchen or flame-thrower, and then strip each item to the last nut and bolt. Badly corroded parts were then sand-blasted and protected against further rusting by covering with a zinc-based paint. Then a long process of re-assembly, priming, filling and painting each part — a labour of love as far as Tony was concerned — ended up with its removal to Elizabeth Castle for public viewing.

Much rubbish has been written and is still perpetrated about the construction and use of the tunnels. Even the author can remember as a schoolboy visitor to the Jersey underground hospital being told by the guide of the 'slave workers who were walled up in the concrete lining of the tunnels'. Other fairy tales state that the purpose of the workings was as gas chambers for exterminating the population.

Guernsey has its equal share of underground tunnels and, possibly because the major publicity in the past has centered on those in Jersey's St. Peter's Valley, those on the island remain largely as they were left in 1945. Over thirty years later the Channel Islands Occupation Society were privileged to be the first people to see inside No. 2 Tunnel at Petit Bouet, now owned by the States Housing Authority. This complex became the final resting place of a considerable amount of captured German war material and special permission was obtained to unseal the entrance in November 1978. One can imagine the excitement of the privileged Society members as they entered a veritable Aladdin's

A collector's dream come true . . . or was it! This is the mouldering condition of the material in the St. Peter's Valley tunnel Ho 2 when it was opened for inspection and clearing out in 1973. Nearly thirty years in a damp atmosphere had wreaked havoc with metal, leather and wood (CIOS/M. Delanoe).

The sea walls, whilst ugly reminders of an invasion defence never tested, have their practical uses — and certainly have an architectural quality of 'beauty'. *Above*: This is the most impressive of all constructed around Longy Bay, Alderney.

Cave. Within the tunnel lay mouldering artillery limbers, rusting field kitchens and searchlights intermingled with smaller items of equipment. The biggest surprise was undoubtedly an intact Freya radar aerial and a chassis, minus the bowl reflector, from a Wuerzburg. However, even the enthusiastic explorers had to accept that their fragile state after so many years corrosion, combined with practical difficulties of removing anything from the site, precluded the possibility of ever recovering the pieces. (It was interesting to observe that while the existence of the equipment was reported fully by the *Jersey Evening Post*, no mention of it was made in a similar article in the *Guernsey Evening Press and Star*).

Another tunnel inspected at the same time was No. 1 at Delancey in Grandes Maisons Road, currently owned by Captain Donald Bisset. The Germans used the area as a tank park and after the war the British Army disposed of a number of light tanks and artillery tractors by sealing them up in the half-completed tunnel. However, all were removed and melted down during the 1953 scrap metal drive and nothing remained to be seen.

When Lieutenant-Colonel A. B. Rogers, the officer commanding the clear-up, commented in 1945 on the fate of those bunkers and other concrete structures which were too massive to demolish, he said that his policy was to remove all woodwork and camouflage netting and then to leave the concrete to weather. Time would do much, he said, and in the years to come he believed that these would not prove to be such eyesores as people had at first expected. Indeed, he prophesied, in many cases they would be no worse than the Martello towers erected in an earlier day.

Thirty-five years later and much of what Colonel Rogers predicted has come true. Although the Islanders wanted to be rid of anything that reminded them of five years of misery, possibly no one realised at the time that the German fortifications would actually have their uses in years to come. The observation tower at La Corbiere is a good example. When Jersey Marine Radio decided to move its operations from the end of Victoria Pier at St. Helier to a new site, the German tower was chosen primarily for its interference-free location. The main control desk was mounted on the roof inside a glass-panelled observation deck to give a complete view of Jersey's western coastline. It was opened in March 1977.

A contrast is provided by the present-day use of Resistance Point Mole Verclut tunnel — that of a vivier or seafood farm. This tunnel had been driven into Gibraltar Rock at St. Catherine's Bay to be used as an ammunition store for a 105mm coastal defence gun mounted at the entrance. For several years after the war it remained empty before the Public Works Department bricked it up. Now it is filled with dozens of tanks containing

Those walls which ran near civilisation had to be breached — like this gap cut beside Third Tower, Beaumont, Jersey (Evening Post).

The end of the Mirus guns — literally! This circle of steel in the Guernsey Occupation Museum at the Forest is all that remains of the once mighty monsters. So the story goes, George Dawson ceremoniously made the first cut and found the Russian steel so hard that a special gas mixture had to be formulated by British Oxygen before Upham's could begin work. The picture *above* shows 'Richard Heaume's' gun being prepared for the melting pot (John Upham).

thousands of lobsters and crabs. Sea water is pumped into the tanks at every high tide and a circulation system oxygenates the water so that the creatures can survive long enough to be eaten!

Another tunnel — Ho 1 in St. Peter's Valley — provides an excellent environment for a mushroom farm.

Many other German constructions have found post-war uses, one of the most important being the anti-tank walls built at the rear of several of the beaches. Whilst marring the pre-war vistas of the islands, their practical use as a sea wall has been tested on many occasions and, as a means of prevention of the erosion of the hinterland, admirably demonstrated in St. Ouen's Bay.

After the war, there was an acute shortage of steel in Britain and one of the first businessmen to reach the Islands was George 'Orange Juice' Dawson. He set up shop in the Manor Hotel, Guernsey and soon made a deal with the States for the purchase of all scrap iron on the island for £25,000. Although, with the benefit of hindsight this now seems a pittance, the States at that time still had to honour Reichsmarks — German currency still being accepted by banks until July 1946. (The final exchange rate was given as 2½d (1p) for ten pfennigs.)

Dawson began a high-powered selling enterprise, selling much of the material before he had bought it. Early casualties were the Mirus guns and on June 23, 1947, the Evans Welding Company of Croydon began the work of reducing the guns to manageable lumps of metal of five tons each. Many residents near the gun-sites expressed regret that the guns were to be removed, feeling that they would be great souvenirs to show future visitors to the island. Nevertheless, money spoke louder than words and all the guns ended up in the melting pot, no doubt still being recycled in motor cars today!

Today, the four gun positions are individually owned and each remains to be seen in a different environment.

The No. 1 gun at Les Rouvets is now on

Chouet Quarry near L'Ancresse Bay in the north of Guernsey (see page 19) was selected by the Army as a dumping ground for much explosive material cleared on the island. *Left:* **In 1945 the tower was not so precariously balanced on the edge of the cliff as it is**

land owned by Lieutenant-Colonel Patrick Wootton and entry is via the Les Rouvets Tropical Vinery. This and No. 4 are the only two positions still left to be seen as they were after the guns were removed.

No. 2 gun-site now lies in the garden of a private house on La Vieille Rue owned by John Cross and No. 3 gun pit forms an attractive sunken playground for La Houguette School.

Probably the easiest to inspect and the most frequently visited position is that of the No. 4 gun in the Rue de l'Arquet, now leased by Richard Heaume. The underground magazines and accommodation are all still open.

Incidentally, Mr. Ronald Mauger (see page 27) spent several happy years after the war acting as a guide to visitors wanting to see the Guernsey defences. When the island's tourist trade restarted in 1947, several of the leading hotels used his services and he conducted up to three tours a day until 1955. 'We had much fun in the bunkers of the flak batteries,' he remembers, 'especially when the ladies struggled up the ladders. In the control room I gathered the party round the telephones still intact but not, of course, in working order. I made some ejaculation then jumped to the telephone on the wall. Cranking the handle madly, I shouted into the transmitter: "Achtung! Achtung! Funfzig britische Flieger sued-west. Schiessen!" (Attention! Fifty British aircraft south-west. Fire!) The audience loved this pantomime.'

Because of the ever-present danger of explosive devices being uncovered, each of the islands has its own permanent bomb disposal officer; Police Sergeant McCord on Guernsey and Mr. Eric S. Walker in Jersey who have been involved in numerous incidents in the post-war years. One area which has revealed much ordnance is the old German training ground at Egypt on Jersey's northern coast. The small hamlet was evacuated and it provided a realistic area or 'Kampfbahn' to practice street fighting, not unlike the present-day simulations provided in many countries to train their police and armies in urban conflict. Egypt was also the danger area for German artillery firing from St. Martin's Church and an unexploded 210mm shell came to light in November 1976.

On Guernsey, large quantities of mines and ammunition were originally dumped in the abandoned, water-filled quarry at Chouet in the far north of the island. When, after some years, there was a local shortage of stone, all this material had to be painstakingly cleared in 1967 prior to the draining and re-opening of the quarry.

The largest cache of explosives dealt with since the war was that discovered in the sunken wreck of the steamer *Arnold Maersk*, a vessel of some 2,500 tons which sank on May 22, 1943 after striking the Grunes aux Dardes rocks between Elizabeth Castle and Noirmont Point off the southern coast of Jersey. The ship had been carrying a cargo of hay, coal, timber and bombs when it foundered in the early hours although all the crew were rescued by boats of the Harbour Defence Group. Some bombs were recovered at the time but the ship was all but forgotten until Tony Titterington discovered the wreck on a diving foray in 1969.

The find was reported to Eric Walker who, from the description given, believed the bombs were 500kg thin-walled aircraft

War leaves behind it a dangerous legacy and the Channel Islands have had their fair share of the post-war 'harvest of steel'. This baby was uncovered at Egypt in November 1976. Mr. Eric Walker and PC Stuart Elliott prepare to detonate the 210mm shell after moving it down the cliff.

today *(right)* **due to subsequent quarrying operations after the explosives were cleared in 1967. The workings are now operated by Ronez Limited, an offshoot of British China Clays Limited (Public Record Office).**

bombs. He immediately contacted the Royal Navy whose experts soon confirmed the identification. It was appreciated that the whole south coast of Jersey was in some danger and if the entire load, estimated at over 60,000lbs exploded, severe damage could be expected to the sea front at St. Aubin's and windows would be broken over a wide area.

Although it was in the middle of the tourist season, a mass exodus had to be avoided if possible (despite the efforts of the national press to do otherwise), and work began to lift the bombs on August 18. As each was raised it was taken some three miles further out to sea and destroyed by firing a charge of plastic explosive taped to each bomb. To protect marine life as much as possible, the bombs were detonated while suspended beneath oil drums but scores of fish were still stunned by the explosions. The last seventeen bombs could not be moved and, after suitable precautions had been taken ashore, they were set off in situ.

The most recent discovery of unexploded bombs occurred in January 1979 when a gardener at Mont Orgueil Castle at Gorey spotted the bottom of a shell poking through the grass in the Grand Battery, visited by hundreds of thousands of tourists since the war. Eric Walker was called in and together with two other bomb disposal officers from the States Police, immediately carried out an investigation of the area. Some nineteen shells of French WWI vintage were uncovered which

the Germans had transported to the castle to act as 'roll bombs' in the event of an invasion. The idea of these weapons was to hang the shells nose down on wires around the walls and to release them onto the heads of troops storming the ramparts.

In view of the size of the find, Eric Walker decided to call in outside assistance and the

The local Press in the Channel Islands has played an important part in educating the public, especially over-enthusiastic schoolboys, into the inherent dangers of all explosive devices. It is never, ever worth the risk of playing with or trying to defuse a thirty-year-old item just to improve one's collection. The author remembers hearing at school in the 1950s of one boy who used to spend all his holidays in Jersey searching for war material and one 'prank' was to throw live grenades, such as this one, over the cliffs at Noirmont to hear the satisfying boom at the bottom. (Come back O'Brien, all is forgiven!)(Evening Post).

Army's bomb disposal unit at Chattenden in Kent was contacted. The act of calling in outside assistance raised some criticism and in order to set local fears at rest, Mr. Walker wrote an open letter to the Press:

'In insisting upon outside experts to deal with the Mont Orgueil Castle roll bombs, the Gorey residents must believe that some lack of skill or diligence has occurred. In view of my long service to the community, perhaps they would consider the following explanation and then judge if this lack of confidence is justified.

'In about 1955, the late Captain G. Dorey, Gardien of the castle for many years, informed me that two roll bombs (old French shells converted to bombs) had been fixed to the wall, and indicated an area of thick undergrowth where he had observed them, abandoned by the Germans. Permission was given to clear this undergrowth and the search revealed two missiles. He did not know of any others.

'In 1977, more were discovered in a different spot on the outside of the castle. A complete search of the eastern face was ordered; this involved clearing much undergrowth and using metal detectors. It proved fruitless despite approximately 700 man-hours of hard work. The Public Works engineer agreed that this search was so thorough that it was unlikely that any missiles remained. This work was carried out by PC Terry Underwood, Public Works staff and myself, PC Stuart Elliot being on holiday. On both these occasions, Army units were invited to assist; officers came over, had a look at this thick bracken and steep slope and declined, although the last visitor offered the suggestion that we put a tractor on the beach and pull some sort of cutter up and down.

'Sadly, at no time did we have any reason to suspect that bombs were concealed inside the castle; perhaps because the Germans' normal disposal procedure was to remove the detonators and then cut the support wires, allowing them to fall and be removed. Recovery upwards would be difficult. Perhaps the Germans were disinclined to drop these from such a height as they had doubts as to the stability of this type of missile. I worked with the German forces during the liberation clearance, sometimes on roll bombs, and also regarded them with some suspicion.

'On the discovery at the Grand Battery, PC Terry Underwood, PC Stuart Elliot and I worked with all possible speed to inspect the new find, determine the number and check the condition. All appeared reasonable. We then fixed slings to allow easy recovery and reburied each one at a known distance to prevent the chance of a mass explosion. One missile would present little risk outside the castle but, of course, the nineteen exploding together would be very serious indeed.

'Having stabilised the position and removed any immediate risk, the Defence Committee considered the options and instructed me to apply for a helicopter as being the safest method. If a helicopter was unavailable, a crane and Land-Rovers were to be used but this would involve some inconvenience to persons living nearby and would be more trying to ourselves.

'If this appears to be an unnecessary fuss, I would explain our reasons. A shell of this type recently exploded at Quen-Plage, France, for no apparent reason at all, and the last one we moved in 1977 began emitting white fumes, this indicating that all was not well. Moving these objects is difficult as they weigh 365lb each. It is said the Chinese liken a bomb to a sleeping dragon; left alone he is quite happy but if disturbed he can become very angry. This seems to sum up the problem. Even after having disposed of some 2,000 shells, mines and bombs in the last twenty-five years, including a hundred of this type, I cannot predict absolutely how this very old and powerful ammunition will behave. It dates

UXB 1979 style. Better than any TV series, residents of Gorey were given a piece of the action when nineteen ex-French WWI shells were discovered buried in Mont Orgueil Castle. Tramped by many thousands of feet since the war, the find provided considerable interest . . . and recriminations. *Above:* David Le Put and Robert Dixon-Cheesman, two home-grown sappers, prepare the shells with strops for lifting (Evening Post).

No. 230 Squadron Puma air transports the baskets of shells out to the beach near the Seymour Tower, three miles from shore (Evening Post).

Major Barry Birch of 49 Explosive Ordnance Disposal based at Chattenden, Kent, veteran of a thousand bangs, supervises Corporal Matulewicz as he performs a last minute hop, step and jump over the shells. In the trade, this is called the dance of the three beehives! (Evening Post).

Left: **2.14 p.m. February 26, 1979 and Barry's big bang signifies the end of the drama.** *Rihgt:* S rv^e yin g t^h e d**ma6e (Evening Post).

from 1917 and uses an explosive discontinued by the major powers at the time of the Crimean War and, although more powerful than TNT, it has the disadvantage of becoming unstable under certain conditions.

'Due to these factors, it would have been quite wrong not to wait for a very low tide and the prospect of better weather, which also allows time for a thorough search inside the castle.

'Last year a number of objects were disposed of, it being believed to be in the public interest that we operated discreetly. This would appear to have some disadvantage, causing some to doubt our experience in this field and to call for outside intervention.

'Several far more difficult tasks have been completed: for instance, the clearance of La Rosiere quarry by States Police Sergeant Ray Medder involving 700 fused mines mixed with twenty tons of barbed wire in four feet of mud down a pit seventy feet deep.

'In view of these facts, I trust the people of Gorey will accept that care was being taken of their old and lovely castle. Everything possible was being done and no other safer solution could be devised by anyone. Clearly blame for this trouble and expense rest with those who concealed such deadly weapons in a public place, well knowing that, as time passed, removing them safely would become more difficult. It was an act of extreme irresponsibility.'

A team from 49 Explosive Ordnance Disposal, commanded by Major Barry Birch was sent to the island and the unit spent the next three weeks thoroughly searching the castle grounds but no other explosives were found.

As removal of the shells by road would have necessitated the evacuation of up to 400 houses in Gorey, Major Birch agreed that the best plan would be to lift them out by helicopter. A Royal Air Force Puma from No. 230 Squadron was brought in from Odiham and on February 26, after the area was cordoned off, the operation began. At low tide the shells were lifted in batches of four to the beach, there being only three hours to complete the move before the sea would cover the spot chosen near the Seymour Tower, some three miles off the south-east corner of the island. The 170kg shells were laid on the beach and a beehive charge stood on each, the whole nineteen being wired together before being simultaneously detonated. Five huge craters were blown in the sand but no damage was caused and another exciting episode was over.

Flashback to 1945. Disposal of the heavy guns by Major Frank Sargent was by the simple expedient of flinging them over the cliffs at Les Landes. For the next episode in this thrilling story, turn the page (Evening Post).

As has been described, after the liberation, immediate measures were taken by British engineers to demilitarise all the Channel Islands and whilst much equipment was dumped at sea, all the heavy guns on Jersey were run over the cliff top at Les Landes. Today a tangled mass of gun barrels and carriages lie at the foot of the cliffs dubbed 'Le Creux aux Canons' and for those with a head for heights it is possible to clamber down to inspect the rusting relics at low tide.

As with all history, one generation's rubbish is another's treasure and many proposals have been made over the years to recover the guns. Mr. Graeme Pitman, the Chief Executive of the Public Works Committee, confirmed that 'we have had many requests to recover the guns from the foot of the cliffs', and the dangers of leaving the guns where they lay were tragically brought home in 1977 when a young enthusiast, Stephen Bird, disappeared while climbing in the area with friends. His jacket and scarf were retrieved but his body was never recovered.

As a consequence the Committee approached Lieutenant Bryan Cooke, OC 59th Independent Commando Squadron, then carrying out demolition work on Alderney, to ask if he would inspect the site and see if the guns could be recovered. The Committee had already decided the job could not be carried out by a private contractor and it was hoped it could be undertaken as a military exercise.

Nothing was done until 1979 when the private company of L. C. Pallot (Tarmac) Limited, at the instigation of the CIOS, succeeded in raising one barrel on Saturday, June 30. The contractors had made a first attempt to raise a barrel the previous weekend but decided that a stronger cable was needed. Even so it took most of the day to winch the ten-ton, 20ft barrel some 275ft up the cliff. At one stage, hydraulic jacks had to be used to lift it clear of some protruding rocks.

Major Sargent must be one of the few people who have satisfied everyone's childhood dreams of smashing things when he ran twenty-five heavy German guns out of town! (Evening Post).

Above: **One of the barrels didn't quite reach the bottom — this is where we found it in April 1979. It was recovered three months later.**

Although unprotected against the elements and salt water for thirty-five years, the barrels at the foot of the cliff are in a remarkable state of preservation (A. Pike).

Once on level ground, the piece was identified as a 150mm K18 brought to Jersey from Guernsey in August 1944. The Occupation Society were most grateful for the generosity of Mervyn and Dudley Pallot in recovering the gun free of charge and for transporting it to Noirmont at no cost.

Later that year in October, Major Sargent made his first visit to Jersey since he left in March 1946. He was enthusiastically received by members of the Society, pleased to meet face to face with the man who disposed of so much that the Society now strives to preserve. As Major Sargent commented at his talk to members, 'Some people now think it inconceivable that we dumped and destroyed all that potentially valuable and interesting equipment but that is with the benefit of hindsight. There was a totally different atmosphere in the island in 1945. The people had suffered bitterly and resented it. Our orders were to dispose of and destroy all enemy equipment, though we did present the States with specimens of all the different types of guns we captured, and this view was reinforced by an instruction I received from a senior States Member who told me forcibly: "We want this island cleansed of the taint of the German occupation".'

None the less, in spite of all his post-war efforts, the Society had a surprise in store for the Major. Taking him out to Noirmont he was proudly shown the K18 recovered only four months previously, the scene being recorded for posterity on film.

Thanks to the Pallot brothers, the barrel we saw half-way down the cliffs is now on display at Noirmont (J. Le Boutillier).

'What did you do in the war, Daddy?' Major Sargent: 'I pushed guns over cliffs son!' (Evening Post).

The whole Les Landes area was handed back by the War Office to the States of Jersey in 1960. The authorities, fully appreciating the value of the area, passed an Act which stated: 'The public has been extremely fortunate to acquire this large area of heathland and long stretch of superb coastline. The committee recommends that it should be preserved for the general enjoyment of the public.'

However, apart from using part of the site as a refuse dump nothing positive emerged until a plan was published in September 1977 outlining its conversion into an amenity area — something which would be of immense value to an overcrowded island. The plan envisaged the construction of car parks to restrict the use of motor vehicles in the area; the creation of a picnic area, nature trail with seating along cliff paths and the conversion of the site where the huge Jubilee bonfire had been lit in an artillery emplacement.

Although action has yet to be taken on the scheme, when we visited the area in the company of the President of the Jersey Branch of the Occupation Society, Peter Bryans, he told us that the Society had plans to preserve the battery and the nearby naval radar tower (featured on our cover). The whole area is wild and windswept and it was considered an ideal setting for scenes for the film *Force 10 from Navarone*. Forty-nine locals were recruited as extras to play either Germans, American Rangers or Commandos, during a three-week filming stint in January 1978. (Other scenes were filmed at La Hague Manor, which acted as a German HQ, and at Grosnez.)

Whilst at Les Landes, Peter showed us another gun barrel (or rather the pieces of it) recently acquired by the Society — and told us a remarkable story. The gun, it seems, was a 22cm K532 (f) which had blown up on August 19, 1944 while firing at a British destroyer. After the war, no one gave a thought to what had happened to the pieces — not even the vigilant Society. In the early 1970s, a new prison was being built on the site of Batterie Roon at La Moye and on February 10, 1979, while preparing their new kitchen garden, the inmates dug up the remains!

In 1953, a renewed scrap drive led to the disposal of all those guns which had been left in situ in 1945 as they were then considered too difficult to remove and had escaped the 1947 scrap metal programme. As it was, the cupolas or front plates had to be cut through in each case to enable the gun barrel to be extracted. As a result, only four coastal guns still remain to be seen in the Channel Islands today — and all of these are on Jersey.

Probably the most interesting is that preserved by the Channel Islands Occupation Society (by the courtesy of the Public Works Committee) at La Corbiere at the southern end of St. Ouen's Bay. The casemate, complete with 10.5cm K 331(f) gun was handed over in 1978 and was first opened to the public on September 3. The 1918-vintage gun was originally a French Model 1913 Schneider field gun, one of eighty-two installed by the Germans in the Islands. With a range of six miles, the Germans considered it an ideal weapon for coastal defence and thirty were emplaced on Jersey of which just the four survive. Since its first opening, the CIOS have replaced the wooden anti-ricochet boarding surrounding the embrasure which was designed to fire across St. Ouen's Bay.

The defences at St. Ouen's were built with the sweat and toil of Russian workers and Lieutenant Georg Kosloff, captured at Smolensk in 1941, confirmed in an interview in 1945 that many of the workers from the nearby Val de la Mare camp who died of ill-treatment were buried where they fell — in the land bordering the picturesque bay — a sobering thought for the thousands of holidaymakers who now frequent the beach.

Most of the anti-tank ditch at St. Ouen's has been filled in and, although the minefields and wire have long been removed, we were very interested to see a group of tetrahedrons still lying in the corner of one field where they had been dumped in 1945.

Another more-official memorial marking the island's wartime past is used by many visitors forty years later, possibly without them realising its true significance. During the occupation, to avoid undue unemployment on Jersey, the States initiated the construction of the New North Road and the project served to utilise local civilian labour who might otherwise have been made to work by the Germans on more warlike constructions. It was finally opened in July 1947 renamed North Marine Drive (now more commonly known as 'La Route du Nord') and was dedicated to all those who had died in two world wars as a 'perpetual reminder of their loyalty and devotion.'

A 22cm K532 (f) dug up by convicts in the vegetable garden of the prison at La Moye — former site of Batterie Roon. We photographed it lying in one of the emplacements of Batterie Moltke at Les Landes (see page 75).

This Sechsschartenturm, Resistance Nest 'Gruene Dune', at Rocquaine, Guernsey, suffered severely in the 1953 scrap drive. The scrap men must have used so much gas in cutting through this thickness of hardened steel that one wonders if it was really worth it.

Another victim of the gas bottle. January 6, 1953 and the Guernsey firm of John Upham Limited begin removing the majority of the steel remaining on the Islands for a London scrap metal dealer. Many rare relics were thus lost — bitterly resented by all those interested in the occupation period. This is Resistance Point Bel Royal, Jersey (Evening Post).

Above: May 1, 1975 and the Touillets Tower bites the dust at Castel, Guernsey. 'Blaster' Bates brought down the former German artillery observation post to provide the site for the island's IBA television mast (Guernsey Press).

No this is not 'Blaster' Bates but Peter Bryans relaxing in his natural habitat of the La Corbiere gun — saved by the CIOS.

CHAPTER SEVEN
The Channel Islands Today

During our visit to Alderney, Colin Partridge showed us these tetrahedra beach obstacles rusting in a field on part of the old Norderney Camp site (see page 101) in the north-east corner of the island where they had been dumped at the end of the war. Colin couldn't help thinking that one would be ideal as a doorstop and Trevor Davenport gave him a hand.

The Channel Islands Occupation Society is a unique organisation formed for the study, research, documentation and preservation of the Islands' wartime history. Today its two branches of Jersey and Guernsey have a combined total of nearly 400 members but in the beginning there were only two.

The Society has its foundations in the boyish enthusiasm of two Guernsey schoolboys, Richard Heaume and John Robinson who, from the middle 1950s, were enthusiastically collecting and researching all they could on the occupation period. Being born after the end of the war, they could not know what the German occupation was really like, but it was a part of everyday conversation and a part of their history. It was their enthusiasm which led them in 1956 to form the 'German Occupation Society of the Channel Islands'. Slowly interest grew and with it came a change of name when, in the mid-1960s, the present Society was born with the objects of studying and recording the years from 1939-1945. Mr. A. D. Priaulx was the first president and Mr. Ken Tough the treasurer while Richard became the social secretary, organising visits and tours of fortifications and talks on various topics concerning the war in the Channel Islands.

In May 1966, Richard opened his museum at the Forest in Guernsey and, with its immaculate captioning and displays, it is a marvellous record of that island's wartime history. Richard admits to an obsession with the Organisation Todt and is an enthusiastic collector of relics concerning 'the Organisation' and, on a recent trip to Alderney, was seen by members to use every effort to wrest a wooden wheelbarrow from a farmer which he claimed was genuine 'OT'!

On May 17, 1971 a Jersey branch was formed with fifteen founder members and in October that year the first combined dinner was held at St. Margaret's Lodge Hotel, Guernsey, together with members of The Alderney Society, a separate body yet having an affinity with the aims of the CIOS. The Jersey delegation included Richard Mayne, who had also opened a German Occupation Museum in 1966 at St. Peter's, Jersey, located in one of the original German bunkers.

The feelings of the membership were expressed at the dinner by Conseiller Raymond Falla (Guernsey's representative on the Island's purchasing committee) who said that, '. . . the part we all played in the occupation must be kept alive and I salute the members of your Society for their perseverance'.

Mr. Ronald Mauger (see page 27) commented how the Society was breaking new ground that night as Jersey and Alderney members were present for the first time. He said he felt strongly that someone should represent Sark the following year for, although that island had no society, it had been occupied by the Germans. Herm and Jethou, however, in his view did not come into it — the latter was never occupied and the former for only eleven weeks by a flak battery.

Mr. Cyd Gardner, replying on behalf of the guests, told several stories of his own experiences. One concerned the landing of British Commandos on Guernsey's south coast in July 1940 and how he met two of them. *(Second Lieutenants Hubert Nicolle and James Symes who were landed back on Guernsey on September 4, 1940 on an intelligence gathering mission. See page 15.)* They gave him a collar-stud and said: 'When we've gone back, open the back of it'. He said he did so and was amazed to find a Commando compass inside.

'That compass', said Mr. Gardner, 'has always been my most treasured possession, that is, until tonight because it now gives me great pleasure to present it to Mr. Richard Heaume for the use of the Society and I've no doubt it will find its way into his museum.'

From that historic meeting, the Channel Islands Occupation Society has gone from strength to strength. A professionally-produced magazine is published annually and the Society has also brought out its own

booklets on specific aspects of the occupation. Regular monthly meetings are held with tours made to visit various sites of interest and speakers give talks on particular aspects of the wartime history and film shows depict life during the occupation. The Society's own 'film', produced with the aid of the Jersey Camera Club, is an excellent 40-minute dual projector slide show *Scars on the Landscape* with synchronised commentary which was instrumental in opening the eyes of the States authorities to the fact that wartime architecture was just as important to the history of the Islands as that of former periods of history. The slide-film was compiled by Michael Ginns, secretary of the Jersey branch. Although Michael was born in England he has lived all his life in the Channel Islands . . . except, that is, when the German authorities decided to deport all Channel Islanders with close English connections to be interned in Germany. Michael was among the 1,186 chosen from Jersey in September 1942. The camps which awaited their pleasure were Internierungslager V at Biberach (with a branch at Schloss Wurzach) in Wurtemburg and Ilag VII at Laufen, Bavaria. After the camp was liberated by the Free French in April 1945, Michael then spent the next five years in the Royal Army Service Corps.

Michael agreed that he had adopted our 'then and now' theme, ideally suited to a dual-projector slide show, merging the 'then' picture into the 'now', all linked with a musical score and commentary spoken by Robin Cox. The combined effect is really excellent and readers should not miss an opportunity to see the show if they visit Jersey.

The whole purpose of *Scars on the Landscape* was to focus attention on the official (and unofficial) vandalism which was robbing Jersey in particular of its wartime past at an increasing rate of knots. In the years following the war, the island fortifications had slowly been whittled away although not without considerable expense. The massive bunker which stood beside the harbour at Gorey cost £3,125 and five weeks' effort in November 1972 using drills and explosives by Demax International before its bulk was swept away — the site is now part of the car park.

As a result of making *Scars on the Landscape* the Jersey Branch was asked by the authorities to prepare a list of wartime fortifications which, in their view, merited preservation. Top of their list was perhaps the most extensive of all Jersey's defence installations at Noirmont Point. There, four 150mm guns had been installed on open platforms controlled from an underground command bunker. The bunker had been constructed by blasting a huge hole in the cliff top some forty feet deep. Walls were fabricated of two-metre-thick reinforced concrete, the work being carried out in this case by conscript labour as well as by local volunteers willing to assist the German war programme. Ventilation units, both power and manually operated, filtered all air and an escape shaft led to the surface.

When the battery was in action, the gunnery officer observed the target through the foremost of the two periscopes which were mounted in the double metal dome that stands at the extremity of Noirmont Point, while the fire-control officer gathered the information supplied by the range finder. After corrections for wind and humidity had been added, this information was fed into a mechanical computer. The range data was then transmitted from the computer by cable to each of the four guns which were mounted on the concrete platforms above ground. The naval gunners had only two tasks to perform: to load the weapons and to keep two pointers on a dial in line with one another by operating the traversing and elevating mechanism. The actual firing of the guns was performed by the fire-control officer from a control panel deep in the bunker itself.

In April 1979, similar 'objets de la guerre' were still to be seen at La Pulente *(above)* and these of the 'Mark II' variety *(below)* overlooking Beauport Bay. Sadly the tetrahedra were all scrapped by the Public Works Department early in 1980 but the hedgehogs have been saved for Noirmont and will be moved there in due course.

Only 3½ examples of the six-loopholed machine gun turret still exist in the Channel Islands. This one at La Mare Mill on St. Ouen's Bay is so complete in its camouflage that it was listed (A13) and categorised of 'special importance' in the historical survey, conducted by the Jersey branch of the CIOS on all the island's fortifications which merited preservation, for the Island Development Committee.

In June 1945 a reporter was allowed to inspect the Noirmont headland for the first time since the surrender. His report on his visit was published in the *Evening Post* on June 11:

'It was obvious in '41 and '42 that big works were going on at Noirmont but the ordinary man in the street could not get there to see it all; even now visitors are not encouraged so let me try and give some idea of what is there although I had not the time to see half of it. The whole of the point is a fortress bristling with heavy guns and honeycombed with underground chambers which form the living quarters and storerooms for the garrison. Those chambers go deep down into the solid rock; no wonder we used to hear and see constant explosions as they blasted their way deeper and deeper into the good Jersey granite. Great steel girders and thick concrete walls hold up the caves and electric light and water are laid on everywhere. Air conditioning keeps the air sweet and massive steel doors to each dug-out, and every room in it, make sure that any attack would be a slow and costly process. At the extreme end of the point is the control room. Here, deep in the rock, protected by concrete and steel, is the nerve centre of the battery with range-finders and all the paraphernalia for firing a battery — all now so much "junk".

'Somewhat to my surprise I found that the guns themselves — 6-inch and 4.7-inch I am told they are — were quite exposed on concrete and steel emplacements with concrete ramps leading to the underground shell rooms. I would have thought that while they were about it they would have placed them in disappearing turrets since, apparently, money was no object and slave labour was cheap. I saw no elaborate furnishings here; everything was strictly utilitarian though I heard rumours of a very elaborate underground "palace" at Portelet with carpets on the floors and luxury furniture looted from island houses.'

After the Germans abandoned the battery, anything that was considered useful was appropriated by the locals and virtually everything that was movable in the shelters, command bunker and the naval observation tower disappeared. In May 1948, Mr. Rolf Bertram led three visitors around the interior of the Noirmont battery without lights and accidently walked through an entrance which had a ten-foot drop the other side. Knocked unconscious, he was lucky that his injuries were not serious, but, as a result, the Public Works Committee had the opening to the command bunker sealed up.

For many years a trip to see the observation tower at Noirmont has been on the itinerary of virtually every tourist visiting Jersey. It has been pictured on post cards and has now become an instantly recognisable feature of the island. As a consequence of this interest shown in the site, the excellent relationship between the CIOS and the Public Works Committee led to the latter handing over the bunker to the Society in 1977 for preparation for public display. Dedicated volunteers removed tons of rubble and one interesting find by Robin Cox was a pistol which some long-forgotten German had tucked up one of the ventilation shafts. A new vandal-proof entrance was constructed over the former open stairway, entry now being through a lockable manhole and thence down a vertical ladder.

The bunker was formally opened on Thursday, June 23, 1977 by the Public Works Committee President, Senator John Averty, who declared in his speech that it was a pleasure to be able to hand over the task of looking after the bunker to such an enthusiastic and capable group of people.

Visitors can inspect the interior of the observation tower and the underground rooms on selected Sunday afternoons during the summer and the Society intends adding a permanent display of military artifacts as time goes on. An early acquisition was an anchor from a German tug which sank near the Dog's Nest rocks outside the harbour. Although the tug was raised, the anchor was left on the rocks until it was retrieved by cadets from the St. Helier Yacht Club in 1951. Two years later it was sold to a Wing Commander Bradbury who used it for many years as a deep water mooring for his yacht. In 1977, it was recovered once again from the sea by the States tug *Duke of Normandy* and thence entered the care of the Society.

Manhandling the one-ton anchor to Noirmont was no mean feat. After restoration on a farm at St. Peter's, it was loaded onto the trailer of Graeme Sty's jeep and transported to the tower where it was lowered down to the door using the winch on Peter Le Conte's Dodge. However, due to the restricted space, it was impossible to manhandle it inside until someone had a brainwave. The cable from the winch was threaded down the former radio antenna tube and thence to the doorway where it was attached to the anchor. The plan worked perfectly and the former German naval anchor was pulled to its final resting place where it can be seen today.

The purposes to which various bunkers can be put has often raised arguments for and against their authorised use. Several on the island have been used by wilder elements for drinking sessions and the like but an application by the Hermitage Youth Club to use the bunker at Parcq de l'Oeillere at St.

Opening of the Noirmont command bunker by Senator John Averty (left) standing with Peter Bryans and Alfred Pipon. The German entrance was a single flight of steps open to the air but these have been covered with a concrete slab for security and one now descends through the lockable manhole cover (when it's open!) (Evening Post).

Operation ANCHOR took place on May 20, 1978 and, by a combination of muscle and ingenuity, it was manhandled inside the Noirmont observation tower (D. Bishop).

The modern-day tourist to the Islands with an interest in the past must be observant and look in all odd corners for still-visible signs of the occupation. Who has resisted the temptation to leave their initials in wet cement? This seemed to be a favourite pastime for the OT and other construction battalions. *Above right:* This inscription can be seen in the remains of the old German stables *(above left)* at the back of the car parking area on the corner of David Place and Victoria Street in the centre of St. Helier.

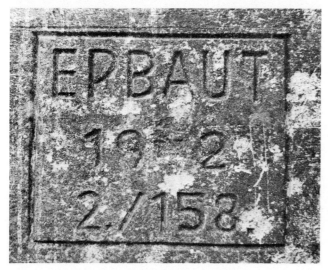 

Above: Unfortunately vandals have obliterated the '4' for some reason in this inscription at Icart, Guernsey, denoting that this particular casemate was built in 1942 by the 2nd Company of the 158 Fortress Engineer Construction Battalion.

Above: One of the most elaborate 'signatures' to be seen in the Islands is this impressive inscription beside Bay View Court, Mont Matthieu, Jersey — so unique that the CIOS survey listed the casemate of special importance for its preservation.

The Russiskaya Osvoboditelnaya Armiya consisted of Russian PoWs who changed sides and fought for Hitler. The carving at Bouley Bay records the passing of one such Russian in 1944. The letters ROA transcribed into the Cyrillic alphabet become POA.

 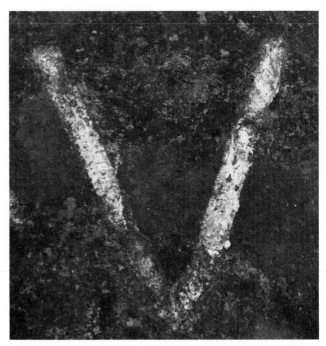

Some painted graffiti has also survived the passing of the years although, for obvious reasons, is much rarer. *Above left:* This V-sign beside the front door of No. 4 Brock Terrace, The Grange, St. Peter Port, whilst no doubt original, has been repainted by a latter-day patriot. *Above right:* That beside the gateway to the Summerland Hotel is undoubtedly genuine WWII-vintage.

Ouen's Bay raised a storm of protest from local residents. Their letter of objection stated that: 'converting the building for use would be material development in a green belt area; there are no toilet facilities; parking is not possible near the bunker; access to the headland is by a dangerous corner; the natural beauty of the headland is likely to be ruined by cars and motor cycles; the youth of the area is already served by the new Communicare Centre; the place would be a death trap if there was a fire and there would be no access for the emergency services.'

A building being put to good use in the St. Ouen's area is Grantez Mill, close by the parish church. The mill was bought by the States in 1911 for £100 and, although the sails had already been taken down, the milling equipment still existed prior to the outbreak of war. It was ideally suited as an observation post and the Germans added an extra floor with a 360-degree field of view and painted conspicuous landmarks with their ranges on the walls. In April 1978, the mill was refurbished by the local scout troop and now provides them with two floors for training and one for games.

An even more practical use for the German bunkers was suggested by a reader of the *Jersey Evening Post* in a letter published in March 1979. This lady thought that with the present state of the world, they should all be cleared out in preparation for a Third World War!

The London Weekend television series *Enemy at the Door* has probably done more than anything else to focus the attention of the general public on the wartime Channel Islands. However, the series received a mixed reception when shown in the Islands for, although fiction, certain factual incidents were portrayed and locals were quick to point out errors, both actual and implied. It also brought up painful memories for many when it dealt with informers, black marketeers and 'Jerrybags' — the Islanders' name for local girls who befriended the Germans. The series was based mainly on Guernsey and because the television company cut corners and filmed several scenes on Jersey, Guernseymen were up in arms. The producer of the second series, Jonathan Alwyn, sympathised with the locals who had been put out but explained that Jersey was blessed with bays and coves more accessible than those on Guernsey. However he said that he hoped to make amends in the second series although commenting that some shooting would have to be done in Jersey for continuity's sake. Sark and Herm were locations used in the re-enactment of the Commando raid (see Chapter Five).

During 1976-77, Colin Partridge (author of *Hitler's Atlantic Wall*, an expert treatise on the subject) and David Kreckeler conducted a survey of all the military structures on Guernsey with a view to making recommendations to the authorities as to which were worthy of preservation.

One of the problems with the retention of wartime constructions on Guernsey has been that, unlike Jersey, most are situated on private land, out of the jurisdiction of the Public Works Committee. Naturally, land owners need, or would like to obtain, a commercial rent for the use of buildings on their land and this has inhibited the growth of the preservation of interesting sites.

However, on February 1, 1979, the naval observation tower at Pleinmont on the southwestern extremity of the island was leased by the Guernsey branch of the CIOS for a period of ten years from its owner, Mr. F. J. Merrien. Immediate plans were made to clean the building and the States Fire Brigade kindly obliged in hosing out the interior. A pair of original gates have been fitted to secure the tower against vandalism and a platform ladder installed to gain access to the top of the

What a pity some officials could not have found somewhere, other than on this original inscription at La Collette, St. Helier (near the power station) to paint a white square!

The 1979 television series 'Enemy at the Door' publicised the role of the Channel Islands during the war to the public at large more than any other book or film. Some scenes were filmed on authentic locations — see page 124. The actors are standing on the orginal German slipway defences at Creux harbour on Sark, which were cut to ground level after the war (London Weekend).

Above: **Part of 'The Raid' episode was filmed outside the Dixcart Hotel (unfortunately no stills were taken) and this cottage on Herm provided a useful, kickable door. The 'Lee-Enfield' rifle is painfully a dummy.** *Below:* **Other parts of the series were not so authentic as to their locations. This realistic-looking sequence was, in reality, filmed at Rochester, Kent (London Weekend).**

tower. It is hoped to install original optical equipment on open days which will give breathtaking views from the top of the tower. The personnel shelter adjoining the tower has also been taken over and this will eventually be used as an exhibition area.

Regrettably, the excellent co-operative spirit which exists between the Occupation Society and the authorities of Jersey (and to a lesser extent in Guernsey) has been, until recently, sadly lacking in Alderney. This lack of official interest has led to most of the wartime sites being left to the ravages of nature, vandals . . . and the British Army! The latter has performed its own brand of destruction on Alderney which is frequently used for training.

Matters came to a head on the island in the autumn of 1977 when a decision was taken to carry out extensive demolition work at Fort Albert. The restoration of the buildings in the courtyard was said to be technically and financially impracticable by the Public Works Committee which stated that before this valuable site within the fort's perimeter could be utilised, the buildings would have to be cleared.

The Army was called in to do the job and, perhaps through an over-enthusiastic use of explosives in the 'training exercise', damage to the outer works resulted. Charges of 'authorised vandalism' were levelled against the Committee which was put well and truly on the spot when it was revealed that the only bridge across the moat had been rendered unsafe and therefore the huge piles of rubble now littering the courtyard could not be removed — leaving an even worse blot on the landscape!

Frustrations in the lack of interest in the preservation of the island's history led to the formation of the Alderney Fortifications Centre by Colin Partridge, Trevor Davenport and Major Tony Hynes in 1979. A lease was sought for the well-preserved original Victorian magazine and custodianship of the outer walls of the fort; the bridge across the moat was rebuilt and by January 1981 work had begun shifting the estimated 2,000 lorry loads of rubble. In the short life of the new organisation, it has also been responsible for a programme to restore the two-storey, underground fortress commander's bunker at Les Rochers and has produced descriptive leaflets for The Alderney Society on three phases of the island's fortifications: pre-Victorian, Victorian and German.

No sooner had the Fort Albert fiasco hit the headlines than a similar issue was raised over the proposed demolition of a pier in Braye harbour.

In 1942, the Germans had worked on the construction of the pier which later became popularly known as the 'German jetty'. After the war, with lack of maintenance, the condition of the steel superstructure of the pier had deteriorated to the point where it was considered unsafe and a plan to demolish it was first put before the island's parliament in 1962. However, no action was taken and it fell further into disrepair and by the 1970s was considered dangerous and a hazard to navigation. Much argument ensued as to the merits of spending some £93,000 of the island's money on its removal, a Mrs. Leda Miller opposing such an idea saying that it would make more sense to spend the same amount of money on casing the steel piles and rebuilding the pier with concrete blocks and hardcore to end up with something of benefit for the future.

Nevertheless, demolition contractors were called in during 1978 and when we visited the harbour in April 1979 the work had been 'completed'. We were told (and were able to see with our own eyes) that the demolition work had only been carried out to the waterline and that far from removing a hazard, all that had been done was to create an even worse danger of jagged underwater steel piles, invisible except at low tide! This unsatisfactory state of affairs remained until January 1981 when a further £50,000 was voted to enable the demolition to be completed.

This is an authentic relic from Herm. Although it doesn't look much like it, this is a Grumman F6F Hellcat which was recovered from the north coast of the island where it had lain for nearly thirty years. The Tostevin brothers transported the remains back with them to Guernsey but it was in such a pitiful condition that we are told nothing now remains.

The German jetty photographed by Ken Tough in June 1978 just before its demolition. The job was only half done and the work resulted in a dangerous underwater obstruction which was only finally cleared in 1981.

One interesting building on Alderney is 'Peacehaven' — an appropriate name for the bungalow in Braye Road where the surrender was signed in 1945 (Colin Partridge).

In spite of, or as a result of, the island's controversies, Alderney now presents the greatest opportunity for the military buff to see, in one place, a microcosm of the entire Atlantic Wall largely as it was left in 1945. Although the guns (and much of the metal work) were removed in the post-war scrap drives, batteries such as Annes at the Giffoine, are amongst the most original on the Islands. The many unsealed bunkers on Alderney allow close inspection but care must be taken as this easy access has also enabled the Army to test various explosive charges — Batterie Annes has suffered in this way although shell hoists and stencilled operating instructions make this an extremely interesting site to visit.

Another place which is a must is the site of Lager Sylt — the only SS concentration camp on British soil during the war. The concrete sentry boxes still remain to mark its extent as does the underground tunnel to the Kommandant's house. The site of the latter can be pinpointed by the fancy stone 'patio' with its magnificent view overlooking the sea. The house was purchased after the war by a Mrs. E. Hunt who had it moved to its present site on Longy Common in late 1946.

As has been described in the earlier chapter on Alderney, the conditions in the four camps on the island resulted in the deaths of many of the inmates. They were mostly buried on Longy Common — a large barren tract of land behind Longy Bay — the latter now ringed with a most impressive anti-tank wall. All the remains were exhumed after the war and taken to France for depositing in the German ossuary at Mont-de-Huisnes. A local family, the Hammonds, constructed a memorial at Longy to all those that had died and were first buried nearby and this was unveiled in 1966.

The first post-war pilgrimage of survivors from the camps by members of the Amicale des Anciens Deportes de l'Ile Anglo-Normande d'Aurigny (Aurigny is the French name for Alderney) was held in 1967. Most of the camp inmates had been French with a sprinkling of Russians, Poles, Spaniards and Jews from all over Europe. A few of those that came through their ordeal stayed on the island after the war and these and other survivors took part in another memorial pilgrimage in 1977. This time, however, numbers were sadly depleted when they attended a wreath-laying ceremony at the Hammond Memorial where a tribute to their sacrifice was paid by the President of the Alderney States, Mr. Jon Kay-Mouat. The last official visit by the various organisations representing the nations whose dead are commemorated by the memorial was in May 1979 — the 35th anniversary of their liberation in May 1944.

Another service marking a different aspect of the island's history was held in the parish church of St. Anne on June 22, 1980 when a special service of remembrance recalled the 40th Anniversary of the evacuation when all but seven residents were transported to England.

One aspect of the wartime architecture of the Channel Islands that the authorities have been pleased to retain (the true purpose of which probably goes unnoticed to the majority of holidaymakers) are the the sea walls. These are, in reality, all German anti-tank barriers and they have completely changed the aspect of some of the bays from their pre-war appearance. For example, St. Brelade's Bay had no wall at all until 1942 when wall No. 6 was built by the Kehl Construction Company between March and October. This precise dating is possible due to the action of one of the workers, whose name is now long forgotten, who left his initials and construction date impressed in the wet concrete of each section as it was completed. It is also possible to pinpoint the provenance of the materials used in its construction: to the west, gritty sand from Grouville (where at least a million tons were removed for various building projects around the island) mixed with aggregate from Ronez and La Crete Quarrys, while the eastern end has been built with sand from the St Brelade's beach itself.

Incidentally, the granite blocks at the eastern end form part of the butts to the old Jersey Militia musketry range established in 1856 but abandoned four years later after the locals protested about inaccurate shooting from Beauport Battery across the far side of the bay. The Germans incorporated a 75mm Pak 40 anti-tank gun casemate into the corner of Le Grouin Point which blended perfectly with the stone used on the butts.

However, as recently as March 1980, the spectre of the Public Works Committee was seen to fall on a short length of seawall at La Haule on St. Aubin's Bay. This isolated piece of concrete was intended to be incorporated into the western end of the anti-tank wall from Beaumont but, although the foundations for the intervening stretch were completed by the OT, the wall above was not. By now the Jersey folk, led by the Channel Islands Occupation

One memorial which cannot any longer be seen on Guernsey is the De Saumarez Monument which stood for over sixty-five years on this plinth in Delancey Park, St. Sampson's. Baron James de Saumarez was a British sailor born in Guernsey on March 11, 1757. He entered the navy at thirteen and was promoted to lieutenant for bravery during the Charleston battle (1776) in the War of the American Revolution and to commander for his part in the battle against the Dutch at the Dogger Bank (1781). Thereafter he helped defeat the French off Dominica in 1782 and was second in command to Nelson in the Battle of the Nile (1798). In short, he was one of Guernsey's greatest sons and his exploits were perpetuated in the erection of the memorial. When the Germans came, they tore it down as it blocked the arc of fire of one of their batteries. After the war, stone from the rubble was taken to Herm to help to rebuild the jetty — a small token gesture to the memory of a great naval man.

No book on the Channel Islands would be complete without a picture of Sark's tiny, two-cell prison. Although rarely used and of no particular significance to the Second World War, we found this picture of a German officer inspecting it and so were able to justify its inclusion. Before the war one could collect the key from the cottage and look inside oneself! (Bundesarchiv).

Society, are well aware of the history attached to seemingly worthless pieces of concrete and the talk of spending £2,000 to demolish it with Cardox (a form of explosiveless demolition which breaks down the concrete by rapidly expanding carbon dioxide) soon brought a flurry of correspondence in the local Press. It was pointed out that it would make much better sense to use the existing German foundations, which had only been covered with a shallow layer of sand, and join the two pieces of wall. Another writer recalled a similar plan some ten years previously to do the self-same thing, recalling that the PWC 'only wanted to blow up the wall because it was there'.

Within a week, the Committee president, Senator John Averty, promised a stay of execution following appeals from the Constable of St. Brelade, Mr. Max de la Haye, and the Occupation Society. As Senator Averty commented: 'We have no hard feelings about this wall at all and it was a very close thing in people's minds whether it was an eyesore or an historic monument'.

Below: **Sark has few wartime relics on display but these guns are to be seen in the grounds of La Seigneurie. The policeman *(above)* is definitely of post-war vintage although we are not so sure about his bicycle! Seconded from the Guernsey force, he is only stationed on the island during the summer to cater for any unruly visitors; crime by the Sarkees being virtually unknown.**

The *Jersey Evening Post* is a virile and investigative organ, always eager to report on matters concerning the history of the island, and its reporters are rarely 'scooped'. Thus, in 1977 a most interesting item appeared in the *Post*, reminiscent of the stories of Japanese soldiers lost in the jungles of the Far East:

'A 54-year-old German soldier has been living in woods in Trinity since the occupation, believing that the Second World War was still being fought. Hans Gruber, dirty and dishevelled, though in good health, was "captured" at 6.30 this morning by a Trinity housewife, Mrs. Mary Ecobichon, while he was attempting to steal food from her secluded farmhouse at Le Champ Chemin, to the east of Les Platons. Before being taken to the General Hospital, where he is now undergoing a variety of tests, Hans took the *JEP* and members of the Trinity Honorary Police to the tiny air raid shelter which has been his home since 1945.

'Nestling deep in the woods at Egypt, less than a quarter of a mile from the cliffs at La Colombiere, the shelter is just eight feet long and five feet high and completely underground. The one entrance, almost completely hidden by overgrown brambles and gorse, is just a hole leading to some steep, slippery steps worn smooth by Hans' steel-tipped boots.

'In an exclusive interview with a German-speaking *JEP* reporter, Hans revealed that the tiny shelter has been his home since February 3, 1945.

'Hans, who was born in Cologne, came to Jersey in 1942 as a private in General Graf von Schmettow's 319th Infantry Division. For the next 2½ years he carried out general patrol duties, mainly on the west coast but, early in 1945, he received special orders from the division's new commander, Majorgeneral Wulf, to act as a look-out on the north coast following a threatened invasion from the sea by British marines. Hans was assigned an area between Belle Hougue Point and Vicard Point and, at the beginning of February 1945, he moved to the air raid shelter which had been built specially for him.

'Incredibly, ever since then, Hans has kept watch for the arrival of British troops, talking to no one and living off the land and food stolen from the few houses in the area. With no fresh orders or radio he has lived in a world of the past, still believing that his homeland was at war with the Allies.

Above: **When tourists hire their cars at Jersey airport, how many realise that Hallmark Hire Cars use an ex-German wartime blister hangar as their garage (Charles E. Brown).**
Below: **Its wooden construction is totally unlike the corrugated iron structures used by the RAF and is a worthy example which we hope will be preserved.**

Still to be seen on Guernsey. *Left:* A 'Ringstand' at the Batterie Mirus gun site, which once mounted a French tank turret — either a Hotchkiss, Renault or Somua. *Right:* This anti-tank obstacle, constructed of railway lines embedded into the sea wall, can still be seen at the eastern end of L'Ancresse Bay.

'His original supplies ran out after just a few months so Hans turned to living off the countryside, snaring rabbits and birds, and collecting berries. He used up most of his ammunition shooting rabbits but still had one bullet when discovered this morning — perhaps saved for a final gesture when the long-awaited invasion arrived.

'Among the rubbish found in the shelter were empty packets of Knackbrot — German hard-tack biscuits — and rotting tins which had once contained German meat.

'Hans' Gewehr 98 rifle was also found in the shelter, along with ammunition pouches, a bayonet, a cut-throat razor, a water bottle and a gas mask — all badly corroded. Hans' uniform, though repaired by him many times, is still usable, and his helmet, which he has spent many hours polishing, is in good condition.

'Before he departed for the hospital, Hans said that he could not believe the war was over as he still heard the V-2 rockets taking off daily — a reference, presumably, to the noise made by BAC One-Eleven jets.

'The television and radio masts which Hans saw going up many years ago — it was, in fact, in 1955 and 1961 respectively — he presumed to be new German communications systems. And what he thought was cheering at a visit by Himmler or Goebbels in 1949 was, in fact, a Royal visit by the then Princess Elizabeth.

'The Constable of Trinity, Mr. Jack Richardson, has taken overall charge of Hans for the time being and has already been in contact with the Home Office and the German Embassy in London. He is also contemplating what further charges, if any, should be brought against Hans who has already been charged with acting in a suspicious manner in Trinity between February 3, 1945, and April 1, 1977.

'Constable Richardson also praised the courage of Mrs. Ecobichon, who used her husband's shotgun to "capture" Hans in her kitchen this morning. "He was pinching my broccoli," was all she would say.'

As stated earlier, the *Post* is a vigorous local paper run by people who still retain a sense of humour — the date of the story: April 1!

Although our more eagle-eyed readers will fault this picture of Hans, when it was reproduced on newsprint at 65-line screen (our ruling is 133 lines to the inch) in the Evening Post, it fooled many people. The newspaper is renowned for pulling pranks on its readers (Jersey Evening Post).

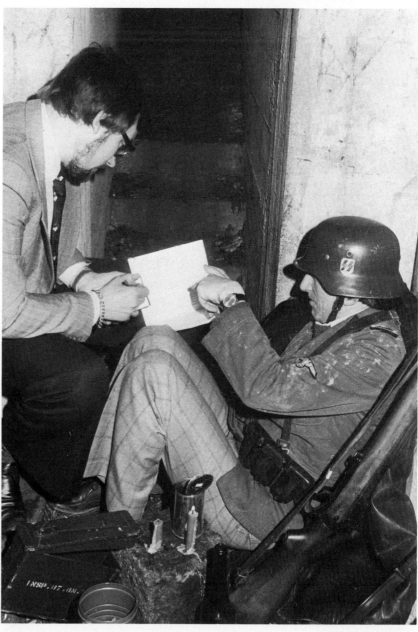

During 1947, Field-Marshal The Viscount Montgomery of Alamein, then Chief of the Imperial General Staff, made extensive overseas tours to the various countries of the New Commonwealth. In May, he slotted in a visit to the Channel Islands and made a triumphant parade in 'Old Faithful' (now at Beaulieu) through St. Peter Port *(above)* on the 23rd followed by a trip to Jersey the following day *(below)*. Although probably not in keeping with the great man's puritanical streak, this was the outstanding Guard of Honour at Samares Manor. The girls came from the Jersey School of Physical Culture (Both H. R. Clayton).

Above: In May 1977, Monty's other Humber 'Victory' (at Coventry Museum) took part in the Military Vehicle Conservation Group Liberation Day tour to Jersey (Evening Post). *Below:* Guernsey came into its own in 1980 with an impressive military vehicle tour organised by Roland and Jill Poole of the MVCG to coincide with the 35th anniversary of the liberation. The big day was May 9 when seventy-five military vehicles, led by the Humber, joined a parade of decorated floats and old vehicles. Who's that in the front seat hiding behind a moustache? Why, 'General' Ken Tough no less! (Guernsey Press).

CHAPTER EIGHT
The Cemeteries

Fort George Military Cemetery, Guernsey. The old burials of the British garrison lie at the back. These pre-1914 graves have individual markers not of the standard Commonwealth War Graves pattern which was only adopted in 1919. In front lie the German interments of the Second World War — originally marked with crosses *(photo below)* but now to be seen with the standard Volksbund Deutsche Kriegsgraeberfuersorge headstones. These are the only German graves still existing in the Channel Islands (CWGC).

Robin Cox, a member of the Jersey Branch of the Occupation Society, has a particular interest in the disposal of the dead. Robin has made an extensive study of wartime burials on the Islands and I am indebted to him for his permission to reproduce here a precis of his research.

By far the most important cemetery in the Channel Islands for the interested visitor is that at Fort George on Guernsey. Not only was the old fortress (since demolished) the major fortification on the island but its cemetery presents a microcosm of all the Islands' military history through the ages. Right from the earliest burials with their individual headstones, through the First World War and on to the Second, the serried ranks of graves, stepped up the picturesque, wooded hillside, both friend and foe, provide a unique location. Only here, in all the Channel Islands, did the German war graves authorities decide to leave their dead where they were originally buried. It was developed as a permanent German war cemetery and an appropriate ceremony marking the event was held on October 5, 1963. Today it is maintained by the States of Guernsey by agreement with the Commonwealth War Graves Commission (CWGC) and the Volksbund Deutsche Kriegsgraeberfuersorge (VDK).

Fort George Cemetery is quite small and most of the German and Allied military dead, and also the foreign labourers working for the Germans, who died on Guernsey, were interred at Le Foulon Cemetery, St. Peter Port, originally an independent burial ground opened in 1858.

After HMS Charybdis was sunk on the night of October 23/24, 1943, twenty-one bodies (one unidentified) were subsequently washed up on Guernsey during the following days. The British sailors were given full military honours by the Germans at their funeral in Foulon Cemetery in November *(photo above)* attended by upwards of 5,000 islanders (Carel Toms Collection). Today an annual service is still held on 'Charybdis Day' in Guernsey in memory of those who lost their lives. *Below:* The temporary markers were replaced after the war when the Commonwealth War Graves Commission laid out the permanent plot.

This was the entrance to the Jerbourg Road German Military Cemetery. This burial ground had been laid out on commandeered private land and it therefore could not be retained as a permanent war cemetery. After the war, the States moved the graves to Foulon and, in 1961, the Volksbund Deutsche Kriegsgraeberfuersorge exhumed all the remains and re-interred them in the ossuary at Mont-de-Huisnes in Normandy. Although ossuaries have been used in the past for mass burial by the French, this is the first time the system has been adopted by the German war graves commission in France.

The sinking of the cruiser *Charybdis* and the destroyer *Limbourne* on the night of October 23/24, 1943 resulted in a number of seamen's bodies being washed ashore on beaches in Guernsey and Jersey. The two ships had been taking part in Operation TUNNEL — a sweep by a force of six destroyers, led by the 5,450-ton light cruiser HMS *Charybdis*, to ambush a German merchant ship being escorted up the English Channel. In the event, it was the British flotilla that was surprised by the escorting German E-boats. The *Charybdis* was hit by two torpedoes . . . and only 107 were rescued from her complement of 569. Also forty-two other seamen lost their lives when HMS *Limbourne* (one of the destroyers) was crippled in a similar fashion. She later had to be sunk by British warships.

Other German military dead were interred in a special military cemetery opened on private land off the Jerbourg Road, St. Martin's, and a similarly special cemetery for the foreign members of the Organisation Todt was opened, again on commandeered private land, at Les Vauxbelets.

During the war, German burials and those of foreign workers in German service at Foulon Cemetery were extensive and these were increased when the States of Guernsey brought in the remains of those exhumed from the Foreign Workers' Cemetery at Les Vauxbelets and the Jerbourg Road German Military Cemetery. All these were exhumed again in 1961 for transfer to the ossuary at Mont-de-Huisnes so that only the British war graves, most in a compact plot, but some scattered throughout the cemetery, remain at Foulon today.

During the Great War, a prisoner-of-war camp had been opened for members of the German forces at Les Blanches Banques, St. Brelade, Jersey. Although the first death took place in the hospital in the barracks in the neighbouring parish of St. Peter and in which parish the corpse was buried, all subsequent fatalities at the camp, caused by either drowning or the epidemic of Spanish Influenza, were buried in the Strangers' section of St. Brelade's Churchyard.

Tended by hotel workers between the wars, this area of the churchyard was the obvious choice for the burials of German troops on Jersey in the Second World War and consequently the first burial of the German military at St. Brelade took place in July 1940, ten days after the arrival of the first German forces. The first interments were located near the graves of those who had died in the Great War but, in February 1942, the first burials took place in two new large blocks, one situated on each side of a hoggin path. In March 1942, the Germans effected a considerable transformation from a simple military cemetery to a 'Heldenfriedhof', or heroes' cemetery. Trees were planted, rough granite paths laid and pairs of wooden gates installed between granite pillars set in the roadside walls of both the cemetery and the Rectory garden. These impressive gates were fitted with large representations of the Iron Cross. The stones of the Great War graves were turned round so that they all faced west, and any of the grave crosses of earlier occupation burials (which had been made of a varnished hardwood by men of the deceased soldiers' units) were replaced by standard black and white painted oak crosses, again depicting the Iron Cross. Fast-growing bushes were planted along the path between the civil and military sections.

In 1943, the SS *Schokland* went down close to the southern shore of Jersey with a high loss of life. The 1,500-ton *Schokland* was a Dutch cargo ship taking Germans on leave to St Malo. Some 200-250 Germans (together with fifteen prostitutes being retired for faithful 'service') were crammed in No. 3 hold in the stern for the short sea crossing when the ship foundered in the early hours of January 5. As the passengers had to negotiate a twenty-foot climb up a vertical ladder and then squeeze through an eighteen-inch-square hatch, it is no wonder that there was a large loss of life. For days afterwards, many bodies were washed ashore and locals recall them being 'stacked like timber' on the Albert Pier before burial. Incidentally, the dead did not include

This exhibit in Castle Cornet shows two of the miniature coffins, about two feet long, used by the VDK to transport the exhumed remains of all their Channel Islands war dead (except those at Fort George) to France.

the captain who swam ashore only to be court-martialled for deliberately sabotaging his ship. It actually hit a submerged rock although there was a report that a Spaniard placed a bomb on board. The ship now lies in seventy-eight feet of water, some 2,000 yards off Noirmont Point, and is now owned by Tony Titterington who bought the salvage rights in 1960.

The gradual burial rate at St. Brelade was jolted unpleasantly with the sinking of the *Schokland* and over the next week twenty-eight spaces were filled. By the autumn of 1943, the first block was nearly full and, aware of the impact of an impressive funeral, the occupying authorities chose the first plot in the new block for the burial of Oberleutnant Zepernick, an efficient and popular officer killed in an RAF raid over northern France whose body was brought back to Jersey for interment. This plot allowed plenty of room for mourners, the firing party and for the wreaths. The new block was full by February 1945, when the first of sixteen bodies was laid in a new area in the Rectory garden.

Above: **St. Brelade's Bay German Military Cemetery as it was at the end of the war. All the remains were also exhumed in 1961** (Societe Jersiaise). *Bottom:* **Today the area is slowly being taken over for parish burials although the old granite path laid out by the German military can still be discerned** *(below).*

Above: **Heroes' Memorial Day, March 21, 1943 (R. H. Mayne).** *Bottom:* **Our 'hero' Robin Cox making his day in April 1979.**

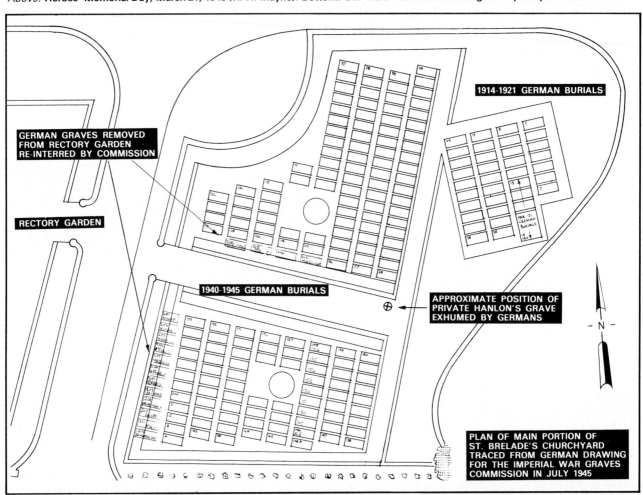

Above: **One of the first steps after the surrender was to replace all the German wartime markers with standard British military white crosses.** *Right:* **Except for these few original wooden markers preserved in Richard Mayne's German Occupation Museum at St. Peter, all the rest went for firewood. New unissued crosses were cut up and made into house nameboards.**

The Strangers' Cemetery in Jersey at the top of Westmount, St. Helier was opened in 1865 as a result of the closure of the earlier strangers' cemetery situated where the Westmount Flats now stand. Cholera and other epidemics had swept through a growing St. Helier in 1794, 1832, 1849, 1856 and 1865, and it was due to deaths from this last epidemic that the need for a new burial ground for soldiers of the garrison and temporary residents in the island was felt.

Prior to 1881, any person wishing to be buried in a parochial churchyard had to be buried with the burial service of the Established Church. Objections by Methodists, Roman Catholics, Jews and Quakers led to the establishment of their own cemeteries where their ministers were able to read their own burial services over the graves. The Strangers' Cemetery fulfilled the function of a free cemetery as it was owned by the States of Jersey, the island's government, and was much used by French Roman Catholics. However, in 1880, a law was passed in England, which had equal force in the Channel Islands, which prohibited Rectors from disallowing visiting ministers reading their services in the parish churchyards.

As a result, strangers to St. Helier were able to be buried in the nearby new Mont-a-l'Abbe Cemetery, which was opened in 1881, and the Strangers' Cemetery became the place of repose only of the very poor or the unknown, which accounted for the very small number of memorial stones as most graves were indicated with temporary wooden markers. As time passed, burials in the cemetery became fewer and fewer until the last interment before the war which took place in 1934.

The place remained neglected until the occupation, when, in 1941, the German forces secretly used it for a small number of unrecorded funerals, probably suicides. These funerals seem to have come as a complete surprise to the cemetery's superintendent, Mr. R. Pallot, who complained to the Constable (or Mayor) in July 1942 that, while taking soundings preparatory to the digging of graves for two Frenchmen who had died from accidental gassing at Bel Royal, some new coffins had been discovered of which no record existed.

With the arrival of the Organisation Todt and its management in late 1941, it was decided to use the Strangers' Cemetery as the burial ground for non-German members of the OT and it was renamed Westmountfriedhof.

The first three persons to be buried there (on Wednesday, February 11, 1942) were Frenchmen who had been found drowned. A

The body of Sang. Hermangiledo, one of the Organisation Todt workers, is laid to rest in the Spanish section of the Strangers' Cemetery at Westmount on April 8, 1942. Note the presence of three uniformed OT officials. Mr. J. B. Le Quesne, the local undertaker (who held the contract for OT burials), stands near the coffin (Francisco Font).

Following the exhumation of all the remains in 1961, the site was chosen for Jersey's first crematorium and the former burial area was landscaped and the wall and gate altered. Stage managed by Margaret Ginns, crematorium Superintendent Roy Edlin takes the place of the cemeteries' superintendent while his staff join the OT (M. Ginns).

fourth Frenchman recovered from the sea was buried eight days later. The next four were Algerians who had died through eating water-hemlock taken from the ditches in St. Peter's Valley, followed by four Spaniards and then two more Frenchmen, the burial of whom brought to light the unrecorded use of the cemetery by the German military.

On August 21, 1942, the first of the Russian impressed labourers, 28-year-old Gregory Gnida, was laid to rest, the first interment in the south-west square. Burials in this plot had increased by sixty-nine by April 1943 with a further four more being added between then and June 30, 1944. These last had died from lingering disease after all the other Russians had been taken back to the French mainland to work on the V-weapon sites. The Russian section of the cemetery was consecrated by Bishop Methode of the Russian Orthodox Church in Paris on March 12, 1943, by which date sixty-two Russians had been buried in the square. (A total of seventy-two Russians were eventually buried in the plot.)

The care with which the Germans divided the cemetery into national sections is demonstrated by the trouble that was caused when a corpse was found floating in the sea near Castle Street on October 14, 1942. After a German inquest, it was decided that it was that of a Russian worker who had escaped from Elizabeth Castle, which at that time was used as a punishment centre by the OT, and the body was duly buried in the Russian section on the 17th.

However, the OT were not satisfied that the body was that of one of their employees and they ordered an exhumation, which took place on the 19th, as did an examination, followed by reburial on the same day in the same grave. Still not satisfied, as to lose a man reflected on the efficiency of the OT paperwork, they ordered a further exhumation and examination on the 21st with the result that a new grave was dug in the French section, into which the remains were placed, apparently vindicating the OT and their personnel section.

At the Strangers' Cemetery, an effect of the sinking of the *Schokland* in January 1943 was to cause the opening of a Dutch section for two of the crew and, within seven days, the burial of three Frenchmen, one Algerian, one Pole, one Belgian and two Russians. They were joined by an unknown seaman who was found on the beach at L'Etacq, and who was buried in a section on his own on the instructions of the Attorney-General who had learned of the Germans' pride in keeping the nationalities separate.

Also as a result of the loss of the *Schokland* and the activity at the German military cemetery at St. Brelade in which the twenty-eight military victims were buried, it was decided to exhume two soldiers who had committed suicide and have them reburied in a new section at Westmount.

The first to be moved was a 32-year-old Obergefreiter who was exhumed and reburied on January 19, 1943. By Liberation Day, fourteen graves had been filled in this section, eight through suicide, one through illness and five through execution, the last taking place on May 8, 1945.

One of the small contingent of Italian volunteers had been buried at Westmount on December 22, 1943 but, upon representation, it was decided that an Italian section should be opened at St. Brelade with the result that the body was transferred there on January 19, 1944.

Prior to the opening of the crematorium in the Strangers' Cemetery grounds in 1961, residents wishing that sort of funeral had to be sent to Guernsey, where such facilities were made available at Foulon Cemetery in 1929. The crematorium was dedicated by the Dean on December 19 and the first cremations took place the following day.

In 1970, the Department of Public Health

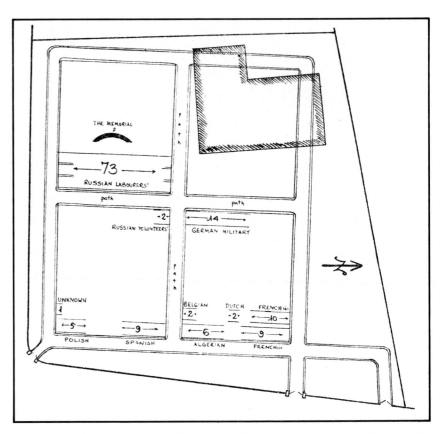

The Strangers' Cemetery on Westmount showing the relative position of the crematorium erected in 1961 and the memorial *(below)* dedicated in 1970 (Evening Post).

erected a memorial on the Russian plot to all the foreign workers who died during the occupation. The granite memorial bears three engraved stones: one from the Soviet Government; one from the French Consulate and one on behalf of the Spanish republican community in Jersey. Each year the Soviet Embassy in London sends a senior representative to the wreath-laying ceremony but the 30th anniversary of the liberation saw the attendance of a special visitor: Mr. Vasily Marempolsky, a former slave worker, who returned from the Soviet Union with two other members of the Soviet War Veterans Committee. Mr. Marempolsky had been deported from his native Ukraine in 1942 when fifteen and he was put to work building railways and erecting the sea wall around St. Aubin's Bay. He was welcomed back by Mr. Vincent Gasulla-Sole, another former slave worker, now resident in Jersey and formerly of the Spanish Republican Army. Mr. Gasulla-Sole had nursed his friend in the camp inmates' hospital (which was located in the Girls' College), when he was admitted with dysentery in 1943. Other wreaths were laid by Colonel Alexei Toujikov of the Central Committee of the Soviet War Veterans Association and a partisan leader during the siege of Leningrad; Captain Vladimir Khuzhokov of the Soviet Embassy in London; a member of the States parliament Deputy Norman Le Brocq, and Mrs. Stella Perkins on behalf of the Jersey Communist Party.

On June 3, 1943 the body of Sergeant Dennis Butlin was picked up at La Pulente and arrangements were made for his funeral at Mont-a-l'Abbe Cemetery, just down the road from the Westmount Strangers' Cemetery. When a second airman, Sergeant Abraham Holden, was recovered at Samares, it was decided to hold a double funeral which took place on June 6. The previous evening the two coffins had 'lain in state' in the hospital chapel and hundreds of islanders had paid them their last respects. Hundreds more lined the route to the cemetery and two lorries had to be laid on to transport all the wreaths.

As regards the foreign workers of the Organisation Todt, prior to October 1941, these were either German workers of the building and contracting firms which had successfully tendered for military work, or French unemployed who had volunteered for work at French labour exchanges and had found themselves sent to Jersey. This situation changed dramatically in the autumn of 1941 when forced and voluntary workers from most European countries arrived. After early repercussions, it was decided that as Italians were allies of the Germans they were thus eligible for burial with the German troops at St. Brelade. This decision led to the exhumation of one Italian soldier from Westmountfriedhof and his reburial at St. Brelade in a small area set apart for Italians.

The first Allied military personnel to be buried in Jersey in the last war were airmen who were washed ashore after being brought down into the sea as a result of aerial engagements. Sergeants Butlin and Holden were found on June 3 and 5, 1943 respectively and were buried in Block O of the new Mont-a-l'Abbe Cemetery. Their bodies were followed by that of an American airman, Sergeant A. Poitras who was found at Bonne Nuit on September 24, 1943 and three days later he too was buried in Block O. Block O was also used for the *Charybdis* victims who were washed up on Jersey.

The sudden arrival of the bodies of so many members of the Allied forces spurred on the Jersey authorities to find a proper military cemetery for Allied dead. The choice fell on part of Howard Davis Park near St. Luke's Church (formerly named Plaisance, after the large house which once stood there) and this land was dedicated by Dean Matthew Le Marinel on November 26, 1943. The first body to be buried there was that of Petty Officer Thomas on November 19, 1943 in grave number 33. The next graves were then filled with thirty coffins which had been exhumed from Mont-a-l'Abbe to which were added the remains of one soldier brought from St. Brelade. He was Private Hanlon, a guard at the prisoner-of-war camp at Les Blanches Banques in the First World War and who had died of disease at the military hospital which had been set up in the then-quite-new Brighton Road School. As a resident in, but a stranger to, the parish of St. Brelade, he had been buried in the old 'strangers' part of the churchyard and in setting out their cemetery in 1942, the Germans had, in fact, laid a path over his grave. After the sinking of the

Above: **This was the first burial of Allied servicemen on Jersey during the Second World War and the Luftwaffe provided bearers and a graveside firing party. (The two deaths had occurred in unconnected incidents) (Evening Post).** *Below:* **When the new Allied War Cemetery was opened in Howard Davis Park in November 1943, the remains were exhumed and re-interred (see page 232) but Robin Cox was able to pinpoint the original grave location, since used for later civilian burials.**

Charybdis, most of the burials at Plaisance were those of the bodies of American sailors or airmen washed ashore.

With the liberation, control of the German military cemetery passed from the Standortskommandantur to the Imperial War Graves Commission. One of their first acts was to transfer the sixteen German coffins from the Rectory garden at St. Brelade across the road into the main cemetery. There they were reburied wherever there was room — along the walls and paths. At the same time, the Swastika which formed the central part of the Iron Cross decoration on the crosses was painted out, but only on those which marked those coffins which had been reinterred. Later all the German crosses were replaced with white British military crosses.

The Americans had all their dead removed from the Howard Davis Park in June 1946 and, in 1949, the French Government arranged for the exhumation of remains for five families who wished to have their relatives brought back from the Strangers' Cemetery. Three more private exhumations took place in the 1950s: an Austrian from Westmount in 1957, a German from St. Brelade and a Belgian, Squadron Leader Gonay, whose RAF plane had crashed on a farm at St. Ouen in June 1944.

Above: **On May 10, 1946, a ceremony of remembrance was held in the Allied War Cemetery in Howard Davis Park to mark the first anniversary of the liberation at which Sir Alexander Coutanche, still Bailiff of Jersey, laid a wreath beside the stone dedicated on November 26, 1943 (Evening Post).** *Below:* **The park today is a haven of peace and beauty — the stone is now surrounded by shrubs (M. Ginns).**

Schoolchildren lined the wall of St. Luke's Church — they had tended the graves during the war (Evening Post).

In 1961, all the Channel Islands' authorities were approached by the Volksbund Deutsche Kriegsgraeberfuersorge — the German equivalent of the Commonwealth War Graves Commission — and it was decided that, apart from those at Fort George in Guernsey, all the remaining military and Organisation Todt war dead should be transferred to a new charnel house at Mont-de-Huisnes in Normandy near Mont St. Michel. This transfer was effected in the latter part of 1961 and marked the construction of the first German tomb of its kind ever to be built in France. The remains of 11,931 dead from the Second World War, previously buried in the Islands and the French Departments of Morbihan, Ille-et-Vilaine, Mayenne, Sarthe, Loir-et-Cher, Indre-et-Loire, Vienne and Indre, have been concentrated there in thirty-four vaults. It was dedicated on September 14, 1963.

Above: **The repatriation of the American servicemen buried in Howard Davis Park via the Victoria Pier in June 1946 has left a rather confusing state of affairs of which the visitor to the cemetery must be made aware. Although not marked as such, the crosses retained by the States, which purport to record American burials, are symbolic only, although this is against CWGC policy** (Evening Post).

On a hillside near Huisnes-sur-Mer lie the remains of the German war dead removed from the Islands (VDK).

Only one execution of a judicial nature was carried out by the Germans in the Channel Islands during the war, that of Francois-Marie Scornet, a young Frenchman who, with a number of his countrymen, landed in Guernsey under the misapprehension that it was the Isle of Wight, where they had hoped to join the Allies. At a military court hearing in Jersey, Scornet took responsibility for the whole venture. He was found guilty and sentenced to death, being executed by a firing squad against a tree in the grounds of St. Ouen's Manor, in the north-west of the island, on March 17, 1941. His body was then transported to town, lying in the mortuary chapel at Almorah Cemetery (opened as 'The St. Helier's General Cemetery' in 1854) while a grave was dug. He was buried quietly in a Roman Catholic holding grave in the extreme north-western corner of the cemetery on the next day, March 18. As it was a holding grave, his coffin was covered by two others in the succeeding months. This was realised by the undertaker, Mr. John Barette Le Quesne, in 1944 and the two additional coffins were removed and buried elsewhere in the cemetery.

Scornet's remains were exhumed in 1945 and with much pomp carried back to France, but even today an annual memorial service is held in the grounds of St Ouen's Manor by the French deportees association.

In December 1940, sixteen young Frenchmen, led by the 20-year-old Francois-Marie Scornet *(right)* **from Ploujean, Finisterre, sailed from the Brittany coast in the SM401 'Bukara' in an attempt to reach England and join the Free French forces. Their knowledge of navigation was minimal and, having sighted land on the 16th, they came ashore boldly singing the Marseillaise. To their horror they were immediately captured by the Germans for what they had taken to be the Isle of Wight was, in actual fact, Vazon Bay, Guernsey. Scornet immediately accepted the blame and, as a result, was taken to Jersey (then the location of the B.d.b.K. HQ) by the Geheimfeldpolizei to stand trial. He was imprisoned in the civilian jail in Patriotic Street, St. Helier. The old prison buildings were demolished in 1975 after the gaol had been moved to a new purpose-built structure at La Moye.** *Below:* **This was the site in April 1979 with work proceeding on an extension to the General Hospital. (Painting of Francois-Marie Scornet by George Campbell.)**

The trial was held in the Old Committee Room of the Royal Court on February 2 and 3. Scornet had bravely taken on himself responsibility for the entire escapade even though he must have known the inevitable result of so doing. Francois-Marie was adjudged guilty of 'wilfully supporting England in the war against the German Empire' (as the Germans claimed he was a released ex-French prisoner-of-war) and he was sentenced to death. He was taken to the nearby Grand Hotel while preparations were made for his execution. *Above:* How many residents today realise that a condemned man was once held in their comfortable hotel?

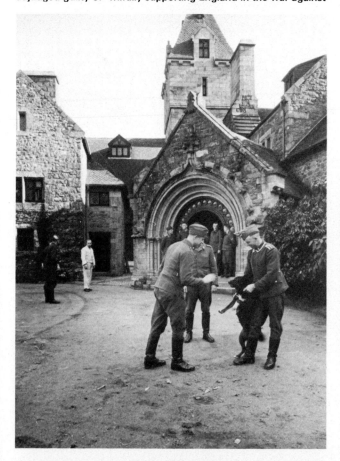

Left: Some 4½ miles away in the parish of St. Ouen, the old Manor House had been requisitioned by the Germans from the de Carteret family in whose possession it had been for generations. It lay in a quiet secluded part of the island and the embankment of La Grande Route de St. Ouen, which ran along the eastern boundary of the estate, formed a convenient stop-butt (Bundesarchiv). *Right:* Today the courtyard in front of the main door has remained almost unaffected by the passage of time. Not so other parts of the house — see overleaf. Photographs by kind permission of Philip Malet de Carteret.

On February 13, while Scornet was languishing in St. Helier, his comrades were taken to Cherbourg on the coastal guard boat 'Bernhard von Tschirsky' and handed over to the GFP for attention. Then on March 6 a disastrous fire occurred at the Manor. The Germans had installed barrack room stoves in the oldest part which dated back to the 12th century. One whole wing and drawing room were completely gutted with the loss of many family heirlooms and much valuable furniture. When 'Albert Speer' in the guise of Michael Ginns (at ease on the sofa in the photo *above right*) visited the house, although much of the restoration work had been done, the drawing room still had to be completed. In spite of the damage, the execution site was not changed. At six o'clock on the morning of Monday, March 17, 1941, Father Mare was summoned to the Grand Hotel where he was told by the Germans that one of his faith required the last rites. He was taken out to a lorry where a young Frenchman was sitting next to his coffin. Taking his place, he was driven to St. Ouen's Manor . . . through the impressive gateway . . . to where the firing squad was waiting.

The stone marks the spot where Scornet died at 8.20 a.m. The tree, in front of which he stood, has since been cut down.

Scornet's body was taken to the chapel at Almorah Cemetery where his coffin rested overnight while a grave was being prepared.

PUBLICATION:

The population is herewith notified, that FRANÇOIS SCORNET, born on May 25th 1919, residing in Plouiean (Department Finistere) has been sentenced

TO DEATH

by the German War Court and has been shot on March 17th, 1941. This had to be done, because of his favouring the actions of the enemy by wilfully supporting England in the war against the German Empire.

German War Court.
March 23rd, 1941.

Above: This was the site of his grave for the next four years before his remains were exhumed *(below)* for returning to France on September 18, 1945.

'My dear parents, my very dear parents. The end of my life is at hand. I am going to die for France facing the enemy bravely. I forgive everyone. Vive la France. Vive Dieu. For the last time I embrace you.' Francois-Marie Scornet's last letter received by his parents after his death. This stone was erected by the Amicale des Anciens Deportes de l'Ile Anglo-Normande d'Aurigny at St. Ouen's Manor on October 15, 1949.

As we have seen in earlier chapters, Alderney had a history all of its own — a darker side of which we may never know or comprehend the complete story. All the suffering endured by the forced labourers made itself visible here on Longy Common where Charles Brown found their overgrown graves. In May 1945 they were marked with simple crosses but in 1961 they were taken with the German war dead to France. Today the area where the former OT cemetery was located is still wild and windswept.

At first, Alderney's military and Organisation Todt dead were buried in St. Anne's Churchyard in the town. However, as the death rate increased, more room had to be found and the military laid out an area off the Longy Road, next to Alderney's Strangers' Cemetery, the centrepiece of which was a large granite stone, finely dressed and carrying a quotation from the Book of Job. In all seventy German soldiers were buried on the island during the period of the occupation. After the ground was cleared in 1961, the memorial stone was lifted over the wall and left resting against the wall in the Strangers' Cemetery.

Almost next to the Strangers' Cemetery is a graveyard for Roman Catholic residents and, on Longy Common itself, a roughly delineated area was set aside in which the OT workers were buried. Today an area of uneven ground, left after the exhumations, defines the boundaries of the burial ground.

Aerial photography shows up much which is invisible or undetectable from the ground. *Above:* No. 541 Squadron photograph of February 1, 1945 of Longy Common with the polygonal-shaped cemetery behind the anti-tank wall constructed around the bay (Crown Copyright). *Below:* Our aerial photograph still shows the exact boundaries of the cemetery.

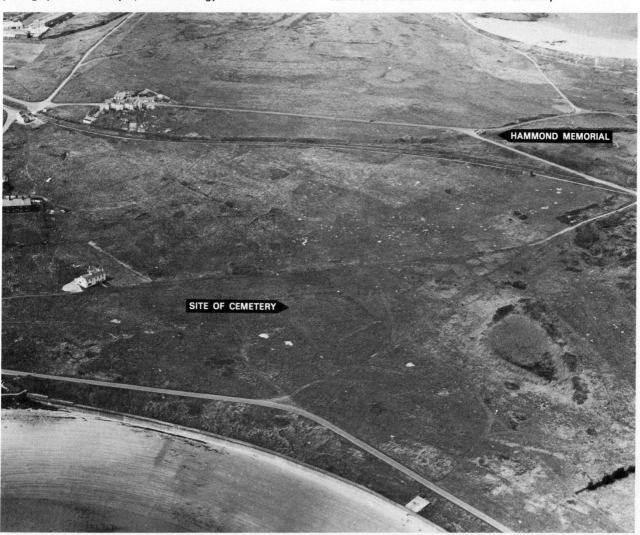

Above: The German military cemetery for the majority of the dead on Alderney was situated at Valongis behind the Strangers' Cemetery (Alderney Journal). *Right:* When the remains were exhumed by the VDK in 1961, the memorial stone was dumped over the wall where we photographed it in 1979. *Below:* The area once used for the burials is now a barren plot (Colin Partridge).

Above: **This impressive memorial to all the foreign workers who died on Alderney was built by the Hammond family in 1966.**

Sark has an interesting history and the military section of its graveyard at St. Peter's Church lies in the north-west corner of the new yard across the road. In a now empty area five Germans were buried but, for some as yet unknown reason, many soldiers who died on Sark were taken across to Guernsey for burial.

It was on Sark that the German Army doctor, who also attended to the islanders' needs, was murdered. The suspicion fell upon his batman, all the more so when he was found dead down a well. The doctor's body was taken to Fort George Cemetery, whereas that of the batman was buried at night in a field belonging to the tenement 'La Connellerie' north of La Seigneurie. It appears that the doctor was, in fact, murdered by a soldier who the MO had refused to pass as unfit for the Russian front and who left the island in a draft the next morning. He confessed to the murder while himself dying and, as a result, the body of the batman was exhumed and buried at Le Foulon in Guernsey but was exhumed again after the war and taken with the others from that cemetery to Mont-de-Huisnes.

In the Sark churchyard are to be seen several noteworthy stones. One is to a Dutch sailor found washed up on a Sark beach and another to Corporal R. Bellamy, one of the two French Commandos who were killed in the raid on Sark in December 1943. (The other, Private Andre Dignac, was exhumed and taken back to France by his family after the war.) There is also the grave of Major John Skelton — a very poignant reminder of the terrible extremes to which people were driven by the introduction of the forced transportation of English residents to Germany in September 1942.

This ordinary stone surround in St. Peter's Churchyard extension marks the grave of Major Skelton who committed suicide rather than be deported with his wife from Sark.

However, to see the most famous grave of all, one must walk across the road. There, in a shady corner beside the church lies a stone simply marked: Sibyl Mary Collings D.S.O., widow of D. J. Beaumont — R. W. Hathaway, 13:1:1884 — 14:7:1974.

German graves marked by the Imperial War Graves Commission at Foulon Cemetery prior to exhumation in 1961 (CWGC).

CHAPTER NINE
The War Graves

The following list has been compiled with the assistance of the Commonwealth War Graves Commission and is based on their published registers for First and Second World War Casualties. The Commission is always very pleased to answer queries on cemeteries and memorials commemorating Commonwealth war dead and can be contacted at 2 Marlow Road, Maidenhead, Berkshire SL6 7DX.

It is probably not generally known that the First World War did not officially end in 1918 nor the Second World War in 1945 although these are the popularly accepted dates. Although Armistice Day was November 11, 1918 the 'Termination of the Present War (Definition) Act' laid down that the First World War would officially cease when an Order in Council under that Act declared the war ended. An Order in Council was duly made declaring August 31, 1921 to be that date. It was therefore decided that Commonwealth service personnel who died within the period August 4, 1914 and August 31, 1921 would be classified as casualties of the 1914-1918 War.

Similarly, as late as 1947, it was not possible to foresee when a date would be arrived at to clearly define the end of the Second World War. Nevertheless peace treaties had been signed with many countries with which the Commonwealth had been at war and it was therefore decided that the participating governments, under the auspices of the Commonwealth War Graves Commission Charter, should be asked to agree the date of December 31, 1947 and that a Supplemental Charter be requested fixing this date for the purpose of commemorating Commonwealth war casualties. The participating Governments agreed and the Supplemental Charter was granted. The reason this date was chosen by the Commission was that it was approximately the same period of time after the surrender of Germany (thirty-one months) as that following the Armistice in the First World War (thirty-two months).

It must also be remembered that for some time after the cessation of hostilities, men serving in the Commonwealth armed forces were still losing their lives in many parts of the world or dying of wounds received prior to the cessation. Many died in their homeland, others in military hospitals overseas. Others were losing their lives either as occupation forces or while fighting in small operations such as in Russia after the First World War or in Palestine after the Second. There were additional casualties suffered from mine clearance and bomb disposal. These then were the reasons for the inclusion of casualties in the CWGC registers of those who died after the generally accepted dates for the termination of both wars.

The war graves in the Channel Islands, as described in Chapter Eight, saw considerable reorganisation in 1961 when the Volksbund Deutsche Kriegsgraeberfuersorge exhumed the remains of German and foreign workers for reburial in the German War Cemetery at Mont-de-Huisnes. This total included eight servicemen and one interned German national from the First World War who were reburied in the German WWI cemetery at Cernay near Mulhouse, France.

In order that this should be a complete listing of all the war dead in the Islands, we have included the 130 dead from the Great War together with all 206 from the Second World War, including 111 German burials, scattered throughout twenty-nine different cemeteries and burial grounds.

GERMAN AND OTHER WAR DEAD REMOVED FROM THE ISLANDS IN 1961 BY THE VOLKSBUND DEUTSCHE KRIEGSGRAEBERFUERSORGE

	Navy	Army	Air Force	OT and others Known	OT and others Unknown	Totals Known	Totals Unknown
Alderney German Cemetery							
German	17	52**	2	—	—	67	4
St. Anne's Churchyard							
German	—	—	—	8*	1****	8	1
Foreign Workers							
(Belgian)	—	—	—	1	—	1	—
(French)	—	—	—	2	—	2	—
(Dutch)	—	—	—	5	—	5	—
(Polish)	—	—	—	1	—	1	—
(Russian)	—	—	—	45	—	45	—
(Spanish)	—	—	—	1	—	1	—
(Yugoslav)	—	—	—	6	—	6	—
(Nationality unknown)	—	—	—	—	3	—	3
Alderney Russian Cemetery							
Foreign Workers							
(French)	—	—	—	4	—	4	—
(Russian)	—	—	—	243	64	243	64
(Nationality unknown)	—	—	—	2	1	2	1
Castel (St. Mary's) Churchyard, Guernsey							
German	—	—	—	1*	—	1	—
Cobo (St. Matthew's) Churchyard, Guernsey							
German	—	—	—	2*	—	2	—
Forest Parochial Cemetery, Guernsey							
German	—	2***	—	1*	—	2	1
Foreign Workers							
(French)	—	—	—	2	1	2	1
St. Andrew's Church Cemetery, Guernsey							
German	—	—	—	1*	—	1	—
St. Brelade's Churchyard, Jersey							
German	28	184***	1	1	—	213	1
St. Helier's (Mont-a-l'Abbe) Strangers' Cemetery, Jersey							
German	1	15	—	—	—	16	—
Foreign Workers							
(Belgian)	—	—	—	2	—	2	—
(French)	—	—	—	17	2	17	2
(Polish)	—	—	—	4	1	4	1
(Russian)	—	—	—	71	2	71	2
(Spanish)	—	—	—	9	—	9	—
(Nationality unknown)	—	—	—	1	—	1	—
St. Peter Port (Foulon) Cemetery, Guernsey							
German	36***	101***	—	1*	1*	136	3
St. Sampson's Churchyard, Guernsey							
German	—	2	—	—	—	2	—
St. Saviour's Parochial Cemetery, Guernsey							
German	—	9	—	—	—	9	—
Sark Parochial Cemetery							
German	1	4	—	—	—	5	—
Vale (Domaille) Church Cemetery, Guernsey							
German	—	—	—	2*	—	2	—
Total	83	369	3	433	76	880	84

* Recorded only as German Forces
** Includes four unknown
*** Includes one unknown
**** When the remains came to be exhumed, one grave could not be located and is now commemorated by a plaque at Mont-de-Huisnes

ALDERNEY

ST. ANNE'S CHURCHYARD

There are six burials from the First War and one from the Second. Five are in the military plot to the east of the church. Up to 1961 there were, in addition, sixty-four foreign workers buried here, making a total of sixty-five burials of the 1939-1945 War. Of these one was Belgian, two were French, five Dutch, one Polish, forty-five Russian, one Spanish, six Yugoslav and three (unidentified) of unknown nationality.

FIRST WORLD WAR

BARKER, Pte. T., 16376. 11th Bn. North Staffordshire Regt. February 28, 1915.
BOULTON, 2nd Mate H. D. SS *Pascal* Mercantile Marine. December 17, 1916. Buried south-west of church.
DOWNES, Pte. Frank, 19107. 4th Bn. North Staffordshire Regt. March 3, 1916. Age 19.
POCOCK, Gnr. A., 8689. Guernsey and Alderney District Establishment, Royal Garrison Artillery. April 19, 1917.
ROBERTS, Pte. Fredrick, 8209. 'C' Coy. 4th Bn. North Staffordshire Regt. Died of sickness, September 28, 1914. Age 33. Son of the late William and Martha Roberts of Hill Top, West Bromwich, Staffs.
SQUIRES, Spr. E. R., 250202. Royal Engineers. June 9, 1921. Son of R. Squires of Lower High Street, Alderney. Buried close to military plot.

SECOND WORLD WAR

ONIONS, Spr. George Edgar, 5059418. 259 Field Coy., Royal Engineers. June 21, 1945. Age 22. Son of Edwin and Beatrice Onions of Birches Head, Stoke-on-Trent. Buried in Row A6, Grave 1.

ALDERNEY ROMAN CATHOLIC CEMETERY

This is sited on the road from St. Anne to Longy. There is one burial from the 1914-1918 war.

COLLINS, Gnr. James, 2303. 109th Coy. Royal Garrison Artillery. Died of sickness, September 29, 1917. Age 55. Husband of the late Mary Collins. Buried in south-west part.

GUERNSEY

COBO (ST. MATTHEW'S) CHURCHYARD, CASTEL

There are three war burials from 1914-1918 and one airman from 1939-1945. At one time there were two men from the German occupying forces but the remains were exhumed in 1961. The war graves are north-east of the church which is situated on the Rue de l'Eglise in the Catel parish near Cobo Bay.

FIRST WORLD WAR

LE COCQ, Sgt. Y. M. F., 3144. 6th Bn. Royal Irish Regt. August 19, 1916.

TORODE, Pte. W. D., 1558. 'A' Coy. 2nd Bn. Royal Guernsey Light Inf. Died of wounds, May 5, 1918. Age 22. Son of Daniel and Alice Torode of Grandes Rocques, Catel, Guernsey.

TOSTEVIN, Pte. W., 2109. 2nd Bn. Royal Guernsey Light Inf. April 6, 1918. Age 19. Son of William Le Prevost Tostevin and Louisa Tostevin of La Lande, Catel, Guernsey.

SECOND WORLD WAR

LE CHEMINANT, Sgt. John Eric, 922801. RAF (VR). Serving in HMS *Daedalus*. November 23, 1945. Age 35. Son of Norman and Ethel May Le Cheminant; husband of Gwendoline Blodwen Le Cheminant of St. Peter Port.

FORT GEORGE MILITARY CEMETERY, ST. PETER PORT

Fort George was the major fortress on the island of Guernsey and lies to the south of St. Peter Port. The cemetery is situated just outside the ramparts in clifftop woodland and contains war graves of both world wars. There are only two Commonwealth burials from the 1939-1945 war, an airman of the Royal Air Force and an airman of the Royal Canadian Air Force. There are also nineteen German sailors, eighty-eight German soldiers and four German Merchant Navy seamen buried here, making a total of 113 burials.

After the First World War, a Cross of Sacrifice and a memorial plaque were erected here in honour of all members of His Majesty's Forces buried in Guernsey, Alderney and Sark. These now serve to commemorate also those 1939-1945 war casualties who are buried on these islands. The Cross of Sacrifice is inscribed 'Their glory shall not be blotted out' and on the memorial plaque are these words: 'The Cross of Sacrifice is erected to the honoured memory of those members of His Majesty's Forces who gave their lives during the wars of 1914-1918 and 1939-1945 and are buried in the islands of Guernsey, Alderney and Sark'.

FIRST WORLD WAR

BLANFORD, Lt. Frank Burrell. 2nd Bn. Royal Guernsey Light Inf. April 25, 1917. Age 39. Son of Thomas and Amy Blanford. Served in the South African Campaign. Born in London. Grave H 170.
BLOOR, Pte. S., 7560. 4th Bn. North Staffordshire Regt. January 3, 1915. Son of William and Martha Bloor of 7 Providence Square, Hanley, Stoke-on-Trent. Grave L 188.
BROWN, Pte. F., 27314. 4th Bn. North Staffordshire Regt. June 23, 1916. Grave N 202.
CLARK, Sgt. N., 9014. 2nd Home Service Bn. North Staffordshire Regt. February 13, 1917. Age 27. Husband of Florence E. S. Clark of 40 Coventry Rd., South Norwood, London. Grave O 210.
CLEWES, Pte. C., 66190. 17th Bn. Royal Defence Corps. September 17, 1917. Age 28. Son of Margaret Tempest (formerly Clewes) of 148 Staffordshire St. A., Sacriston, Durham, and the late Matthew Clewes. Grave O 208.
FITZGERALD, Pte. A., 8128. 4th Bn. North Staffordshire Regt. January 11, 1915. Age 22. Son of William Henry and Mary Ann Fitzgerald of 15 Bright St., Hanley; husband of S. J. Tellright (formerly Fitzgerald) of High St., Hanley, Staffs. Grave M 199.
JONES, Pte. M., 24593. 2nd Home Service Bn. North Staffordshire Regt. July 20, 1917. Grave O 211.
JONES, Pte. Samuel, 19359. 4th Bn. North Staffordshire Regt. January 17, 1916. Grave M 194.
JONES, Pte. Thomas, 27383. 2nd Home Service Bn. North Staffordshire Regt. July 11, 1917. Age 37. Son of Thomas and Sarah Jones of 71 Margaret St., Treherbert (Rhondda), Glam. Grave O 207.
LASAUCE, Pte. A., 623. 2nd Bn. Royal Guernsey Light Inf. June 23, 1917. Grave O 209.
LINDLEY, Pte. J., 7719. 2nd Bn. Yorkshire Regt. August 14, 1914. Age 30. Son of Alfred Lindley of 12 Airley Top, Huddersfield. Grave L 191.
LOMAX, Pte. J., 27340. 4th Bn. North Staffordshire Regt. Died of sickness, July 12, 1916. Age 24. Son of Emma Lowe (formerly Lomax) of 10 Firs Rd., Carrington Lane, Ashton-on-Mersey, Manchester, and the late James Lomax. Grave N 201.
MASCALL-THOMPSON, Capt. Cecil. (Hon. Maj., Cdg. RASC in Guernsey). Royal Army Service Corps, formerly The Rifle Brigade. January 21, 1916. Age 66. Son of Julius Henry Thompson, RN; husband of Antoinette Sarah Mascall-Thompson (nee Stedman) of 172 Clapham Rd., Clapham, London. Educated at Eton. Grave H 109.
MORAN, L/Cpl. J., 8381. 4th Bn. North Staffordshire Regt. February 15, 1915. Grave M 198.
MURPHY, Pte. M., 7111270. 3rd Bn. Royal Irish Regt. July 4, 1921. Grave Q 220.
OSBORNE, Pte. H., 574. 2nd Bn. Royal Guernsey Light Inf. November 17, 1918. Grave Q 222.
PEARSE, CQMS W., 26468. Royal Engineers. January 1, 1918. Grave P 216.
SALT, Pte. J. J., 63000. 17th Bn. Royal Defence Corps. November 28, 1917. Grave P 217.
SLEETH, Pte. M., 66309. 17th Bn. Royal Defence Corps. January 20, 1918. Grave P 214.
THRESHER, Sgt. Thomas Richard, 5578. 2nd Bn. Royal Irish Regt. November 4, 1918. Age 35. Husband of Mabel S. Thresher of 'Strathblane', 11 Belgrave St., Petersham, Sydney, Australia. Grave P 212.
TOSTEVIN, Pte. S., 2513. 2nd Bn. Royal Guernsey Light Inf. November 2, 1918. Grave P 213.
WILSON, L/Cpl. G., 2324. 2nd Bn. Royal Guernsey Light Inf. November 7, 1918. Age 43. Son of John and Jane Wilson; husband of Christiana Wilson. Born at Bingley, Yorks. Grave Q 223.
YATES, Pte. A. J., 8382. 4th Bn. North Staffordshire Regt. December 6, 1914. Grave L 190.

SECOND WORLD WAR (BRITISH)

BEACH, Sgt. (Flt. Eng.) William, 1567112. RAF (VR). No. 61 Sqdn. January 28, 1944. Age 21. Son of William and Caroline Beach of Westerhope, Newcastle-on-Tyne. Grave 32.
BIDDLECOMBE, Flt. Sgt. (Pilot) Conrad Peter Vivian, R/164649. RCAF. No. 226 (RAF) Sqdn. June 19, 1944. Age 22. Son of Mr. and Mrs. Conrad Herbert Biddlecombe, and stepson of Elizabeth Biddlecombe of New York City, USA. Grave 34.

SECOND WORLD WAR (GERMAN)

BANCK, Major Bruno, November 19, 1942. Age 55. Grave 85.
BASTIAN, Gfr. Karl, March 28, 1942. Age 32. Grave 60.
BEETZ, Uffz. Hans-Joachim, September 3, 1943. Age 22. Grave 113.
BEHRENS, Gefr. Franz, February 18, 1943. Age 20. Grave 74.
BIELESBERGER, Leut. Josef, July 12, 1942. Age 20. Grave 51.
BIEROTH, Gefr. Alois, February 4, 1942. Age 35. Grave 67.
BLASCHEK, Major F. R., November 7, 1941. Age 36. Grave 143.
BLEYER, Uffz. August, October 4, 1942. Age 28. Grave 42.
BLOCH, O.T.Mann Peter, March 9, 1943. Age 41. Grave 96.
BLUM, Feldw. Albert, January 20, 1942. Age 47. Grave 64.

BOES, O.Gefr. Peter, October 16, 1942. Age 26. Grave 73.
BOHM, Pion. Hubert, February 7, 1941. Age 30. Grave 159.
BUCHMANN, O.T.Mann Franz, September 17, 1942. Age 35. Grave 41.
BUECHOLD, Masch.Gfr. Julius, June 21, 1943. Age 21. Grave 112.
BUCKREUS, O.Pion. Johann, February 4, 1942. Age 35. Grave 68.
CHRISTIANI, Generalmajor Horst, November 19, 1942. Age 49. Grave 87.
DALL, Heizer Fritz, January 2, 1942. Age 22. Grave 71.
DAVIDS Masch.Mat.d.l. Elso, November 7, 1943. Age 43. Grave 116.
DIESINGER, Gefr. Friedrich, January 28, 1942. Age 32. Grave 70.
DOERNACH, Gfr. Wilhelm, June 6, 1943. Age 41. Grave 109
DREES, Gfr. Richard, September 10, 1940. Age 29. Grave 168.
DREYER, Leut. Ulrich, September 23, 1941. Age 21. Grave 152.
ESSINGER, Gfr. Heinrich, October 4, 1942. Age 30. Grave 43.
FIEDLER, O.Pion. Johann, May 11, 1942. Age 34. Grave 62.
FISCHER, Uffz. Albert, June 8, 1943. Age 23. Grave 108.
FRIEDRICH, O.Gfr. Walfried, November 12, 1943. Age 40. Grave 117.
FROMME, Leut. Anton, October 17, 1940. Age 23. Grave 156.
FUERST, O.Feldw.Vtr. Ludwig, November 19, 1942. Age 60. Grave 86.
GERASCH, Masch.Gfr. Karl, August 2, 1942. Age 22. Grave 37.
GIERTH, Schtz. Fritz, March 20, 1942. Age 19. Grave 58.
GOEBEL, Asst.Arzt. Dr. August, April 29, 1942. Age 29. Grave 61.
GOEBGES, Fk.O.Gfr. Franz, May 14, 1943. Age 21. Grave 101.
GESELLE, Kanonier Fritz-Paul, June 12, 1943. Age 39. Grave 110.
GRAJEWSKI, Pol.Wachtm. Erich, February 26, 1943. Age 38. Grave 88.
HANKE, Feldw. Helmut, January 17, 1942. Age 23. Grave 146.
HASSDENTEUFEL, NSKK/Strm. Alois, May 25, 1943. Age 38. Grave 107.
HERMANN, Arbeiter Ludwig, May 27, 1942. Age 40. Grave 45.
HIRT, Gfr. Manfred, January 17, 1942. Age 19. Grave 148.
HINKEL, Hptm. Johann, March 28, 1943. Age 46. Grave 95.
HIRSCHMILLER, Strm.Gefr. Gerhard, November 19, 1942. Age 18. Grave 81.
HOCHT, Gfr. Ernst, March 11, 1942. Age 29. Grave 55.
HOFMANN, Uffz. Johannes, November 8, 1942. Age 23. Grave 76.
IMKE, Uffz. Friedel, December 28, 1941. Age 23. Grave 145.
JANAUSCH, Matrose Franz, September 8, 1941. Age 19. Grave 150.
JUERGENS, O.T.Mann Heinrich, May 8, 1943. Age 57. Grave 91.
JUST, Masch.O.Gfr. Franz, May 20, 1943. Age 21. Grave 103.
KALTENMAIER, Mtr.Gfr. Josef, March 5, 1942. Age 21. Grave 63.
KAMINIAREK, O.T.Mann Johann, May 17, 1943. Age 49. Grave 89.
KELLER, Leut. Willi, November 19, 1940. Age 42. Grave 157.
KEMPER, Mtr.O.Gefr. Erwin, November 19, 1942. Age 21. Grave 82.
KERSCHKE, Mtr.O.Gfr. Georg, May 14, 1943. Age 20. Grave 98.
KIEFER, Uffz. Albert, October 3, 1941. Age 22. Grave 153.
KLEBE, Mar.Art. Martin, July 7, 1941. Age 21. Grave 161.
KLEIN, Matr. Emil, August 2, 1942. Age 30. Grave 36.
KLUTTIG, Uffz. Albin, March 7, 1942. Age 26. Grave 54.
KONKUS, Gfr. George, August 22, 1942. Age 34. Grave 38.
KOCK, O.Polier Heinrich, April 10, 1943. Age 42. Grave 94.
KRIEGEL, Arbeiter Michael, June 10, 1942. Age 57. Grave 46.
KURZ, Uffz. Helmut, August 9, 1940. Age 25. Grave 166.
KURZ, Leut. Otto, January 21, 1942. Age 35. Grave 65.
LANDGRAF, Gfr. Heinz, April 12, 1943. Age 31. Grave 93.
LORENZ, Gfr. Willi, April 23, 1943. Age 23. Grave 92.
LUCKNER, Schtz. Helmut, March 19, 1942. Age 33. Grave 57.
MEWES, Gfr. Konrad, January 17, 1942. Age 21. Grave 147.
MOHR, O.Pion. Karl, February 21, 1942. Age 39. Grave 66.
MOLITOR, Leut. Peter, October 17, 1940. Age 26. Grave 155.
MUEHLBACH, Gfr. Anton, August 9, 1940. Age 19. Grave 164.
MUEHLHANS, Gfr. Gerhard, August 9, 1940. Age 22. Grave 165.
MUELLER, Gefr. Erich, December 24, 1941. Age 32. Grave 69.
NAUJOKAT, O.Gefr. Erich, September 6, 1942. Age 28. Grave 40.
NEIDIG, O.Schtz. Johann, July 24, 1942. Age 33. Grave 50.
NIEDERMEYER, Kanonier Anton, August 9, 1940. Age 26. Grave 162.
NITZ, Uffz. Gunter, September 25, 1941. Age 23. Grave 154.
OPITZ, Mar.Art. Willi, April 7, 1941. Age 20. Grave 160.
OSWALD, O.Gfr. Peter, October 4, 1942. Age 35. Grave 44.
RECKTENWALD, O.Gfr. Jakob, July 5, 1942. Age 36. Grave 119.
RICHTER, Polier Alfred, July 5, 1942. Age 33. Grave 48.
RICHTER, Mtr.O.Gfr. Karl, May 14, 1943. Age 20. Grave 100.
RITSCHER, Wachtm. Georg, May 14, 1943. Age 28. Grave 97.
ROEBEN, Feldw. Karl, June 15, 1942. Age 25. Grave 47.
ROEHLICH, Schtz. Fritz, October 13, 1942. Age 30. Grave 72.
RONNEBURG, Schtz. Willy, March 27, 1942. Age 34. Grave 59.
ROSELT, O.Gfr. Friedrich, November 11, 1943. Age 35. Grave 115.
ROTHMEIER, Gfr. Johann, October 26, 1943. Age 38. Grave 114.
RUPP, Schtz. Hans, March 12, 1942. Age 33. Grave 56.
SAUR, Gefr. Gustav, February 7, 1943. Age 33. Grave 80.
SCHADOWSKI, O.T.Mann Bernhard, November 7, 1942. Age 57. Grave 75.
SCHAUDER, Masch.O.Gfr. Gustav, May 14, 1943. Age 24. Grave 99.
SCHERER, O.Schtz. Erwin, November 12, 1942. Age 28. Grave 77.
SCHMIDT, Ober.Leut. Dr. Friedrich, July 11, 1942. Age 39. Grave 49.
SCHMIDT, O.Gfr. Georg, December 3, 1941. Age 28. Grave 144.
SCHNEIDER, Kanonier Heinz, August 9, 1940. Age 19. Grave 163.
SCHNEIDER, Pion. Karl, February 7, 1941. Age 28. Grave 158.
SCHNEIDER, Heizer Paul, January 19, 1942. Age 29. Grave 142.
SCHNEIDER, Gfr. Werner, August 23, 1942. Age 31. Grave 39.
SCHELLENBERGER, Pionier Wilhelm, August 27, 1941. Age 29. Grave 149.
SCHUBERT, Strm.Gfr. Walter, May 14, 1943. Age 19. Grave 102.
SCHULZ, M.A.Maat. August, November 13, 1943. Age 48. Grave 118.
SEIDEL, O.Gefr. Oskar, December 25, 1942. Age 23. Grave 79.
SIPPEL, Gfr. Heinrich, August 23, 1943. Age 37. Grave 167.
SCHNUECKER, O.Gfr. Willi, May 24, 1943. Age 29. Grave 104.
STEINBRINK, Gefr. Heinrich, November 28, 1942. Age 31. Grave 78.
STEVENS, Mtr.O.Gfr. Hermann, June 20, 1943. Age 28. Grave 111.
TIMEFEI, Schiffskoch Simon, May 23, 1943. Age 26. Grave 106.
TOMANEO, O.T.Mann Vincenc, August 16, 1942. Age 49. Grave 52.
TREUDE, Gren. Wilhelm, May 16, 1943. Age 33. Grave 90.
TUCHEN, Matrose Erich, September 8, 1941. Age 18. Grave 151.
VAN KEMP, Masch.Gefr. Hans, November 19, 1942. Age 21. Grave 83.
VAN LIESHOUT, Kapitaen Johannes, May 23, 1943. Age 44. Grave 105.
VOLKENANNDT, Uffz. Ernst, March 4, 1942. Age 28. Grave 53.
WACKES, O.Gefr. Hilmar, November 19, 1942. Age 34. Grave 84.

ST. PETER PORT (ST. JOHN'S) CHURCHYARD

This cemetery is located on the north side of the town in St. John's Road. There are four men buried here who served in the 1914-1918 war.

BROUARD, Cpl. John, 774. 2nd Bn. Royal Guernsey Light Inf. Died of tetanus, June 14, 1917. Age 24. Son of the late Harry and Patty Mary Brouard; husband of Clara F. Brouard of La Planque Farm, Water Lanes, St. Peter Port, Guernsey. Buried north of the church in Grave 248.

CLARK, Maj. Wilfrid Thomas De Lacey, Royal Marine Light Inf. February 24, 1919. Age 34. Son of the Rev. Henry and Ada Blanche Clark of Guernsey; husband of Kathleen Ethel Clark (nee Foote) of 'Ashburton', Lower Rohais, Guernsey. Buried east of the church.

ENGLAND, Gnr. William Henry, 192119. 81st Siege Bty. Royal Garrison Artillery. March 29, 1918. Age 21. Son of Frederick and Margaret England of 'Ashbrook', St. Sampson's, Guernsey. Buried south-east of the church.

JENKINS, Whlr. A., 89237. 5th 'C' Reserve Bde. Royal Field Artillery. July 11, 1916. Age 19. Son of Mr. and Mrs. J. Jenkins of Surprise Cottage, Le Grand Bouet, St. Peter Port, Guernsey. Buried east of the church.

ST. PETER PORT (CANDIE ROAD) CHURCH CEMETERY

This is situated on the north-west side of St. Peter Port in Candie Road. It was laid out in 1831 and belongs to the parish of St. Peter Port. A memorial was erected at St. Peter Port Cemetery to the men of Guernsey who fell in the First World War. There are four 1914-1918 burials.

GALPIN, Spr. E. G., WR/40964. 329th Quarry Coy. Royal Engineers. November 25, 1918. Age 36. Son of William R. Galpin. Grave BX 87.

MADELL, L/Cpl. R., 5998850. 1st Bn. Essex Regt. May 14, 1921. Son of Mrs. A. Madell of 12 Mount Durand, St. Peter Port. Grave EX 44.

ROWSELL, Gnr. A. C. B., 192239. 'C' Bty. 36th Fire Command. Royal Garrison Artillery. January 11, 1920. Grave FX 29.

SHEPHERD, Pte. W. C. T4/094157. H.T. Royal Army Service Corps. January 18, 1920. Grave FE 5Y.

ST. PETER-IN-THE-WOOD PAROCHIAL CEMETERY

The church is located in Rue de l'Eglise at Les Buttes, five miles south-west of St. Peter Port. Here lie three men from the 1914-1918 war.

LE MOIGNE, Pte. P. G., 2593. Royal Guernsey Light Inf. June 17, 1916. Age 18. Son of Mrs. Adelaide Le Moigne. Grave D 20.

LE PROVOST, Pte. J., 167. 2nd Bn. Royal Guernsey Light Inf. February 26, 1920. Grave C 6.

ROBERT, Pte. Henry, 172. 2nd Bn. Royal Guernsey Light Inf. Died of wounds, April 18, 1918. Age 31. Son of Frederick and Louisa Robert of La Cite, St. Peter-in-the-Wood, Guernsey. Grave D 29.

ST. MARTIN'S PAROCHIAL CEMETERY

This parochial cemetery is located at La Rue des Frenes, half-a-mile west of St. Martin's Church. There are five burials from the 1914-1918 war and one from 1939-1945.

FIRST WORLD WAR

FRAMPTON, CSM Algernon Herbert Charles, 169. 2nd Bn. Royal Guernsey Light Inf. June 14, 1920. Age 35. Son of Herbert Charles and Kate Bowes Frampton; husband of Georgina Emily Frampton of Albata, Grand Bouet, Guernsey. Grave I 67.

HILL, 2nd Lt. Norman Ernest Albert. West Yorkshire Regt. Died of pneumonia, March 27, 1915. Age 28. Son of John Rowland Hill and Frances Hill of Bebington, Cheshire, and Guernsey. A tea planter in Ceylon. Grave F 2.

ST. PETER PORT (FOULON) CEMETERY

This cemetery, which belongs to the States of Guernsey, is situated a mile due west of the town and was opened in 1858. It contains the graves of thirty-two members of His Majesty's Forces who either died while serving on the island or who were washed ashore on the coast including four 1914-1918 war graves. The total includes one unknown naval casualty of the Second World War.

During the years of the German occupation, burials in this cemetery of members of the occupying forces, and of foreign workers whom they brought to the island, were numerous. After the war the States of Guernsey Board of Administration transferred to this large burial ground the remains of more than forty foreign workers from the Foreign Workers Cemetery at Les Vauxbelets, so that the ground they occupied, which had been commandeered without payment, could

FIRST WORLD WAR

AYRES, Pte., Frank Rowland, 3455. 6th Bn. Royal Irish Regiment. May 13, 1919. Son of P. P. and R. Ayres of Primrose Cottage, Victoria Avenue, Banques, Guernsey. Grave O 16

BLAISE, Gnr. Emile, 89317. Royal Field (formerly Royal Garrison) Artillery, July 2, 1920. Son of Ives Blaise of La Ramee, St. Peter Port. Grave 1814.

DAVID, Cadet C., 2224. 2nd Bn. Royal Guernsey Light Inf. Died of influenza, December 6, 1918. Age 19. Son of John and Annie Margaret David of Chaumont, Rohais, Guernsey. Grave A 16

MOYLAN, Ch. Yeo. Sigs. Thomas Joseph, 7619. HMAS *Australia* Royal Australian Navy, February 16, 1919. Age 36. Son of Thomas Moylan and his wife, Mary Kelly; husband of Eva Moylan of Delhi House, 112 Manners Rd., Southsea, Hants. Born in Guernsey. Grave H 22.

SECOND WORLD WAR

BOOTH, Ldg. Sto. Frederick, D/KX. 93025. RN. HMS *Charybdis*. October 23, 1943. Age 24. Son of Alfred and Mary Booth of Pitsmoor, Sheffield. Grave 38.

BRADFORD, Mechanician 2nd Cl. Frank D/K. 62154. RN. HMS *Charybdis*. October 23, 1943. Aged 38. Son of William John and Kate Jane Bradford of Torpoint, Cornwall; husband, of Ellen Patricia (Nellie) Bradford of Torpoint. Grave 54.

BUNN, AB. Louis Leslie, C/JX. 351947. RN. HMS *Charybdis*. October 23, 1943. Age 19. Son of Walter Louis and Maria Sarah Bunn of Hethersett, Norfolk. Grave 37.

CLAYTON, Boy 1st Cl. Donald Jeffrey, D/JX. 246429. RN. HMS *Charybdis*. October 23, 1943. Age 17. Son of Ella Clayton, and foster-son of Amy Barrington of Little Breinton, Herefordshire. Grave 52.

LE HURAY, Pte. C. R., 1079. 1st Bn. Royal Guernsey Light Inf. March 7, 1917. Grave I 40.

TAYLOR, Sgt. Charles Cecil, 8894. 1st Bn. Border Regt. March 11, 1919. Husband of H. Taylor of Le Hurel, Saints Road, St. Martin's. Grave A 28.

WHITEHEAD, Gnr. William, 15345. No. 2 Depot, Royal Garrison Artillery. September 5, 1914. Age 34. Husband of Marguerite Marthe Whitehead of La Haye, St. Martin's, Guernsey. Grave I 62.

SECOND WORLD WAR

MULHOLLAND, Lt. John Desmond, 107259, MC. The Duke of Cornwall's Light Infantry. September 3, 1945. Age 26. Son of Flt. Lieut. J. E. Mulholland, RAF, and of Dorothy Mulholland of St. Martin's. Barrister-at-Law. Grave D 118.

be returned to the owner. They also transferred remains from the Jerbourg Road German Military Cemetery, St. Martin, the land for which had been similarly commandeered. These burials were all exhumed in 1961, the German war dead totalling 139.

The graves of the Royal Naval casualties, men of the crew of HMS *Charybdis*, which was sunk by the enemy in the English Channel on October 23, 1943 were made by the British Legion in Guernsey. It would appear that because of lack of space, it was customary to prepare graves for up to four burials. Thus the Joint Grave 27. (The Commonwealth War Graves Commission acquired the freehold in 1950.) Three of the naval graves and one of a soldier are in a small reserve plot lettered AM. This is separated from the rest of the cemetery by a grey granite wall on the eastern side, a path on the northern and western sides and a low hedge on the southern side. The other war graves are located elsewhere in the cemetery.

CLAYTON, Ord. Tel. William, D/JX. 341388. RN. HMS *Charybdis*. October 23, 1943. Aged 21. Son of James and Emily Clayton of Wigan, Lancashire. Grave 56.

DOBSON, Ldg. Supply Asst. William John (Bill), D/MX. 68963. RN. HMS *Charybdis*. October 23, 1943. Age 27. Son of William Henry Hamilton Dobson and Bessie Jane Dobson; husband of Phyllis Mary Dobson of Notting Hill, London. Grave 40.

GILLINGHAM, Dvr. William Clifford, T/10690118. Royal Army Service Corps. November 24, 1946. Age 44. Son of William and Eliza Gillingham; husband of Bertha Lilian Gillingham of St. Peter Port. Buried in Section AG, Grave 33.

HALL, Gnr. William James, 1547282. 77 Bty., 44 Lt. AA Regt., Royal Artillery. January 16, 1940. Age 29. Son of John and Margaret Hall of Lisburn, Co. Antrim, Northern Ireland. Buried in Section AN, Grave 116.

HARPER, ERA 4th Cl. Thomas Gordon, D/MX. 102904. RN. HMS *Charybdis*. October 23, 1943. Son of Dr. Thomas Harper and Margaret Harper of Stranraer, Wigtownshire; husband of Eleanor Harper of Balderton, Nottinghamshire. Grave 53.

HERBERT, Coder, John Wilson, D/JX. 342996. RN. HMS *Charybdis*. October 23, 1943. Age 20. Son of George W. Herbert and Elsie M. Herbert of Hereford. Grave 50.

JOHNS, Dvr. Edgar, T/5568312. Royal Army Service Corps. January 13, 1946. Grave 34.

JONES, AB Frank, D/J. 22492. RN. HMS *Charybdis*. October 23, 1943. Age 46. Son of John and Henrietta Jones of St. Thomas, Exeter. Grave 51.

JOUHNING, Pte. Walter Henry, 13006880. Pioneer Corps. June 9, 1947. Age 41. Husband of E. F. Jouhning of St. Peter Port. Buried in Section DD/BB, Joint Grave 27.

KANE, AB Joseph Hugh, D/JX. 363846. RN. HMS *Charybdis*. October 23, 1943. Age 20. Son of Thomas and Margaret Kane of Sandyford, Glasgow. Grave 36.

LAWSON, Ord. Sea. David Connal, D/JX. 421609. RN. HMS *Charybdis*. October 23, 1943. Age 18. Son of George Hogg Lawson and Mary Quin Lawson of Springburn, Glasgow. Grave 42.

LE FRIEC, Sgt. Yves, 7811128. The King's Own Scottish Borderers. February 16, 1947. Age 49. Son of Yves Marie and Mariane Le Friec; husband of Ellen (Nellie) Le Friec of Croutes. Buried in Section AG, Grave 95.

MACDONALD, Ord. Sea. John Donald, P/JX. 518696. RN. HMS *Charybdis*. October 23, 1943. Age 18. Son of Donald and Williamina Macdonald of Brora, Sutherlandshire. Grave 49.

MAIDMENT, Ord. Sea. John, D/JX. 246207. RN. HMS *Charybdis*. October 23, 1943. Age 17. Son of Edwin William and Eva Gertrude Maidment of Weston-super-Mare, Somerset. Grave 46.

MAY, Ldg. Supply Asst. Kenneth Rich, D/MX. 67815. RN. HMS *Charybdis*. October 23, 1943. Age 33. Son of Robert Phillips May and Mary Ellen May of St. Austell, Cornwall. Grave 57.

MORGAN, AB John Rees, D/JX. 230394. RN. HMS *Charybdis*. October 23, 1943. Age 23. Son of Howel Rees Morgan and Mabel Elizabeth Morgan of Crickhowell, Brecknockshire. Grave 39.

MURPHY, Ord. Sea. Patrick, D/JX. 416730. RN. HMS *Charybdis*. October 23, 1943. Son of James and Margaret Murphy of Westport, Co. Mayo, Irish Republic. Grave 55.

OZANNE, Sgt. Milford, S/126190. Royal Army Service Corps. November 3, 1947. Age 39. Buried in Section DD/BB, Joint Grave 27.

ROBERTS, Marine Clifford Ernest, PLY/X. 4181. Royal Marines. HMS *Charybdis*. October 23, 1943. Age 19. Son of Ernest Charles and Amy Maude Roberts of Maulden, Bedfordshire. Grave 45.

SOMERS, Sto. 1st Cl. Henry, D/KX. 134952. RN. HMS *Charybdis*. October 23, 1943. Grave 48.

WHITE, Supply Asst. John Edward, D/MX 68473. RN. HMS *Charybdis*. October 23, 1943. Age 24. Son of Henry Edward and Frances Alice White of Hounslow, Middlesex. Grave 41.

WILLIAMS, Second Offr. William Baker, Merchant Navy, SS *Roebuck* (Weymouth). June 13, 1940. Age 37. Son of Mr. and Mrs. J. W. Williams of St. Peter Port; husband of Olive Williams of St. Peter Port. Buried in Section FF/EE2, Grave 74.

YOUNG, Sto. 2nd Cl. Alfred Thomas, D/KX. 162164. RN. HMS *Charybdis*. October 23, 1943. Age 17. Son of Thomas Henry and Florence Mabel Young of Sutton, Surrey. Grave 47.

ST. SAMPSON'S CHURCHYARD

This burial ground lies between St. Sampson's Harbour to the north and a disused quarry to the south. Some years ago there was a landslip in the churchyard which left a number of graves, including seven war graves, in a dangerous situation where access is impossible. Under the circumstances, the Commonwealth War Graves Commission decided to commemorate the war casualties on a special memorial panel fixed to the perimeter wall. There were two German graves until they were exhumed in 1961.

LE HURAY, Pte. Henry Edward, 251. 2nd Bn. Royal Guernsey Light Inf. Died of pneumonia, December 6, 1918. Age 25. Son of Edward Le Huray and Ellen Shaw, his wife; husband of Emily Elizabeth Torode (formerly Le Huray) of Route Militaire, St. Sampson's, Guernsey. Born in Guernsey. Grave 98.

LE PREVOST, Sgt. Frederick George, 9278. 'D' Coy. 1st Bn. Devonshire Regt. Died of pneumonia, February 18, 1919. Age 25. Son of Mrs. Julia Le Quelenec of Banks, St. Sampson's, Guernsey. Grave 73.

LE QUELENEC, L/Sgt. Francis Henry, 12. 2nd Bn. Royal Guernsey Light Inf. Died of meningitis, May 5, 1917. Age 42. Husband of Julia Le Quelenec of Banks, St. Sampson's, Guernsey. Born in Jersey. Grave 73.

MAHY, Pte., William Blondfield, G/3362. 9th Bn. Royal Sussex Regt. November 26, 1915. 1912 Plot, Row 3, Grave 1.

NICOLLE, Pte. Douglas, 1167. 1st Bn. Royal Guernsey Light Inf. March 19, 1917. Age 36. Son of James William and G. Nicolle of Baubigny Cottage, St. Sampson's, Guernsey. Grave 329.

PERRODOU, Pte. Leon, 49. 1st Bn. Royal Guernsey Light Inf. Died of pneumonia, July 12, 1919. Age 19. Son of Francoise and Jean Marie Perrodou of Kimberley Terrace, Sandy Hook, L'Islet, Guernsey. Grave 54.

SANDS, Pte. W. J., 3788. 2nd Bn. Royal Guernsey Light Inf. November 1, 1918. Age 18. Son of Mr. W. Sands of Trafalgar Square, Vale, Guernsey. Grave 304.

WALDEN, Cpl. George Edward, WR/28617. 321st Quarry Coy. Royal Engineers. March 5, 1919. Age 52. Son of Thomas and Fanny Walden; husband of Louisa A. Walden of 9 Victoria Avenue, Banks, Guernsey. Grave 108.

VALE (DOMAILLE) CHURCH CEMETERY

This churchyard adjoins the church of St. Michel-du-Valle in the far north of Guernsey. Both burials are from the 1914-1918 war. Two German burials were exhumed in 1961.

MARTEL, Pte. Ernest, 1650. 2nd Bn. Royal Guernsey Light Inf. March 30, 1919. Age 27. Son of F. and Harriet Martel of Les Pequeries Vale, Guernsey; husband of W. Gillingham (formerly Martel) of Albert Cottage, Les Pequeries' Vale, Guernsey, Grave 80.

TOSTEVIN, Spr. T., WR/28603. Royal Engineers. January 7, 1920. Age 30. Son of Nicholas Tostevin of Le Houmet, Vale, Guernsey; husband of Blanche C. Tostevin of 16 Paris St., St. Peter Port, Guernsey. Grave 81.

VALE PAROCHIAL CEMETERY

This cemetery adjoins the Vale Churchyard. All three burials are of men who served in 1914-1918.

FALLA, Spr. Alfred John, WR/28460. 321st Quarry Coy. Royal Engineers. January 18, 1920. Age 28. Son of James John Falla and Elizabeth Mary Gaudion, his wife; husband of Maybliss Linda Falla of Monica, Vale Rd., Guernsey. Grave 210.

GIRARD, Spr. W., WR/28473. Royal Engineers. March 11, 1919. Grave 254.

LE PREVOST, Pte. Albert John, 2188. 'D' Coy. 2nd Bn. Royal Guernsey Light Inf. Died of meningitis, November 21, 1918. Age 19. Son of Thomas Dickson Le Prevost and Mary Judith Le Prevost of La Haize, Vale, Guernsey. Grave 55D.

JERSEY

GROUVILLE (LA CROIX) CEMETERY

This cemetery is situated in La Rue a Don at Grouville and there is just one burial here from 1914-1918.

CAIN, Pte. A. B., 85. Garrison Bn. Royal Jersey Militia. May 31, 1919. Husband of Hilda Rose Cain of Fort d'Auvergne, Havre-des-Pas, Jersey. Buried in north-east part, Grave 162.

ST. BRELADE'S CHURCH CEMETERY

One man who served in the 1939-1945 war is buried in the church cemetery located behind St. Brelade's Bay.

HEMSLEY, Spr. Charles William, 3757455. 660 Constr. Coy., Royal Engineers. April 15, 1947. Age 58. Son of Jesse and Mary Ann Hemsley; husband of Maria Hemsley of St. Aubin. Plot A, Grave 61.

ST. BRELADE'S CHURCHYARD

The cemetery is situated to the north of the church and here lie four soldiers from the 1914-1918 war. There was also originally one other First War burial, that of Private Hanlon in the Strangers' Plot at the north end of the churchyard, together with seven German graves. As the plot was used by the Germans for 1939-1945 burials. the Germans exhumed the remains of Private Hanlon and reinterred them in St. Helier's War Cemetery *(see page 232).*

There were originally 215 burials of the 1939-1945 war in this churchyard made up of one RAF airman; twenty-eight German sailors; 184 German soldiers (one unidentified); one German airman and one German Merchant Navy seaman. All the German war dead were exhumed in 1961.

FIRST WORLD WAR

DE LA COTE, Pte. F. A., 29080. 6th Bn. Dorsetshire Regt. April 3, 1918. Buried on the north-west side, Grave 550.

FONEY, Gnr. P. E. F., 31059. 5th Reserve Bde. Royal Field Artillery. December 10, 1918. Age 29. Son of Philip and Mary Foney of Mont Les Vaux, St. Aubin, Jersey; husband of Vera Burgoyne. Buried on the north-west side, Grave 500.

LE SAUVAGE, 2nd Lt. Ernest Davies. 1st Bn. Dorsetshire Regt. and Royal Flying Corps. Mentioned in Despatches. Killed while flying, May 30, 1916. Age 19. Son of Ernest P. M. Le Sauvage of The Lodge, Beaumont, Jersey. Buried on south side of east-west plot, Grave 612

WINDEBANK, Boy E. W., 19463. 3rd Bn. York and Lancaster Regt. February 10, 1919. Buried in north-east part of main plot, Grave 463.

SECOND WORLD WAR

HULTON, Pilot Offr. (Pilot) Henry Stephen Penton, 33428. RAF. No. 18 Sqdn. March 21, 1940. Age 20. Son of Lt-Col. Henry Horne Hulton, and of Isobel Hope Millicent Hulton of Beaumont. Buried on north side, Grave 452.

Above: The imposing entrance gates at St. Brelade's Heldenfriedhof (Societe Jersiaise). *Bottom:* The pillars were retained but the gates were replaced soon after the war.

Below: There were a few private exhumations before those in 1961. Oberleutnant Otto Winter was removed by his family in July 1956 from Grave No. 194 for return to the Fatherland.

ST. CLEMENT'S CHURCHYARD

Situated on St. Clement's Inner Road at St. Clement, here lie two men, one from each of the First and Second World Wars.

FIRST WORLD WAR

DURELL, Shipt. 2nd Cl. L. E. H. PO/M. 5454. RN. HMS *Cator*. July 15, 1918. Buried north-west of church near north boundary.

SECOND WORLD WAR

PORTEOUS, Maj. Kenneth St. Clare, 230818. Royal Electrical and Mechanical Engineers. August 11, 1945. Age 33. Son of Ernest George and Jane Louise Porteous of Samares; husband of Florence Clarice Porteous. Grave 100

ST. HELIER'S (ALMORAH) CEMETERY

This cemetery at La Pouquelaye, St. Helier covers an area of six acres and belongs to the town of St. Helier. It contains twenty-four war graves from 1914-1918 (including those of a Belgian soldier), and originally one interned German civilian, and two burials from 1939-1945.

A memorial can also be seen to the 862 men of Jersey who lost their lives in the Great War.

FIRST WORLD WAR

ABBOTT, Pte. Herbert John, 9764. Depot (Aldershot), Royal Army Medical Corps. Died May 19, 1918 of sickness following wounds (gas) received in France. Age 22. Son of Mabel Mary Abbott of 53 Great Union Rd., St. Helier, Jersey, and the late Thomas Joseph Abbott. Born in Lancashire. Plot O, Row 9, Grave 31.

BEERE, L/Cpl. Alfred Henry, 12812. 351st Coy. Royal Defence Corps. May 31, 1916. Age 45. Son of the late George and Anne Beere. Born at Cowes, Isle of Wight. Plot A, Row 3, Grave 8.

BOOKER, Cpl. Fredrick, 8591. 4th Bn. South Staffordshire Regt. October 3, 1914. Age 30. Son of Isaac and Ann Booker of Walsall; husband of Sarah Booker of 103 Old Birehills, Walsall, Staffs. Plot Q, Row 5, Grave 5.

BRIERLEY, Pte. Guy, 892. Jersey Garrison Bn., Royal Guernsey Light Inf. Died of pneumonia, October 31, 1918. Age 34. Son of Eli and Mary Ann Brierley of Lees, Oldham; husband of Miriam Brierley of 31 Wesley St., Failsworth, Manchester. Plot M, Row 6, Grave 6.

BROWN, Pte. F. W., 772. *See* LEAMON, the true family name.

CATELINET, Sgt. W. E. 278353. London Electrical Engineers, Royal Engineers. Accidentally killed, May 30, 1919. Age 26. Son of Philip George and Ellen Catelinet of Jersey; husband of Gertrude Catelinet of 17½ Windsor Rd., St. Helier, Jersey, Plot M, Row 4, Grave 22.

CAWLEY, Officer's Std. 3rd Cl. G. R., L/6572. RN. HMS *Q36* March 30, 1917. Grave 10/15A

CORNIERE, Pte. Peter Francis, 31560. Depot, Hampshire Regt. Died of wounds, October 19, 1917. Age 40. Son of Peter Corniere and Eugenie Le Monnier, his wife; husband of Jeanne Francoise Corniere of 18 Phillip's St., St. Helier, Jersey. Plot F, Row 8, Grave 8.

COWEY, CSM T., 279110. Royal Garrison Artillery, attd. Permanent Staff, Royal Jersey Artillery. June 17, 1918. Plot A, Row 1, Grave 68.

CRENAN, Rfn. Joseph, 9469. 7th Bn. Depot, Royal Irish Rifles, formerly 7th Bn. June 26, 1917. Son of Mrs. Jane Crenan of Mill House, Trinity Rd., St. Helier, Jersey. Plot H, Row 2, Grave 19.

DE STE. CROIX, Bmdr. P. W., 69709. 179th Siege Bty., Royal Garrison Artillery. September 25, 1917. Plot F, Row 8, Grave 1.

GRIFFIN, Pte. R. G. P., TR/8/5383. 34th Training Reserve Bn. June 30, 1917. Age 18. Son of Henry and Ellen Griffin of 9 Sand St., St. Helier, Jersey. Plot H, Row 6, Grave 22.

HALE, Pte. Charles, 9909. 4th Bn. South Staffordshire Regt. December 29, 1914. Age 40. Son of Joseph and Elizabeth Hale; husband of Esther Jane Hale of 72 Tantara St., Walsall, Staffs. Born at Rotherham. Plot Q, Row 5, Grave 5.

HERESEY, Sgt. John Adolphus, 510656. Canadian Army Service Corps. January 24, 1919. Age 39. Husband of Ellen Heresey of 9A Wineva Avenue, Toronto, Canada. Plot P, Row 4, Grave 13.

LEAMON (served as BROWN), Pte. F. W., 772. Royal Jersey Garrison Bn., Channel Islands Militia. March 27, 1920. Plot Q, Row 3, Grave 21.

LE CLEARE, Rfn. Louis Joseph, 10180. Depot, Royal Irish Rifles, formerly 7th Bn. Died May 7, 1918 of pneumonia following wounds (gas). Age 18. Son of Eugene and Maria Le Cleare of 6 Mayfield Cottages, St. Helier, Jersey. Plot A, Row 1, Grave 70.

LE CORNU, Pte. J. E., 43957. Depot, Hampshire Regt. October 28, 1918. Age 37. Husband of Hester Elizabeth Le Cornu of 30 Great Union Rd., St. Helier, Jersey. Plot N, Row 8, Grave 17.

LYDON, Pte. J., 8826. 4th Bn. South Staffordshire Regt. October 19, 1914. Plot Q, Row 5, Grave 5.

NICOLLE, Pte. J. F., EMT/46512. Royal Army Service Corps. September 13, 1919. Husband of E. E. Nicolle of 48 Kensington Place, St. Helier, Jersey. Plot C, Row 9, Grave 34.

RALPH, Sgt. Ernest, 5144. 4th Bn. South Staffordshire Regt. October 12, 1914. Age 37. Husband of Edna May Ralph of 6 Dudley St., Wednesbury, Staffs. Plot Q, Row 5, Grave 5.

TAYLOR, Dvr. A. G., T2/017888. H.T. Royal Army Service Corps. September 9, 1919. Age 39. Husband of E. E. Taylor of 19 Cannon St., St. Helier, Jersey. Plot G, Row 2, Grave 10.

VILLARD, Able Seaman J. F., 231087/PO, RN. HMS *Birmingham* August 5, 1916. Grave I/4D.

WHITMORE-SEARLE, 2nd Lt. B. 10th attd. 4th Bn. South Staffordshire Regt. October 3, 1915. Plot M, Row 8, Grave 5.

WRIGHT, Pte. W., 10492. 4th Bn. South Staffordshire Regt. March 7, 1916. Age 46. Husband of A. Wright of 5/5 Scotland St., Parade, Birmingham. Plot Q, Row 3, Grave 18.

SECOND WORLD WAR

FAIERS, Pte, Gordon Robert, 19154433. General Service Corps. July 5, 1947. Age 19. Son of Cecil Henry Faiers and of Marie Perida Faiers (nee Arthur) of St. Helier. Grave 24/19V.

WALKER, Capt. Ian Patrick Goold, 160177. 1st Northamptonshire Yeomanry and The Queen's Bays (both Royal Armoured Corps). July 6, 1945. Age 31. Son of Maj. George Goold Walker, DSO, MC, and Lilian Sophie Walker of L'Etacq. Grave 22/13F.

ST. HELIER'S (MONT-a-L'ABBE) NEW CEMETERY

The cemetery is located at the top of Tower Road in St. Helier. There are burials here from both the First and Second World Wars. Here also lies the only female war casualty of the Armed Services to be buried in the Channel Islands. Sixteen German war dead have been exhumed.

FIRST WORLD WAR

BAUDAINS, Pte. George, 998. Royal Jersey Garrison Bn., Channel Islands Militia. October 25, 1918. Age 27. Son of George Baudains; husband of Jane Mabel Baudains (nee Le Brun) of 3 Richmond Place, First Tower, Jersey. Plot X, Row 9, Grave 48.

DE LA LANDE, Rfn. Arthur, 4099. 'D' Coy. 7th Bn. Royal Irish Rifles. May 19, 1916. Age 20. Son of Mary Elizabeth De La Lande. Born in Jersey. Plot P, Row 9, Grave 17.

GLENDEWAR, Spr. George Ambrose, 189706. Royal Engineers. July 7, 1917. Age 34. Son of Orlando John Glendewar; husband of Maud G. B. Glendewar of 29 Cannon St., St. Helier, Jersey. Born in Jersey. Plot U, Row 9, Grave 52.

HARVEY, Pte. Harold Joseph, 1290. 'D' Coy. 2nd Bn. Royal Jersey Militia. Died of meningitis, September 5, 1915. Age 19. Son of Joseph and Louisa Harvey of Wesley Cottage, Dieg Rd., St. Saviour's, Jersey. Plot T, Row 10, Grave 32.

LARBALESTIER, Bmdr. B., 87192. 5th Bde. Canadian Field Artillery. December 2, 1916. Section V, Grave 5, North part.

LE LION, Dvr. W. G., 20922. 75th Field Coy. Royal Engineers. November 23, 1918. Age 29. Son of Eugene Bienaime Le Lion of 7 Pomona Rd., St. Helier, Jersey. Plot R, Row 9, Grave 24.

RODDY, Maj. Edwin Louis. 1st Bn. Cheshire Regt. Died of sickness, July 3, 1919. Age 45. Son of Col. P. Roddy, VC, and Margaret Roddy; husband of Josephine Morgan Roddy (formerly Pressey) of 14 Howers Place, Cambridge. Plot T, Grave 174.

SYVRET, Air Mech. Second Class. Francis Philip, 80548. 23rd Aircraft Repair Shop, RAF. June 24, 1918. Age 35. Son of Phillip Edward and H. De G. Syvret of St. Brelade's, Jersey; husband of Louisa Elizabeth Syvret of Bermuda Villa, Aquila Rd., St. Helier, Jersey. Plot X, Grave 73.

SECOND WORLD WAR

BANKS, Dvr. William Alfred, T/14402737. Royal Army Service Corps. May 1, 1946. Age 19. Son of William Henry Louis and Louise Mary Banks of St. Helier. Section M, Grave 72.

BENNETT, Cpl. Ernest William, 330257. RAF. April 23, 1947. Age 44. Son of Ernest and Louise Mary Bennett; husband of Amy Harriet Bennett of St. Helier. Section S, Grave 50.

CLIFTON, Pte. Alfred, 5437254. The Green Howards (Yorkshire Regt.). June 13, 1947. Age 27. Son of Susan Ann Clifton of St. Helier. Section S, Grave 168.

DE LA HAYE, L/Cpl. Francis Charles, 5732241. The Dorsetshire Regt. July 11, 1947. Age 27. Son of Francis Charles and Clara De La Haye; husband of Stella Mary De La Haye of Clowne, Derbyshire. Section O, Grave 168.

HOUSSEL, Marine Francis Horace, PO/X. 3440. Royal Marines. HMS *Nelson*. April 15, 1940. Age 20. Son of Mr. and Mrs. F. J. Houssel of St. Helier. His brother Denis Francis also died on service. Section T, Grave 41.

JENKINS, Gnr. James, 1422405. *See* TARDIVEL, Peter Marie.

LE SAUX, Pte. George E., 5441193. Army Catering Corps. March 18, 1946. Age 30. Son of John Peter and Mary Margaret Le Saux of St. Helier. Section U, Grave 43.

NORMAN, PO Albert John, P/J. 7098. RN. HMS *Victory*. July 27, 1945. Age 52. Son of John Philip and Alice Jane Norman of St. Helier; husband of Margaret Norman of St. Helier. Section U, Grave 159.

PIPON, Pte. John Alfred, 7600584. Royal Army Ordnance Corps. August 3, 1942. Age 38. Son of Alfred Edward and Louisa Pipon of St. Helier; husband of Albertine Josephine Pipon (nee Brugano), of Gorey. Section S, Grave 111.

RABET, Wren Simone Yolande, 29702. Women's Royal Naval Service. HMS *St. Angelo*. October 21, 1947. Age 21. Daughter of Francis Ernest and Laurentine Albertine Rabet of St. Helier. Section O, Grave 142.

TARDIVEL, Peter Marie (served as JENKINS, Gnr. James, 1422405). 10 Bty., 7 HAA Regt., Royal Artillery. December 5, 1947. Age 48. Section S, Grave 35.

VALLER, Sgt. (Air Gnr.) Thomas Edward, 3051950. RAF (VR). March 18, 1946. Age 20. Son of Edward and Ada Valler; husband of Patricia Winifred Leigh Valler of St. Saviour's. Section S, Grave 169.

ST. HELIER'S (MONT-a-L'ABBE) OLD CEMETERY

This part of the cemetery is located south-east of the New Cemetery across the road. It was opened in 1855.

CORNISH, Pte. F. A., 111255. Depot, Tank Corps. April 24, 1919. North part, Plot F, Grave 264.

INGRAM, Pte. Arthur J., 11964. Royal Marine Light Inf. January 20, 1919. Age 33. South part, Plot I, Grave 68.

TEMPLE, 1st Mate Frederic. SS *Clan Mackenzie* Mercantile Marine. Drowned as a result of an attack by an enemy submarine, March 5, 1918. Age 42. Son of Mr. and Mrs. Henry W. Temple; husband of Bessie Ingram Temple of Prospect House, Crow Wood, Widnes, Lancs. Born in Liverpool. Plot D, Grave 158.

ST. HELIER'S WAR CEMETERY

This cemetery in the Howard Davis Park adjoins St. Luke's Church and is close to the south-eastern boundary of St. Helier. Entry is via La Route du Fort. It is an enclosure on land ceded to the States of Jersey by the owner as a memorial to his son who was killed during the First World War. This portion of the park was dedicated on November 26, 1943 as a burial place for British and Allied war casualties, the ground having been consecrated by clergy of the Church of England, the Roman Catholic Church and Free Churches. Existing war graves were moved into it from St. Helier's (Mont-a-l'Abbe) New Cemetery and, in all, fifty-two British and Allied servicemen were buried here.

In June 1946, the American authorities moved the graves of eleven members of their forces to the United States Military Cemetery at Blosville, France; a Belgian airman, too, has been exhumed and repatriated. There now rest here forty men of the Forces of the United Kingdom, thirty-eight of whom belonged to the Royal Navy and two to the Royal Air Force. Only twelve of the sailors are identified; they belonged to the crew of HMS *Charybdis* sunk by the enemy in the English Channel on October 23, 1943 and their bodies were washed ashore.

There is also one 1914-1918 war grave here, transferred by the Germans from a part of St. Brelade's Churchyard which they took over for the burial of their own service dead.

FIRST WORLD WAR

HANLON, Pte., G., 12864. 351st Protection Coy. Royal Defence Corps. November 11, 1916.

SECOND WORLD WAR

BUTLIN, Sgt. Dennis Charles, 547514. RAF. May 1, 1943. Grave 1.

CAMERON, Ordnance Mechanic 4th Cl. David L., D/MX. 75978. RN HMS *Charybdis*. October 23, 1943. Age 23. Son of Isaiah and Mary J. Lewis Cameron of Falkirk, Stirlingshire. Grave 28.

DRAKE, Musician James, X. 1930. Royal Marine Band. HMS *Charybdis*. October 23, 1943. Grave 25.

DUNCAN, AB. Hector, D/JX. 257943. RN. HMS *Charybdis*. October 23, 1943. Age 35. Son of John and Isabella Duncan of Dundee; husband of Helen Baxter Duncan of Dundee. Grave 26.

HARROLD, Ord. Sea. Kenneth Ivan, D/JX. 417624. RN. HMS *Charybdis*. October 23, 1943. Age 18. Son of Leonard and Mabel Harrold of Melksham, Wiltshire. Grave 42.

HOLDEN, Sgt. (Obs.) Abraham, 591390. RAF. No. 431 (RCAF) Sqdn. April 11, 1943. Age 22. Son of Shepherd and Florence Holden; husband of Elsie Holden of Oswaldtwistle, Lancashire. Grave 4.

MADDEN, ERA 4th Cl. Ivor William, D/MX. 55184. RN. HMS *Charybdis*. October 23, 1943. Age 22. Son of William Thomas Madden and Ethel Madden of Eastney, Portsmouth. Grave 27.

MILTON, Supply PO Arthur L., D/MX. 54131. RN. HMS *Charybdis*. October 23, 1943. Age 26. Son of Leonard James Milton and Maud Eveline Milton of Swansea. Grave 31.

REYNOLDS, AB. Mervyn, D/JX. 178507. RN. HMS *Charybdis*. October 23, 1943. Age 24. Son of Ernest Reynolds, and of Lillie Maud Reynolds, of Carnmenellis, Cornwall. Grave 32.

RILEY, AB. John Bernard, D/JX. 176204. RN. HMS *Charybdis*. October 23, 1943. Age 24. Son of John and Mary Ann Riley of Manchester. Grave 39.

THOMAS, Regulating PO Frank, D/MX. 74113. RN. HMS *Charybdis*. October 23, 1943. Age 31. Son of Gomer and Maud Louise Thomas; husband of Josephine Thomas of Ford, Plymouth. Grave 33.

TOOZER, AB. Arthur S. C., D/JX. 285701, RN. HMS *Charybdis*. October 23, 1943. Age 36. Son of George Enock and Miriam Toozer of Cardiff; husband of Minnie Toozer of Splott, Cardiff. Grave 24.

WELLENS, ERA. 4th Cl. Thomas, D/MX. 74832. RN. HMS *Charybdis*. October 23, 1943. Age 26. Husband of Annie Wellens of Waterloo, Ashton-under-Lyne, Lancashire. Grave 29.

WILLIS, AB. Arthur W., P/JX. 299074. RN. HMS *Charybdis*. October 23, 1943. Age 23. Son of Arthur W. Willis and Charlotte Willis of Custom House, Essex. Grave 30.

The present-day graves of Sergeants Butlin and Holden in St. Helier's War Cemetery. They were the first to be buried in Howard Davis Park — see page 213. The oak for all the crosses in the cemetery was donated by Lady Stewart, of Grainville Manor, St. Saviour's Hill in memory of her only son killed in the First World War. At the special request of the States on behalf of the people of Jersey, the Commission agreed to retain the wooden crosses in preference to the standard headstone for as long as they could be adequately maintained. To date, quite a number have already been replaced from the same stock of wood.

ST. JOHN'S NEW CHURCHYARD

Located beside the parish church, here lies one soldier from the 1914-1918 war.

RONDEL, Pte. Clarence Philip, 8/40130. 53rd Bn. Devonshire Regt. Died of pneumonia, November 2, 1918. Age 18. Son of John George and Louisa Rondel (nee Hotton), of La Fontaine, St. John's, Jersey. Buried in north plot near west boundary.

ST. LAWRENCE'S CHURCHYARD

The two war graves from the 1914-1918 war are in the north-west part of the churchyard which lies beside the parish church.

LANGLOIS, Pte. Philip John, 474. 'D' Coy. 1st West Bn. Royal Jersey Militia. February 2, 1915. Age 28. Son of Ph. G. Langlois of Le Passage, St. Lawrence, Jersey. Grave 68.

SYVRET, Lt. S. de B. Royal Scots Fusiliers. May 14, 1918. Grave 33.

ST. MARTIN'S CHURCHYARD

There are four war graves in the churchyard laid out beside the parish church, three from the First War and one from the Second.

FIRST WORLD WAR

BENDELL, Lieut. Albert, MBE, Royal Navy. April 26, 1920. Age 54. Son of John Payne Bendell; husband of Amelia Matilda Bendell (nee Renouf) of Ebenezer House, Trinity, Jersey. North-east of the church. Grave 58.

PAYN, Pte. J. de C. B., 6971. 1st Regt. South African Inf. March 4, 1916. Buried near north-east corner of churchyard. Grave N 38.

ROBERTS, Pte. H. G., 668. Royal Jersey Garrison Bn., Channel Islands Militia. October 29, 1918. Buried in north-west corner. Grave 146.

SECOND WORLD WAR

DEAN, Sgt. Thomas Charles Harold, 910713. RAF. (VR). March 10, 1946. Age 38. Son of William and Annie Maude Dean; husband of Margaret Dean of Faldouet. South part, Grave 183.

ST. MARY'S CHURCHYARD

This cemetery beside the parish church contains one burial from 1914-1918.

LE CORNU, Pte. Philip Francis, 1054442. 14th Bn. Canadian Inf. (Quebec Regt.). Died September 14, 1918 of wounds received August 15, 1917. Age 24. Son of Philip Francis Le Cornu and Mary A. Seager, his wife. Born in Jersey. Buried south-east of the church.

ST. PETER'S CHURCHYARD

This cemetery located adjacent to the parish church has two 1914-1918 war graves. It originally contained that of a German naval stoker until exhumed.

DUTTON, 2nd Lt. G. A. 4th Bn. South Staffordshire Regt. May 5, 1916. Buried west of church.

ELLIS, Pte. C., 26782. 6th Coy. Royal Army Medical Corps. July 23, 1918. Buried near west entrance. Grave 559.

Gefreiter Helmut Lehr, died July 8, 1940, the first German burial at St. Brelade's (CIOS/K. Pigeon).

ST. OUEN'S CHURCHYARD

The churchyard lies by the parish church. There are five graves here, all from the 1914-1918 war. Three of the graves lie against the southern boundary.

ECOBICHON, Pte. S. W., 1026. Royal Jersey Garrison Bn. Channel Islands Militia. October 19, 1918. Age 30. Husband of Lilla Marie Amy Ecobichon of Chandos Granty, St. Ouen, Jersey. Grave 45.

GIONTA, Rfn. J. L., 8701. 4th Bn. Royal Irish Rifles. September 24, 1920. Buried in north-west corner. Grave 168.

GODFRAY, Sgt. Francis George, 109. Royal Jersey Garrison Bn., Channel Islands Militia. April 24, 1919. Age 24. Son of Philip Godfrey. Grave 44.

MAUGER, Gnr. Philip Edward, 108134. Machine Gun Corps (Inf.). March 2, 1919. Age 26. Son of Edward Mauger and Jane Pigeon, his wife; husband of Florence Louisa Dimmick (formerly Mauger) of La Fosse-au-Bois, St. Ouen, Jersey. Buried north of church. Grave 228.

TURNER, Pte. J. F., 114055. Royal Army Medical Corps. Died of meningitis, June 23, 1919. Age 20. Son of Francis Turner and Eugenie Le Fevre, his wife, of L'Etacq, St. Ouen, Jersey. Grave 12.

ST. PETER'S (ST. MATTHEW'S) ROMAN CATHOLIC CHURCHYARD

In spite of its ambiguous title, this churchyard lies near Six Roads in St. Lawrence. Here lie four men who died while serving in the 1914-1918 war.

MINIER, Gnr. John Francis, 66537. 'B' Siege Depot, Royal Garrison Artillery. August 20, 1917. Age 27. Son of Francois Rene Minier of Middle Hill, Ville au Bas, St. Lawrence, Jersey. Buried north of church.

TREUSSARD, L/Cpl. Peter Donald, 27025. 1st Bn. South Staffordshire Regt. Died of wounds, June 6, 1917. Age 40. Son of Francois Treussard; husband of Marie Francoise Treussard of Elmside Cottage, St. Peter, Jersey. Buried north of church.

URVOY, Pte. A. P., 46514. 3rd Bn. Wiltshire Regt. November 1, 1918. Buried near north-west corner of church.

VELER, L/Sgt. John Rene, 12744. MM. 'F' Coy. 6th Bn. Dorsetshire Regt. Died of pneumonia, October 30, 1918. Age 25. Son of Peter and Anne Veler of Villa Les Abeilles, Au Rosais, St. Servan-sur-Mer, Ille-et-Vilaine, France. Born in Jersey. Buried in north-west corner.

ST. SAVIOUR'S CHURCHYARD

The cemetery is located next to St. Saviour's Church in St. Saviour's Hill. Here rest two servicemen, both from the First World War. There was originally the grave of a German civilian, interned on the island in 1914-1918, until exhumed.

FIRST WORLD WAR

DU HEAUME, Capt. H. T. Royal Army Medical Corps, August 5, 1916. Age 55. Son of Philippe du Heaume. Born in Jersey. Buried in the old ground north-west of church.

LE GALLAIS, Lt. R. W. Royal Flying Corps. Killed while flying, September 15, 1917. Age 19. Son of M. F. H. Le Gallais and J. Le Gallais of 22 Rue Louis Hymans, Brussels. Born in Jersey. Buried in the old ground south of church.

ST. SAVIOUR'S PAROCHIAL CEMETERY

This burial ground is located adjacent to the churchyard. There is one burial from the Second World War.

SECOND WORLD WAR

UPTON, Gnr. Denis Edward, 1797247. Royal Artillery. October 3, 1946. Age 26. Husband of Lilian Maud Upton of St. Helier. Row 15, Grave 77.

TRINITY CHURCHYARD

There is one burial from 1914-1918 in the churchyard which is located beside the parish church.

GRAHAM, Sto. 1st Cl. S. R. K/20381. RN. HMS *Woolwich*. December 1, 1918. Age 23. Son of Samuel James Graham of 34 Hill St., St. Helier, Jersey. Buried north of east end of church.

SARK

ST. PETER'S CHURCHYARD

St. Peter's lies near the centre of the island. There is only one British serviceman buried on Sark and he served in the 1914-1918 war. There were also five German war dead buried on the island until exhumed in 1961. There is also the grave of one Dutch seaman and one French Commando (the other has been exhumed *see page 167*).

CARRE, Pte. Helier, 7661. Royal Guernsey Light Inf. November 13, 1918. Age 18. Son of Helier and Harriet Carre of La Colinette, Sark. Buried north of church in old ground.

CHAPTER TEN
The War Museums

After the war there was a surfeit of war material on the Islands, so much in fact that its historical value in later years was not appreciated at the time. Much was scrapped or dumped at sea or just left to rot in the sealed-up tunnels.

However, due to the foresight of Mr. Ralph Mollet, Secretary to La Societe Jersiaise and with the help of its Keeper, Mr. F. S. Clements, a military museum was opened at La Hougue Bie near Grouville in 1946.

La Hougue Bie takes its name from an involved once-upon-a-time-legend concerning a monster and a French baron whose body is purported to be buried in the dolmen under the large mound, now topped by a small chapel. The Germans constructed an underground shelter and battalion HQ in the grounds and this is the building which was converted into a museum. However, as La Societe's interests span all the periods in Jersey's history, an associated exhibition hall covers archaeology, geology, transport and agricultural history.

One of the earliest members of the Channel Islands Occupation Society was Richard Mayne and in 1966 he opened his Bunker Museum at St. Peter's. He had been able to build up his collection of uniforms, weapons and other artefacts fairly easily in the early post-war years as many people just wanted to be rid of anything German. Although interest in militaria has grown out of all proportion in the intervening years, nowadays Richard is still given relics which come to light, like the old German stove found in 1976 in the attic of the Ommaroo Hotel still with a half-burned photograph of a German officer among the ashes. A circular wooden practice target was recently found being used as a dustbin lid at Highlands College and an anti-aircraft gun dug up from a field at Fliquet.

The bunker museum consists of six rooms and a central corridor and is now associated with the adjacent Jersey Motor Museum run by Michael Wilcock. One of the most significant exhibits is a collection of letters sent to the Geheimfeldpolizei HQ informing on civilians selling goods on the black market or revealing the owners of hidden radio sets. The letters on display were never delivered as they were intercepted by a worker in the Jersey Post Office, but one wonders how many were successful in reaching their destination? As the inscription over the display proclaims: 'This is what war does to people'.

Elizabeth Castle, on its detached rock about three-quarters-of-a-mile from the shore at St. Helier, is another location of occupation relics including some of the items taken from the St. Peter's Valley tunnel in 1973.

The first Second World War museum to be opened in the Channel Islands was La Hougue Bie in an original German headquarters in 1946. The museum has many specimens handed over by the British Army after the clear up in 1945.

As its name suggests, the castle was built in the reign of the first Queen Elizabeth and, by the outbreak of the First World War, was a modern army barracks occupied by the Artillery and later by the South Staffordshire Regiment. In 1922 the British government offered to sell the castle to the States of Jersey for £1,500 to which they agreed, the hand over of the keys being concluded with all due ceremony on May 21, 1923.

When the Germans took over in 1940, they fortified the castle, adding new works which

Part of the display of militaria in Elizabeth Castle in the ammunition room of the 105mm gun covering the harbour entrance (States of Jersey Public Works Department).

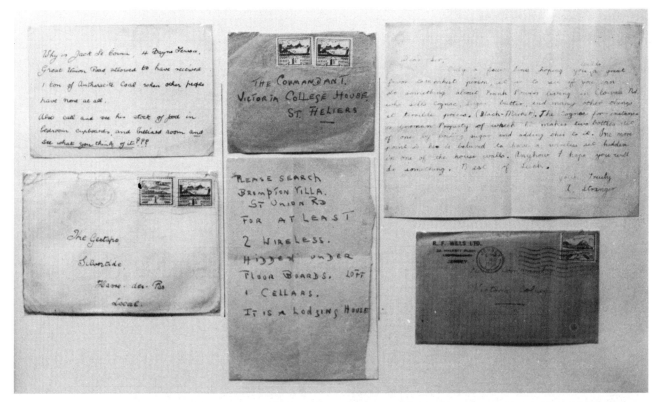

One of the most revealing, although pitifully tragic, of all the war exhibits in the Channel Islands is this one at Richard Mayne's German Occupation Museum at St. Peter's, Jersey. Destined for the German authorities, these letters from potential informers were fortunately intercepted by a Post Office employee; nevertheless they show how low some people will sink in time of war.

Above: This 20mm cannon on a naval-type mounting complete with shield was dug up from a field at Fliquet in the parish of St. Martin in 1971 and is now on display outside the museum. *Right:* Richard Mayne also has one of the rare examples of the Enigma Model C encoding machines used by all the German forces during WWII. The machine was invented in the 1920s by Arthur Scherbius who died long before he saw his creation put into use. It was the output from this machine that the Allies decoded using Ultra based at Bletchley Park, Buckinghamshire.

Richard's professionally-printed captions for each item on display put many official museums in the shade.

made it the cornerstone of the the harbour defences for five years. It also served as a punishment prison for recalcitrant members of the OT. The work of clearing up the castle was carried out in 1945-46 by a party of German prisoners-of-war and when it was handed back to the States on May 29, 1946 it was in a better state of preservation than it had been prior to the war.

It takes a little over ten minutes to walk out to the castle at low tide along the causeway and the old German road can still be seen on the right-hand side. To the left once stood a line of telegraph poles carrying communications from the mainland. The long, sloping track leading to the concrete shelter immediately inside the main entrance was the site of a 600mm searchlight which overlooked the harbour. Nearby is an anti-tank gun position dated 1944 and two bunkers containing 105mm French guns, restored complete with range markings.

Three other buildings constructed by the Germans on Jersey, while not qualifying as museums in the accepted sense of the word, are nevertheless the repositories of relics and well worth a visit. These are the battle headquarters at l'Aleval; the German underground military hospital in St. Peter's Valley and the observation tower and artillery command bunker at Noirmont which is to become a permanent point of display for occupation ephemera owned or on loan to the Jersey branch of the Channel Islands Occupation Society.

The best museum for Second World War relics, as far as presentation and quality of exhibits is concerned, is undoubtedly the Occupation Museum at the Forest, Guernsey. This is the brainchild of Richard Heaume, the schoolboy founder of the island's branch of the CIOS. Richard is a 1944 model and, although too young to have remembered the occupation, traces his interest in military hardware to the middle fifties. In those pre-replica days, interest in German equipment was limited to a few collectors and much could still be easily acquired. Richard confesses to being one of the early troglodytes obtaining many items from Guernsey's military tunnels. By the 1960s, these sources began to dry up and a new era began called 'the house to house'. This brought to light many items, from attics, from cellars and from gardens

Left: 'Genuine OT' is Richard Heaume's claim for this wooden wheelbarrow prised from an Alderney farmer. *Right:* This is one of the 305mm Mirus shells.

This horse-drawn ambulance is of WWII vintage having seen service in Guernsey and, post-war, on Sark (Carel Toms Collection).

One of the Renault FT17 turrets which were once commonplace in the Channel Islands. Both 'Richards' have examples on display. This one came from Les Nicolles, the Forest and can be seen at the Guernsey Occupation Museum as can the 37mm Flak salvaged from a wreck off Jethou in 1977.

Richard Heaume's poignant scenario brings home to the visitor the hardships faced during the occupation by all the Islanders.

and garages. Farmers sheds also proved profitable storehouses of long forgotten relics.

In 1956, Richard, then aged twelve, jointly formed the embryo of the present Occupation Society and has remained ever since one of its staunchest supporters. Ten years later he decided to put his collection on public display and he converted two cottages for the purpose next to his parents' farm.

Nevertheless, with all Richard's sense of history, he admits to failing to purchase one unique occupation relic. After the war all the German military horses were sold off and whilst some 300 went into the cooking pot, another 300 went to farmers for agricultural work. (One very real problem was that they only knew German words of command!) In 1968, Paddy, the last surviving German horse died and with it Richard's chance to have a real exhibit faded into dust. Stuffed, it would have looked ideal between the shafts of Richard's horse-drawn ambulance. This was the only purpose-built vehicle to be constructed on the island during the German occupation of Guernsey. It had been presented by the St. John Ambulance Brigade to Sark after the war where it provided sterling service until replaced by a horse-drawn caravan. In 1973, following protracted negotiations, the owner finally agreed to exchange the wagon for a set of horse harnesses and, in September, a team of volunteers led by Richard brought the ambulance back to Guernsey. Richard comments that, 'bringing it down Harbour Hill without horses proved quite a frightening experience!' The vehicle was restored at Mallett's Coachworks where it was originally built and put on show in his museum in 1974.

Various wreckage from aircraft brought down on the island is also on display including a wooden Spitfire or Hurricane propeller trawled up off St. Martin's Point in 1978; a Typhoon propeller from the sea off Herm and one from a Heinkel He 111 off Jethou.

Richard also organises a Fortress Tour of the island's defences which includes Batterie Mirus, the Naval Signals HQ for the whole of the Channel Islands at St. Jacques (where the German raid on Granville in March 1945 was planned) and several of the observation towers. All this and cream teas . . .

Guernsey's equivalent to Elizabeth Castle is Castle Cornet, similar in many ways although more easily accessible along Castle Pier. The

In 1947, King George VI gave Castle Cornet to the people of Guernsey as a token of their loyalty during two World Wars. Besides the German Occupation Museum, the castle contains five other collections and two picture galleries (see also pages 40-43 and 46).

castle has had a chequered history dating back to 1150 and the reign of Henry II. It has seen many battles, two sieges (one of over eight years) and was partly destroyed when the powder magazine exploded in 1672. Prior to WWII it was occupied by a detachment of the Royal Garrison Artillery although not garrisoned as such.

Like its sister in Jersey, Castle Cornet was heavily fortified by the Germans for harbour defence and much of the alteration work survives today. There is a fine museum of artillery weapons and military exhibits including the historic rum barrel from HMS *Beagle* on which the German articles of surrender were signed.

Alderney does not possess a war museum as such, but the Museum of the Alderney Society is housed in the old Town School, built in 1790, which was renovated by Society volunteers and opened in April 1972. This fine, small museum contains exhibits from all periods of the island's history, including two displays of the German Occupation. These contain small arms, insignia, trenching tools, sign boards, a gunnery range board, field library box, publications, newspapers, radio transmitter and tuner, field telephone, sandals and a striped uniform of one of the inmates of the Sylt camp, baskets for 88mm Flak shells and a mint 20mm Flak 38 barrel amongst other items. The museum also holds a small selection of German documents and photographs, all found in Alderney, and one of the most interesting exhibits is a 1:2000 scale relief model of the eastern half of the island made from pre-war maps and wartime aerial reconnaissance, showing the German defences. This model was made for the office of Combined Operations in 1944 and presented to the museum in 1971 by the RAF.

The Alderney Library, run entirely by volunteer helpers with the aid of a States grant, is located on the first floor of the Island Hall, the former residence of the island's hereditary governors until the 19th century. The closed collection contains an album of photographs taken in Alderney between early 1941 and mid-1942 by Sonderfuehrer Hans Herzog of the Nebenstelle of FK 515. These were presented to Mr. Colin Benfield in the

This 105mm gun came from Brehon Tower, a Victorian sea-fort built in 1855-56 on a rock midway between Guernsey and Herm, which was converted into a flak tower. Although valuable as part of the harbour defences, this isolated rock was, perhaps, the most unpopular posting in the Channel Islands. The gun was dumped in the sea at the end of the war and recovered by Roger Berry in August 1971.

early 1960s by Herzog and later donated to the Alderney Library, and contain a wide selection of photographs covering military life in Alderney, both on and off duty.

Mention has already been made of the Alderney Fortifications Centre, dedicated to the preservation of the island's military architecture of all periods, which is now in the process of clearing and restoring the German Fortress Commander's bunker at Les Rochers. This will not be a museum, in the

accepted sense, but it is intended to be open, on completion, to conducted parties.

Fortress Tours have been successfully introduced in 1980, with separate trips for the German and Victorian fortifications which are such a splendid feature of the island's landscape. These operate twice weekly during the season, and throughout the year on demand.

The Alderney Museum display includes a blue and white striped concentration camp suit which was found under the floorboards of a barrack room at Fort Albert.

CHAPTER ELEVEN
Your Holiday Hotel

When a country is taken over in war, one of the first demands is for accommodation for the invading forces. Hotels provide an easy, immediate solution to the problem and, as there are few tourists in an occupied country, they usually remain in enemy use. Being a centre of tourism, both before the war and today, the Channel Islands are well provided with hotels which saw varied histories. Particularly for visitors to the Islands, relaxing in the comfort of their surroundings, this brief run-down of the uses to which each hotel was put by the Germans may provide some intriguing thoughts.

On Jersey, one of the most interesting is, perhaps, the **Metropole Hotel** in Roseville Street, St. Helier which was used as the German Military Headquarters (Befehlshaber der britische Kanalinseln) with Oberst Graf Rudolf von Schmettow in residence until July 1941. Thereafter this position was filled by Generalmajor Erich Mueller with a new headquarters in Guernsey (not in a hotel — see Chapter One) and von Schmettow remained at the Metropole solely as Inselkommandant of Jersey.

The **Pomme d'Or Hotel** near the Weighbridge in St. Helier was probably the next in line of seniority as it became the HQ of the Hafenkommandant Kanalinseln (and Hafenkommandant Jersey when each island got its own 'Hako'), the Hafenkommandant being the Senior Naval Officer. In 1945 the hotel saw a change of management when Colonel William Robinson, the Allied Military Commander, took it over as his HQ.

The Standortkommandant or Garrison Commander used the **Ritz Hotel** at Colomberie, St. Helier for his HQ for a while in 1941 but for the rest of the time it served as billets until it was abandoned in 1945.

The quartering of troops was an important role for the island's hotels; the **Royal Hotel**, David Place, St. Helier and the **Demi des Pas Hotel** at The Dicq, St. Clement were both used for the purpose. Offices and billets for the Kriegsmarine were located in the **Yacht Hotel** in Caledonia Place, St. Helier.

Other hotels were taken over in the neighbourhood for defence positions, i.e. **Bouley Bay Hotel** for troops manning Resistance Point Bouley Bay; the **Prince of Wales Hotel** in St. Ouen's for men at Strongpoint Greve de Lecq; **High Tower Hotel** (now **Sands Disco**) on the Five Mile Road for Resistance Point High Tower, and **L'Etacquerel Hotel** for troops

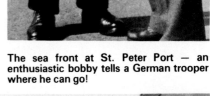

The sea front at St. Peter Port — an enthusiastic bobby tells a German trooper where he can go!

If you would like to stay in the German Military headquarters, go to the Metropole; extended in 1979/80. (R. H. Mayne and M. Ginns).

manning Strongpoint Etacquerel. Similar situations prevailed at Resistance Point Fort Henry at Grouville, where the nearby **Grouville Hall Hotel** provided easy accommodation, and at Gorey where the **British-Elfine Hotel** (now **The Moorings**) was used by the Germans at Mont Orgueil and Gorey Harbour. **Portelet Hotel** was a convenient billet for troops manning defences in the St. Brelade's area while the **Star Hotel** (now closed) in Wharf Street, St. Helier was used for those stationed at posts on and around the harbour.

Hotels made convenient buildings for combined headquarters and quartering and the **Gloster Hotel** on Gloster Terrace, Rouge Bouillon, St. Helier was taken over by Infantry Regiment 582 and the **Aberfeldy Hotel** in Old St. John's Road by Engineer Battalion 319. The HQ for Machine Gun Battalion 16 was located in **St. Peter's Hotel** (then called the **Alexandra Hotel**) while the

The German Naval HQ with its massive bunker (left) on the corner of Conway Street. Above: The two-storey construction took six months to demolish in 1956 (J. Parke and Societe Jersiaise).

Being a pre-war holiday retreat, there were plenty of hotels on the Islands to provide accommodation for troops. The Royal Hotel in David Place, St. Helier, became a billet while the Ommaroo was used by officers of 'the organisation'.

The Moorings Hotel
and Restaurant
GOREY

1st REGISTER

Telephone: 0534 53633
Telex: 4192085

An attractive small hotel situated at the foot of the ancient castle of Mont Orgueil and overlooking the beautiful sandy bay of Grouville. The hotel is open all the year round and is centrally heated throughout. Fitted with the latest Fire Alarm System. All rooms are en suite and have TV. Its famous Restaurant is a favourite rendezvous of the local inhabitants and offers superb English and Continental cuisine.

Write to resident manager for illustrated brochure and tariff

---

unit's Second Company used **La Moye Hotel** on the Route Orange at St. Brelade.

The **Grand Hotel** at West Park, St. Helier became the HQ of Festungspionerstab XIV (Fortress Engineer Staff 14) from August 1941 and the present **Ocean Hotel** on Westmount provided accommodation for gunners of the 105mm harbour-blocking Batterie Endrass of Army Coastal Artillery Regiment 1265 (although it was not then a hotel, simply Westmount House). Another building falls in a similar category — that of the **Mont de la Rocque Hotel** at St. Brelade which was the location of the First Battalion of Artillery Regiment 319.

The **Palace Hotel** at Bagatelle, St. Saviour had an interesting history as prior to the outbreak of war it was one of Jersey's foremost hotels. The 150-room Palace enjoyed a panorama over the entire bay and began the war as a Wehrmacht communications centre and later became an officers' training school. Much of the planning for the German 'Commando' raid on the American-occupied French port of Granville on the night of March 8/9, 1945 was carried out here but, on the eve of the assault, a small fire broke out in a ground floor room — used rather unwisely as a cordite store. Because of the secrecy surrounding the forthcoming raid, it was imperative to 'protect' the plans which lay scattered throughout the hotel so the Germans preferred to stand by and the fire was allowed to increase its hold. Instead of calling in the civilian fire brigade, the Kommandant decided to have his own men place demolition charges on all three floors to blow out a section of the hotel to stop the fire spreading along the building. Explosives were placed in position but the Germans were literally 'playing with fire' and, not surprisingly, the whole hotel exploded before their very eyes. Many buildings in St. Saviour's were damaged in the blast and nine soldiers mortally wounded. Although the firemen were now permitted to try to dowse the blaze, it was a dangerous job as there were repeated detonations as small arms ammunition exploded. The shell of the building later had to be demolished because of its dangerous condition, the site now being occupied by the Palace Close housing estate.

The Headquarters for the Organisation Todt in Jersey (Abschnittsbauleitung Julius) was located at the **Portland Hotel** situated at No. 19 Midvale Road in St. Helier while OT officers were billeted in the **Ommaroo Hotel** on Havre des Pas after it was requisitioned on December 4, 1941.

Rest and recreation was another important factor to be considered for sustaining the morale of the occupying forces and both officers and men had facilities provided in various hotels. The Offiziersheim — which

Hotels provided convenient accommodation for troops manning nearby defences. The Moorings Hotel was then named the British-Elfine Hotel.

Guest House Register Grade "A"

"L'Etacquerel"

Telephone: (0534) 82492

L'Etacq, St. Ouen's Bay Jersey, C.I.

A small country Hotel on the north-west point of the island, with a panoramic view over St. Ouen's Bay, which is reached from the bottom of our grounds.

We assure a happy holiday in relaxing surroundings with good English food.

Pleasant lounge bar with Sea Views. Open to Guests and their Friends.

All bedrooms have H. & C., are comfortably furnished and include central heating, making it ideal for early and late holidays. (No hidden extras.)

Bed, Breakfast and Evening Meal. Terms £7·00 to £10·50 daily. 5% service charge
Open March to October
RESIDENTIAL LICENCE   SEPARATE TV LOUNGE   AMPLE CAR PARKING SPACE

Write, enclosing unaffixed stamp, or telephone for Illustrated Brochure to Resident Manageress:
Maureen Bentley

The Germans left the L'Etacquerel a legacy: the deepest underground bunker in Jersey.

was a cross between a British Army NAAFI, a reception centre and a rest home — was located on the Havre des Pas in the **Fort d'Auvergne** while the **British Hotel** (now closed) on Broad Street seems to have been used as a conventional hotel for visiting high-ranking personnel.

NCOs and ordinary ranks had their Soldier's Club in the **Mayfair Hotel** in St. Saviour's Road where the men were looked after by nurses of the German Red Cross led by Sister Elizabeth Bergmann. In 1971, Sister Bergmann returned with a party of her former staff to find the hotel considerably enlarged and modernised.

What better place to spend your holiday than the former Offiziersheim in St. Helier?

NOW UNDER THE SAME MANAGEMENT AS THE NORFOLK LODGE HOTEL

Fort d'Auvergne Hotel

HAVRE-DES-PAS, ST. HELIER, JERSEY, C.I.   Telephone: (STD 0534) 22950

The hotel is built on the site of an old Fort and is bounded by the sea on two sides, yet is only minutes from the town centre. Your comfort and enjoyment is our responsibility, and if you are ready for a break, away from the cares of today—for that ideal honeymoon—to celebrate some special anniversary—or to have a holiday with friends or your family—we are at your service. The hotel is centrally heated.

Colour Brochure will be sent to you with pleasure by our Resident Manager
A SECOND REGISTER ✦✦✦ HOTEL

WHERE A WARM WELCOME AWAITS YOU . . . .

Above: 'Danke schoen fuer alles . . . kommen Sie zurueck . . . Danke, Danke . . .' One can just picture the conversation outside the main Soldatenheim in Jersey — or perhaps it was a curt: 'On your bike!' (Bundesarchiv).

St. Saviour's Road, St. Helier, some forty years later. The Mayfair Hotel has seen considerable change over the intervening years and, apart from the chimney, there was little to line up when we took our comparison.

Left: The Alexander Hotel once sported a machine gun position on its roof (R. H. Mayne). *Right:* Now renamed St. Peter's Hotel. Richard Mayne's museum was opened in the rear in 1966 and the conservatory demolished in 1973 (M. Ginns).

St. Brelade's Bay Hotel was Soldatenheim II and was unusual in that it accommodated all ranks, with officers on one side of the main corridor and NCOs and ORs on the other.

The Organisation Todt had their OT-Heim in the **Beaufort Hotel** in Green Street although this was then an elegant Victorian building and not the modern concrete and glass structure in use by modern-day tourists since 1950.

For those service personnel desiring other forms of pleasure, the ladies of the **Victor Hugo Hotel** at Greve d'Azette, St. Clement no doubt proved accommodating although the OT had their own brothel in the **Abergeldie Hotel** on Havre des Pas until the OT demolished it themselves (in a fit of pique?) on April 28, 1943 to make way for the new railway line to the east. Services were then transferred to the **Norman House Hotel** (now the **Lido Bay Hotel**) at First Tower.

For those who contracted certain diseases associated with these hotels, treatment was provided at the **Merton Hotel** which from December 1942 became a Luftwaffe hospital.

Some hotels had a very much more sinister connection and **Silvertide** (no longer a hotel) on Havre des Pas was the headquarters of the Geheimfeldpolizei (note that the unit on the island was the Secret *Field* Police, not the Gestapo).

By 1945, the number of civilians convicted by the authorities of crimes exceeded the space available in the civil prison and the **Chelsea Hotel** in Gloucester Street was taken over as 'holding' accommodation until a small barred room was free in the main 'hotel'.

The Islands are well known for their holiday camps and, with the addition of a barbed wire fence, they became very convenient prison camps. The **Grouville Holiday Camp** was initially used by the British authorities as an internment camp for alien civilians and after July 1, 1940 the Germans used it for British troops trapped on the island. When these men were transferred to France, it was then used to billet Spanish OT workers. The site of this camp is now covered by a modern housing estate.

Jersey Holiday Camp (now the **Jersey Holiday Village**) at Portelet was used as quarters for the naval gunners manning Batterie Lothringen at Noirmont Point and the **Jersey Jubilee Holiday Camp** (now Pontin's Holiday Village) at Plemont, St. Ouen, housed German sailors for a time from German ships laid up in Jersey's harbours during the winter of 1944-45.

Transformation at the Portland — the Organisation Todt HQ in Jersey.

CHELSEA HOTEL
JERSEY, CHANNEL ISLANDS

SECOND REGISTER ♦♦♦♦   TELEPHONE: (0534) 30241

SITUATED ONE MINUTE FROM SEA FRONT AND TOWN CENTRE
* LARGE BALLROOM
* SNACK BAR
* SPACIOUS LOUNGES
* RECREATION ROOM
* TELEVISION LOUNGE
* LIFT
* RADIO, INTERCOM, BABY-LISTENING AND SHAVER POINTS IN ALL BEDROOMS
* FULLY LICENSED ATTRACTIVE CARNIVAL AND TUDOR BARS
* COACH EXCURSIONS DAILY
* MOST BEDROOMS WITH PRIVATE BATH OR SHOWER AND W.C.
* ACCOMMODATION FOR 250 GUESTS
* HEATED SWIMMING POOL
* ATTRACTIVE GARDEN
* SOLARIUM

ONE OF THE LONGEST ESTABLISHED HOTELS IN THE CHANNEL ISLANDS. UNDER THE SAME FAMILY OWNERSHIP FOR OVER 60 YEARS. NOT OPEN TO THE GENERAL PUBLIC

Write for Colour Brochure and special early and late season holiday details to DEPT. I.J., CHELSEA HOTEL, JERSEY, Channel Islands
A BINNINGTON COMPANY

The Chelsea Hotel, situated in Gloucester Street, St. Helier, immediately opposite the Jersey General Hospital, was initially taken over for conversion into the German Army and Naval hospital (utilising two different sections of the front of the hotel) with the larger, rear part taken over for troop billeting. Double bunks were installed in the dining room and ballroom and other public rooms as well as in the 100 bedrooms. Towards the end of the occupation, one part of the building was altered, having slits cut in the bedroom doors and barbed wire fixed over the windows, to convert it into a temporary prison although the war ended before it could be used.

PORTLAND HOTEL

2nd Register   *Telephone:* (STD 0534) 30842

Midvale Road, St. Helier, Jersey, C.I.

Residential Licence
Attractive Cocktail Bar
Heated Swimming Pool
Separate Television Lounge
Radio in Bedrooms
Bedrooms with Private Bath
Some large Family Rooms available
Excellent Cuisine
Friendly Atmosphere
Garden and Large Car Park
Dancing one night a week during season
Coach Trips arranged from Hotel
Close to shops, cinemas, and within easy distance of the sea

BED, BREAKFAST AND EVENING DINNER
No Service Charge is added to your account
Under the personal supervision of the Proprietors: Mr. and Mrs. R. Blaby

Left: One of the popular conceptions of the German officer in wartime is that he always drives around in an open touring car. This party lives up to that image as they dismount outside the former British Hotel in Broad Street, St. Helier (Bundesarchiv). *Right:* Equally ubiquitous is this Morris — perhaps to the Germans it typifies the English! The hotel is now a merchant bank.

Silvertide: then the Geheimfeldpolizei HQ, now a St. Helier guest house.

It is difficult to imagine that the plush 'First Register' 4-star St. Brelade's Bay Hotel was once a German Soldatenheim. (M. Ginns and R. H. Mayne).

The Merton in its heyday as a Luftwaffe hospital (R. H. Mayne).

The Savoy, then the Channel Islands Hotel, was the Guernsey Kommandantur — complete with its AA recommendation and bobby!

GRANGE LODGE HOTEL

FULLY LICENSED   **OPEN TO NON-RESIDENTS**

A.A.   R.S.A.C.   R.A.C.

Position ideal, standing in 3 acres of grounds in the residential part of St. Peter Port. Bus stand 50 yards from Hotel to all parts of the Island. Excellent cuisine. Moderate inclusive terms. Central heating, hot and cold in all bedrooms. Also rooms with private bathrooms. Spacious Lounge and Dining Room, also Lounge Bar.
Television in all bedrooms.
Swimming Pool   Television   Car Park
Terms and Brochure on Application to Resident Manager
Telephone: **Guernsey 25161**
Telegrams: **Grange Lodge, Guernsey**

★

ESTABLISHED UNDER THE SAME PROPRIETORSHIP FOR OVER 40 YEARS

Grade

THE MANOR HOTEL

Established 1932

Telephone: **Guernsey 0481 37788** (2 lines)   Telegrams: **Manor Hotel, Guernsey**

PETIT BÔT, GUERNSEY

A.A. and R.A.C. Hotel   Fully Licensed

THE MANOR HOTEL, set in an unrivalled position facing south, is now Guernsey's largest and most up-to-date country hotel. Its situation on the south side of the island, overlooking Petit Bôt Valley and Bay, is acknowledged by visitors to be unique. The bay, a photograph of which appears in our Brochure, is reached by a good road. It is surrounded by beautiful gorse- and heather-covered cliffs, which in the early Spring and late Autumn are a blaze of colour. Conditions are ideal for Bathing, Recreation, Sun Bathing and Boating.
The Hotel has 56 bedrooms all with H. & C., bedside lights, and radio. Electric razor plugs. Some of the bedrooms have private bathrooms. Excellent cuisine and choicest wines. Cocktail bars, lounges and television. A modern passenger lift serves all floors.

Fully inclusive terms from £8·50 to £10·00 per person daily, according to season. No service charge

Illustrated Brochure on application to: **Mr. S. G. MARSHALL**

Grade

Venue for most of the official meetings between the civilian and German authorities from 1941 to 1945 was Feldkommandantur 515 Nebenstelle at the Grange Lodge Hotel in The Grange, St. Peter Port, Guernsey.

Although the aerodromes on the Channel Islands did not play a major part in the Battle of Britain, 'Pikas' (Ace of Spades) Geschwader, JG53 equipped with Messerschmitt Bf 109s, stayed at the Manor Hotel for a time.

If you are staying on Guernsey then the most important, as has been described in earlier chapters, is the **Grange Lodge Hotel**. This was taken over completely for living and office accommodation for Feldkommandantur 515. (The German command unit could be considered roughly the equivalent of a Civil Affairs Unit in the British Army.)

The **Royal Hotel**, on the Esplanade, St. Peter Port, although it featured prominently in many photographs of the period, was never a headquarters as such although it was taken over by the Germans from the very first day of the occupation. They used the hotel for accommodation for officers in the island temporarily or who did not form part of a military unit stationed in Guernsey. The local manager, Mr. Mentha, who was of German extraction, was retained together with certain of the pre-war staff. The restaurant was also a favourite venue for senior officers dining out. Billeting charges had to be met by the States of Guernsey at a negotiated rate.

One of the hotels with considerable historical significance concerning the German operations in the Channel Islands was **La Colinette Hotel** at St. Jacques (and the **Bon Air Private Hotel** opposite). Both became the headquarters of Seeko-KI — the naval signals station for the Islands — and powerful transmitting equipment was installed which could communicate with Berlin. It was from here that the first stage of the 'Channel Dash' of the Scharnhorst and Prinz Eugen on February 12-13, 1942 was monitored until the ships were further north up the Channel. Initially the radio masts were mounted on top of the hotel until more secure bunkers were constructed close by in 1943, after which the hotels became billets for the signals staff.

The **Hotel de Normandie**, just off the High Street in St. Peter Port was first taken over by German HQ personnel in June 1941. The licensee, Mr. A. Isler, was, however, permitted to remain in residence in one part of the hotel until March 19, 1942 when he was evicted. The building was used for billeting for NCOs and men of the Divisional headquarters and mess and recreation rooms were provided.

During the Battle of Britain, when the aerodromes on the Channel Islands were used by the Luftwaffe more as advanced landing grounds, the **Manor Private Hotel** in the valley above Petit Bot Bay was the headquarters of the Luftwaffe fighter unit stationed at the airport. When this left the island, a small HQ unit remained for a time and the hotel was permanently kept in a state of readiness for instant use throughout the occupation. When the Marshall family returned after the war, one of their first guests (at £3 17s. 6d. per week) was George Dawson on the island for his scrap metal drive — *see page 183.* (Michael Marshall incidentally has written two excellent books on Sark: *Hitler Invaded Sark* and another entitled simply *Sark*. He is also an ardent worshipper of Anders Lassen and gave us much help with our research on Operation BASALT.)

Another hotel taken over by the Luftwaffe during the early days of the occupation was **Le Chene Hotel** in the parish of the Forest but it was little used and eventually a Flak unit occupied it intermittently.

Before the war, you could stay at the Old Government House for 7/6 per night. The Germans converted it into a Soldatenheim for both officers and NCOs.

Above: **Background action is the ivy-covered Red Lion Hotel which became an OT mess. Foreground action is an MG34 in a Tobrukstand sited at the rear of the 47mm anti-tank bunker at Hougue a la Perre (see page 22) (Royal Court).** *Below:* **First positions, please! Ken Tough, Colin Partridge and Richard Heaume on their chalk marks for the re-make of this scene during our visit to Guernsey in 1979.**

Many of the island's hotels performed the role of billets, among these being the **Bella Luce Hotel**, St. Martin's; **Captain's Hotel**, St. Martin's (infantry, marine and artillery units); **Hotel de France**, St. Peter Port; **Hubits Hotel**, St. Martin's (infantry); **La Piette Hotel**, overlooking the sea at St. Peter Port (navy and naval artillery); **Queens Hotel**, St. Martin's; **Rockmount Hotel**, Cobo Bay; **Victoria Hotel** on the Esplanade (naval) and **St. Saviour's Hotel** which was commandeered in November 1942 and the licensee Charles Rose evicted. It was thereafter occupied by artillery troops.

The Organisation Todt took over the **Red Lion Hotel**, on the sea front at St. Peter Port near Tramsheds corner, in late 1941 for use as a private lounge and bar. The staff were retained, the OT paying the wages and running costs. Both officers and NCOs used it for overnight accommodation.

Another St. Peter Port hotel under new management was the **Richmond Hotel** (now called the **Duke of Richmond**) which was requisitioned early in 1941 and used to house military administrative staff, mainly officers. **Moore's Queens Hotel**, today **Moore's Central Hotel** (next to the Hotel de France) was taken over completely in April 1941 for use by the German Customs Police. **Old Government House**, despite its grand sounding name, was only used by German officers in the early days, it being converted into a Soldatenheim for the remainder of the war. Officers and NCOs could spend their leave and off-duty periods at the hotel.

Present-day visitors to the **Albany Hotel** in Queens Road (then called the **Burnham Court Hotel**) will probably be interested to learn that in former days it was the Geheimfeldpolizei headquarters.

Several hotels suffered worse fates: **Old Mill House Hotel** at Petit Bot was demolished for what the Germans described as 'military reasons', as was **Petit Bot Hotel** itself. The real reason was that both were situated beside the Martello tower and after Operation AMBASSADOR, the Germans feared the hotels might be used by a raiding party. **Sunnyside Hotel** at Vazon was also demolished early in the occupation.

L'Eree Hotel, at L'Eree in the parish of St. Pierre-du-Bois on the west coast, was used until 1942 as a German billet with the licensee in residence, was partially converted into a strongpoint and thereafter occupied by artillery units and Panzer Jaeger (mobile anti-tank gunners).

Another 'strongpoint' hotel was **Houmet du Nord** where the main portion was occupied at first by a German infantry unit. In September 1941, the proprietor was evicted and the whole area converted into a strongpoint and a 105mm gun position constructed alongside *(see page 25).*

The **White Hart Hotel** at St. Peter Port also underwent major reconstruction and became a strategic bunker on a corner of the Esplanade overlooking the sea with a 47mm anti-tank gun installed on the ground floor.

The **Albion Hotel**, beside the Town Church, was used as a naval billet until October 1942 when the owner Mr. F. Etor was evicted and it was taken over by German naval engineering workshop staff. Two months later, Mr. Hughes was evicted from the **Beaulieu Hotel** (now the **Carlton Hotel**) and the German infantry took over.

The same fate befel the **Gouffre Hotel**, at Le Gouffre and the **Grandes Rocques** at Cobo Bay when the licensee was kicked out to enable an artillery unit to make the hotel their HQ and accommodate their NCOs and ORs.

One hotel which was spared the rigours of an enforced takeover was the **Prince of Wales**

Above: The close proximity of the Martello tower at Petit Bot Bay, on the southern coast of Guernsey, was the reason for the demolition of Old Mill House Hotel and Petit Bot Hotel seen here in this pre-war photo (N. Grut). *Below:* Petit Bot in 1979 with the Manor Hotel used by the Luftwaffe in 1940 a little way inland up the valley.

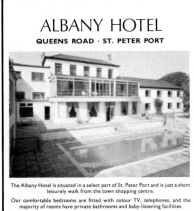

Although now sporting a different name, this was the GFP headquarters on the island.

Above: The entire ground floor of the White Hart Hotel, located in a strategic position on an important junction on the Esplanade, was converted into a fortified bunker and has recently been totally rebuilt *(below)* (Royal Court).

in St. Peter Port — mainly because of its very limited living accommodation. It kept going without interference from the German forces and the licensee even managed to operate his bar, albeit with meagre supplies. For the first two years of the occupation, he also kept the restaurant open.

The **Imperial Hotel** at Pleinmont was also able to keep its bar open for civilians and the proprietor together with his wife and brother were allowed to remain in the domestic accommodation until 1942 when it was taken over by the German infantry.

The owner of **L'Ancresse Lodge** at L'Ancresse Vale was also allowed to stay until the end of 1941, together with German troops mainly occupied in light artillery and weapon maintenance, until they later took over completely.

Civilians were also allowed to share the bar of the **Wayside Cheer Hotel** at Grandes Rocques with German officers or partake of afternoon tea although the menu was very restricted. Later the Germans occupied the premises entirely and, when the British arrived in May 1945, they kindly allowed

The Crown Hotel was taken over as the German Naval Headquarters and it was from this building that Vizeadmiral Huffmeier set out to sign the surrender — see page 45 (R. H. Mayne). Although no longer a hotel, its present use is in keeping with its former role.

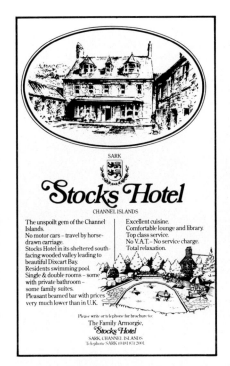

Swimming Pool and Gardens

A corner of the Dining Room

Hotel Entrance with Harbour in background

Alderney's Grand Island Hotel
Welcomes you to a perfect Island holiday

The Grand Island has long been established as Alderney's leading holiday hotel. Reasonably priced and offering excellent value, its location, amenities and services are second to none on the Island. Ideally situated, with panoramic views over the harbour, it is close to the village and within easy reach of the golf course and Alderney's sandy beaches. Good food and wines are served at extended, unhurried meal-times. The centrally heated bedrooms are well appointed, all with radio and most with balcony and private bathroom. The Grand Island's own amenities include a solar heated swimming pool in a delightful setting of gardens and terraces, tennis court, games room, two lounges and a colour TV room. There is dinner dancing most nights and a friendly and informal atmosphere prevails, with the hotel so constructed as to ensure peace and quiet for those who wish.

Tel: Alderney (STD 048 182) 2848
A.A., R.A.C., and B.T.H.A. RECOMMENDED

Stocks Hotel on Sark was used for billeting for a period before being abandoned and stripped of its furniture.

them to remain as it became an internment centre for German officers and NCOs awaiting transfer to British PoW camps!

The 'first and the last' could be an ideal title for the **Channel Islands Hotel** and the **Crown**. The former (now retitled the **Savoy**), situated next to the Grand Hotel on the Esplanade, was taken over in the first few days of the occupation and became the Kommandantur. Then, just a couple of hundred yards down the road, in what is now the Royal Yacht Club building, Vizeadmiral Friedrich Huffmeier, had his Naval Headquarters and it was from here, through a doorway long since blocked up, that he emerged to surrender the Channel Islands on May 9, 1945.

Alderney, as we have seen, was a vast military camp with no real civilian life as such. The Kommandant occupied the **Connaught Hotel** (then a private house) on Connaught Square opposite the Vicarage in St. Anne while at 'Air Force Corner' on the Butes, the Luftwaffe took over the **Grand** with billets and messing facilities at the **Belle Vue**. Naval personnel occupied the **Sea View Hotel** at Braye while the **Victoria Hotel** (in Victoria Street, St. Anne) was a canteen for voluntary workers on the island.

As has been described *(see page 125)*, the **Bel Air Hotel** was burnt down during the occupation on Sark. The first Germans to arrive on Sark stayed at the hotel and a spark from a stove set the thatched roof ablaze.

The annexe to the **Dixcart Hotel,** owned by Mrs. Pickthall, was used for billeting until Operation BASALT *(see page 148)* when the joint manageresses, Miss Page and Miss Duckett, were interned. The nearby **Stocks Hotel** had also been used as a billet by harbour personnel but the Commando raid on

An amazing coincidence occured when I came to write this chapter — less than six hours later, the Grand Island Hotel in Alderney, the island's major hotel with 82 beds, caught fire and was completely gutted in the early hours of March 20, 1981 (Colin Partridge).

Sark changed everything. After 1942 the Germans retreated into the 20th century version of the stockade and they created a fortified compound, defended by mines and wire, centered on Le Manoir, the Kommandant's quarters. To furnish their new accommodation, **Stocks Hotel** was stripped of furniture, the Germans going so far as to pull down the **Bungalow Hotel** on Little Sark to provide building materials.

I am indebted to Michael Ginns on Jersey, Vernon Le Maitre on Guernsey, 'Bunny' Pantcheff on Alderney and Mrs. Sheila Falle on Sark for their valuable assistance in compiling this chapter.

APPENDIX

The Channel Islands Occupation Review

As Editor of *After the Battle* magazine, I have been closely involved with the military history of the Second World War for the past decade. Having studied campaigns in various areas and visited many battlefields during this period, I have no hesitation in stressing the importance of the work being done by the Channel Islands Occupation Society. Not only has the Society grown in stature in its ten-year life but its publishing programme is gradually covering every facet of the war in the Islands. The yearly Review, alternately published by the Guernsey and Jersey Branches, contains a broad spread of articles and, whilst this book has been designed specifically to cover the war years in our own 'then and now' style, more detailed coverage on fortifications, tunnels, civilian life or any other aspect can usually be found in past Reviews. Therefore, although regrettably not all are still in print, it may be helpful to include a complete contents list of all the Society's editions published during its first ten years.

As a long-standing overseas member of the Channel Islands Occupation Society, I look forward to the next ten years and hope that many readers interested in the wartime history of the Islands will support the CIOS by taking out membership. Further information may be obtained from the secretaries of either branch whose addresses are given below.

Guernsey publishes the 'even' years and copies may be had from: K. H. Tough Esq., 'Gladclift', Ruettes Brayes, St. Martin's, Guernsey C.I.
Jersey publishes the 'odd' years and copies may be had from: W. M. Ginns Esq., 'Rangistacey', Rue des Sablons, Grouville, Jersey C.I.

1973 Now out of print but a one-off photocopied version (of reasonable quality) can be prepared to special order for £1.50

CONTENTS
German Tunnels in Jersey by Michael Ginns. A comprehensive look at the many tunnels, whether planned or completed, which the Germans excavated to form impregnable, bomb proof shelters. In all, nineteen were planned.
What was the RAD? by Michael Ginns. A brief description of the organisation of the Reichsarbeitsdienst (RAD) or State Labour Service, units of which served in the Channel Islands in 1941-42.
The German Naval Signals Organisation in the Channel Islands by Frank Wilson. A most interesting account of the naval signals organisation and the naval command structure based on information given by the former naval Signals Officer, Herr Willi Hagedorn.
Hitler's Fortification Order. October 20, 1981, marks the fortieth anniversary of the famous decision by Hitler to turn the Channel Islands into impregnable fortresses. This is a translation of that order.

Patrie — The Story of a Railway Engine by Michael Ginns. The life story of a metre-gauge steam locomotive which came to Jersey in 1942 to open the then newly laid OT railway.
Wartime Shipping Losses in CI Waters: A list of all known marine casualties between 1940 and 1945, compiled by Richard Mayne and John Wallbridge.

1974 Now out of print but a photocopied version is available at £2.50

CONTENTS
Liberation Day Reconnaissance by Rex Ferbrache. An account of the first liberation troops ashore in Guernsey by someone who took part in the operation.
Forgotten Islanders by Richard Mayne. The impression is sometimes given that the German occupation of the Channel Islands was not all that bad. By comparison with what happened elsewhere, this is doubtless true but occasionally the vicious side of Nazism showed its face and this list gives the names of all those Channel Islanders from Jersey who found themselves inside concentration camps. Many never returned.

Batterie Mirus Command Centre by David Kreckeler. A detailed account, with a ground plan, of the command bunker of the largest artillery battery in the Channel Islands.
Guide to Vazon Bay Defences by Colin Partridge and David Kreckeler.
L'Angle Tower by Colin Partridge. A detailed study of this most unusual artillery direction-finding tower by Colin Partridge who has spent fourteen years researching the Atlantic Wall.
Anti-Tank Walls in Jersey. Hitler was concerned that the wide, flat and sandy beaches of Jersey would be ideal for tank landing craft. Thus he ordered that they be sealed by the massive concrete walls that survive to this day. Michael Ginns traces their history and tells where they may be found.

1975 Out of print but a photocopied version is available at £1.50.

CONTENTS
Left Behind. A poignant account of the feelings of those Channel Islanders who remained at home in 1940 gleaned from some undelivered letters which were discovered in 1974.
Cezembre by Denis Holmes. An account of the famous siege of this islet which lies outside the harbour of St. Malo. Compiled from the accounts given by some of the Germans who participated in August 1944.
People who escaped from Jersey during the Occupation compiled by Richard Mayne. Many sought to escape from life under German rule. Some succeeded, some ended up in jail . . . others died trying.
German Artillery in the Channel Islands by Michael Ginns. A comprehensive account of the manner in which the Germans lavished more artillery protection on the Islands than anywhere else in Western Europe.
Infantry Strongpoints in Jersey by Peter Bryans. A listing of every defensive infantry position in Jersey together with the weapons mounted.
A List of churches in Guernsey during the Occupation and how they fared, compiled by Fred Martin.

1976 Still available at £1.00 per copy.

CONTENTS

What was the ROA? by Michael Ginns. A description of the Russiskaya Osvoboditelnaya Armiya (Russian Liberation Army) consisting of Russian PoWs who changed sides and fought for Hitler. Units of these renegades were employed in the Channel Islands from the autumn of 1943 onwards.

The Weapons of the German Soldier in the Channel Islands by Terry Gander. The author, a renowned expert on all WWII weapons, describes the many types of personal weapons with which the German soldier in the Channel Islands was armed.

Snow Joke by John Bouchere. By Christmas 1944 everyone in the Channel Islands — occupiers and occupied alike — faced starvation. In typically humorous fashion the author describes the fate of his family's Xmas 1944 dinner — a small saucepan of beans! But it could not have been funny at the time.

Guide to Perelle Bay Defences by Colin Partridge and David Kreckeler. A detailed examination of the defences of another of Guernsey's bays.

The Sinking of HMS Charybdis and HMS Limbourne by John Wallbridge. A description of the action in which two British warships were destroyed in an engagement with German E-boats off the Channel Islands.

Leaflets dropped in the Channel Islands during WWII by John Goodwin and Richard Mayne. A very detailed listing and description of all the propaganda leaflets dropped by the RAF.

The German Military Communication Network in Guernsey by David Kreckeler.

1977 still available at 70p per copy.

CONTENTS

The Fuel Shortage Solution — Holzgaz! by John Bouchere. During the course of his apprenticeship as a motor mechanic during the occupation, the author helped to convert many lorries to run on wood, charcoal or anthracite.

Resistance Point La Rocque A by Peter Bryans. A detailed description of the defences of La Rocque Harbour.

A Tour of Alderney's German Defences by Michael Ginns. The author takes the reader — be he a student of fortifications or merely an interested tourist — on a guided tour of the island of Alderney. There is much to see!

Machine Guns used in the Channel Islands by Terry Gander. With the author's usual expertise, a detailed account of all types of machine guns employed by the Germans is given.

1978 Out of print. A photocopied version is available at £2.50.

CONTENTS

Escapes from Guernsey and Alderney during the Occupation by David Kreckeler. An account of several daring escapes with reference to official sources.

Wartime Quarries in Jersey by Michael Ginns. To produce the millions of cubic metres of concrete consumed in the fortifications, the Germans required an even larger quantity of aggregate. This article lists all the quarries in Jersey which the Germans enlarged or reopened.

The States of Guernsey 1939-45 by Ken Tough. An account of how the States (the island's governing body) functioned during the war.

German Mortars used in the Channel Islands. A descriptive account by Terry Gander.

MV Dorothea Weber by John Wallbridge. An account of the adventures of a small auxiliary schooner which the Germans employed on the supply run to the Channel Islands.

1979 copies still available at 90p each.

CONTENTS

The La Corbiere Gun by Terry Gander. A description of the 10.5cm K 331(f) coastal defence gun which the Jersey Branch of the Occupation Society is restoring in a bunker on the south west corner of the island.

The Commando Raid on Sark 1942 by Michael Ginns. Culled from contemporary official German reports, this article describes the Commando raid on Sark which resulted in prisoners-of-war in Germany and Canada being handcuffed.

A Boy at War by John Bouchere. The Channel Islands were the only place in occupied Europe where civilian radio sets were confiscated. The author recounts how, with a friend, he 'liberated' a couple of sets from the store where they were impounded.

A Wehrmacht Geologist in Jersey by Dr. Clive Bishop. An account of the activities of Professor Kluepfel, an eminent geologist who served with Fortress Engineer Command 14 in Jersey from 1941-44. He was responsible for the siting of tunnels, camps and other installations, depending upon the suitability of the terrain, the supply of water, etc.

Panzers in Guernsey by Richard Heaume. The activities of Panzer Battalion 213 and its Char B Renault tanks in Guernsey 1942-45.

1980 copies still available at 95p per copy.

CONTENTS

Channel Islands Ro Ro by John Wallbridge. The landing craft and other amphibious craft used by the liberation forces in 1945.

The Story of the Palace Hotel Explosion by Denis Holmes. At long last the true account of how this once famous hotel met its end in 1945. The details were supplied by a former German officer who was there at the time.

The 22cm Kanone K 532(f) by Terry Gander. An historical account of a captured French gun, one of four in Jersey, which blew up in August 1944 and is now preserved by the Occupation Society at Les Landes.

Cross-Channel E-boat Raid 1943 by David Kreckeler.

1981 just out. At the time of writing the price is estimated at £1.10

CONTENTS

Karl Greier — Reluctant Soldier by Margaret Ginns. The true story of an Austrian hairdresser who came to Jersey in 1930 and, having married an English girl, became a reluctant conscript into the German Army in 1943. Captured by the Russians, he passed three unpleasant years in Soviet prison camps before release. His attempts to return to Jersey were almost as frustrating as his captivity.

The Electricity Supply in Jersey during the Occupation by Leslie Hawkley.

The Kommandant by Michael Ginns. An attempt to explain the different types of Kommandant (e.g. Feldkommandant; Standortkommandant; Inselnkommandant) who served in the Channel Islands and whose various functions have caused confusion ever since!

The 4.7cm Pak 36(t) by Terry Gander. An authoritative account dealing with the unusual Czech fixed anti-tank gun which was to be found in casemates all along the Atlantic Wall.

Also available from the Jersey Branch:-
Jersey Besieged 1944 which contains seventy photos from official German sources depicting the state of the island's fortification when Jersey was cut off from the outside world. This is a *limited* edition and the price is £3.00 per copy.